The Geodynamics of the Aegean and Anatolia

Geological Society books refereeing procedures

The Society makes every effort to ensure that the scientific and production quality of its books matches that of its journals. Since 1997, all book proposals have been refereed by specialist reviewers as well as by the Society's Books Editorial Committee. If the referees identify weaknesses in the proposal, these must be addressed before the proposal is accepted.

Once the book is accepted, the Society Book Editors ensure that the volume editors follow strict guidelines on refereeing and quality control. We insist that individual papers can only be accepted after satisfactory review by two independent referees. The questions on the review forms are similar to those for *Journal of the Geological Society*. The referees' forms and comments must be available to the Society's Book Editors on request.

Although many of the books result from meetings, the editors are expected to commission papers that were not presented at the meeting to ensure that the book provides a balanced coverage of the subject. Being accepted for presentation at the meeting does not guarantee inclusion in the book.

More information about submitting a proposal and producing a book for the Society can be found on its web site: www.geolsoc.org.uk.

It is recommended that reference to all or part of this book should be made in one of the following ways:

From: Taymaz, T., Yilmaz, Y. & Dilek, Y. (eds) 2007. *The Geodynamics of the Aegean and Anatolia*. Geological Society, London, Special Publications, **291**.

Piper, D. J. W., Pe-Piper, G., Perissoratis, C. & Anastasakis, G. 2007. Distribution and chronology of submarine volcanic rocks around Santorini and their relationship to faulting. *In*: Taymaz, T., Yilmaz, Y. & Dilek, Y. (eds) *The Geodynamics of the Aegean and Anatolia*. Geological Society, London, Special Publications, **291**, 99–111.

GEOLOGICAL SOCIETY SPECIAL PUBLICATION NO. 291

The Geodynamics of the Aegean and Anatolia

EDITED BY

T. TAYMAZ
Istanbul Technical University, Turkey

Y. YILMAZ
Kadir Has University, Istanbul, Turkey

and

Y. DILEK
Miami University, Ohio, USA

2007
Published by
The Geological Society
London

THE GEOLOGICAL SOCIETY

The Geological Society of London (GSL) was founded in 1807. It is the oldest national geological society in the world and the largest in Europe. It was incorporated under Royal Charter in 1825 and is Registered Charity 210161.

The Society is the UK national learned and professional society for geology with a worldwide Fellowship (FGS) of over 9000. The Society has the power to confer Chartered status on suitably qualified Fellows, and about 2000 of the Fellowship carry the title (CGeol). Chartered Geologists may also obtain the equivalent European title, European Geologist (EurGeol). One fifth of the Society's fellowship resides outside the UK. To find out more about the Society, log on to www.geolsoc.org.uk.

The Geological Society Publishing House (Bath, UK) produces the Society's international journals and books, and acts as European distributor for selected publications of the American Association of Petroleum Geologists (AAPG), the Indonesian Petroleum Association (IPA), the Geological Society of America (GSA), the Society for Sedimentary Geology (SEPM) and the Geologists' Association (GA). Joint marketing agreements ensure that GSL Fellows may purchase these societies' publications at a discount. The Society's online bookshop (accessible from www.geolsoc.org.uk) offers secure book purchasing with your credit or debit card.

To find out about joining the Society and benefiting from substantial discounts on publications of GSL and other societies worldwide, consult www.geolsoc.org.uk, or contact the Fellowship Department at: The Geological Society, Burlington House, Piccadilly, London W1J 0BG: Tel. +44 (0)20 7434 9944; Fax +44 (0)20 7439 8975; E-mail: enquiries@geolsoc.org.uk.

For information about the Society's meetings, consult *Events* on www.geolsoc.org.uk. To find out more about the Society's Corporate Affiliates Scheme, write to enquiries@geolsoc.org.uk.

Published by The Geological Society from:
The Geological Society Publishing House, Unit 7, Brassmill Enterprise Centre, Brassmill Lane, Bath BA1 3JN, UK

(*Orders*: Tel. +44 (0)1225 445046, Fax +44 (0)1225 442836)
Online bookshop: www.geolsoc.org.uk/bookshop

British Library Cataloguing in Publication Data

A catalogue record for this book is available from the British Library.

ISBN 978-1-86239-239-7

Typeset by Techset Composition Ltd., Salisbury, UK

Printed by MPG Books Ltd, Bodmin, UK

Distributors

North America
For trade and institutional orders:
The Geological Society, c/o AIDC, 82 Winter Sport Lane, Williston, VT 05495, USA
Orders: Tel +1 800-972-9892
Fax +1 802-864-7626
E-mail gsl.orders@aidcvt.com

For individual and corporate orders:
AAPG Bookstore, PO Box 979, Tulsa, OK 74101-0979, USA
Orders: Tel +1 918-584-2555
Fax +1 918-560-2652
E-mail bookstore@aapg.org
Website http://bookstore.aapg.org

India
Affiliated East-West Press Private Ltd, Marketing Division, G-1/16 Ansari Road, Darya Ganj, New Delhi 110 002, India
Orders: Tel +91 11 2327-9113/2326-4180
Fax +91 11 2326-0538
E-mail affiliat@vsnl.com

Contents

The geodynamics of the Aegean and Anatolia: introduction

T. TAYMAZ[1], Y. YILMAZ[2] & Y. DILEK[3]

[1]*Department of Geophysical Engineering, İstanbul Technical University, Maslak, TR–34469, İstanbul, Turkey (e-mail: taymaz@itu.edu.tr)*

[2]*Kadir Has University, Fatih, İstanbul, Turkey*

[3]*Department of Geology, Miami University, Oxford, OH 45056, USA*

The complexity of the plate interactions and associated crustal deformation in the Eastern Mediterranean region is reflected in many destructive earthquakes that have occurred throughout its recorded history, many of which are well documented and intensively studied. The Eastern Mediterranean region, including the surrounding areas of western Turkey and Greece, is indeed one of the most seismically active and rapidly deforming regions within the continents (Fig. 1). Thus, the region provides a unique opportunity to improve our understanding of the complexities of continental tectonics in an actively collisional orogen. The major scientific observations from this natural laboratory have clearly been helping us to better understand the tectonic processes in active collision zones, the mode and nature of continental growth, and the causes and distribution of seismic, volcanic and geomorphological events (e.g. tsunamis) and their impact on societal life and civilization. The tectonic evolution of the Eastern Mediterranean region is dominated by the effects of subduction along the Hellenic (Aegean) arc and of continental collision in eastern Turkey (Anatolia) and the Caucasus. Northward subduction of the African plate beneath western Turkey and the Aegean region is causing extension of the continental crust and volcanism in the overlying Aegean extensional province. Eastern Turkey has been experiencing crustal shortening and thickening as a result of northward motion of the Arabian plate relative to Eurasia and the attendant post-collisional magmatism (Taymaz *et al.* 1990, 1991*a*, *b*; McClusky *et al.* 2000, 2003; Dilek & Pavlides 2006, and references therein; Fig. 2). The resulting combination of forces (the 'pull' from the subduction zone to the west and 'push' from the convergent zone to the east) is causing the Turkish plate to move southwestward, bounded by strike-slip fault zones: the North Anatolian Fault Zone (NAFZ) to the north and the East Anatolian Fault Zone (EAFZ) to the south. Interplay between dynamic effects of the relative motions of adjoining plates thus controls large-scale crustal deformation and the associated seismicity and volcanism in Anatolia and the Aegean region (Taymaz *et al.* 2004).

Regional synthesis

Given its location in the Alpine–Himalayan orogenic belt, and at the collisional boundary between Gondwana and Laurasia, the geological history of the Aegean region and Anatolia involves the Mesozoic–Cenozoic closure of several Neotethyan oceanic basins, continental collisions and subsequent post-orogenic processes (e.g. Sengör & Yılmaz 1981; Bozkurt & Mittwede 2001; Okay *et al.* 2001; Dilek & Pavlides 2006; Robertson & Mountrakis 2006). The opening of oceanic branches of Neotethys commenced in the Triassic and they closed during the Late Cretaceous to Eocene time interval. The closure of Neotethyan basins is recorded by several suture zones (e.g. Vardar, Izmir–Ankara–Erzincan, Bitlis–Zagros, Intra-Pontide, Antalya sutures), along which Jurassic–Cretaceous ophiolites and mélanges are exposed (e.g. Sengör & Yılmaz 1981; Robertson & Dixon 1984; Dercourt *et al.* 1986; Stampfli 2000; Okay *et al.* 2001; Parlak *et al.* 2002; Elmas & Yılmaz 2003; Parlak & Robertson 2004; Robertson & Ustaömer 2004; Robertson *et al.* 2004*a*, *b*; Stampfli & Borel 2004; Bagcı *et al.* 2005, 2006; Dilek *et al.* 2005; Çelik *et al.* 2006; Dilek & Thy 2006; Parlak 2006, and references therein). The polarity of subduction, the timing of ocean basin opening and closure, and the location of Neotethyan suture zones remain somewhat controversial. The destruction of oceanic basins was also accompanied and followed by: (1) Cretaceous to early Palaeocene arc magmatism (e.g. Pontide arc: Okay & Sahintürk 1997; Yılmaz *et al.* 1997); (2) development of accretionary-type forearc basins (e.g. Haymana–Polatlı Basin; Koçyiğit 1991; Tuz Gölü Basin, Görür *et al.* 1998); (3) late Palaeocene to Miocene and younger post-collisional magmatism (Aldanmaz *et al.* 2000; Keskin 2003; Boztuğ *et al.* 2004, 2006; Karslı *et al.* 2004; Aslan 2005; Innocenti *et al.* 2005; Altunkaynak & Dilek 2006); (4) the development of several blueschist belts (e.g. Late

From: TAYMAZ, T., YILMAZ, Y. & DILEK, Y. (eds) *The Geodynamics of the Aegean and Anatolia.* Geological Society, London, Special Publications, **291**, 1–16.
DOI: 10.1144/SP291.1 0305-8719/07/$15.00

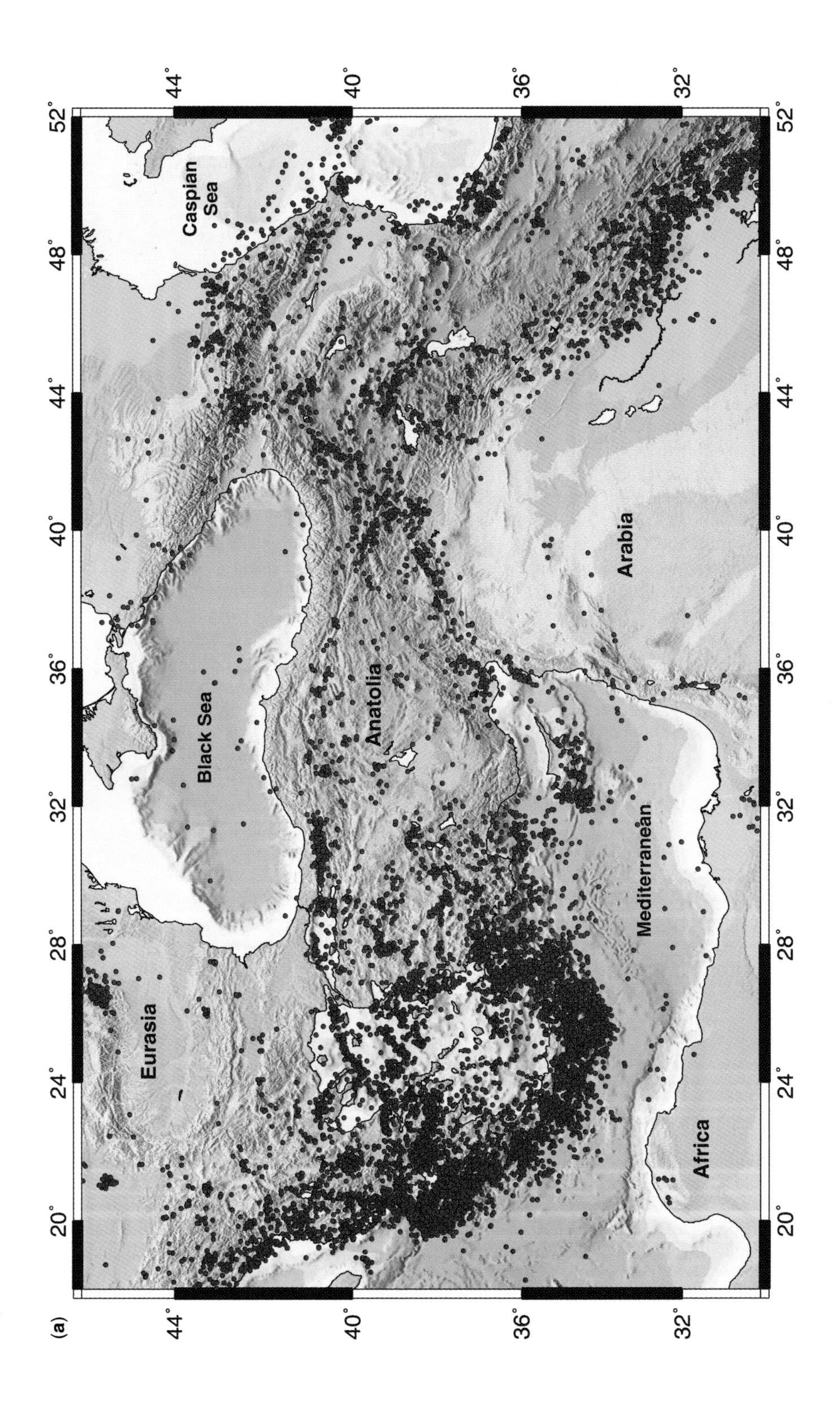
(a)
Caspian Sea
Black Sea
Anatolia
Eurasia
Mediterranean
Arabia
Africa
44°
40°
36°
32°
20°
24°
28°
32°
36°
40°
44°
48°
52°

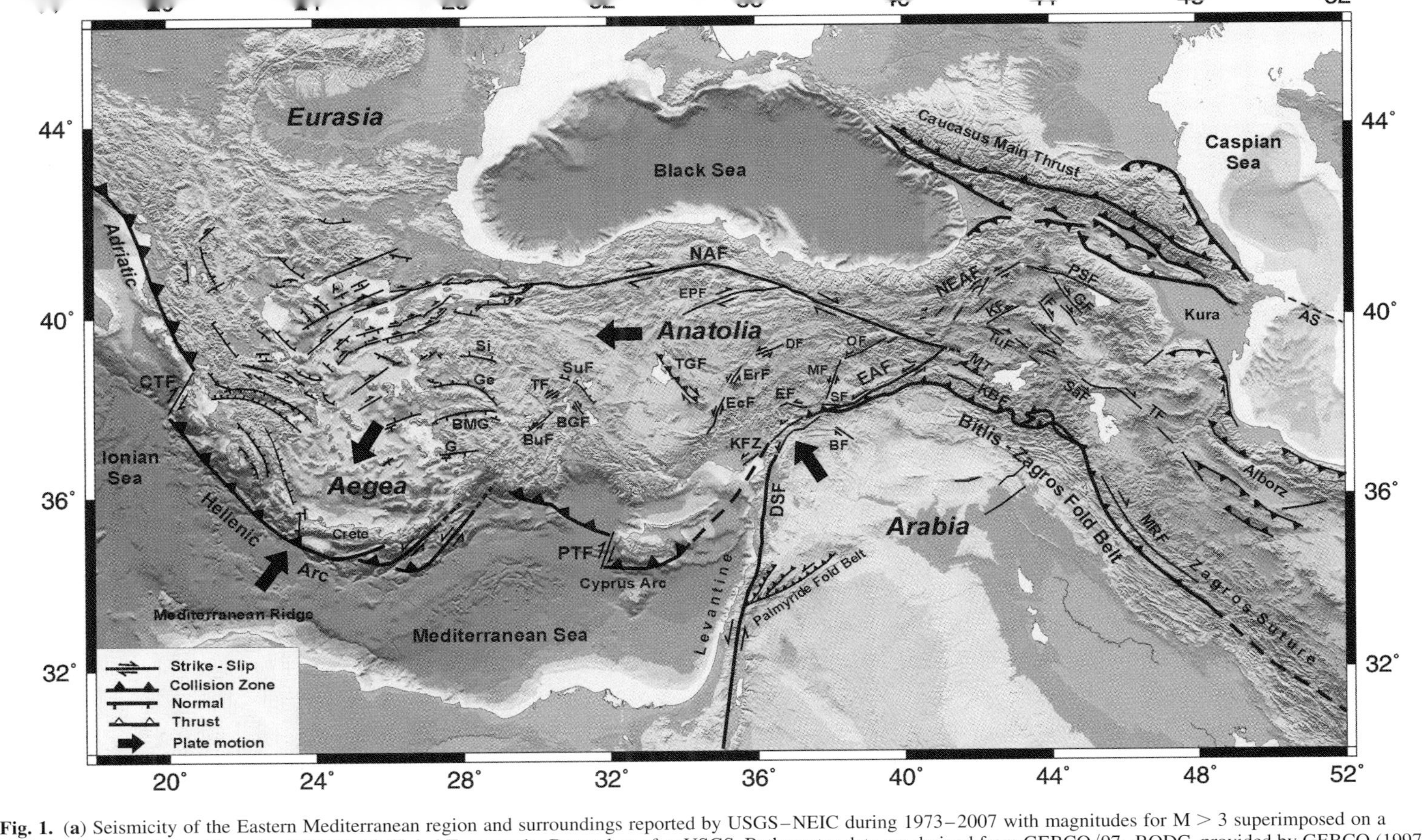

Fig. 1. (**a**) Seismicity of the Eastern Mediterranean region and surroundings reported by USGS–NEIC during 1973–2007 with magnitudes for M > 3 superimposed on a shaded relief map derived from the GTOPO-30 Global Topography Data taken after USGS. Bathymetry data are derived from GEBCO/97–BODC, provided by GEBCO (1997) and Smith & Sandwell (1997*a*, *b*). (**b**) Summary sketch map of the faulting and bathymetry in the Eastern Mediterranean region, compiled from our observations and those of Le Pichon & Angelier (1981), Taymaz (1990), Taymaz *et al.* (1990, 1991*a*, *b*); Şaroğlu *et al.* (1992), Papazachos *et al.* (1998), McClusky *et al.* (2000) and Tan & Taymaz (2006). Large black arrows show relative motions of plates with respect to Eurasia (McClusky *et al.* 2003). Bathymetry data are derived from GEBCO/97–BODC, provided by GEBCO (1997) and Smith & Sandwell (1997*a*, *b*). Shaded relief map derived from the GTOPO-30 Global Topography Data taken after USGS. NAF, North Anatolian Fault; EAF, East Anatolian Fault; DSF, Dead Sea Fault; NEAF, North East Anatolian Fault; EPF, Ezinepazarı Fault; PTF, Paphos Transform Fault; CTF, Cephalonia Transform Fault; PSF, Pampak–Sevan Fault; AS, Apsheron Sill; GF, Garni Fault; OF, Ovacık Fault; MT, Muş Thrust Zone; TuF, Tutak Fault; TF, Tebriz Fault; KBF, Kavakbaşı Fault; MRF, Main Recent Fault; KF, Kağızman Fault; IF, Iğdır Fault; BF, Bozova Fault; EF, Elbistan Fault; SaF, Salmas Fault; SuF, Sürgü Fault; G, Gökova; BMG, Büyük Menderes Graben; Ge, Gediz Graben; Si, Simav Graben; BuF, Burdur Fault; BGF, Beyşehir Gölü Fault; TF, Tatarlı Fault; SuF, Sultandağ Fault; TGF, Tuz Gölü Fault; EcF, Ecemiş Fau; ErF, Erciyes Fault; DF, Deliler Fault; MF, Malatya Fault; KFZ, Karataş–Osmaniye Fault Zone.

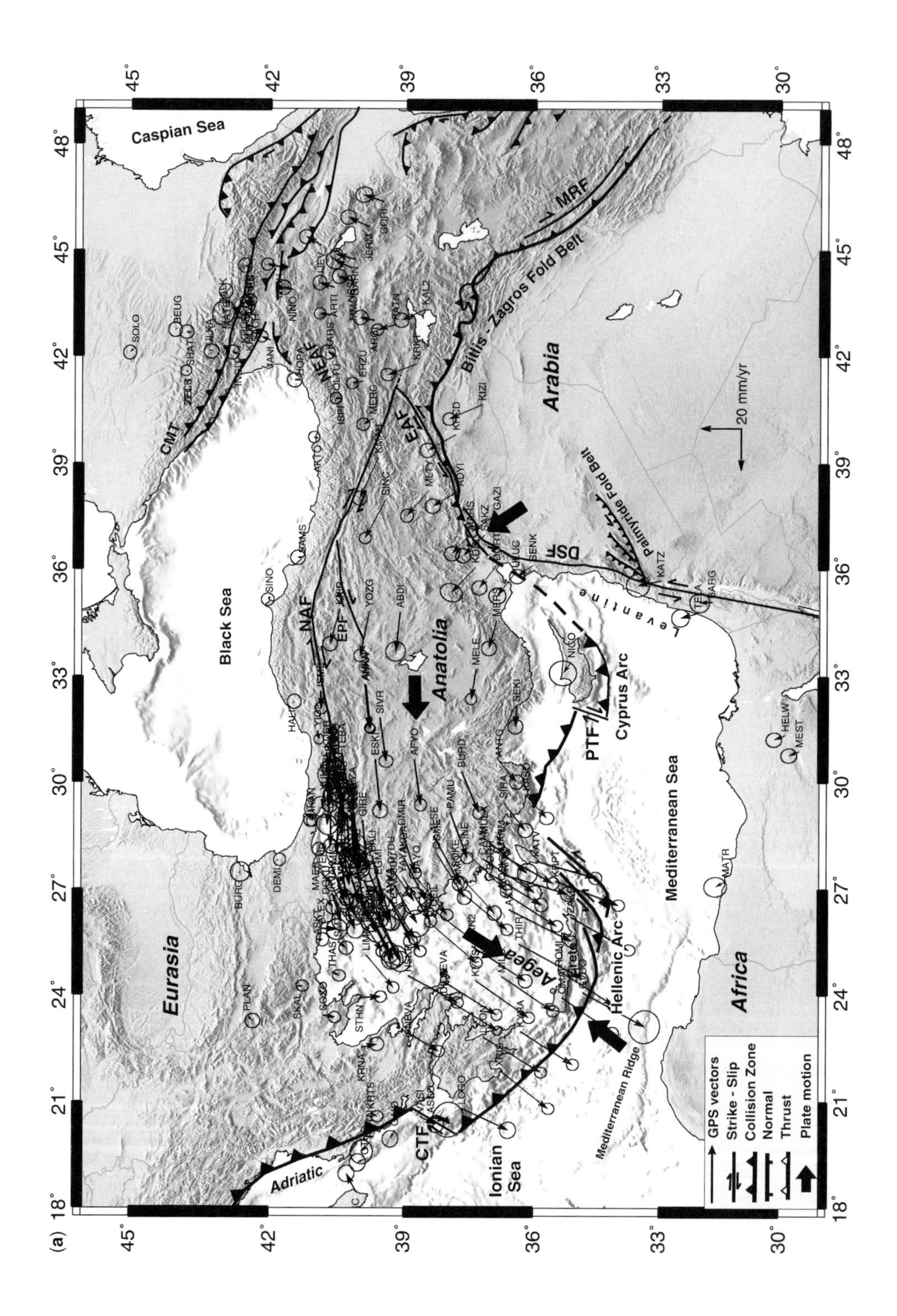
(a)
Caspian Sea
Black Sea
Eurasia
Anatolia
Arabia
Africa
Mediterranean Sea
Ionian Sea
Adriatic
Aegean
Hellenic Arc
Cyprus Arc
Mediterranean Ridge
Levantine
Bitlis - Zagros Fold Belt
Palmyride Fold Belt
Crete
NAF
EAF
NEAF
EPF
DSF
MRF
CMT
CTF
PTF
20 mm/yr
GPS vectors
Strike - Slip
Collision Zone
Normal
Thrust
Plate motion
18°
21°
24°
27°
30°
33°
36°
39°
42°
45°
48°

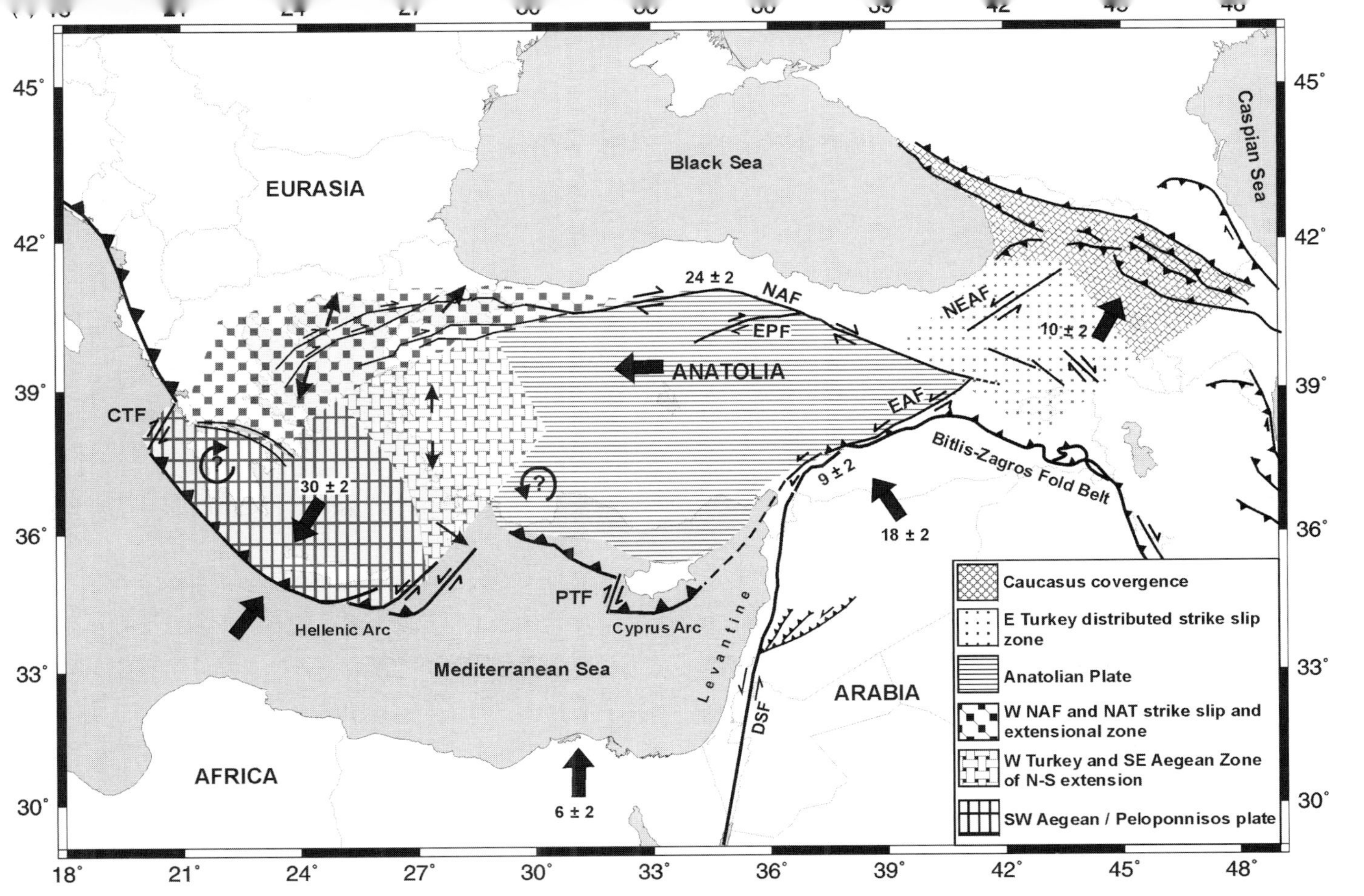

Fig. 2. (**a**) GPS horizontal velocities and their 95% confidence ellipses in a Eurasia-fixed reference frame for the period 1988–1997 superimposed on a shaded relief map derived from the GTOPO-30 Global Topography Data taken after USGS. Bathymetry data are derived from GEBCO/97–BODC, provided by GEBCO (1997) and Smith & Sandwell (1997*a*, *b*). Large arrows designate generalized relative motions of plates with respect to Eurasia (in mm a^{-1}) (recompiled after McClusky *et al.* 2000). NAF, North Anatolian Fault; EAF, East Anatolian Fault; DSF, Dead Sea Fault; NEAF, North East Anatolian Fault; EPF, Ezinepazarı Fault; CTF, Cephalonia Transform Fault; PTF, Paphos Transform Fault; CMT, Caucasus Main Thrust; MRF, Main Recent Fault. (**b**) Schematic map of the principal tectonic settings in the Eastern Mediterranean. Hatching shows areas of coherent motion and zones of distributed deformation. Large arrows designate generalized regional motion (in mm a^{-1}) and errors (recompiled after McClusky *et al.* (2000, 2003). NAF, North Anatolian Fault; EAF, East Anatolian Fault; DSF, Dead Sea Fault; NEAF, North East Anatolian Fault; EPF, Ezinepazarı Fault; CTF, Cephalonia Transform Fault; PTF, Paphos Transform Fault.

Cretaceous Tavşanlı Zone in Turkey: Okay *et al.* 1998; Sherlock 1999; Çamlıca metamorphic belt in NW Turkey: Okay & Satır 2000, and references therein; Eocene–Oligocene Cycladic blueschist belt in the central Aegean: Altherr *et al.* 1979; Avigad & Garfunkel 1989, 1991; Okrusch & Bröcker 1990; Jolivet *et al.* 1994, 2003; Avigad *et al.* 1997; Bröcker *et al.* 2004; Ring *et al.* 2001; Trotet *et al.* 2001; Bröcker & Pidgeon 2007; Lycian Nappes and Menderes Massif: Oberhänsli *et al.* 2001; Okay 2001; Rimmelé *et al.* 2003, and references therein; Bolkar Mountains in the Central Taurides: Dilek & Whitney 1997); (5) high- to low-grade metamorphism affecting larger areas.

The nappe translation and burial of large areas beneath advancing ophiolite nappes has resulted in regional metamorphism and consequent formation of crustal-scale metamorphic massifs, such as the Rhodope Massif, Strandja Massif, Cycladic Massif, Menderes Massif and Central Anatolian Crystalline Complex (Şengör *et al.* 1984; Whitney & Dilek 1997; Bozkurt & Oberhänsli 2001*a*, *b*; Okay *et al.* 2001; Gautier *et al.* 2002; Whitney *et al.* 2003; Şengün *et al.* 2006; Bozkurt 2007, and references therein).

The closure of oceanic basins resulted in crustal thickening and subsequent post-orogenic extension and magmatism in the west (Aegean extensional system) and collisional intracontinental convergence in eastern Turkey and the Caucasus that still prevail in the region. The present-day configuration of the Aegean region is therefore the manifestation of three major structures: (1) the Hellenic–Cyprian subduction zone; (2) the dextral North Anatolian fault system (NAFS); (3) the sinistral East Anatolian fault system (EAFS). Along the Hellenic–Cyprian trenches the African plate is subducting NNE beneath the Anatolian plate at varying rates causing lithospheric tearing and intra-plate deformation (Dilek 2006). The NAFS and EAFS are world-class examples of intracontinental transform fault systems that intersect at a continental triple junction in northeastern Turkey (e.g. Bozkurt 2001; Şengör *et al.* 2005). The continuum of deformation along the NAFS and EAFS has resulted in the WSW extrusion of the intervening Anatolian plate onto the Eastern Mediterranean lithosphere, accompanied by its counter-clockwise rotation, between the converging Eurasian and Arabian plates (Rotstein 1984). The sinistral Dead Sea fault system (DSFS) facilitates the northward motion of Arabia and also plays an important role in the active tectonics of the region.

Subsequent to a series of continental collisions and the demise of the Neotethyan seaways, the Aegean region experienced roughly NNE–SSW-oriented extension since the latest Oligocene to Early Miocene times (Dilek 2006, and references therein). This region, the Aegean extensional system (AES), covers a large area that includes Greece, Macedonia, Bulgaria, Albania and SW Turkey and forms one of the most spectacular and best-studied continental extensional regions. The cause of the onset of extension is controversial and may be (1) slab retreat along the Aegean subduction zone and consequent back-arc extension, (2) collapse of an overthickened crust, (3) westward escape of Anatolia along its plate boundaries, the NAFS and EAFS, or (4) differential rates of convergence between NE-directed subduction of the African plate relative to the hanging-wall Anatolian plate; that is, rapid southwestward movement of Greece relative to Anatolia (e.g. McKenzie 1978; Dewey & Şengör 1979; Le Pichon & Angelier 1981; Rotstein 1984; Şengör *et al.* 1985; Şengör 1979, 1987; Dewey 1988; Jackson & McKenzie 1988; Kissel & Laj 1988; Taymaz *et al.* 1990, 1991*a*; Seyitoğlu & Scott 1991, 1992; Taymaz & Price 1992; Bozkurt & Park 1994; Meulenkamp *et al.* 1994; Taymaz 1996; Saunders *et al.* 1998; Thomson *et al.* 1998; Koçyiğit *et al.* 1999; Bozkurt 2000, 2003; McClusky *et al.* 2000, 2003; Yılmaz *et al.* 2000; Okay 2001; Doglioni *et al.* 2002; Purvis & Robertson 2004; Sato *et al.* 2004; Seyitoğlu *et al.* 2004; Seyitoğlu *et al.* 2004; Bozkurt & Sözbilir 2004, 2006; Purvis *et al.* 2005; and references therein).

The AES is currently under the influence of forces exerted by northward subduction of the African plate beneath the southern margin of the Anatolian plate along the Hellenic–Cyprean trenches and dextral slip on the North Anatolian fault system. The continental extension has expressed itself in two distinct structural styles:

(1) Rapid exhumation of deep-burial metamorphic rocks in the immediate footwall of currently low-angle brittle–ductile normal faults (detachment fault and metamorphic core complexes). The footwall deformation preserves evidence for a progressive transition from ductile to brittle where mylonites are overprinted by breccias and, in turn, by cataclasites. Exhumation was accompanied by synchronous deposition of continental red clastic sediments in the basin(s) located in the detachment hanging walls.

(2) Late stretching of crust and a consequent graben formation along Plio-Quaternary high-angle normal faults (the modern phase of extension or rift mode). Several core complexes (e.g. Rhodope, Cycladic, Kazdağ, Menderes, Niğde core complexes: Lister *et al.* 1984; Dinter & Royden 1993; Gautier *et al.* 1993, 1999; Bozkurt & Park 1994; Gautier & Brunn 1994; Dinter *et al.* 1995; Vandenberg & Lister 1996; Whitney & Dilek 1997; Hetzel *et al.* 1998; Jolivet & Patriat 1999; Jolivet & Faccenna 2000; Lips *et al.* 2001; Bonev & Stampfli 2003; Ring *et al.* 2003; Gessner *et al.* 2004;

Beccaletto & Steiner 2005; Bonev 2006; Bonev *et al.* 2006*a*, *b*; Bozkurt *et al.* 2006; Bozkurt 2007; Régnier *et al.* 2007, and references therein) and overprinting approximately east–west-trending grabens (e.g. Gulf of Corinth, Büyük Menderes and Gediz grabens) therefore form the most prominent elements of the AES.

The Aegean region is therefore considered as a perfect natural laboratory to study mechanisms of core-complex formation, synchronous basin evolution and subsequent graben formation during its post-collisional extensional tectonic evolution. The papers in this book shed some light on various aspects of this extensional tectonics of the Aegean region, but there are still many contentious issues concerning the origin, timing and evolution of Neogene crustal extension in this broad zone of convergence between Africa and Eurasia (see Taymaz & Price 1992; Taymaz 1993; Taymaz *et al.* 2004; Bozkurt & Mittwede 2005; Dilek & Pavlides 2006, and references therein for details).

The Aegean region is also characterized by widespread post-collisional magmatism expressed by extensive volcanic sequences, hypabyssal intrusions and granitoid bodies (Fytikas *et al.* 1976, 1984; Altherr *et al.* 1982; Bingöl *et al.* 1982; Innocenti *et al.* 1984, 2005; Güleç 1991; Seyitoğlu *et al.* 1992, 1997; Hetzel *et al.* 1995*a*, *b*; Richardson-Bunbury 1996; Ercan *et al.* 1997; Yılmaz *et al.* 2001; Işık *et al.* 2003; Erkül *et al.* 2005; Ring & Collins 2005; Tonarini *et al.* 2005; Yücel-Öztürk *et al.* 2005; Aldanmaz 2006; Altunkaynak & Dilek 2006; Bozkurt *et al.* 2006; Pe-Piper & Piper 2006; Dilek & Altunkaynak 2007, and references therein). The extant data suggest that there may have been close temporal and spatial relationships between magmatism and subduction roll-back processes and/or Neogene continental extension in the Aegean region, where the age of volcanic activity becomes younger southwards. There are good examples of synextensional granites emplaced into the footwall rocks of detachment faults (i.e. Simav and Alasehir detachment faults), providing crucial evidence for the age of core-complex formation. Therefore, geochronology and thermochronological studies have recently concentrated on these granitoid bodies in the region (e.g. Ring & Collins 2005; Thompson & Ring 2006).

This introduction is aimed at presenting a synoptic overview of the regional geology and geophysics based on the existing literature, as well as outlining the results of recent literature on existing controversies about the tectonic and geodynamic evolution of the Aegean region. The geology of this region has been reviewed in a series of recent special publications, providing in-depth coverage of the extant data and models, and readers are referred to these publications for additional information (Robinson 1997; Gourgaud 1998; Bozkurt & Rowbotham 1999*a*, *b*; Durand *et al.* 1999; Bozkurt *et al.* 2000; Bozkurt & Mittwede 2001, 2005; Aksu *et al.* 2002, 2005; Akıncı *et al.* 2003; Taymaz *et al.* 2004; Şengör *et al.* 2005; Bozkurt 2006; Dilek & Pavlides 2006; Robertson & Mountrakis 2006).

Research themes

This Special Publication includes a wide range of contributions, illustrating both the diversity of study regions being actively researched and of techniques now available to investigate crustal deformation. It also complements the recent compilations on this region as listed above. Coverage ranges from the Levantine region in the east to SW Bulgaria in the west, with emphasis on the Aegean extensional province and the adjacent western part of the North Anatolian Fault Zone as well as the Hellenic and Cyprean subduction zones. We have grouped papers into the following key themes.

The Aegean Sea and the Cyclades

Katzir *et al.* review the tectonic position and field relations of major ultramafic occurrences in the Cyclades and document in detail the petrography and chemical compositions of ultramafic and associated rocks on the islands of Evvia, Naxos, Tinos and Skyros. They then discuss the origin and mode of emplacement of these rocks and the orogenic evolution of the Cyclades. Widespread serpentinization of most of the ultramafic rocks suggests denudation prior to reburial causing Alpine metamorphism. Relict mantle assemblages and mantle-like oxygen isotope ratios from Naxos meta-peridotites are attributed to the emplacement of these mantle rocks onto a continental margin via collision and subsequent high-pressure (HP) metamorphism (M_1) at 550–650 °C and ≥14 kbar. The meta-basites of the Skyros and Evvian mélanges record M_1 temperatures of 450–500 °C and 400–430 °C, respectively. Thus, from Evvia south-eastwards progressively deeper (i.e hotter) levels of the subducted plate are exposed. Interestingly, temperatures of the M_2 overprint also increase from Evvia through Skyros to Naxos. The diverse P–T paths of the Cycladic blueschists are predicted by thermal modelling of tectonically thickened crust unroofed either by erosion or by uniform extension.

Mehl *et al.* present detailed structural data from the islands of Tinos and Andros documenting the exhumation of HP metamorphic rocks in the Cyclades. The data are consistent with localization of deformation and its progressive evolution whereby early ductile fabrics are superimposed by low-angle semi-brittle shear planes and, in turn, by steeply dipping late brittle structures. The authors also confirm the role of boudinage formation in localizing ductile–brittle transition and emphasize the continuum of strain from ductile to brittle domains during exhumation. One of the main conclusions of the paper is that the strain localization process depends on both rheological stratification and compositional heterogeneity.

Pe-Piper & Piper document the occurrence of Miocene igneous rocks on the island of Samos as part of a Late Miocene–Quaternary back-arc setting in the Aegean Sea. Three groups of Late Miocene igneous rocks are differentiated: (1) an intrusive complex of monzodiorite and minor granites; (2) potassic trachytes and minor rhyolite; (3) bimodal rhyolites and basalts. New K–Ar ages combined with existing geochronology and biostratigraphy suggest an ages of 10–11 Ma for the first two groups and 8 Ma for the bimodal volcanic rocks. Radiogenic isotope and trace element compositions suggest partial melting of an enriched garnet lherzolite mantle source for the origin of monzodiorite and basalt. The authors show that trachyte and monzodiorite rocks may have evolved by fractional crystallization of a parental magma similar to that of the younger basalt. Emplacement and eruption of the monzodiorite, minor granites, potassic trachytes and rhyolite are attributed to regional extension and listric faulting, whereas the younger basalt extrusion was probably associated with north–south strike-slip faulting that provided pathways for different types of mantle melts.

Piper *et al.* interpret marine seismic reflection profiles from around Santorini to show the distribution of active faults and the occurrence of submarine volcanic rocks interfingering with stratified basinal sediments in the south Aegean arc. Two distinct phases of recent volcanism appear to have taken place in the area: the 1.6 ka and 0.65–0.55 Ma Akrotiri episodes. Accordingly, the ages of subsurface submarine volcanic horizons of Santorini (lower and upper volcanic units) are estimated as latest Pliocene and the younger Akrotiri episode. Because Santorini is located at the intersection of several fault sets of different orientations (east–west, ENE–WSW and NE–SW) and different ages, Late Neogene basin subsidence and volcanism are interpreted to have resulted from changing fault patterns associated with the collision of the African and Aegean–Anatolian plates.

Bonev & Beccaletto document structural evidence on the latest Oligocene to Present extensional tectonics within a back-arc setting in the north Aegean above the Hellenic subduction zone. The data come from two distinct locations: eastern Rhodope–Thrace of Bulgaria–Greece and the Biga Peninsula of NW Turkey. The structural data from the metamorphic rocks are consistent with top-to-the-NNW–SSE- to NE–SW-directed extension in dome-shaped core complexes in the footwalls of low-angle detachment faults. The results of this study combined with the available literature from other parts of the Aegean region suggest that the extensional history in the region comprises syn- and post-orogenic episodes during the Paleocene–Eocene and the latest Oligocene–Early Miocene, respectively. The former event was attributed to gravitationally induced hinterland-directed exhumation of the orogenic stack during the closure of the Vardar Ocean, whereas the latter was the consequence of widespread back-arc extension. The recognition of southward migration of extension and magmatism from the Rhodope complex in the north to the present position of the Hellenic trench in the south supports subduction roll-back processes that have prevailed in the region since Late Cretaceous time.

Georgiev *et al.* report the results of recent global positioning system (GPS) campaigns aimed at monitoring and studying the active deformation in SW Bulgaria. The analyses of GPS data for the 1996–2004 period provide firm evidence for active faulting in the region. The region is divided, based on geology and geodetic data from 38 GPS sites, into five blocks of homogeneous kinematic behaviour with average motions varying between 1.3 and 3.4 mm a^{-1}. The rate of motion for the whole region is *c.* 1.8 $\pm$ 0.7 mm a^{-1} in a N154° direction (to the SSE) with respect to the stable Eurasia; this result correlates well with the geological data on neotectonic motions in SW Bulgaria.

The Hellenic and the Cyprus arcs region

Karagianni & Papazachos present a database of regional earthquakes recorded by a portable broadband three-component digital station and a shear velocity model of the crust and uppermost mantle beneath the Aegean area using simultaneous inversion of Rayleigh and Love waves. The results are consistent with strong lateral variations of the S-wave velocities for the crust and uppermost mantle in the Aegean. The authors confirm the presence of thin crust (<28–30 km) for the whole Aegean Sea region and even thinner (20–22 km) crust in the southern and central Aegean Sea. On the other hand, the crust on land is much thicker, around 40–45 km in western Greece and a mean of 35 km in the rest of the country. A significant sub-Moho upper mantle low-velocity zone (LVL mantle) identified in the southern and central Aegean Sea correlates well with the high heat flow in the mantle wedge above the subducted slab and with the related active volcanism in the region.

Meier *et al.* investigate the structure and dynamics of the plate boundary in the area of Crete by receiver function, surface wave and micro-seismicity using temporary seismic networks, and summarize the results with special emphasis on their implications for geodynamic models. The authors then propose that the island of Crete represents a horst structure in the central forearc of the retreating Hellenic subduction zone. The reported properties of the lithosphere and the plate interface beneath Crete are attributed to extrusion of material from a subduction channel, driving differential uplift of the island by several kilometres since about 4 Ma.

Yolsal *et al.* inspect historical tsunamis known to have occurred in the Eastern Mediterranean Sea region identified from verified catalogues in three groups and correlate them with the seismogenic zones such as the Hellenic and the Cyprus arcs, the left-lateral strike-slip Dead Sea Fault and the Levantine rift. The authors conduct numerical simulations involving the initiation and propagation of tsunami waves as series of large sea-waves of extremely long wavelength and period generated by an impulsive undersea disturbances or activity near the coasts (i.e. earthquake-induced tsunamis). The authors then compute water surface elevation distributions and theoretical arrival times (i.e. calculated travel times) for the Paphos, Cyprus earthquake of 11 May 1222 and for the Crete earthquake of 8 August 1303, which are known to be the largest and well-documented tsunamigenic events in the region. The authors confirm that the coastal topography, sea bottom irregularities and near-shore bathymetry are crucial components in tsunami wave simulations, and they further suggest that improvement of the resolution of bathymetric maps, particularly for the details of the continental shelf and seamounts, would facilitate a better understanding of tsunami generation and tsunami-prone mechanisms.

Structural complexities associated with strike-slip faulting in Anatolia

Ergin *et al.* report on the influences of the Late Quaternary tectonics and sea-level changes on sedimentation in the Sea of Marmara, as observed in the Sarköy Canyon in the western part of this sea. They present the results of detailed sedimentological work on several sediment cores collected from this submarine canyon. The work is also supported by the interpretation of seismic section profiles and ^{14}C dating of base sections in the sediment cores. The dated sediments (12 ka BP) marked the shift of depositional environment from lacustrine to the present marine conditions. The change of grain size from sand- to gravel-sized particles at the base to siliciclastic mud upwards in the succession is interpreted to mark changes in the Pleistocene–Holocene conditions. The widespread occurrences of faults, synsedimentary structures and submarine slides or slumps interpreted on seismic profiles form the most important records of active tectonics in the canyon and prove once more the major role of faulting and associated deformation on sedimentation in the Sea of Marmara.

Taymaz *et al.* investigate the seismotectonics of the North Anatolian Fault (NAF) in the vicinity of the Orta–Çankırı region (central Turkey) by analysing a moderate-sized ($M_w = 6.0$) earthquake that occurred on 6 June 2000. The authors correlate source rupture characteristics of this event with those obtained from the field mapping (neotectonic) and geodetic (InSAR) studies. The authors then discuss the faulting in this anomalous earthquake in relation to the local geometry of the main strike-slip system (NAF), and speculate that this event may not be a reliable guide to the regional strain field in NW central Turkey. The authors tentatively suggest that one possible explanation for the occurrence of the 6 June 2000 Orta–Çankırı earthquake could be localized clockwise rotations as a result of shearing of the lower crust and lithosphere.

Gürsoy *et al.* stress the importance of travertine occurrences in the study of active faulting, as these deposits are commonly linked to earthquake activity during which geothermal reservoirs are reset and activated by earthquake fracturing. They study the palaeomagnetic record of three travertine fissures in the Sıcak Çermik geothermal field near Sivas in central Anatolia to understand the ambient field at the time of deposition and to identify cycles of secular variation of the geomagnetic field, with the aim of estimating the rate of travertine growth. The travertines are dated by the U–Th method and vary in age between 100 and 360 ka. The authors analyse sequential samples collected from the margins (earliest deposition) to the centres (last deposition) of fissure travertines and conclude, based on the assumption that these cycles record time periods of 1–2 ka, that travertine layers identify resetting of the geothermal system by earthquakes with magnitudes of 4.5–5.5 at every 50–100 years. Travertine precipitation appears to have occurred at rates of 0.1–0.3 mm a^{-1}. The data are also consistent with the occurrence of major earthquakes (M *c.* 7.5) at approximately every 10 ka.

The majority of the papers in this thematic book were presented at the International Symposium on the Geodynamics of Eastern Mediterranean: Active Tectonics of the Aegean, held at the Kadir

Has University, İstanbul, Turkey, during 15–18 June 2005. This meeting was organized in memory of Professor Kâzım Ergin (1915–2002), a source of pride for the İstanbul Technical University. Kâzım Ergin (Mehmet Kâzım Ergin), known to his colleagues and students as Kâzım Hoca, was a Turkish geophysicist whose theoretical and experimental research contributed to many aspects of solid earth geophysics (Taymaz 2002, 2004). He was also an important figure in advancing the teaching of geosciences in Turkey in the decades after World War II, both as an instructor and as an administrator. Ergin served in high-level administrative capacities in various institutions. After the establishment by the government in 1963 of the Scientific and Technical Research Council of Turkey (TÜBİTAK), he was one of the early appointees to its engineering research group. He was eventually elected chairman of the Scientific Board of TÜBİTAK, a capacity in which he served until he retired in 1979. Ergin also served as a director of the Istanbul Technical University; as a member of the NATO Science Committee Executive Council and of its Scientific Board; as a member of the Executive Council of the European Science Foundation (where he was Turkey's first representative); and as an Executive Council rapporteur for the UNESCO Working Group on Seismicity and Seismotectonics. He died on 24 November 2002 on Teachers' Day, an annual holiday in Turkey. He shall always be remembered as one of the pioneering figures in the development of Earth Sciences in Turkey, for his individual contributions as a university teacher and administrator, and for his influence on his colleagues and students.

The symposium was sponsored by Kadir Has University, the Scientific and Technological Research Council of Turkey (TÜBİTAK), the Turkish Academy of Sciences (TT/TÜBA-GEBİP/2001-2-17), the British Council, the Geological Society of London, the Alexander von Humboldt (AvH) Foundation, the Turkish Petroleum Corporation (TPAO), and Gemini-Club Tourism. The editors would like to thank the members of the Organizing Committee and the staff and students at Kadir Has University who ensured the smooth running of the June 2005 symposium. Thanks are due to J. Turner (Series Editor) for his continuous encouragement, help and comments during the preparation of this volume, to the Geological Society Publishing House for editorial work, and to Angharad Hills for her continuous help at every stage of production of this volume. We are grateful to E. Bozkurt, C. Yaltırak and S. Yolsal for their help with editorial work and with the preparation of individual chapters in the book. Critical scholarly evaluation of scientific papers published in this Special Publication was no small task. We are most grateful to the referees for their dedicated and objective work, constructive criticism and suggestions, which collectively improved the quality of this book and helped us maintain high scientific standards. We finally thank the contributors to this book for their time and effort, and active participation in producing this exciting volume on the geodynamics of the Aegean and Anatolia.

References

AKINCI, Ö., ROBERTSON, A. H. F., POISSON, A. & BOZKURT, E. (eds) 2003. The Isparta Angle, SW Turkey—its role in the evolution of Tethys in the Eastern Mediterranean Region. *Geological Journal*, **38**, 191–394.

AKSU, A. E., YALTIRAK, C. & HISCOTT, R. N. (eds) 2002. Quaternary paleoclimatic–paleoceanographic and tectonic evolution of the Marmara Sea and environs. *Marine Geology*, **190**, 1–552.

AKSU, A. E., HALL, J. & YALTIRAK, C. (eds) 2005. Miocene to Recent tectonic evolution of the Eastern Mediterranean: new pieces of the old Mediterranean puzzle. *Marine Geology*, **221**, 1–440.

ALDANMAZ, E. 2006. Mineral–chemical constraints on the Miocene calc-alkaline and shoshonitic volcanic rocks of western Turkey: disequilibrium phenocryst assemblages as indicators of magma storage and mixing conditions. *Turkish Journal of Earth Sciences*, **15**, 47–73.

ALDANMAZ, E., PEARCE, J. A., THIRLWALL, M. F. & MITCHELL, J. 2000. Petrogenetic evolution of late Cenozoic, post-collision volcanism in western Anatolia, Turkey. *Journal of Volcanology and Geothermal Research*, **102**, 67–95.

ALTHERR, R., SCHLIESTEDT, M., OKRUSCH, M. *ET AL.* 1979. Geochronology of high-pressure rocks on Sifnos (Cyclades, Greece). *Contributions to Mineralogy and Petrology*, **70**, 245–255.

ALTHERR, R., KREUZER, H., WENDT, I. *ET AL.* 1982. A late Oligocene/early Miocene high temperature belt in the Attic–Cycladic crystalline Complex (S. E. Pelagonian, Greece). *Geologische Jahrbuch*, **E23**, 97–164.

ALTUNKAYNAK, Ş. & DILEK, Y. 2006. Timing and nature of postcollisional volcanism in western Anatolia and geodynamic implications. *In*: DILEK, Y. & PAVLIDES, S. (eds) *Post-collisional Tectonics and Magmatism in the Mediterranean Region and Asia. Geological Society of America, Special Papers*, **409**, 321–351.

ASLAN, Z. 2005. Petrography and petrology of the calc-alkaline sarıhan granitoid (NE Turkey): an example of magma mingling and mixing. *Turkish Journal of Earth Sciences*, **14**, 183–207.

AVIGAD, D. & GARFUNKEL, Z. 1989. Low-angle faults above and below a blueschist belt: Tinos Island, Cyclades, Greece. *Terra Nova*, **1**, 182–189.

AVIGAD, D. & GARFUNKEL, Z. 1991. Uplift and exhumation of high-pressure metamorphic terrains—the example of the Cycladic blueschist belt (Aegean Sea). *Tectonophyics*, **188**, 357–372.

AVIGAD, D., GARFUNKEL, Z., JOLIVET, L. & AZANÒN, J. M. 1997. Back-arc extension and denudation of Mediterranean eclogites. *Tectonics*, **16**, 924–941.

BAĞCI, U., PARLAK, O. & HÖCK, V. 2005. Whole-rock and mineral chemistry of cumulates from the Kizildag

(Hatay) ophiolite (Turkey): clues for multiple magma generation during crustal accretion in the southern Neotethyan ocean. *Geological Magazine*, **69**, 53–76.

BAĞCI, U., PARLAK, O. & HÖCK, V. 2006. Geochemical character and tectonic environment of ultramafic to mafic cumulate rocks from the Tekirova (Antalya) ophiolite (southern Turkey). *Geological Journal*, **41**, 193–219.

BECCALETTO, L. & STEINER, C. 2005. Evidence of two-stage extensional tectonics from the northern edge of the Edremit Graben (NW Turkey). *Geodinamica Acta*, **18**, 283–297.

BINGÖL, E., DELALOYE, M. & ATAMAN, G. 1982. Granitic intrusions in western Anatolia: a contribution to the geodynamic evolution of this area. *Eclogae Geologicae Helvetiae*, **75**, 437–446.

BONEV, N. 2006. Cenozoic tectonic evolution of the eastern Rhodope Massif (Bulgaria): basement structure and kinematics of syn- to postcollisional extensional deformation. *In*: DILEK, Y. & PAVLIDES, S. (eds) *Post-collisional Tectonics and Magmatism in the Mediterranean Region and Asia.* Geological Society of America, Special Papers, **409**, 211–235.

BONEV, N. G. & STAMPFLI, G. M. 2003. New structural and petrologic data on Mesozoic schists in the Rhodope (Bulgaria): geodynamic implications. *Comptes Rendus Géosciences*, **335**, 691–699.

BONEV, N., BURG, J.-P. & IVANOV, Z. 2006*a*. Mesozoic–Tertiary structural evolution of an extensional gneiss dome—the Kesebir–Kardamos dome, eastern Rhodope (Bulgaria–Greece). *International Journal of Earth Sciences*, **95**, 318–340.

BONEV, N., MARCHEV, P. & SINGER, B. 2006*b*. $^{40}Ar/^{39}Ar$ geochronology constraints on the Middle Tertiary basement extensional exhumation, and its relation to ore-forming and magmatic processes in the eastern Rhodope (Bulgaria). *Géodinamica Acta*, **19**, 265–280.

BOZKURT, E. 2000. Timing of extension on the Büyük Menderes Graben, western Turkey and its tectonic implications. *In*: BOZKURT, E., WINCHESTER, J. A. & PIPER, J. D. A. (eds) *Tectonics and Magmatism in Turkey and the Surrounding Area.* Geological Society, London, Special Publications, **173**, 385–403.

BOZKURT, E. (ed.) 2001. Neotectonics, seismicity and earthquakes in Turkey. *Geodinamica Acta*, **14**, 1–212.

BOZKURT, E. 2003. Origin of NE-trending basins in western Turkey. *Geodinamica Acta*, **16**, 61–81.

BOZKURT, E. (ed.) 2006. Metamorphic terranes of the Aegean region. *Geodinamica Acta*, **19**, 249–453.

BOZKURT, E. 2007. Extensional vs contractional origin for the Southern Menderes Shear Zone, southwest Turkey: tectonic and metamorphic implications. *Geological Magazine*, **144**, 191–210.

BOZKURT, E. & MITTWEDE, S. K. 2001. Introduction to the geology of Turkey—a synthesis. *International Geology Review*, **43**, 578–594.

BOZKURT, E. & MITTWEDE, S. K. 2005. Introduction: evolution of Neogene extensional tectonics of western Turkey. *Geodinamica Acta*, **18**, 153–165.

BOZKURT, E. & OBERHÄNSLI, R. (eds) 2001*a*. Menderes Massif: structural, metamorphic and magmatic evolution. *International Journal of Earth Sciences*, **89**, 679–882.

BOZKURT, E. & OBERHÄNSLI, R. 2001*b*. Menderes Massif (western Turkey): structural, metamorphic and magmatic evolution—a synthesis. *International Journal of Earth Sciences*, **89**, 679–708.

BOZKURT, E. & PARK, R. G. 1994. Southern Menderes Massif: an incipient metamorphic core complex in western Anatolia, Turkey. *Journal of the Geological Society, London*, **151**, 213–216.

BOZKURT, E. & ROWBOTHAM, G. (eds) 1999*a*. Advances in Turkish Geology, Part I: Tethyan Evolution and Fluvial–Marine Sedimentation. *Geological Journal*, **34**, 3–222.

BOZKURT, E. & ROWBOTHAM, G. (eds) 1999*b*. Advances in Turkish Geology, Part II: Magmatism, Micropalaeontology and Basin Evolution. *Geological Journal*, **34**, 223–319.

BOZKURT, E. & SÖZBILIR, H. 2004. Tectonic evolution of the Gediz Graben: field evidence for an episodic, two-stage extension in western Turkey. *Geological Magazine*, **141**, 63–79.

BOZKURT, E. & SÖZBILIR, H. 2006. Evolution of the large-scale active Manisa fault, southwest Turkey: implications on fault development and regional tectonics. *Geodinamica Acta*, **19**, 427–453.

BOZKURT, E., WINCHESTER, J. A. & PIPER, J. D. A. (eds) 2000. *Tectonics and Magmatism in Turkey and the Surrounding Area.* Geological Society, London, Special Publications, **173**.

BOZKURT, E., WINCHESTER, J. A., MITTWEDE, S. K. & OTTLEY, C. J. 2006. Geochemistry and tectonic implications of leucogranites and tourmalines of the Southern Menderes Massif, southwest Turkey. *Geodinamica Acta*, **19**, 363–390.

BOZTUĞ, D., JONCKHEERE, R., WAGNER, G. A. & YEĞINGIL, Z. 2004. Slow Senonian and fast Palaeocene–early Eocene uplift of granitoids in the central eastern Pontides, Turkey: apatite fission–track results. *Tectonophysics*, **382**, 213–228.

BOZTUĞ, D., ERÇIN, A. İ., KURUÇELIK, M. K., GÖÇ, D., KÖMÜR, İ. & İSKENDEROĞLU, A. 2006. Geochemical characteristics of the composite Kaçkar batholith generated in a Neo-Tethyan convergence system, Eastern Pontides, Turkey. *Journal of Asian Earth Sciences*, **27**, 286–302.

BRÖCKER, M. & PIDGEON, R. T. 2007. Protolith ages of meta–igneous and metatuffaceous rocks from the Cycladic blueschist unit, Greece: results of a reconnaissance U–Pb zircon study. *Journal of Geology*, **115**, 83–98.

BRÖCKER, M., BIELING, D., HACKER, B. & GANS, P. 2004. High-Si phengite records the time of greenschist-facies overprinting: implications for models suggesting mega-detachments in the Aegean Sea. *Journal of Metamorphic Geology*, **22**, 427–442.

ÇELIK, O. F., DELALOYE, M. & FERAUD, G. 2006. Precise Ar-40–Ar-39 ages from the metamorphic sole rocks of the Tauride Belt Ophiolites, southern Turkey: implications for the rapid cooling history. *Geological Magazine*, **143**, 213–227.

DERCOURT, J., ZONENSHAIN, L. P., RICOU, L.-E. *ET AL.* 1986. Geological evolution of the Tethys belt from the Atlantic to the Pamirs since the Lias. *Tectonophysics*, **123**, 241–315.

DEWEY, J. D. 1988. Extensional collapse of orogens. *Tectonics*, **7**, 1123–1139.

DEWEY, J. F. & ŞENGÖR, A. M. C. 1979. Aegean and surrounding region: complex multiplate and continuum tectonics in a convergent zone. *Geological Society of America Bulletin*, **90**, 84–92.

DILEK, Y. 2006. Collision tectonics of the Mediterranean region: causes and consequences. *In*: DILEK, Y. & PAVLIDES, S. (eds) *Post-collisional Tectonics and Magmatism in the Mediterranean and Asia*. Geological Society of America, Special Papers, **409**, 1–13.

DILEK, Y. & ALTUNKAYNAK, S. 2007. Cenozoic crustal evolution and mantle dynamics of post-collisional magmatism in western Anatolia. *International Geology Review*, **49**, 431–453.

DILEK, Y. & PAVLIDES, S. (eds) 2006. *Post-collisional Tectonics and Magmatism in the Mediterranean and Asia*. Geological Society of America, Special Papers, **409**.

DILEK, Y. & THY, P. 2006. Age and petrogenesis of plagiogranite intrusions in the Ankara mélange, central Turkey. *Island Arc*, **15**, 44–57.

DILEK, Y. & WHITNEY, D. L. 1997. Counterclockwise *PTt* trajectory from the metamorphic sole of a Neo-Tethyan ophiolite (Turkey). *Tectonophysics*, **280**, 295–301.

DILEK, Y., SHALLO, M. & FURNES, H. 2005. Rift-drift, seafloor spreading, and subduction tectonics of Albanian ophiolites. *International Geology Review*, **47**, 147–176.

DINTER, D. A. & ROYDEN, L. 1993. Late Cenozoic extension in northeastern Greece: Strymon valley detachment system and Rhodope metamorphic core complex. *Geology*, **21**, 45–48.

DINTER, D. A., MACFARLANE, A. M., HAMES, W., ISACHSEN, C., BOWRING, S. & ROYDEN, L. 1995. U–Pb and $^{40}Ar/^{39}Ar$ geochronology of the Symvolon granodiorite: implications for the thermal and structural evolution of the Rhodope metamorphic core complex, northeastern Greece. *Tectonics*, **14**, 886–908.

DOGLIONI, C., AGOSTINI, S., CRESPI, M., INNOCENTI, F., MANETTI, P., RIGUZZI, F. & SAVAŞÇIN, Y. 2002. On the extension in western Anatolia and the Aegean Sea. *Journal of Virtual Exploration*, **8**, 169–183.

DURAND, B., JOLIVET, L., HORVÁTH, F. & SÉRRANE, M. (eds) 1999. *The Mediterranean Basins: Tertiary Extension Within the Alpine Orogen*. Geological Society, London, Special Publications, **156**.

ELMAS, A. & YILMAZ, Y. 2003. Development of an oblique subduction zone—tectonic evolution of the Tethys suture zone in southeast Turkey. *International Geology Review*, **45**, 827–840.

ERCAN, T., SATIR, M., SEVIN, D. & TÜRKECAN, A. 1997. Batı Anadolu'daki Tersiyer ve Kuvarterner yaşlı kayaçlarda yeni yapılan radyometrik yaş ölçümlerinin yorumu. *Bulletin of Mineral Research and Exploration Institute (MTA) of Turkey*, **119**, 103–112 [in Turkish with English abstract].

ERKÜL, F., HELVACI, C. & SÖZBİLİR, H. 2005. Stratigraphy and geochronology of the Early Miocene volcanics in the Bigadiç borate basin, western Turkey. *Turkish Journal of Earth Sciences*, **14**, 227–253.

FYTIKAS, M., GUILIANO, O., INNOCENTI, F., MARINELLI, G. & MAZZUOLI, R. 1976. Geochronological data on recent magmatism of the Aegean Sea. *Tectonophysics*, **31**, 29–34.

FYTIKAS, M., INNOCENTI, F., MANETTI, P., MAZUOLI, R., PECCERILLO, A. & VILLARI, L. 1984. Tertiary to Quaternary evolution of volcanism in the Aegean region. *In*: DIXON, J. E. & ROBERTSON, A. H. F. (eds) *The Geological Evolution of the Eastern Mediterranean*. Geological Society, London, Special Publications, **17**, 687–700.

GAUTIER, P. & BRUN, J.-P. 1994. Ductile crust exhumation and extensional detachments in the central Aegean (Cyclades and Evvia Islands). *Geodinamica Acta*, **7**, 57–85.

GAUTIER, P., BRUN, J.-P. & JOLIVET, J. 1993. Structure and kinematics of Upper Cenozoic extensional detachment on Naxos and Paros (Cyclades islands, Greece). *Tectonics*, **12**, 1180–1194.

GAUTIER, P., BRUN, J.-P., MORICEAU, R., SOKOUTIS, D., MARTINOD, J. & JOLIVET, L. 1999. Timing, kinematics and cause of Aegean extension: a scenario based on comparison with simple analogue experiments. *Tectonophysics*, **315**, 31–72.

GAUTIER, P., BOZKURT, E., HALLOT, E. & DIRIK, K. 2002. Pre-Eocene exhumation of the Niğde Massif, Central Anatolia, Turkey. *Geological Magazine*, **139**, 559–576.

GEBCO 1997. *General Bathymetric Chart of the Oceans, Digital Version*. CD-ROM. British Oceanographic Data Centre, Birkenhead.

GESSNER, K., COLLINS, A. S., RING, U. & GÜNGÖR, T. 2004. Structural and thermal history of poly-orogenic basement: U–Pb gecohronology of granitoid rocks in the southern Menderes Massif, western Turkey. *Journal of the Geological Society, London*, **161**, 93–101.

GÖRÜR, N., TÜYSÜZ, O. & SENGÖR, A. M. C. 1998. Tectonic evolution of the Central Anatolian basins. *International Geology Review*, **40**, 831–850.

GOURGAUD, A. (ed.) 1998. Volcanism in Anatolia. *Journal of Volcanology and Geothermal Research*, **85**, 1–537.

GÜLEÇ, N. 1991. Crust–mantle interaction in western Turkey: implications from Sr and Nd isotope geochemistry of Tertiary and Quaternary volcanics. *Geological Magazine*, **123**, 417–435.

HETZEL, R., PASSCHIER, C. W., RING, U. & DORA, O. O. 1995*a*. Bivergent extension in orogenic belts: the Menderes Massif (southwestern Turkey). *Geology*, **23**, 455–458.

HETZEL, R., RING, U., AKAL, C. & TROESCH, M. 1995*b*. Miocene NNE-directed extensional unroofing in the Menderes Massif, southwestern Turkey. *Journal of the Geological Society, London*, **152**, 639–654.

HETZEL, R., ROMER, R. L., CANDAN, O. & PASSCHIER, C. W. 1998. Geology of the Bozdağ area, central Menderes Massif, SW Turkey: Pan-African basement and Alpine deformation. *Geologische Rundschau*, **87**, 394–406.

INNOCENTI, F., KOLIOS, N., MANETTI, P., MAZZUOLI, R., PECCERILLO, A., RITA, F. & VILLARI, L. 1984. Evolution and geodynamic significance of Tertiary orogenic volcanism in northeastern Greece. *Bulletin of Volcanology*, **47**, 25–37.

INNOCENTI, F., AGOSTINI, S., DI VINCENZO, G., DOGLIONI, C., MANETTI, P., SAVAŞÇIN, M. Y. & TONARINI, S. 2005. Neogene and Quaternary volcanism in western Anatolia: magma sources and geodynamic evolution. *Marine Geology*, **221**, 397–421.

IŞIK, V., TEKELI, T. & SEYITOĞLU, G. 2003. Ductile–brittle transition along the Alaşehir detachment fault and its structural relationship with the Simav detachment fault, Menderes Massif, western Turkey. *Tectonophysics*, **374**, 1–18.

JACKSON, J. & MCKENZIE, D. 1988. The relationship between plate motions and seismic moment tensors and rates of active deformation in the Mediterranean and Middle East. *Geophysical Journal*, **93**, 45–73.

JOLIVET, L. & FACCENNA, C. 2000. Mediterranean extension and the Africa–Eurasia collision. *Tectonics*, **19**, 1095–1106.

JOLIVET, L. & PATRIAT, M. 1999. Ductile extension and the formation of the Aegean Sea. *In*: DURAND, B., JOLIVET, L., HORVÁTH, F. & SÉRRANE, M. (eds) *The Mediterranean Basins: Tertiary Extension within the Alpine Orogen*. Geological Society, London, Special Publications, **156**, 427–456.

JOLIVET, L., BRUN, J.-P., GAUTIER, P., LALLEMANT, S. & PATRIAT, M. 1994. 3-D kinematics of extension in the Aegean from the Early Miocene to the present, insights from the ductile crust. *Bulletin de la Société Géologique de France*, **65**, 195–209.

JOLIVET, L., FACCENNA, C., GOFFÉ, B., BUROV, E. & AGARD, P. 2003. Subduction tectonics and exhumation of high-pressure metamorphic rocks in the Mediterranean orogen. *American Journal of Science*, **303**, 353–409.

KARSLI, O., AYDIN, F. & SADIKLAR, M. B. 2004. Magma interaction recorded in plagioclase zoning in granitoid systems, Zigana Granitoid, eastern Pontides, Turkey. *Turkish Journal of Earth Sciences*, **15**, 287–306.

KESKIN, M. 2003. Magma generation by slab steepening and breakoff beneath a subduction–accretion complex: an alternative model for collision-related volcanism in Eastern Anatolia, Turkey. *Geophysical Research Letters*, **30**, 8046.

KISSEL, C. & LAJ, C. 1988. Tertiary geodynamical evolution of the Aegean arc: a palaeomagnetic reconstruction. *Tectonophysics*, **146**, 183–201.

KOÇYİĞİT, A. 1991. An example of an accretionary fore-arc basin from central Anatolia and its implications for the history of subduction of Neo-Tethys in Turkey. *Geological Society of America Bulletin*, **103**, 22–36.

KOÇYİĞİT, A., YUSUFOĞLU, H. & BOZKURT, E. 1999. Evidence from the Gediz Graben for episodic two-stage extension in western Turkey. *Journal of the Geological Society, London*, **156**, 605–616.

LE PICHON, X. & ANGELIER, J. 1981. The Aegean Sea. *Philosophical Transactions of Royal Society of London, Series A*, **300**, 357–372.

LIPS, A. L. W., CASSARD, D., SÖZBILIR, H. & YILMAZ, H. 2001. Multistage exhumation of the Menderes Massif, western Anatolia (Turkey). *International Journal of Earth Sciences*, **89**, 781–792.

LISTER, G. S., BANGA, G. & FEENSTRA, A. 1984. Metamorphic core complexes of cordilleran-type in the Cyclades, Aegean Sea, Greece. *Geology*, **12**, 221–225.

MCCLUSKY, S., BALASSANIAN, S., BARKA, A. *ET AL*. 2000. Global positioning system constraints on plate kinematics and dynamics in the eastern Mediterranean and Caucasus. *Journal of Geophysical Research*, **105**, 5695–5719.

MCCLUSKY, S., REILINGER, R., MAHMOUD, S., BEN-SARI, D. & TEALEB, A. 2003. GPS constraints on Africa (Nubia) and Arabia plate motions. *Geophysical Journal International*, **155**, 126–138.

MCKENZIE, D. 1978. Active tectonics of the Alpine–Himalayan belt: the Aegean Sea and surrounding regions. *Geophysical Journal of the Royal Astronomical Society*, **55**, 217–254.

MEULENKAMP, J. E., VAN DER ZWAAN, G. J. & VAN WAMEL, W. A. 1994. On Late Miocene to recent vertical motions in the Cretan segment of the Hellenic arc. *Tectonophysics*, **234**, 53–72.

OBERHÄNSLI, R., PARTZSCH, J., CANDAN, O. & ÇETINKAPLAN, M. 2001. First occurrence of Fe–Mg-carpholite documenting a high-pressure metamorphism in metasediments of the Lycian nappes, SW Turkey. *International Journal of Earth Sciences*, **89**, 867–873.

OKAY, A. İ. 2001. Stratigraphic and metamorphic inversions in the central Menderes Massif: a new structural model. *International Journal of Earth Sciences*, **89**, 709–727.

OKAY, A. İ. & ŞAHINTÜRK, Ö. 1997. Geology of the Eastern Pontides. *In*: ROBINSON, A. G. (ed.) *Regional and Petroleum Geology of the Black Sea and Surrounding Region*. Association of American Petroleum Geologists, Memoirs, **68**, 291–311.

OKAY, A. İ. & SATIR, M. 2000. Coeval plutonism and metamorphism in a latest Oligocene metamorphic core complex in northwest Turkey. *Geological Magazine*, **137**, 495–516.

OKAY, A. İ., HARRIS, N. B. W. & KELLEY, S. P. 1998. Exhumation of blueschists along a Tethyan suture in northwest Turkey. *Tectonophysics*, **285**, 272–299.

OKAY, A. İ., TANSEL, İ. & TÜYSÜZ, O. 2001. Obduction, subduction and collision as reflected in the Upper Cretaceous–Lower Eocene sedimentary record of western Turkey. *Geological Magazine*, **138**, 117–142.

OKRUSCH, M. & BRÖCKER, M. 1990. Eclogites associated with high-grade blueschists in the Cycladic archipelago, Greece: a review. *European Journal of Mineralogy*, **2**, 451–478.

PAPAZACHOS, B. C., PAPADIMITRIOU, E. E., KIRATZI, A. A., PAPAZACHOS, C. B. & LOUVARI, E. K. 1998. Fault plane solutions in the Aegean Sea and the surrounding area and their implication. *Bollettino di Geofisica Teorica ed Applicata*, **39**, 199–218.

PARLAK, O. 2006. Geodynamic significance of granitoid magmatism in the southeast Anatolian orogen: geochemical and geochronogical evidence from Goksun–Afsin (Kahramanmaras, Turkey) region. *International Journal of Earth Sciences*, **95**, 609–627.

PARLAK, O. & ROBERTSON, A. 2004. The ophiolite-related Mersin Melange, southern Turkey: its role in the tectonic–sedimentary setting of Tethys in the Eastern Mediterranean region. *Geological Magazine*, **141**, 257–286.

PARLAK, O., HÖCK, V. & DELALOYE, M. 2002. The supra-subduction zone Pozantı–Karsantı ophiolite, southern Turkey: evidence for high-pressure crystal fractionation of ultramafic cumulates. *Lithos*, **65**, 205–224.

PE-PIPER, G. & PIPER, D. J. W. 2006. Unique features of the Cenozoic igneous rocks of Greece. *In*: DILEK, Y. & PAVLIDES, S. (eds) *Post-collisional Tectonics and Magmatism in the Mediterranean Region and Asia.* Geological Society of America, Special Papers, **409**, 259–282.

PURVIS, M. & ROBERTSON, A. H. F. 2004. A pulsed extension model for the Neogene–Recent E–W-trending Alaşehir Graben and the NE–SW-trending Selendi and Gördes basins, western Turkey. *Tectonophysics*, **391**, 171–201.

PURVIS, M., ROBERTSON, A. H. F. & PRINGLE, M. 2005. Ar-40–Ar-39 dating of biotite and sanidine in tuffaceous sediments and related intrusive rocks: implications for the Early Miocene evolution of the Gördes and Selendi basins, W Turkey. *Geodinamica Acta*, **18**, 239–253.

RÉGNIER, J. L., MEZGER, J. E. & PASSCHIER, C. W. 2007. Metamorphism of Precambrian–Palaeozoic schists of the Menderes core series and contact relationships with Proterozoic orthogneisses of the western Çine Massif, Anatolide belt, western Turkey. *Geological Magazine*, **144**, 67–104

RICHARDSON-BUNBURY, J. M. 1996. The Kula volcanic field, western Turkey: the development of a Holocene alkali basalt province and the adjacent normal faulting. *Geological Magazine*, **133**, 275–283.

RIMMELÉ, G., JOLIVET, L., OBERHÄNSLI, R. & GOFFÉ, B. 2003. Deformation history of the high-pressure Lycian nappes and implications for tectonic evolution of SW Turkey. *Tectonics*, **22**, 1007–1029.

RING, U. & COLLINS, A. S. 2005. U–Pb SIMS dating of synkinematic granites: timing of core complex formation in the northern Anatolide belt of western Turkey. *Journal of the Geological Society, London*, **162**, 1–10.

RING, U., WILLNER, A. & LACKMANN, W. 2001. Stacking of nappes with different pressure–temperature paths: an example from the Menderes nappes of western Turkey. *American Journal of Science*, **301**, 912–944.

RING, U., JOHNSON, C., HETZEL, R. & GESSNER, K. 2003. Tectonic denudation of a Late Cretaceous–Tertiary collisional belt: regionally symmetric cooling patterns and their relation to extensional faults in the Anatolide belt of western Turkey. *Geological Magazine*, **140**, 421–441.

ROBERTSON, A. H. F. & DIXON, J. E. 1984. Introduction: aspects of the geological evolution of the Eastern Mediterranean. *In*: DIXON, J. E. & ROBERTSON, A. H. F. (eds), *The Geological Evolution of the Eastern Mediterranean.* Geological Society, London, Special Publications, **17**, 1–74.

ROBERTSON, A. H. F. & MOUNTRAKIS, D. (eds) 2006. *Tectonic Development of the Eastern Mediterranean Region.* Geological Society, London, Special Publications, **260**.

ROBERTSON, A. H. F. & USTAÖMER, T. 2004. Tectonic evolution of the intra-Pontide suture zone in the Armutlu Peninsula, NW Turkey. *Tectonophysics*, **381**, 175–209.

ROBERTSON, A. H. F., ÜNLÜGENÇ, U. C., İNAN, N. & TASLI, K. 2004*a*. The Misis–Andirin complex: a mid-Tertiary mélange related to late-stage subduction of the southern Neotethys in S Turkey. *Journal of Asian Earth Sciences*, **22**, 413–453.

ROBERTSON, A. H. F., USTAÖMER, T., PICKETT, E. A. *ET AL.* 2004*b*. Testing models of Late Palaeozoic–Early Mesozoic orogeny in Western Turkey: support for an evolving open-Tethys model. *Journal of the Geological Society, London*, **161**, 501–511.

ROBINSON, A. G. (ed.) 1997. *Regional and Petroleum Geology of the Black Sea and Surrounding Region.* Association of American Petroleum Geologists, Memoirs, **68**.

ROTSTEIN, Y. 1984. Counterclockwise rotation of Anatolian block. *Tectonophysics*, **108**, 71–91.

ŞAROĞLU, F., EMRE, Ö. & KUŞÇU, İ. 1992. *Active Fault Map of Turkey, 2 sheets.* Maden ve Tetkik Arama Enstitüsü, Ankara.

SATO, T., KASAHARA, J., TAYMAZ, T., ITO, M., KAMIMURA, A., HAYAKAWA, T. & TAN, O. 2004. A study of microearthquake seismicity and focal mechanisms within the Sea of Marmara (NW Turkey) using ocean bottom seismometers (OBSs). *Tectonophysics*, **75**, 181–241.

SAUNDERS, P., PRIESTLEY, K. & TAYMAZ, T. 1998. Variations in the crustal structure beneath western Turkey. *Geophysical Journal International*, **134**, 373–389.

ŞENGÖR, A. M. C. 1979. The North Anatolian Transform Fault: its age, offset and tectonic significance. *Journal of the Geological Society, London*, **136**, 269–282.

ŞENGÖR, A. M. C. 1987. Cross-faults and differential stretching of hanging walls in regions of low-angle normal faulting: examples from eastern Turkey. *In*: COWARD, M. P., DEWEY, J. F. & HANCOCK, P. L. (eds) *Continental Extensional Tectonics.* Geological Society, London, Special Publications, **28**, 575–589.

ŞENGÖR, A. M. C. & YILMAZ, Y. 1981. Tethyan evolution of Turkey: a plate tectonic approach. *Tectonophysics*, **75**, 181–241.

ŞENGÖR, A. M. C., SATIR, M. & AKKÖK, R. 1984. Timing of tectonic events in the Menderes Massif, western Turkey: implications for tectonic evolution and evidence for Pan-African basement in Turkey. *Tectonics*, **3**, 693–707.

ŞENGÖR, A. M. C., GÖRÜR, N. & ŞAROĞLU, F. 1985. Strike-slip deformation, basin formation and sedimentation: strike–slip faulting and related basin formation in zones of tectonic escape. *In*: BIDDLE, K. T. & CHRISTIE-BLICK, N. (eds) *Strike–slip Faulting and Basin Formation.* Society of Economic Paleontologists and Mineralogists, Special Publications, **37**, 227–264.

ŞENGÖR, A. M. C., TÜYSÜZ, O., İMREN, C. *ET AL.* 2005. The North Anatolian Fault: a new look. *Annual Review of Earth and Planetary Sciences*, **33**, 37–112.

ŞENGÜN, F., CANDAN, O., DORA, Ö. O. & KORALAY, E. 2006. Petrography and geochemistry of paragneisses in the Çine submassif of the Menderes Massif, western Anatolia. *Turkish Journal of Earth Sciences*, **15**, 321–342.

SEYITOĞLU, G. & SCOTT, B. C. 1991. Late Cenozoic crustal extension and basin formation in west Turkey. *Geological Magazine*, **128**, 155–166.

SEYITOĞLU, G. & SCOTT, B. C. 1992. The age of the Büyük Menderes Graben (west Turkey) and its tectonic implications. *Geological Magazine*, **129**, 239–242.

SEYITOĞLU, G., SCOTT, B. C. & RUNDLE, C. C. 1992. Timing of Cenozoic extensional tectonics in west Turkey. *Journal of the Geological Society, London*, **149**, 533–538.

SEYITOĞLU, G., ANDERSON, D., NOWELL, G. & SCOTT, B. C. 1997. The evolution from Miocene potassic to Quaternary sodic magmatism in western Turkey: implications for enrichment processes in the lithospheric mantle. *Journal of Volcanology and Geothermal Research*, **76**, 127–147.

SEYITOĞLU, G., IŞIK, V. & ÇEMEN, İ. 2004. Complete Tertiary exhumation history of the Menderes Massif, western Turkey: an alternative working hypothesis. *Terra Nova*, **16**, 358–363.

SHERLOCK, S. 1999. ^{40}Ar–^{39}Ar and Rb–Sr geochronology of high-pressure metamorphism and exhumation history of the Tavşanlı Zone, NW Turkey. *Contributions to Mineralogy and Petrology*, **137**, 46–58.

SMITH, W. H. F. & SANDWELL, D. T. 1997*a*. *Measured and estimated seafloor topography (version 4.2)*. World Data Centre-A for Marine Geology and Geophysics Research Publication, **RP-1**.

SMITH, W. H. F. & SANDWELL, D. T. 1997*b*. Global seafloor topography from satellite altimetry and ship depth soundings. *Science*, **277**, 1957–1962.

STAMPFLI, G. M. 2000. Tethyan oceans. *In*: BOZKURT, E., WINCHESTER, J. A. & PIPER, J. D. A. (eds) *Tectonics and Magmatism in Turkey and Surrounding Area*. Geological Society, London, Special Publications, **173**, 1–23.

STAMPFLI, G. M. & BOREL, J. D. 2004. The TRANSMED transects in space and time: constraints on the paleotectonic evolution of the Mediterranean domain. *In*: CAVAZZA, W., ROURE, F. M., SPAKMAN, W., STAMPFLI, G. M. & ZIEGLER, P. A. (eds) *The TRANSMED Atlas – The Mediterranean Region from Crust to Mantle*. Springer, Berlin, 53–80.

TAN, O. & TAYMAZ, T. 2006. Active tectonics of the Caucasus: earthquake source mechanisms and rupture histories obtained from inversion of teleseismic body waveforms. *In*: DILEK, Y. & PAVLIDES, S. (eds) *Post-collisional Tectonics and Magmatism in the Mediterranean and Asia*. Geological Society of America, Special Papers, **409**.

TAYMAZ, T. 1990. *Earthquake source parameters in the Eastern Mediterranean Region*. PhD thesis, University of Cambridge.

TAYMAZ, T. 1993. The source parameters of Çubukdağ (Western Turkey) earthquake of 11 October 1986. *Geophysical Journal International*, **113**, 260–267.

TAYMAZ, T. 1996. S–P-wave traveltime residuals from earthquakes and lateral heterogeneity in the upper mantle beneath the Aegean and the Hellenic Trench near Crete. *Geophysical Journal International*, **127**, 545–558.

TAYMAZ, T. 2002. Obituary: Kâzım Ergin. *Turkish Journal of Earth Sciences*, **11**, 247–250.

TAYMAZ, T. 2004. Geophysicists: Kâzım Ergin. *EOS Transactions, American Geophysical Union*, **85**, 14.

TAYMAZ, T. & PRICE, S. 1992. The 1971 May 12 Burdur earthquake sequence, SW Turkey: a synthesis of seismological and geological observations. *Geophysical Journal International*, **108**, 589–603.

TAYMAZ, T., JACKSON, J. A. & WESTAWAY, R. 1990. Earthquake mechanisms in the Hellenic Trench near Crete. *Geophysical Journal International*, **102**, 695–731.

TAYMAZ, T., JACKSON, J. A. & MCKENZIE, D. P. 1991*a*. Active tectonics of the North and Central Aegean Sea. *Geophysical Journal International*, **106**, 433–490.

TAYMAZ, T., EYIDOĞAN, H. & JACKSON, J. A. 1991*b*. Source parameters of large earthquakes in the East Anatolian Zone (Turkey). *Geophysical Journal International*, **106**, 537–550.

TAYMAZ, T., WESTAWAY, R. & REILINGER, R. (eds) 2004. Active Faulting and Crustal Deformation in the Eastern Mediterranean Region. *Tectonophysics*, **391**, 1–374.

THOMSON, S. N. & RING, U. 2006. Thermochronologic evaluation of postcollision extension in the Anatolide orogen, western Turkey. *Tectonics*, **25**, paper number TC3005.

THOMSON, S. N., STÖCKHERT, B. & BRIX, M. R. 1998. Thermochronology of the high-pressure metamorphic rocks of Crete, Greece: implications for the speed of tectonic processes. *Geology*, **26**, 259–262.

TONARINI, S., AGOSTINI, S., ONNOCENTI, F. & MANETTI, P. 2005. δ^{11}B as tracer of slab dehydration and mantle evolution in western Anatolia Cenozoic magmatism. *Terra Nova*, **17**, 259–264.

TROTET, F., JOLIVET, L. & VIDAL, O. 2001. Tectono-metamorphic evolution of Syros and Sifnos islands (Cyclades, Greece). *Tectonophysics*, **338**, 179–206.

VANDENBERG, L. C. & LISTER, G. S. 1996. Structural analysis of basement tectonites from the Aegean metamorphic core complex of Ios, Cyclades, Greece. *Journal of Structural Geology*, **18**, 1437–454.

WHITNEY, D. L. & DILEK, Y. 1997. Core complex development in central Anatolia, Turkey. *Geology*, **25**, 1023–1026.

WHITNEY, D. L., TEYSSIER, C., FAYON, A., HAMILTON, M. A. & HEIZLER, M. 2003. Tectonic controls on metamorphism, partial melting, and intrusion: timing and duration of regional metamorphism and magmatism in the Niğde Massif, Turkey. *Tectonophysics*, **376**, 37–60.

YILMAZ, Y., TÜYSÜZ, O., YIĞITBAŞ, E., GENÇ, Ş. C. & ŞENGÖR, A. M. C. 1997. Geology of tectonic evolution of the Pontides. *In*: ROBINSON, A. G. (ed.)

Regional and Petroleum Geology of the Black Sea and Surrounding Region. Association of American Petroleum Geologists Memoirs, **68**, 183–226.

YILMAZ, Y., GENÇ, Ş. C., GÜRER, Ö. F. *ET AL.* 2000. When did the western Anatolian grabens begin to develop? *In*: BOZKURT, E., WINCHESTER, J. A. & PIPER, J. D. A. (eds) *Tectonics and Magmatism in Turkey and the Surrounding Area.* Geological Society, London, Special Publications, **173**, 353–384.

YILMAZ, Y., GENÇ, Ş. C., KARACIK, Z. & ALTUNKAYNAK, S. 2001. Two contrasting magmatic associations of NW Anatolia and their tectonic significance. *Journal of Geodynamics*, **31**, 243–271.

YÜCEL-ÖZTÜRK, Y., HELVACI, C. & SATIR, M. 2005. Genetic relations between skarn mineralization and petrogenesis of the Evciler Granitoid, Kazdağ, Çanakkale, Turkey and comparison with world skarn granitoids. *Turkish Journal of Earth Sciences*, **14**, 225–280.

The geodynamic evolution of the Alpine orogen in the Cyclades (Aegean Sea, Greece): insights from diverse origins and modes of emplacement of ultramafic rocks

Y. KATZIR[1], Z. GARFUNKEL[2], D. AVIGAD[2] & A. MATTHEWS[2]

[1]*Department of Geological and Environmental Sciences, Ben Gurion University of the Negev, Beer Sheva 84105, Israel (e-mail: ykatzir@bgu.ac.il)*

[2]*Institute of Earth Sciences, The Hebrew University of Jerusalem, Jerusalem 91904, Israel*

Abstract: The Alpine orogen in the Cyclades, wherein both high-pressure metamorphic rocks and ultramafic rocks co-occur, is a key area in studying the emplacement of mantle rocks into the crust. Within the Cyclades three distinct ultramafic associations occur: (1) HP–LT ophiolitic mélanges of the Cycladic Blueschist Unit (CBU) on Evia and Syros; (2) meta-peridotites associated with migmatized leucogneisses on Naxos, which represent the deepest exposed levels of the CBU; (3) a greenschist-facies metamorphosed dismembered ophiolite juxtaposed on top of the CBU by an extensional detachment on Tinos. Most of the Cycladic ultramafic rocks were serpentinized prior to Alpine metamorphism, suggesting denudation prior to reburial. The Naxos meta-peridotites preserve, however, relict mantle assemblage and mantle-like oxygen isotope ratios, and thus indicate direct emplacement from the mantle into an underthrust continent during collision and HP metamorphism (M_1). Thus conditions for M_1 in the Naxos leucogneiss core are constrained by ultramafic assemblages to 550–650 °C and $\geq$14 kbar. Mafic blocks of the ophiolitic mélanges in the NW Cyclades span a wide range of chemical compositions indicating derivation from variable oceanic settings and sequential events of alteration and metasomatism. Given the comparable geochemical heterogeneity in the Syros and Evian mélange intervals, the garnet-bearing meta-basites of the Syros mélange record higher M_1 temperatures (450–500 °C) than the garnet-free epidote blueschists of the Evian mélanges (400–430 °C). It follows that going southeastwards from Evia progressively deeper (i.e. hotter) levels of the subducted plate are exposed. Correspondingly, temperatures of the M_2 overprint also increase from pumpellyite-bearing assemblages on southern Evia, through greenschists on Syros to upper-amphibolite, sillimanite-bearing gneisses on Naxos. The diverse *P–T* paths of the CBU form an array wherein the deeper a rock sequence is buried, the 'hotter' is its exhumation path. Such a pattern is predicted by thermal modelling of tectonically thickened crust unroofed by either erosion or uniform extension.

The occurrence of dense ultramafic rocks, peridotites, the prime constituent of the Earth's mantle, at the surface of the continents requires significant vertical mobility. It is thus not surprising that orogenic belts where continents have collided and vast tectonic movements have taken place host most of the relatively rare peridotites. High-pressure metamorphic rocks best record the vertical movements involved in orogenesis: eclogites and blueschists mostly comprise surface-derived rocks, thus implying a full tectonic cycle of burial and exhumation. Orogenic segments where both high-pressure and ultramafic constituents occur in proximity are thus key areas in answering a fundamental question: how are mantle-derived rocks incorporated into the subduction–exhumation sequence of surficial rocks?

A partial answer to the puzzle of displacement of ultramafic rocks into the crust is given by ophiolite suites representing occasional portions of oceanic plates that escaped destruction at subduction zones and were carried onto the foreland of an adjacent continent. Based on their tectonic setting two major types of ophiolites were distinguished (Moores 1982; Coleman 1984; Wakabayashi & Dilek 2003): (1) ophiolites that occur as thick thrust sheets that rest upon passive margin substrate and are commonly associated with high-temperature metamorphic aureoles at their base (e.g. the Semail ophiolite, Oman; the Pindos ophiolite, Greece); (2) ophiolite bodies that occur as blocks within tectonic blueschist mélange as part of an accretionary prism (e.g. the Franciscan mélange). The differences in the manner of occurrence and tectonic context of the two types reflect their origin and mode of emplacement: Tethyan-type ophiolites formed by thrusting of an oceanic lithosphere slab upon passive continental margin sequences whereas upheaval of oceanic fragments within the accretionary prism of an active margin gave rise to Cordilleran-type ophiolites.

Within the Hellenic segment of the Alpine orogenic belt a major Tethyan-type ophiolite

From: TAYMAZ, T., YILMAZ, Y. & DILEK, Y. (eds) *The Geodynamics of the Aegean and Anatolia.* Geological Society, London, Special Publications, **291**, 17–40.
DOI: 10.1144/SP291.2 0305-8719/07/$15.00

emplacement occurred in mid- to late Jurassic times. The 'Eohellenic' ophiolites are interpreted as originating from a Mesozoic Neo-Tethyan oceanic basin, the Pindos Ocean, and subsequently thrust northeastwards onto the Pelagonian passive continental margin (Fig. 1; Robertson *et al.* 1991; Smith 1993). Deep-water sedimentation continued, however, in the Pindos basin until its final closure in the early Tertiary (Jones & Robertson 1991). Within the Cycladic Massif of the Aegean Sea (Fig. 1), a Tertiary high-pressure orogenic segment that lies to the SE of the Hellenides, thin remnants of the Eohellenic ophiolites occur on the island of Paros (Papanikolaou 1980). However, most of the ophiolites in the Cyclades are regionally metamorphosed at variable conditions, they are highly attenuated and dismembered, and are bounded and dissected by low-angle tectonic contacts. The Cycladic ultramafic rocks are associated with a great variety of country rocks including leucogneisses of continental basement origin (the Main Ultramafic Horizon on Naxos; see below), thus raising questions concerning the provenance of peridotites. The diversity in field relations, metamorphic grade and tectonic position of ultramafic rocks makes the Alpine orogen in the Cyclades an attractive terrain to address the questions of their origin and emplacement. Moreover, in a complicated poly-metamorphosed orogenic segment such as the Cyclades, the relative sluggishness of metamorphic reactions in ultramafic rocks turns them into potential preservers of pre- and early metamorphic evolution invariably effaced by later events in other rocks.

In this paper we review the tectonic position and field relations of major ultramafic occurrences in the Cyclades and examine in detail the petrography and chemical compositions of ultramafic and associated rocks. Thus, their origin and mode of emplacement are unveiled and new constraints on the orogenic evolution of the Cyclades are set.

Regional geological setting

The Cycladic Massif

The Cycladic Massif (Fig. 1) records an Alpine orogenic cycle of collisional thickening, collapse and reworking by back-arc extension (Avigad *et al.* 1997). This cycle is well represented by the metamorphic evolution of the dominant tectonic unit of the Massif, the Cycladic Blueschist Unit (CBU, Lower tectonic Unit), and by its position within the orogenic nappe-pile. The CBU underwent regional eclogite- and blueschist-facies metamorphism during Late Cretaceous to Eocene compression (M_1 metamorphism) (Altherr *et al.* 1979; Andriessen *et al.* 1979; Maluski *et al.* 1981, 1987; Wijbrans & McDougall 1988; Wijbrans *et al.* 1990; Bröcker *et al.* 1993; Bröcker & Enders 1999, 2001; Tomaschek *et al.* 2003; Putlitz *et al.* 2005). Today the Cycladic Blueschist Unit forms part of the highly attenuated crust of the Aegean Sea in an extensional back-arc environment. The tectonic sequence of thickening followed by back-arc extension was suggested as an efficient mechanism for rapid exhumation and preservation of high-pressure metamorphic rocks in the Cyclades (Lister *et al.* 1984; Gautier & Brun 1994; Jolivet & Patriat 1999; Trotet *et al.* 2001*a*). Avigad *et al.* (1997) questioned this notion and showed that back-arc extension in the Aegean lagged behind a significant part of the exhumation of the high-pressure metamorphic rocks. Evidence for the initiation of the Aegean back-arc extension in the earliest Miocene is plentiful and includes: (1) activation of normal-sense detachments that juxtaposed sedimentary and low-pressure metamorphic rocks (Upper Tectonic Unit) on top of exhumed high P–T rocks (CBU) (Lister *et al.* 1984; Faure *et al.* 1991; Lee & Lister 1992; Gautier *et al.* 1993); the earliest detachment was observed on Tinos, where both units were intruded by an 18 Ma granite (Avigad & Garfunkel 1989); (2) abundance of north-dipping ductile extensional fabrics in Early Miocene overprinting assemblages (Urai *et al.* 1990; Buick 1991; Gautier & Brun 1994; Jolivet & Patriat 1999); (3) beginning of sedimentation in extension-related basins (Dermitzakis & Papanikolaou 1981; Sanchez-Gömez *et al.* 2002); (4) regional extension-controlled granitic plutonism (Altherr & Siebel 2002).

Placing the Early Miocene initiation of extension within the P–T–t path of the CBU provides convincing evidence for a time gap between exhumation and extension in the Cyclades. Most of the petrological studies of the Cycladic high-pressure rocks concentrated on the islands of Sifnos, Syros and Tinos, where the best-preserved eclogites occur. The three islands form a NE–SW linear array that hereafter will be termed the Central Eclogite Axis (CEX; Fig. 1). The P–T–t paths of the CBU in the CEX include Eocene eclogite-facies metamorphism at 15 ± 3 kbar and 450–500 °C (M_1) that was followed by a 23–21 Ma greenschist-facies overprint (M_2) at similar temperatures and 5–7 kbar (Matthews & Schliestedt 1984; Evans 1986; Schliestedt 1986; Dixon & Ridley 1987; Schliestedt *et al.* 1987; Okrusch & Bröcker 1990; Avigad *et al.* 1992; Bröcker *et al.* 1993, 2004; Bröcker & Franz 1998). Hence, the Early Miocene onset of back-arc extension in the Aegean was coeval with a greenschist-facies metamorphic event (M_2) that overprinted high-pressure rocks in the CEX. It is thus evident that, at the time of overprinting, the Cycladic blueschists and

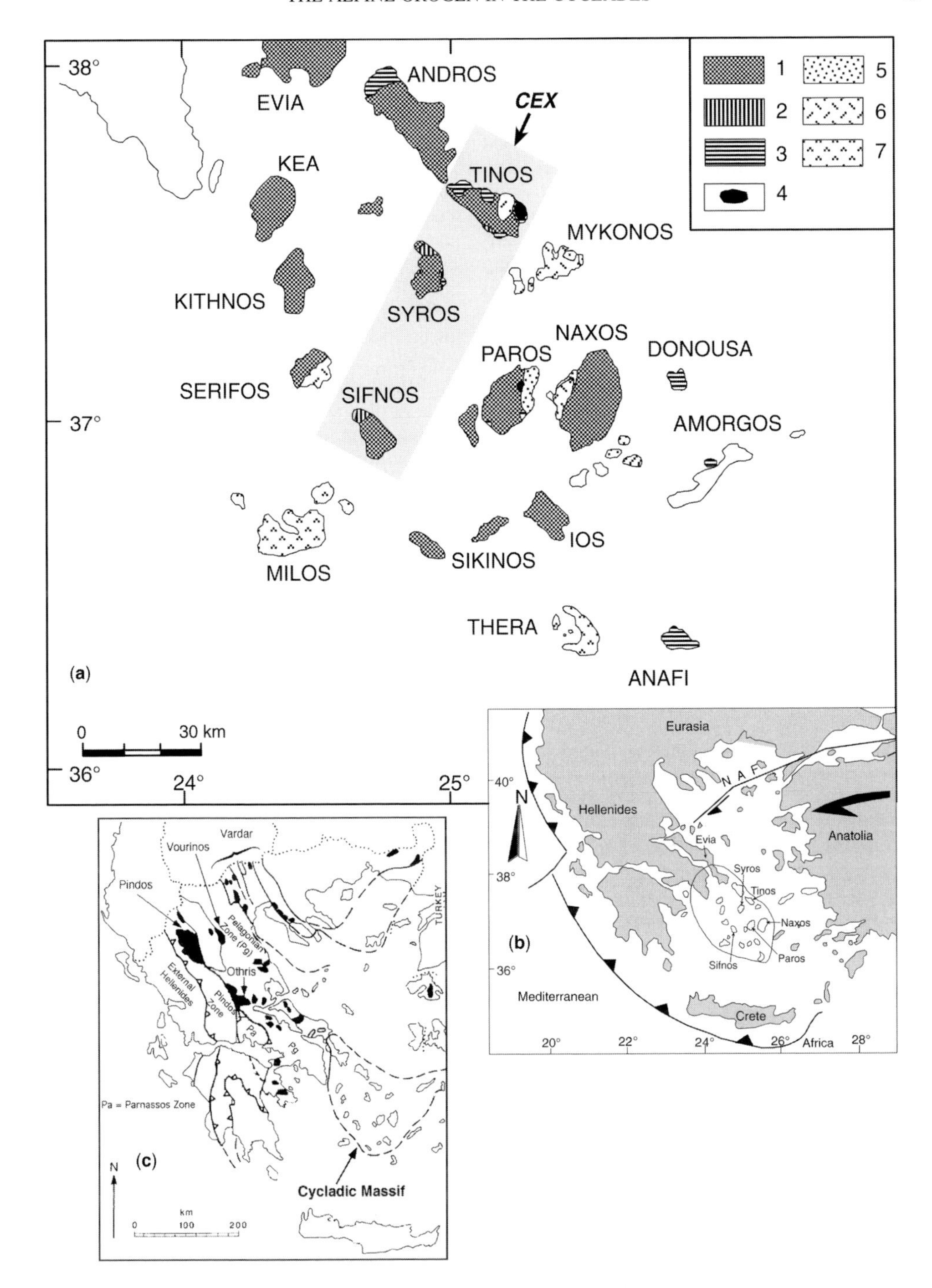

Fig. 1. (**a**) Geological map of the Cycladic Massif (after Avigad & Garfunkel 1991). The Cyclades Blueschist Unit (CBU, the Lower Unit): 1, Eocene high-pressure rocks overprinted at greenschist-facies conditions in the Early Miocene; 2, Eocene eclogite-facies rocks. The Upper Unit: 3, low-pressure metamorphic rocks, mostly of Late Cretaceous age; 4, ophiolites; 5, Early Miocene clastic sediments. Cenozoic igneous rocks: 6, Miocene granitoids; 7, Pliocene to recent volcanic rocks. CEX, Central Eclogite Axis. (**b**) Location map for the Cycladic Massif showing major tectonic features of the Aegean region: back-arc extension in the Aegean is promoted by the southward retreat of the Hellenic Trench and accommodates right-lateral motion on the North Anatolian Fault (NAF). (**c**) Map of Greece showing the geotectonic zones and the locations of main outcrops of Eohellenic ophiolites (from Smith 1993).

eclogites had already isothermally decompressed from their maximum burial depths to much shallower crustal levels. New thermodynamic analysis indicates, however, significant cooling during decompression for Syros and Sifnos eclogites, and suggests variable tectonic scenarios for the early exhumation of the Blueschist Unit (Trotet *et al.* 2001*b*; Schmädicke & Will 2003). Studying the spatial distribution and variation of the $P-T$ paths in the Cyclades can lead to better understanding of the exhumation-related tectonic processes in the time interval between high-pressure metamorphism in Cretaceous–Eocene times (M_1) to overprinting in the Early Miocene (M_2). Thus a major objective of this research is to study the petrological evolution of ultramafic and associated rocks of the Lower tectonic Unit both to the NW (Evia) and to the SE (Naxos) of the CEX. Additionally, a dismembered ophiolite of the Upper Unit that was juxtaposed on top of the blueschists on Tinos is studied to provide insight into processes that affected the overburden that covered the CBU during exhumation.

Ultramafic rock associations

The Alpine orogen in the Cyclades comprises three main tectono-metamorphic units separated by low-angle faults (Fig. 1; Dürr *et al.* 1978*a*; Avigad & Garfunkel 1991). The two upper units contain ultramafic rocks. The dominant Cycladic Blueschist Unit (CBU, the Lower Unit), the evolution of which is discussed above, is delimited from above by flat-lying normal faults that omit a significant part of the overburden that covered it since Early Miocene times (Lister *et al.* 1984; Ridley 1984*a*; Avigad & Garfunkel 1989; Gautier & Brun 1994; John & Howard 1995; Patriat & Jolivet 1998; Bröcker & Franz 1998; Ring *et al.* 2003). These detachments juxtaposed the Upper tectonic Unit, which comprises lithologically diverse thin remnants of the overburden, on top of the CBU. Tectonic windows on Evia and Tinos expose the lowermost weakly metamorphosed Basal Unit (Almyropotamos Unit), which consists of a thick platform carbonate sequence topped by Eocene flysch, beneath the blueschists (Dubois & Bignot 1979; Avigad & Garfunkel 1989). The apparent inverted age and metamorphic sequences indicate that the tectonic contact between the Lower and Basal Units is a major thrust fault (Katsikatsos *et al.* 1986; Matthews *et al.* 1999; Shaked *et al.* 2000).

The CBU is characterized by a change in protoliths from a sequence of continental origin in the central Cyclades (Naxos and Paros) to an oceanic or basinal sequence in the NW Cyclades (Syros, Tinos, Andros) and southern Evia. The Lower Unit on Naxos consists of pre-Alpine granitic and quartzo-feldspathic precursors overlain by a Mesozoic sedimentary sequence dominated by bauxite-bearing shelf carbonates (Jansen & Schuiling 1976; Feenstra 1985; Keay *et al.* 2001). Its counterpart in the NW Cyclades consists mainly of Mesozoic quartz-poor clastic and volcanic protoliths with moderate to minor amounts of carbonates that probably accumulated in a deep basin. Strikingly, ultramafic rocks are associated both with the leucogneisses and the overlying platformal sediments on Naxos and with the basinal sequences of the NW Cyclades, where they are associated as mélange constituents with other ophiolitic lithologies.

The Upper Unit comprises varied sequences of low- to medium-pressure metamorphic rocks and unmetamorphosed rocks. Metamorphic sequences include amphibolite- to greenschist-facies basites, acidites, pelites and marbles on the small islands of Donousa, Nikouria and Anafi on the eastern edge of the Cyclades and on Syros and Tinos (Dürr *et al.* 1978*a*; Reinecke *et al.* 1982; Maluski *et al.* 1987; Patzak *et al.* 1994; Ring *et al.* 2003; Beeri, Y., pers. comm.). High-pressure mineralogy was never found in the metamorphic rocks of the Upper Unit and metamorphism was dated as Late Cretaceous (Dürr *et al.* 1978*b*; Reinecke *et al.* 1982; Maluski *et al.* 1987; Altherr *et al.* 1994; Patzak *et al.* 1994). Both grade and age of metamorphism indicate that the rocks of the Upper Unit were at shallow crustal levels since the Late Cretaceous and escaped the Eocene high-pressure metamorphism that affected the CBU. Rare meta-serpentinites occur on Anafi; however, the largest ultramafic exposure occurs in the Upper Unit of Tinos: a metamorphosed dismembered ophiolite composed of mafic and ultramafic slices (Katzir *et al.* 1996). The tectono-stratigraphic position of the Tinos dismembered ophiolite and its sliced structure greatly differs from those of its counterpart ultramafic associations in the CBU, suggesting a distinctive tectonic setting for its deformation and metamorphism and possibly for its origin.

Three major ultramafic associations are thus focused on: (1) HP–LT ophiolitic mélanges in the NW Cyclades (Syros and Evia); (2) meta-peridotites associated with migmatized leucogneisses on Naxos; and (3) metamorphosed dismembered ophiolite in the Upper Unit of Tinos.

In addition to comparing their Alpine metamorphic evolution, we also use the field relations, pre-metamorphic textural and mineralogical relicts, and whole-rock chemical composition to shed light on the petrogenesis of the igneous constituents of the diverse ultramafic associations and on the tectonic setting in which they were assembled.

HP–LT ophiolitic mélanges (NW Cyclades)

In the basinal association of the NW Cyclades ultramafic rocks occur as sheets and lenses of meta-serpentinite or as meta-serpentinitic envelopes to metabasites within HP–LT ophiolitic mélanges. The two major sequences of ophiolitic mélange in the CBU occur on Syros and on southern Evia (Blake *et al.* 1981). On Syros the mélange interval occurs in the uppermost part of the Blueschist Unit and includes 'knockers' of metagabbro enclosed in a meta-greywacke sequence (Bonneau *et al.* 1980; Dixon & Ridley 1987; Bröcker & Enders 1999, 2001; Tomaschek *et al.* 2003). Thin and highly metasomatized serpentinite envelopes occur at the contacts between the metabasite blocks and the enclosing metasediments. On southern Evia, however, two intervals of ophiolitic mélange occur (Fig. 2). The Ochi ophiolitic mélange occurs at the upper levels of the Blueschist Unit, and according to its position, thickness and composition might be correlated to the Syros mélange (Table 1). The Tsaki ophiolitic mélange occurs at the base of the Lower Unit immediately above the basal thrust. Unlike the Syros mélange, both mélange intervals on Evia include large bodies of meta-serpentinite (up to several hundred metres in diameter).

Origin

In Table 1 the major pre-metamorphic characteristics of the ophiolitic mélange on Syros and Evia are described.

On both Syros and Evia, gabbros preserve a partial record of igneous and hydrothermal mineral assemblages and textures that formed in the oceanic crust, prior to high-pressure metamorphism. However, such preservation in the other mélange components is unique to Evia. The ultramafic protoliths of the meta-serpentinite lenses in the Evian mélanges probably represent two different levels in the oceanic lithosphere: whereas the cumulate wehrlitic lens and the gabbro-hosted wehrlite dyke of Ochi represent the mantle–crust transition zone, the bastite-bearing serpentinite of Tsaki was probably derived from mantle harzburgite. The thin, intensively sheared Syros serpentinites bear no igneous mineralogical or textural relicts, thus their origin is not disclosed. Likewise, the occurrence of metavolcanic blocks in the Evian mélanges is evident by the clearly observed pillowed structure and vesicles. On Syros, however, a volcanic origin for some of the metabasite blocks cannot be demonstrated unambiguously (Seck *et al.* 1996).

Further characterization of the tectonic setting in which the igneous protoliths of the blocks in the mélange were formed is possible by geochemical analysis (Seck *et al.* 1996; Katzir *et al.* 2000). Mineralogical, geochemical and isotope evidence indicates that the mélange constituents had been hydrothermally altered in an oceanic environment and possibly also metasomatized during subduction (Dixon & Ridley 1987; Katzir *et al.* 2000; Putlitz *et al.* 2000; Bröcker & Enders 2001; Tomaschek *et al.* 2003). The rare earth elements (REE) are considered immobile during fluid–rock interaction and are thus useful tools in studying the origin of Evia and Syros metabasites. Chondrite-normalized REE patterns given in Figure 3 show significantly higher REE abundances for both Ochi gabbros and basalts relative to their Tsaki equivalents, indicating different sources. In either of the mélange intervals on Evia the REE contents of metabasalts are significantly higher than those of the associated metagabbros. Basalts are particularly light REE (LREE)-enriched relative to gabbros, whereas heavy REE (HREE) contents of some of the metagabbros are almost equivalent to those of metabasalts. Assuming that the blocks in each mélange interval were assembled from a single source terrane, and given the lack of relict cumulate texture and positive Eu anomalies in Evian gabbros, the most plausible explanation for the depletion in incompatible elements in gabbros relative to the adjacent basalts is the loss of residual melt by compaction. REE abundances in metabasalts are thus a more reliable indicator of the tectonic setting of the source areas of the Evian mélanges. Ochi metabasalts are highly fractionated and LREE enriched: the La content is up to 120 times chondrite values and $(La/Yb)_{CN}$ values range from five to 10. These features indicate an enriched mantle source for Ochi basalts and are comparable with those of enriched mid-ocean ridge basalt (E-MORB) (Sun & McDonough 1989). Consistently lower REE abundances and less fractionated patterns are observed for the Tsaki metabasalts: La abundances are 20–30 times chondrite and $(La/Yb)_{CN}$ values are 2–3. These values are characteristic of normal MORB (N-MORB) to transitional MORB (T-MORB), or of tholeiitic basalts from back-arc setting. Syros metagabbros are the least REE-enriched rocks; however, REE abundances of other HP–LT metabasites of the Syros mélange are highly variable. The slightly upward-convex REE patterns of the Syros metagabbros were interpreted as representing a mixture of REE-poor cumulate phases and N-MORB intercumulus melt (Seck *et al.* 1996). Because of the lack of pre-metamorphic relicts, it is difficult to define the protoliths of Syros glaucophanites and eclogites. However, glaucophanites are fine to medium grained and highly foliated, whereas eclogites (and particularly garnet glaucophanites) show variable grain size and texture

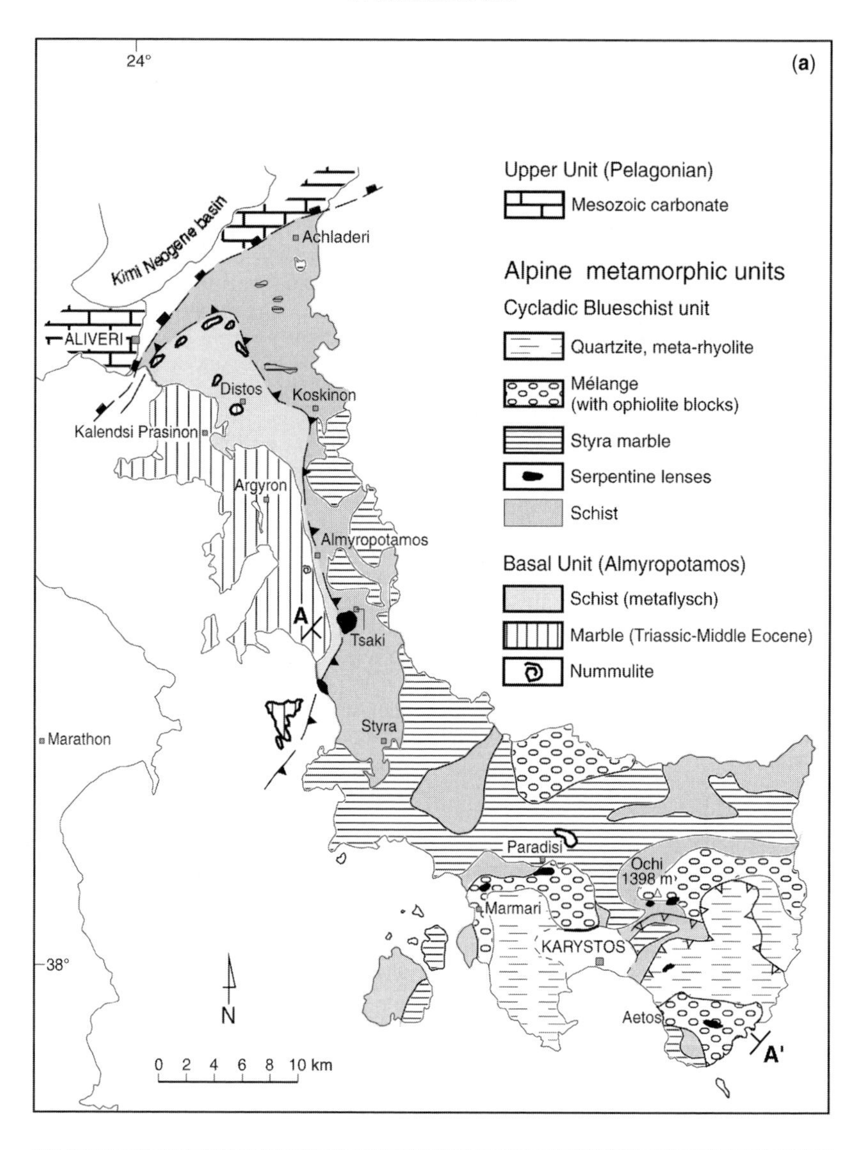

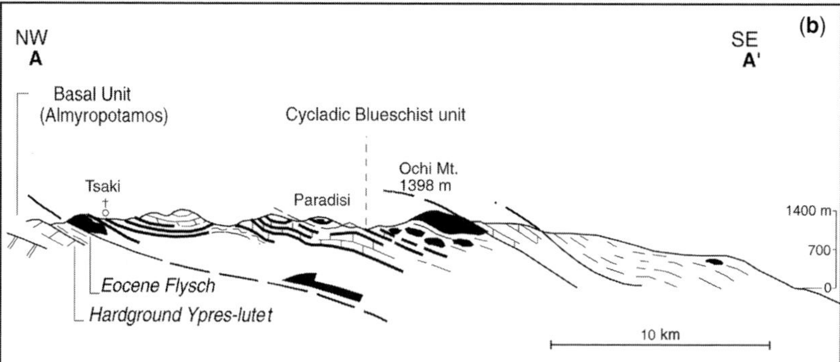

Fig. 2. Geological map (**a**) and cross-section (**b**) of southern Evia after Jacobshagen (1986), showing the tectonic contacts at the base and the top of the Cycladic Blueschist Unit and several sub-units within it.

Table 1. *Lithological and textural characteristics of HP–LT ophiolitic mélanges in the Cyclades with emphasis on pre-metamorphic features*

		Syros	Ochi	Tsaki
Lithological components	Blocks	Metagabbro (≤700 m), meta-acidite (granophyre?), clast-supported meta-igneous breccia of sedimentary origin	Metagabbro (≤200 m), meta-serpentinized wehrlite, metabasalt	Meta-serpentinite (≤500 m), metagabbro, metabasalt
	Matrix	Pelitic schist, meta-tuffites, metasomatized serpentinite, rare meta-chert, meta-ironstone	Semi-pelitic schist, meta-chert, meta-ironstone	Pelitic schist
Underlying sequence		Schist–marble (meta-flysch)	Semi-pelitic schist, meta-tuffite	Basal Unit (Almyropotamos)
Overlying sequence		Pelitic–psammitic schist, marble	Marble, quartzite, volcanogenic schist, meta-rhyolite	Pelitic schist
Pre-metamorphic relicts		Gabbro: pseudomorphs of actinolite after cpx (sub-sea-floor hydrothermal alteration) including rare augite cores; cumulate texture preserved by topotactic growth of metamorphic minerals (gln, ep) over igneous precursors (cpx, plg); intrusive relations between basic and acid rocks	Gabbro: aug, aug replaced by hbd Wehrlite: cumulate texture of cpx poikiloblasts enclosing mesh-textured serpentine after olivine Basalt: pillows zoned from omp and vesicle-rich core to gln-rich and non-vesicular rims Wehrlite dyke in gabbro	Serpentinite: bastite pseudomorphs after opx Gabbro: coarse-grained hbd Basalt: vesicular; sub-ophitic texture; Ti-aug cores in omp
Metamorphic textures related to pre-metamorphic features		Flaser gabbro: porphyroclasts of act or omp Metasomatic rinds at ultramafic contacts: omphacitite, jadeitite	Flaser gabbro: porphyroclasts of aug	Basalt: vesicles filled with various metamorphic assemblages

act, actinolite; aug, augite; cpx, clinopyroxene; ep, epidote; gln, glaucophane; hbd, hornblende; omp, omphacite; opx, orthopyroxene; plg, plagioclase.
Data from Syros and Evia are based on the observations of Dixon & Ridley (1987) and Katzir *et al.* (2000), respectively.

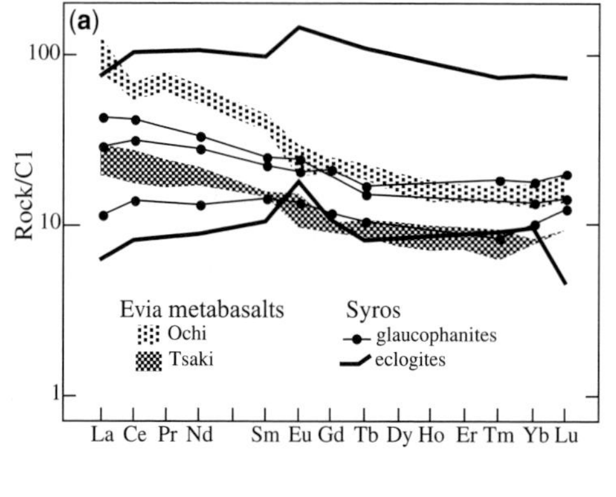

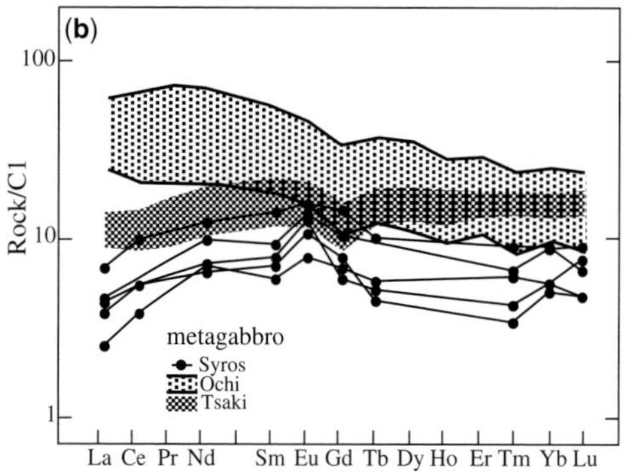

Fig. 3. Chondrite-normalized REE patterns of (**a**) metabasalts and (**b**) metagabbros from the HP–LT ophiolitic mélanges on Syros and southern Evia. Data for Syros and Evia are from Seck *et al.* (1996) and Katzir *et al.* (2000), respectively.

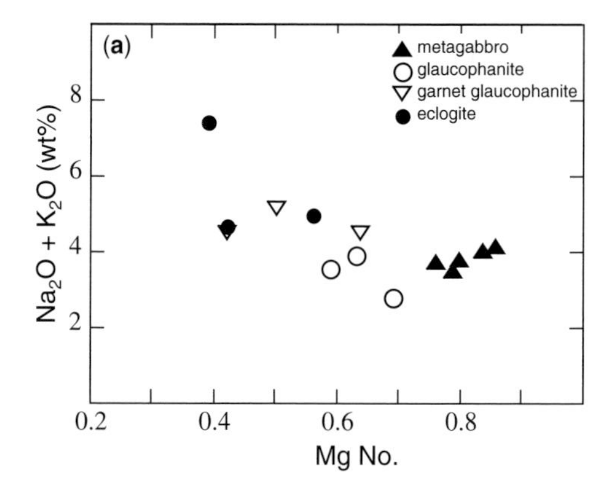

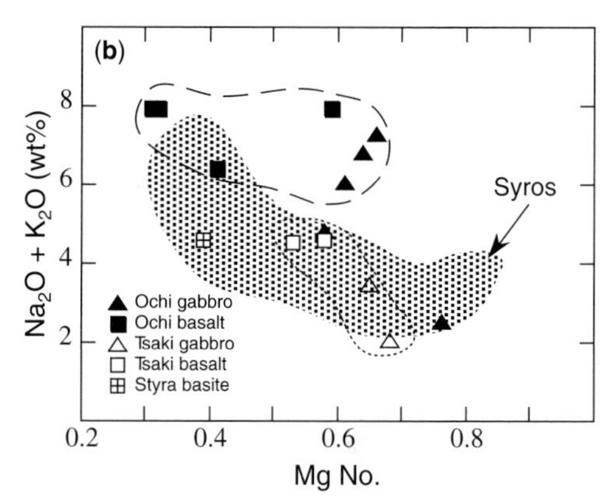

Fig. 4. Mg-number (atomic ratio (Mg/(Mg + Fe)) v. total alkali content ($Na_2O + K_2O$ wt%) in metabasites from (**a**) Syros and (**b**) Evia HP–LT ophiolitic mélanges. Data for Syros and Evia are from Seck *et al.* (1996) and Katzir *et al.* (2000), respectively.

including massive to foliated rocks (Seck *et al.* 1996). Based on textural criteria and the resemblance of their REE patterns to those of the Tsaki metabasalts, the Syros glaucophanites may have been derived from similar, moderately enriched basaltic protoliths. The textural variability of Syros eclogites and garnet glaucophanites is also reflected in their REE patterns. The coarser-grained, massive members have low to moderate total abundances of REE and positive Eu anomalies that suggest cumulate gabbro precursors. However, the most REE-rich eclogites (100 times chondrite) are foliated, possibly representing highly differentiated basalts. To sum up, the basaltic components of the NW Cyclades ophiolitic mélanges span a wide range of REE compositions. The REE patterns are characteristic of variable oceanic environments including N-, T- and E-MORB. These may represent either different segments of a single spreading centre or adjacent small oceanic basins.

Whereas REE abundances are considered as attained during igneous crystallization and retained through post-magmatic events, major element composition is susceptible to hydrothermal and metasomatic modifications. In Figure 4 the alkali element contents of Syros and Evia metabasites are plotted against Mg-number (Mg/(Mg + Fe) atomic ratio). In all three mélange sequences alkali element and iron enrichments are positively correlated. These correlations were interpreted in various ways. Based on mineralogical and geochemical variations between core and rim of pillow basalts in the Evian mélanges, Katzir *et al.* (2000) attributed the Na–Fe trends to various degrees of hydrothermal alteration in the oceanic crust. On Syros, however, where monomineralic reaction rinds (glaucophane, omphacite, actinolite and chlorite) between blocks and serpentinite envelopes are well developed, high-pressure metasomatism was seen as responsible for Na and Fe variations, and in some cases for extreme desilicification (Dixon & Ridley 1987). According to Seck *et al.* (1996), metasomatism in the Syros mélange is rather limited, and Fe-rich eclogites represent strongly differentiated basalts that crystallized in small magma chambers. Regardless of whether the major element variations in Syros and Evia metabasites are the result of igneous differentiation, oceanic hydrothermal alteration, synmetamorphic fluid–rock interaction or some combination of these processes, a major conclusion may be drawn: in all three mélange intervals wide ranges of major element compositions exist (SiO_2 46–55 wt%; Mg-number 0.31–0.76; $Na_2O + K_2O$ 4.5–8 wt%) allowing the crystallization of diverse, yet comparable metamorphic assemblages.

Formation of mélanges

The occurrence of glaucophane and omphacite metasomatic reaction zones at the serpentinite–mafic blocks and serpentinite–metasedimentary matrix contacts on Syros indicates that juxtaposition of the mélange components preceded peak M_1 metamorphism (Dixon & Ridley 1987). On Evia chlorite and tremolite blackwalls envelop the ultramafic bodies and follow their current geometry, thus also indicating pre- to syn-M_1 incorporation of the igneous bodies into the enclosing sedimentary sequence. The boudin shape of most blocks, including the largest ones, suggests that the last stage of the evolution of the mélanges involved significant flattening by extensional strain during metamorphism. However, the process by which the mélange components were initially assembled cannot be unequivocally deduced. The Evian and Syros mélanges may have formed either as olistostrome horizons within thick sedimentary flysch sequences (Dixon & Ridley 1987; see also Mukhin 1996) or as shear zones that separate distinct thrust sheets (Bröcker & Enders 2001). A combined scenario of a structurally weak serpentinite-bearing stratigraphic layer that later developed into a shear zone and was reshaped as a tectonic mélange is also possible (Bröcker & Enders 2001). The Evian and Syros mélanges differ totally, however, from Franciscan-type trench mélanges: unlike their Franciscan counterparts, which form sequences several kilometres thick exposed regionally, ophiolitic mélanges in the Cyclades are rare and occur in thin horizons. Also, the metamorphic conditions recorded in the igneous blocks and sedimentary matrix of the Cycladic mélanges are homogeneous, whereas the Franciscan-type complexes are characterized by high variability of metamorphic temperatures and pressures in the blocks, often included in non- or weakly metamorphosed clastic sediments. Thus, models suggested for the formation of trench mélanges such as return laminar flow in the subduction trench are definitely inapplicable to the HP–LT ophiolitic mélanges in the Cyclades.

High-pressure metamorphism (M_1)

Early field and petrographic studies in the Cycladic Blueschist Unit suggested an increase in M_1 metamorphic grade from southern Evia southeastwards through Andros to the Central Eclogite Axis (Blake *et al.* 1981; Bonneau & Kienast 1982). Southern Evia was considered a lower-grade blueschist terrane because of the occurrence of lawsonite and the absence of garnet in glaucophane-rich metabasites. In comparison, garnet is abundant in Syros eclogites and glaucophanites, and the occurrence of epidote–white mica pseudomorphs after lawsonite confines its stability to pre-peak M_1 prograde conditions (Ridley 1984*b*; Putlitz *et al.* 2005). A temperature range of 450–500 °C was determined for M_1 on Syros by Fe–Mg exchange thermometry of garnet and omphacite (Okrusch & Bröcker 1990). Similar temperatures were obtained by cation exchange thermometry for Sifnos and Tinos eclogites (Schliestedt 1986; Okrusch & Bröcker 1990) and confirmed by oxygen isotope thermometry on Sifnos and Syros (Matthews & Schliestedt 1984; Matthews 1994; Putlitz *et al.* 2000). More recent studies have argued, however, for higher peak M_1 temperatures: ≤ 580 °C (Trotet *et al.* 2001*b*) and 550–600 °C (Schmädicke & Will 2003) (Table 2). The first estimate is based on multi-equilibrium approach using TWEEQ whereas the latter uses the same cation thermometry as in earlier studies. We find no reason to prefer the recent, higher-temperature estimates to the previous ones. On the contrary, peak M_1 temperatures of 450–500 °C have been advocated by numerous petrological and isotope studies on a variety of rock compositions (Evans 1986; Schliestedt 1986; Schliestedt & Okrusch 1988; Matthews 1994; Putlitz *et al.* 2000, 2005; see Table 2).

Our observations on southern Evia showed that the dominant high-pressure assemblage in Ochi and Tsaki metabasites is epidote–glaucophane–omphacite, which defines the M_1 grade as epidote blueschist (Katzir *et al.* 2000). Relict lawsonite enclosed in various high-pressure and retrograde minerals records an earlier, prograde part of the P–T path. The reaction lawsonite + jadeite = zoisite + paragonite + quartz + H_2O is thus a lower temperature limit for the peak M_1 assemblages on both Syros and southern Evia. Notwithstanding, the absence of garnet in Evian blueschists is either temperature or composition dependent. It has been shown above that Evian and Syros metabasites span a rather wide and comparable range of compositions. This includes SiO_2, MgO, FeO and CaO, the activities of which strongly affect the garnet-in reaction in blueschists of the haplobasaltic system: chlorite + quartz + epidote = garnet + actinolite + H_2O (Evans 1990). Thus, peak M_1 temperatures on

Table 2. *Temperature and pressure estimates for the M_1 (Eocene) and M_2 (Miocene) metamorphic events in the Central Eclogite Axis of the Cyclades (Sifnos and Syros) and on southern Evia*

	Sifnos			Syros			Southern Evia		
	P–T	References	Method	*P–T*	References	Method	*P–T*	Reference	Method
M_1									
T (°C)	470 ± 30	1, 2, 3	ce, pd	450–500	3, 7	ce, pd	400–430	10	pd
	480 ± 25	4	oxi	500 ± 30	8	oxi	400	11	pd
	≤580	5	pd	450	9	pd			
	550–600	6	ce	≤580	5				
P (kbar)	15 ± 3	1, 3	pd	14–20	3, 7	pd	>12	10	pd
	14–18	2	pd	≤20	5	pd	>10–12	12	pd
	20	5, 6	pd						
M_2									
T (°C)	≤450	13, 14	pd	370–430	9		300–350	10	pd
	400–500	15	oxi	≤520	5				
	380–440	6	pd						
P (kbar)	5–7	13, 14	pd	6–9	9		4–8	10	pd
	9	6	pd	≤11	5				

References: 1, Schliestedt (1986); 2, Evans (1986); 3, Okrusch & Bröcker (1990); 4, Matthews (1994); 5, Trotet *et al.* (2001*b*); 6, Schmädicke & Will (2003); 7, Ridley (1984*b*); 8, Putlitz *et al.* (2000); 9, Putlitz *et al.* (2005); 10, Katzir *et al.* (2000); 11, Reinecke (1986); 12, Shaked *et al.* (2000); 13, Schliestedt & Matthews (1987); 14, Avigad *et al.* (1992); 15, Matthews & Schliestedt (1984). ce, cation exchange thermometry; oxi, oxygen isotope thermometry; pd, phase diagram calculations.

Evia were not high enough to allow garnet to crystallize in metabasites. The preservation of relict igneous minerals and lawsonite in Evia metabasites also indicates that M_1 temperatures were lower than for the CEX; they were estimated at 400–430 °C (Katzir *et al.* 2000).

The lower and upper pressure limits on Syros are defined, respectively, by the occurrence of $Jadeite_{92}$ + quartz in meta-acidites and of paragonite instead of omphacite–kyanite in eclogites (Ridley 1984*b*; Schliestedt *et al.* 1987; Okrusch & Bröcker 1990). At 470 °C the corresponding reactions of jadeite + quartz = albite and paragonite = jadeite + kyanite+ H_2O define a pressure range of 12–20 kbar (Schliestedt 1986; Schliestedt & Okrusch 1988; Okrusch & Bröcker 1990). The maximum jadeite content of sodic clinopyroxene in Evian meta-acidites is equal to that of their counterparts on Syros, indicating similar minimum M_1 pressures of ≥12 kbar (Schliestedt *et al.* 1987; Katzir *et al.* 2000).

There are two possible ways to interpret the distribution of the M_1 *P–T* estimates in the NW Cyclades and on southern Evia. In the first interpretation, peak pressures, like temperatures, were lower on southern Evia compared with the CEX. Consequently, Evian blueschists were subducted to shallower levels than the Cycladic eclogites, and both terranes experienced similar *P–T* trajectories. Alternatively, Evian blueschists and Cycladic eclogites represent the same depth interval of the subducted plate; however, the latter remained longer at deep crustal levels, which allowed prolonged heating.

Metamorphic overprint (M_2)

The M_2 metamorphic overprint occurred throughout the CEX at greenschist-facies conditions. A typical M_2 assemblage in metabasites includes albite–chlorite–epidote–phengite–calcic amphibole. The presence of barroisitic amphibole in some greenschists and the scarcity of biotite in metapelites constrain M_2 conditions to 6–7 kbar and ≤450 °C, respectively (Schliestedt *et al.* 1987; Bröcker *et al.* 1993). Oxygen isotope thermometry in Sifnos (quartz–epidote; quartz–phengite) and Tinos (quartz–magnetite) greenschist-facies rocks gave temperatures of 400–500 °C and 440–470 °C, respectively (Matthews & Schliestedt 1984; Bröcker *et al.* 1993). The temperature ranges of the M_1 high-pressure metamorphism and the M_2 greenschist overprint overlap, thus indicating that within the M_1–M_2 time interval the high-pressure rocks of the CEX experienced isothermal decompression. This two-point isothermal path was further established by a detailed petrological study of a Sifnos rock sequence that described several stages of metamorphic equilibration at successively decreasing pressures: relict eclogite-facies rocks transformed into albite-bearing, garnet-free epidote blueschist to be finally equilibrated at greenschist-facies conditions (Avigad *et al.* 1992). Recent studies claimed to show, however, that high-pressure rocks of the CEX experienced significant cooling during decompression (Table 2). Based on local equilibria in chlorite–phengite-bearing assemblages, Trotet *et al.* (2001*b*) argued that eclogites at the top of the CBU on Sifnos and Syros cooled during decompression whereas the lower, strongly overprinted part of the section decompressed isothermally. THERMOCALC-based petrogenetic grids calculated for the well-preserved eclogites and for the underlying greenschists on Sifnos show, however, decompressional cooling for both sequences (Schmädicke & Will 2003). The temperatures estimated for the retrograde M_2 overprint in both studies do not appreciably differ from the 'traditional' petrological and isotope thermometry estimates: they all converge at 400–500 °C (Table 2). The apparent cooling postulated for the Cycladic eclogites mostly stems from new peak temperatures assigned to the M_1 event: 550–600 °C. These estimates disagree with numerous well-established petrological and isotope studies (see above) and thus isothermal decompression still seems the most probable path for exhumation in the CEX.

The M_2 assemblage of metabasites in the Evian mélanges, actinolite–pumpellyite–epidote–chlorite–albite–phengite, is characteristic of the pumpellyite–actinolite facies (Banno 1998). Phase diagram calculations using mineral compositions in Evian metabasites limit this paragenesis to pressures of 4–8 kbar and temperatures of 300–350 °C (Table 2; Katzir *et al.* 2000). The M_2 assemblage in Tsaki metagabbros differs, however, from its Ochi counterpart: it includes sodic augite, which indicates lower temperatures relative to the clinopyroxene-free Ochi assemblage (e.g. Maruyama & Liou 1985).

By studying mineral assemblages in ophiolitic mélanges, a clear distinction between the metamorphic histories of the Lower tectonic Unit of Evia and the CEX can be made: whereas Syros eclogites ($T = 450–500$ °C) decompressed isothermally through greenschist-facies conditions, decompression of the lower-grade Evian blueschists ($T = 400–430$ °C) involved cooling to pumpellyite–actinolite-facies conditions ($T = 300–350$ °C). Moreover, Tsaki metabasites at the base of the section on Evia were overprinted at lower temperatures compared with their Ochi equivalents that occur 2 km upsection. This decrease in M_2 temperatures downwards towards the base of the Blueschist Unit can be explained by the underthrusting of the Late Eocene sedimentary rocks of the Basal Unit (Almyropotamos platform), which either caused cooling by conduction or resulted in a more prolonged retrograde metamorphism impelled by the infiltration of fluids from below.

Leucogneiss-associated peridotites (Naxos)

Both the protoliths and the metamorphic evolution of the sequence that hosts ultramafic horizons in the Lower Unit of Naxos are very different from the NW Cycladic mélanges. On Naxos an Early Miocene (18 Ma; Andriessen *et al.* 1979; Wijbrans & McDougall 1988; Keay *et al.* 2001). Barrovian-type overprint (M_2) occurred during extension and exhumation of former high-pressure rocks (M_1) and almost totally effaced their assemblages and fabrics. Naxos (Fig. 5) is a mantled gneiss dome whose core consists of migmatites formed during M2 metamorphism of quartzofeldspathic ortho- and para-gneisses (Buick 1988; Pe-Piper *et al.* 1997; Keay *et al.* 2001). Overlying the leucogneiss core (Buick 1988) is a 7 km thick metasedimentary envelope dominated by siliciclastic schists and gneisses in its lower part ('Lower Series' of Jansen & Schuiling 1976) and by meta-bauxite-bearing marbles in its upper part ('Upper Series'). The grade and intensity of M_2 metamorphism increase with increasing structural depth from greenschist-facies rocks containing relict M_1 assemblages at the top of the sequence to upper amphibolite-facies rocks in the core. The approximate range of M_2 temperatures spanned by the Barrovian facies series is 400–700 °C at pressures of 5–7 kbar (Jansen 1977; Feenstra 1985). Further petrological studies refined the peak M_2 conditions that have occurred at the leucogneiss core to 6 ± 2 kbar and 670–700 °C (Buick & Holland 1989, 1991; Katzir *et al.* 1999, 2002; Matthews *et al.* 2003). However, remnants of the former high-pressure mineralogy (M_1) have not been found in the leucogneiss core of Naxos, thus its exhumation $P–T$ path remains speculative. A first view of the pre-M_2 evolution of the Naxos core was made possible by petrological and oxygen isotope study of peridotite lenses hosted by the silicic gneisses (Katzir *et al.* 1999, 2002).

Field relations and M_2 metamorphism

Ultramafic horizons occur at four structural levels within the metamorphic sequence of Naxos (Fig. 5): (1) the Main Ultramafic Horizon (MUH) occurs within upper amphibolite-facies rocks at the transition from the leucogneiss core to the Lower Series; it is composed of 1–10 m sized lenses of massive to moderately foliated medium- to coarse-grained meta-peridotites; (2) the Agia ultramafic horizon occurs in NW Naxos within sillimanite-grade rocks of the Lower Series; (3) sporadic ultramafic bodies occur within staurolite–kyanite-grade rocks at the transition from the Lower to the Upper Series; and (4) within greenschist-facies rocks of the Upper Series.

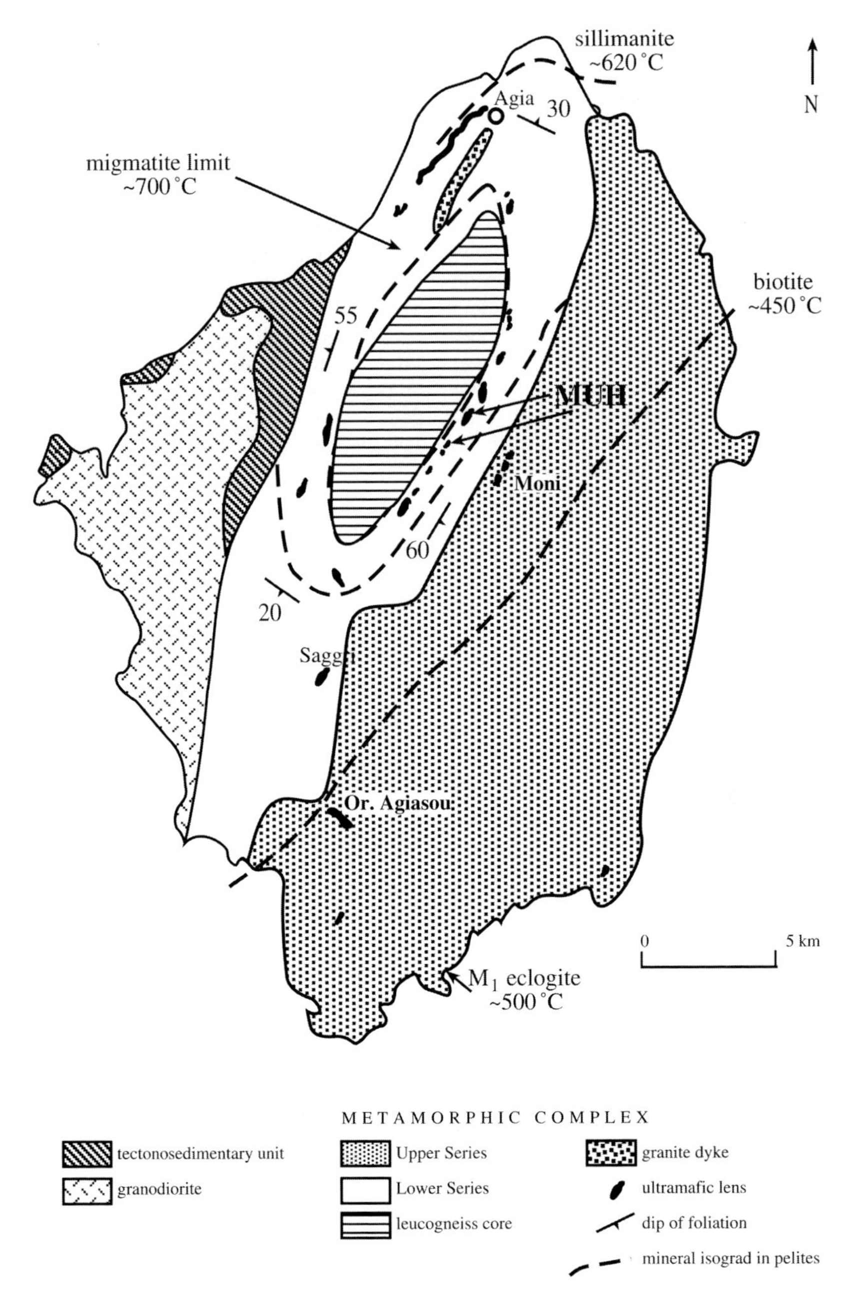

Fig. 5. A simplified geological map of Naxos after Jansen & Schuiling (1976) showing the metamorphic complex of the CBU (subdivided), the tectonosedimentary Upper Unit and a 12 Ma granodiorite. Ultramafic rocks (in black) occur within the metamorphic dome at four structural levels: (1) the Main Ultramafic Horizon (MUH) occurs at the transition from the leucogneiss core to the overlying Lower Series; (2) the Agia ultramafic horizon within the Lower Series; (3) at the transition from the Lower to Upper series (Moni and Saggri exposures); (4) within the Upper Series (Ormos Agiasou exposure). The M_2 mineral isograds mapped by Jansen & Schuiling (1976) and modified by Feenstra (1985) and Buick (1988) are shown. The occurrence of particularly well-preserved M_1 high-pressure rocks at the top of the section on SE Naxos is shown (Avigad 1998).

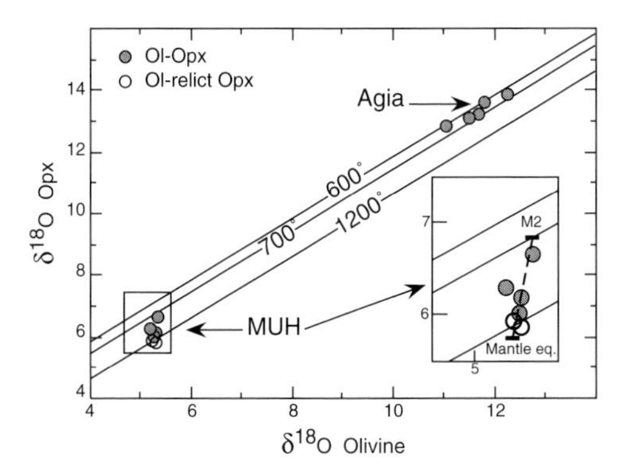

Fig. 6. Plot of $\delta^{18}O$ (orthopyroxene) v. $\delta^{18}O$ (olivine) in meta-peridotites of the MUH and the Agia ultramafic horizon on Naxos. Isotherms are calculated according to Rosenbaum *et al.* (1994). The analytical uncertainty is smaller than the symbol size.

A non-deformed talc–enstatite assemblage and extremely high $\delta^{18}O$ values of olivine and orthopyroxene (11–14‰; Fig. 6) indicate that recrystallization of the Agia meta-peridotites occurred during post peak M_2 infiltration of silica-rich fluids from pegmatites (Katzir *et al.* 2002; Matthews *et al.* 2003). This fluid-flow event erased the former mineralogy of the Agia peridotites, thus precluding further investigation of their origin and emplacement. Notwithstanding, both the MUH meta-peridotites and the ultramafic bodies that occur at the base and within the Upper Series are structurally concordant and isofacially metamorphosed with their host rocks. A synkinematic olivine–orthopyroxene–hornblende–chlorite–spinel assemblage is dominant in the MUH peridotites. Anthophyllite–talc and antigorite–talc schists characterize the two ultramafic occurrences of the Upper Series, respectively. Temperature estimates of the ultramafic assemblages agree well with those of the host M_2 Barrovian series: *c.* 700 °C in the sillimanite schist-associated MUH, *c.* 580 °C in anthophyllite–talc schists enclosed by staurolite–kyanite-grade pelites and <450 °C in antigorite–talc schists of the biotite zone. Highly foliated metasomatic reaction zones are well developed at the ultramafic–felsic rocks interface: phlogopite–actinolite–anthophyllite in the MUH and chlorite–tremolite in the structurally higher located bodies. The variable mineralogy of the metasomatic zones is mostly dependent on the chemical potentials of various ions, but is also consistent with increase in temperature going downsection from the Upper Series to the leucogneiss core.

Field and petrological evidence thus indicates that the Naxos meta-peridotites have experienced the Barrovian M_2 metamorphism with their host rocks.

Pre-M_2 evolution

Relict phases in ultramafic rocks of different structural levels on Naxos indicate two distinct pre-M_2 histories. In the upper two ultramafic bodies mesh textured serpentine associated with fine-grained magnetite is overgrown by the M_2 talc-bearing assemblages. The occurrence of early serpentine fabrics overprinted by M_2 mineralogy indicates that the Upper Series ultramafic rocks were first denuded and hydrated prior to Alpine metamorphism either on the sea floor or during ophiolite emplacement. In contrast, the MUH peridotites show no signs of early serpentinization and instead preserve their mantle oxygen isotope signature ($\delta^{18}O$ (Ol) = 5.2‰) and some of their original mantle assemblage. Large, moderately deformed porphyroclasts of olivine and orthopyroxene are randomly oriented within the M_2 recrystallized matrix. The pre-kinematic orthopyroxenes have high-Al_2O_3 and high-CaO cores (up to 5.5% and 0.9%, respectively) and contain exsolution lamellae of Cr-spinel. Other relicts of the pre-M_2 peridotite include rare grains of high-Al green spinel (60 wt% Al_2O_3). Aluminium and Ca-in-Opx thermometry of relict orthopyroxene, olivine and spinel yields temperatures of *c.* 1050 °C (Katzir *et al.* 1999). Oxygen isotope Opx–Ol thermometry in the MUH meta-peridotites gives a bimodal distribution of temperatures, grouped at 700 °C and 1200 °C, and indicates partial oxygen exchange during M_2 superimposed on previous mantle fractionation (Fig. 6; Katzir *et al.* 2002).

Because emplacement of peridotites into shallow crustal levels invariably involves serpentinization it is concluded that the MUH peridotites were directly transported from the mantle and tectonically interleaved with the continental crustal section of Naxos at depth. Probably, the incorporation of the ultramafic rocks into the upper crustal section occurred while the latter was underthrust and buried to great depth during Alpine collision and high-pressure metamorphism (M_1). Their emplacement at the base of the orogenic wedge is inferred to have involved isobaric cooling from temperatures of *c.* 1050 °C within the spinel lherzolite field to eclogite-facies temperatures higher than 500 °C, the upper stability limit of serpentine. Further support for the deep origin of the Naxos core is given by down-section extrapolation: M_1 pressures of ≥ 12 kbar and temperatures of ≤ 500 °C were estimated for relict eclogite at the top of the section on SE Naxos, corresponding to a regional average gradient of *c.* 15 °C km^{-1} (Fig. 5; Avigad 1998). Bearing in mind that the present-day 7 km thick section separating SE Naxos from the leucogneiss core underwent post-M_1 ductile thinning (Buick 1991), minimum pressures of *c*. 14 kbar and temperatures of 600 °C may be estimated for M_1 at the core.

A dismembered ophiolite (Tinos Upper Unit)

On the island of Tinos several slices of the Upper Unit were tectonically juxtaposed over the CBU before the emplacement of an 18 Ma old granite, which intrudes the contact between them (Fig. 7). A detailed study of the Upper Unit occurrences revealed that they consist mostly of a dismembered ophiolite sequence (Katzir *et al.* 1996). This includes tens-of-metres thick slices of strongly sheared phyllite (metabasalt), mostly at the base, sheared and massive serpentinites and metagabbros (Fig. 8). These slices reach a total thickness of up to 200–300 m. The strong slicing greatly reduced the thickness of the ophiolite and also disrupted the typical order of the various lithologies.

In spite of their common oceanic affinity, the Tinos dismembered ophiolite and the NW Cyclades HP–LT ophiolitic mélanges differ in major aspects of their evolution.

(1) Lower than igneous oxygen isotope ratios in relict oceanic actinolite of metagabbros from the Syros mélange (3–5‰; Putlitz *et al.* 2000) are similar to ratios measured in modern oceanic gabbros and in the intact, fully developed ophiolite suites of Oman and Troodos (Heaton & Sheppard 1977; Gregory & Taylor 1981). These isotope ratios indicate high-temperature interaction with seawater during the generation of new oceanic crust. In contrast, hornblende of sub-sea-floor hydrothermal origin in metagabbros of the Tinos

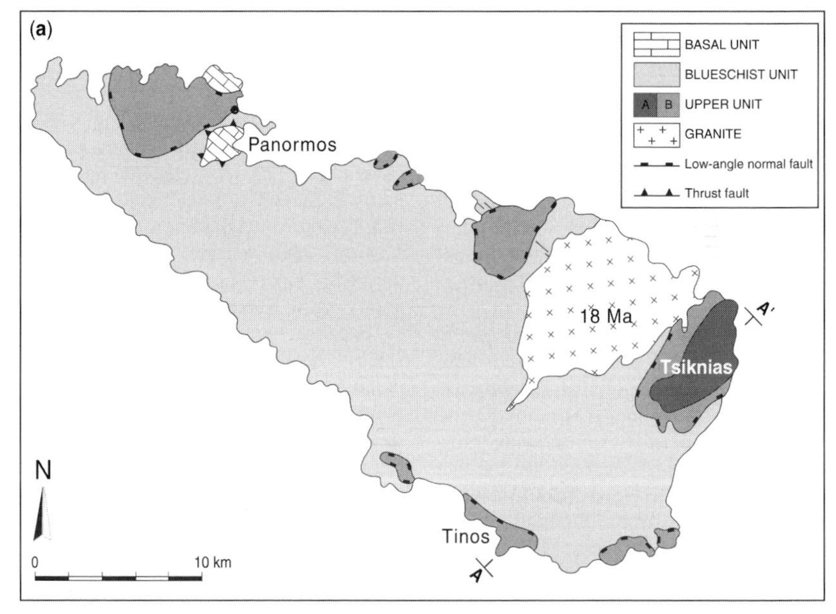

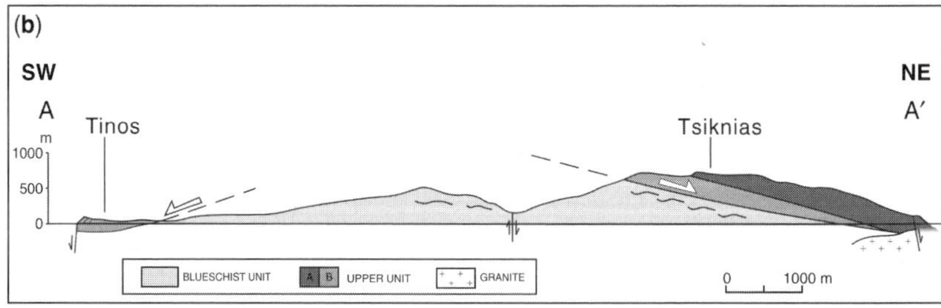

Fig. 7. (**a**) Geological map of Tinos (after Avigad & Garfunkel 1989). The Upper Unit includes slices of a dismembered ophiolite: serpentinites and metagabbros (A, dark grey) overlying mafic phyllites (B, bright grey). (**b**) A structural cross-section (A–A′ in (a)) of Tinos showing the main detachment surface that separates the Upper Tectonic Unit in the hanging wall from the underlying CBU. An 18 Ma granite that intrudes both tectonic units near Tsiknias gives an upper time constraint for their juxtaposition.

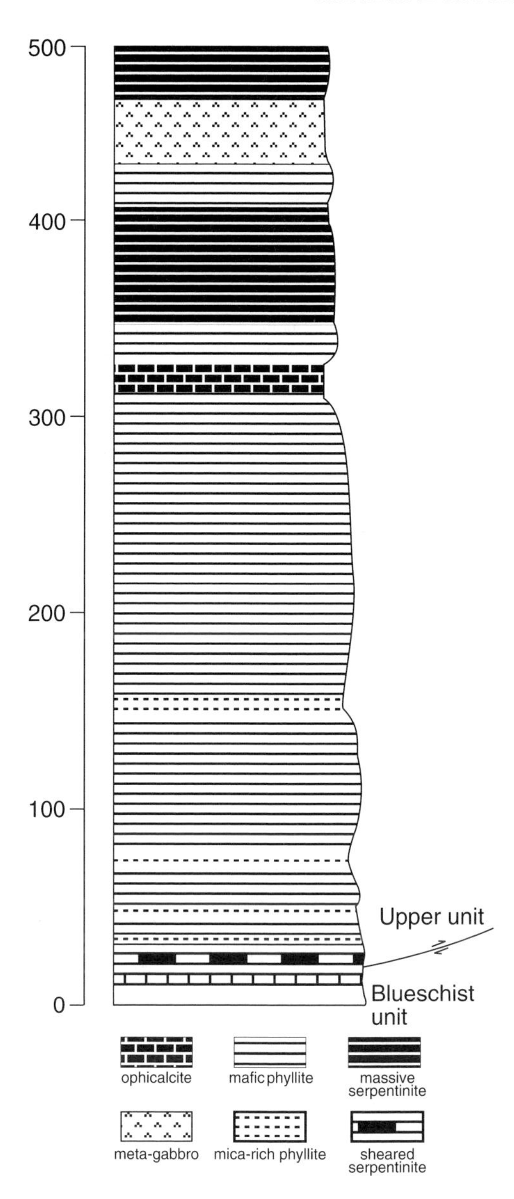

Fig. 8. General lithological columnar section of the dismembered ophiolite of the Upper Unit on Tinos (after Katzir *et al.* 1996; Zeffren *et al.* 2005). On the basis of textural and geochemical observations, mafic phyllites were interpreted as metabasalts.

Upper Unit has higher than igneous $\delta^{18}O$ values (5.5–7.5‰, Putlitz *et al.* 2001). The high $\delta^{18}O$ values of the hornblendes are compatible with interaction of oceanic gabbros with seawater that had previously been enriched in $^{18}O/^{16}O$ by isotope exchange at high temperatures. Isotope exchange is thought to have occurred by deep penetration of seawater during early, hot intra-oceanic thrusting (Putlitz *et al.* 2001). Thus, the Syros and Tinos metagabbros represent oceanic floor segments that have undergone different sea-floor alteration: the Syros metagabbros represent 'typical' sea-floor alteration whereas the Tinos metagabbros represent a different environment, affected by tectonic disturbance.

(2) In contrast to the random occurrence of ophiolitic lithologies in the mélanges, the Tinos ophiolite consists of coherent slices, each slice dominated by a single oceanic lithology and bounded by a low-angle tectonic contact (Fig. 8). The sequence of slices does not follow the primary order of the ophiolite suite, but sometimes repeats or reverses the original order. Dissection of the primary oceanic crust by low-angle reverse faults can account for the disturbed order of the oceanic lithologies. However, reverse faulting cannot account for the absence of primary components in the Tinos ophilolite (deep-sea sediments), or for the significant reduction in its thickness. Omission of major portions of the ophiolite suite requires, instead, a post-thrusting phase of normal faulting with flat-lying faults that cut out substantial portions of the original slice pile. Such extension agrees well with the strong slicing within the ophiolite and with the normal faulting at the base of the Upper Unit (Avigad & Garfunkel 1989, 1991; Patriat & Jolivet 1998).

(3) Unlike the HP–LT mélanges of the NW Cyclades, the Tinos ophiolite was never buried to great depths. However, it was metamorphosed at greenschist-facies conditions in the course of the Alpine orogenesis. Greenschist-facies mineralogy overprints the early oceanic fabrics in all major rock types of the ophiolite and is associated with penetrative deformation in the phyllites and sheared serpentinites and gabbros. Some of the metamorphic mineral growth in the Upper Unit exposure of Mt. Tsiknias, including crystallization of olivine neoblasts in serpentinite, was considered to occur by contact metamorphism induced by the intrusion of the granite at 18 Ma (Fig. 8; Stolz *et al.* 1997). However, this cannot account for a major penetrative regional metamorphism observed in all rock types and exposures of the Tinos Upper Unit. Given the field evidence for early oceanic thrusting, Katzir *et al.* (1996) suggested that metamorphism was induced by continued thrusting and piling of nappes that have created the necessary overburden to cause greenschist-facies conditions. Based on K–Ar ages on amphiboles extracted from an amphibolite-facies slice that constitutes the topmost part of the Upper Unit on Tinos (Patzak *et al.* 1994), a Late Cretaceous age was assigned to the metamorphic event that affected the entire sequence of the Upper Unit (Katzir *et al.* 1996). Late Cretaceous ages were also determined

for Upper Unit gneisses and amphibolites on several other islands in the Cyclades (Reinecke *et al.* 1982; Maluski *et al.* 1987; Altherr *et al.* 1994). The first geochronological results for the phyllites of the Upper Unit were obtained by Rb–Sr dating, which yielded ages between 21 and 92 Ma (Bröcker & Franz 1998). The youngest age was obtained for a sample collected near the contact with the CBU. Recently obtained $^{40}Ar/^{39}Ar$ ages on synkinematic white micas from phyllites at the base of the Upper Unit were concentrated in the Oligocene–Miocene (31–21 Ma; Zeffren *et al.* 2005). These ages suggest that the rocks represent an extensional shear zone that operated in the Late Oligocene and juxtaposed the Upper Unit on top of the partially exhumed high-pressure rocks of the CBU. It is thus concluded that the lower part of the Upper Unit experienced recrystallization and age resetting in a time interval that corresponds to the M_2 overprint in the CBU. The upper part of the dismembered ophiolite records, however, an older, pre-Tertiary metamorphic history.

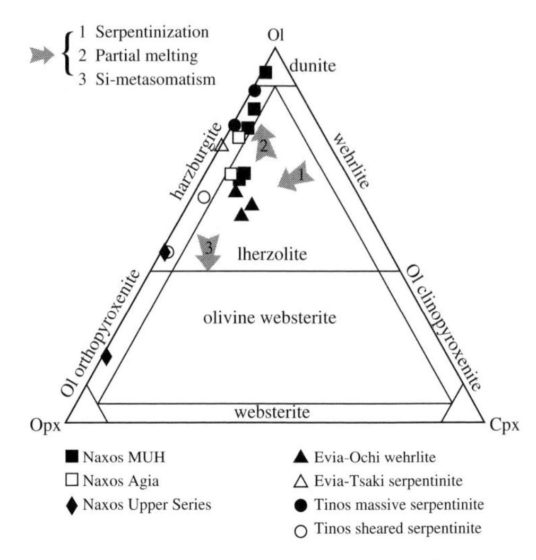

Fig. 9. Spinel peridotite norm compositions of the ultramafic rocks of the Cyclades plotted on the ultramafic classification diagram. Arrows indicate compositional trends resulting from serpentinization, partial melting and Si-metasomatism.

Discussion

Assembly and evolution of the ultramafic associations

The origin and manner of assembly of the major ultramafic rock-bearing associations in the Cyclades are diverse. A clear distinction exists between the Naxos association, where thin ultramafic horizons occur within continental quartzofeldspathic gneisses (MUH) and continental platform metasediments (the Agia and Upper Series horizons), and the other two associations, where ultramafic rocks are involved with oceanic crust. The oceanic associations are further distinguished by their manner of assembly: ophiolitic olistotrome intervals within basinal sedimentary sequences in the Lower Unit compared with sliced dismembered ophiolite in the Upper Unit. The ultramafic rocks of the Lower Unit, both on Naxos and in the NW Cyclades, were incorporated into their respective platformal and basinal host sequences prior to or during the high-pressure M_1 metamorphism. Since then they have experienced with their host rocks the whole cycle of Alpine Tertiary orogenesis including compression, exhumation and metamorphic overprint during extension. The ultramafic slices of the Tinos Upper Unit, however, escaped most of the Tertiary tectonometamorphic cycle, but could not avoid the extensional overprint that resulted in their final juxtaposition against other ophiolitic slices in the Early Miocene (Zeffren *et al.* 2005).

The combined effect of the processes that shaped the Cycladic ultramafic rocks since their derivation from the mantle, through polymetamorphism to denudation and serpentinization (or vice versa) is reflected in their whole-rock geochemical compositions. Partial melting in the mantle, metasomatism by Si-rich fluids derived from crustal host rocks, and serpentinization have distinct geochemical signatures that are clearly seen on major element variation diagrams. Spinel peridotite norms of the Cycladic ultramafic rocks were calculated and plotted on the peridotite classification diagram (Fig. 9). The peridotites of the MUH on Naxos form a roughly vertical trend going from the lherzolite field into the harzburgite and dunite fields. This trend and the negative linear correlations of SiO_2, CaO and Al_2O_3 with MgO (Fig. 10) are compatible with mantle depletion caused by partial melt extraction. The MUH samples with the lowest MgO contents are very close in composition to primitive upper mantle estimates (Hart & Zindler 1986; Hofmann 1988), whereas the others could form by different degrees of partial melting. The compositions of Agia meta-peridotites plot close to the depletion trend defined by the MUH, suggesting a common origin. Excluding the Naxos MUH and Agia rocks, most other ultramafic rocks plot along the Ol–Opx join of the triangular diagram (Fig. 9). This is a clear indication of serpentinization, which always involves major loss of Ca. Even the Ochi wehrlites, which still have significant Ca

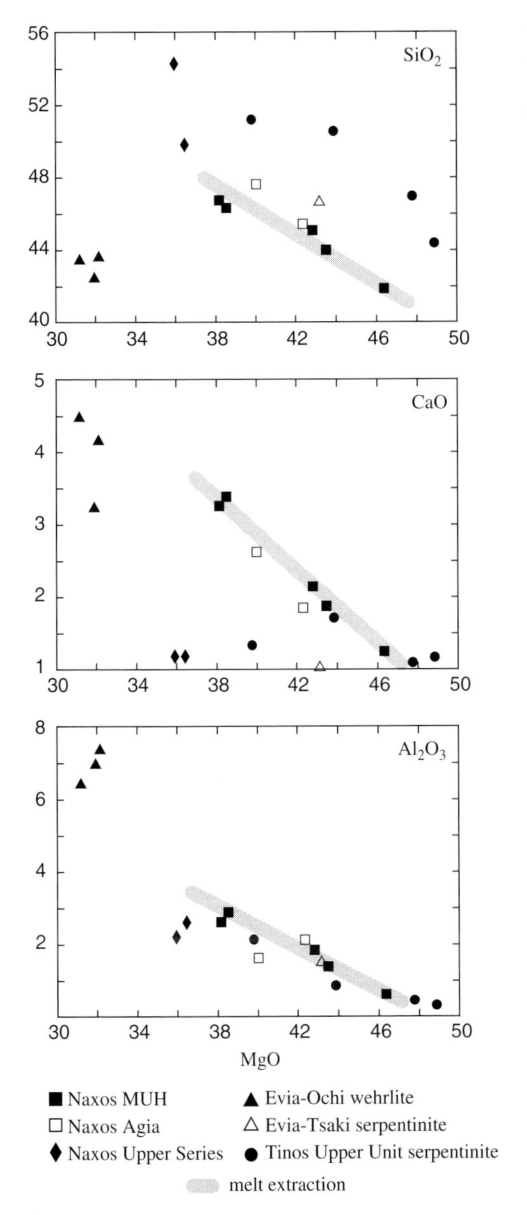

Fig. 10. Variation diagrams of SiO_2, CaO and Al_2O_3 v. MgO for the ultramafic rocks of the Cyclades.

contents, plot within the lherzolite field far to the left of their original compositions, as a result of a high degree of serpentinization. The geochemical modifications during serpentinization, including significant to complete Ca loss and increase in the Si/Mg ratio, are also illustrated in the variation diagrams (Fig. 10). The Tsaki, Naxos Upper Series and Tinos Upper Unit meta-serpentinites have almost no CaO regardless of the MgO content. For a given MgO content they have higher SiO_2 than the Naxos meta-peridotites, which have not experienced serpentinization. Except for the Ochi wehrlites, which are not true mantle peridotites, all the other rocks form a general negative correlation trend on the Al_2O_3–MgO variation diagram. This is not surprising, as Al behaves as a conservative element during serpentinization and its content may be used as a qualitative criterion of mantle fertility. The Naxos Upper Series serpentinites plot on the extreme left side of the peridotite classification diagram; however, their norms, which have less than 50% olivine, cannot be accounted for by serpentinization alone (Fig. 9). Their relatively high SiO_2 contents indicate that after Ca loss in serpentinization, they were silica enriched by externally derived fluids. The occurrence of synkinematic talc in their M_2 metamorphic assemblages and the blackwalls at their contacts with the host quartzo-feldspathic rocks suggest that they were metasomatized by Si-rich fluids derived from metamorphic reactions in the adjacent country-rocks. A similar scenario may apply to one sample of Tinos serpentinite taken from a thin, highly sheared talc-rich horizon at the base of the Upper Unit.

Geochemical trends thus indicate a fundamental difference between the Naxos MUH meta-peridotites and the rest of the ultramafic rocks in the Cyclades. Whereas the latter were first exhumed and serpentinized at near-surface conditions prior to Alpine metamorphism, the MUH meta-peridotites were incorporated at depth into the orogenic wedge, thus avoiding low-temperature hydration.

P–T *paths in the Cyclades*

The equilibrium *P–T* conditions for blueschists and eclogites usually lie at higher pressures and lower temperatures than most normal geothermal gradients. Thermal modelling suggests that only exceptionally rapid uplift can prevent reinstatement of an equilibrium geothermal gradient with consequent heating and destruction of the high-pressure mineral assemblages (England & Thompson 1984; Thompson & Ridley 1987). Abundant field and thermometric data from the Cyclades and many other high-pressure orogenic belts show, however, that high *P–T* rocks decompressed adiabatically with almost no heating, or even with cooling (Ernst 1988; Dunn & Medaris 1989; Platt 1993). In particular, the petrological and geochronological studies on Sifnos, Syros and Tinos have shown that the Cycladic eclogites decompressed isothermally at *c.* 450–500 °C from 15 ± 3 kbar at *c.* 50–45 Ma to 5–7 kbar at *c.* 23–21 Ma. Extensional tectonism has thus been suggested as responsible for the rapid unroofing and exhumation of eclogites in the Cyclades (Lister *et al.* 1984; Jolivet & Patriat 1999; Trotet *et al.* 2001*a*).

The new petrological analysis of metamorphic assemblages in ultramafic and associated rocks on Evia and Naxos allows redrawing of the *P–T* paths of the Lower tectonic Unit away from the CEX and offers a broader perspective on the exhumation processes in the Cyclades (Fig. 11). Because of the occurrence of pre-Alpine basement in a lowermost structural position, the leucogneiss core of Naxos is considered as the deepest exposed levels of the Alpine orogen in the Cyclades. However, its *P–T* path remains speculative, as anatexis and deformation during exhumation (M_2) totally destroyed any evidence for its evolution during collision and M_1 metamorphism. The occurrence of non-serpentinized fertile spinel lherzolites, probably representing the subcontinental mantle, within the leucogneiss core supports its deep origin. The peridotites were interleaved with the subducted upper crustal section during M_1 and can thus serve as indicators of peak M_1 temperatures. The lack of serpentinization and the preservation of relict >1000 °C mantle assemblage indicate that the peridotites cooled to temperatures of *c.* 500–650 °C during M_1 (Katzir *et al.* 1999). Comparable temperatures and pressures >14 kbar are given by down-section extrapolation from relict eclogites on SE Naxos (Avigad 1998). Given the well-constrained temperatures of 670–700 °C calculated for the M_2 high amphibolite overprint (Buick & Holland 1989, 1991), a *P–T* path of decompressional heating emerges for the leucogneiss core on Naxos (Fig. 11).

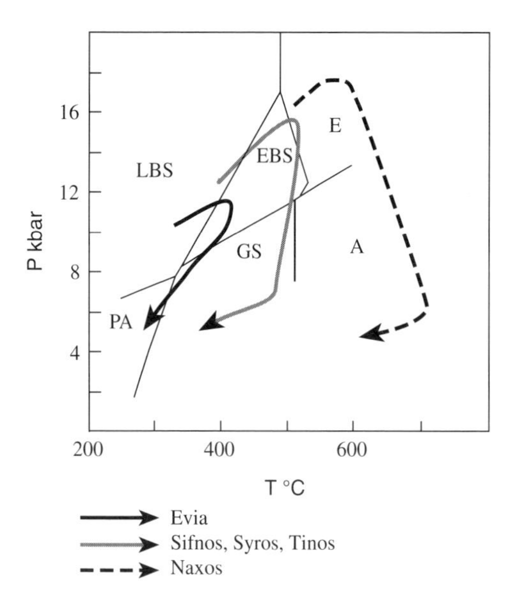

Fig. 11. Pressure–temperature paths for the Cycladic Blueschist Unit (CBU) in the Central Eclogite Axis (Tinos, Syros and Sifnos), on Naxos and on southern Evia plotted on the metamorphic facies diagram of Evans (1990). Abbreviations for metamorphic facies: A, amphibolite; E, eclogite; EBS, epidote blueschist; GS, greenschist; LBS, lawsonite blueschist; PA, pumpellyite–actinolite.

The high-pressure metamorphic rocks of the Cyclades show diverse *P–T* paths (Fig. 11). Whereas eclogites on Sifnos, Syros and Tinos decompressed isothermally, blueschists on Evia cooled and the deepest-buried, peridotite-associated gneisses on Naxos were heated during decompression. The pattern formed by the *P–T* paths in the Cyclades is not random: the higher the maximum pressure is, the 'hotter' is the exhumation path. This pattern was predicted by thermal modelling of thickened continental crust unroofed by erosion (England & Richardson 1977). The *P–T* path of any single metamorphic rock is governed by two concurrently competing processes: temperature increase by conductive relaxation and decompression by erosion of the orogenic pile. Assuming relatively rapid and uniform thickening of the crust, the magnitude of temperature increase is dependent on the time that elapses before the rock is exhumed sufficiently to be affected by the proximity of the cold upper boundary (England & Thompson 1984). Thus the deeper the rock is buried within the orogenic pile, the longer will be the period during which its temperature increases, and the hotter will be the overprint during exhumation. Although developed for erosion, the assumptions and heat transfer equations of the model hold regardless of the actual mechanism responsible for bringing the rocks nearer the surface. Thus the Eocene to Miocene removal of *c.* 30 km of rock overburden implied by the diverse *P–T–t* paths of the Cycladic Blueschist Unit could have been accomplished by erosion, uniform 'pure shear' extension, or a combination of both. However, according to the England & Thompson model, cooling during decompression is possible only if exhumation of the underthrust rocks began very early, simultaneously with the onset of conductive heating. Mid-Eocene flysch that tops the Basal Unit in the Almyroptamos tectonic window of southern Evia indicates such early synorogenic erosion. Generally, however, thick sequences of Eocene or younger clastic sediments are not exposed in the Cyclades. An efficient mechanism that can account for the decompressional cooling of the Evian HP–LT rocks is accretion of relatively cold rocks underneath them (Rubie 1984). Underthrusting of the Basal Unit beneath the Blueschist Unit can also account for the inverted M_2 temperature gradient within the Blueschist Unit towards the basal thrust on Evia (Katzir *et al.* 2000). Cooling by conductive heat transfer is also indicated by thermometry of marbles across the basal thrust on Tinos (Matthews *et al.* 1999). Notwithstanding, the cooling during decompression inferred by Trotet *et al.* (2001*b*) for the uppermost part of the Blueschist Unit in the CEX was attributed to early exhumation by non-uniform, 'simple-shear' extension manifested by deep ductile shear zones (Ruppel *et al.* 1988; Ruppel 1995).

In summary, no single $P-T$ path should be assigned to the Cycladic Blueschist Unit. Instead, a common $P-T$ trajectory for the M_1 high-pressure metamorphism can be drawn. Going southeastwards from southern Evia, progressively deeper levels of the subducted plate are exposed. Correspondingly, temperatures of the M_2 overprint also increase from pumpellyite-bearing assemblages on southern Evia, through greenschists on the CEX, to upper-amphibolite, sillimanite-bearing gneisses on Naxos. The array of enveloping $P-T$ loops thus given is compatible with exhumation by uniform attenuation of the crust accomplished by either erosion or extension. Post-M_1 thrusting of the Blueschist Unit on top of the Basal Unit exerts excess cooling at its base on southern Evia. Restricted cooling at the top of the Blueschist Unit on Sifnos and Syros was explained by local 'simple shear'-type extension that resulted in early exhumation.

Provenance of the Cycladic ultramafic rocks

The relative positions of the Cycladic Massif and adjacent terraines on both sides of the Aegean Sea at the beginning of the Neogene are shown in Figure 12 (from Garfunkel 2004). This palaeogeographical reconstruction, which corrects for the Neogene extension and block rotation in the Aegean Sea, is used as a guide for names and locations of major rock units and terranes in the following discussion.

Several models integrating the tectonometamorphic evolution of the Cycladic Massif into the Alpine history of the Hellenides in continental Greece have been proposed (Biju-Duval *et al.* 1977; Bonneau 1984; Papanikolaou 1987). All of the models emphasize two orogenic events: (1) the Eohellenic event of Late Jurassic age that involved emplacement of ophiolites from the Pindos Ocean over the Pelagonian continental margins (Smith 1993); (2) the Meso-Hellenic final closure of the Mesozoic Pindos basin by subduction and eventual collision of the Apulian and Pelagonian microplates in Early to Middle Eocene times (Robertson & Dixon 1984). This collisional event is thought to have caused the high-pressure metamorphism of rock sequences of the underthrust Apulian plate, including the Pindos deep-water sediments (possibly the NW Cyclades) and the Apulian continental basement and platform (possibly Naxos; Blake *et al.* 1981; Papanikolaou 1987).

Diverse origins and modes of emplacement have been deduced in this study for the ultramafic rocks of the Cyclades. However, the field relations and history of the majority of ultramafic occurrences in the Cyclades hardly fit the above two-stage evolution of the Hellenides. The simplest rocks to correlate are the serpentinites of the Paros Upper Unit, which are covered by transgressive Barremian limestone and thus probably represent the Eohellenic event. Apparently, the high $P-T$ ophiolitic

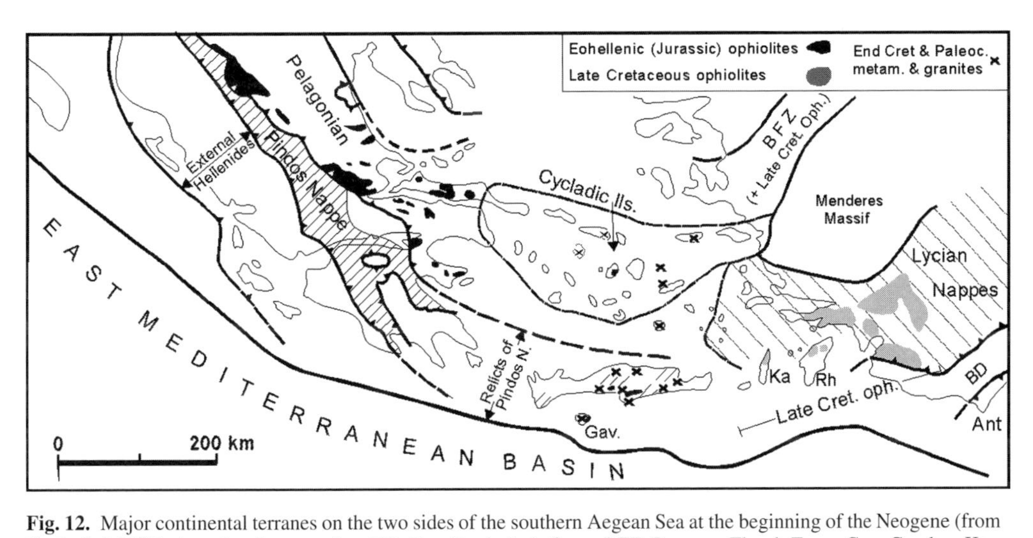

Fig. 12. Major continental terranes on the two sides of the southern Aegean Sea at the beginning of the Neogene (from Garfunkel 2004). Ant, Antalya complex; BD, Bey Daglari platform; BFZ, Bornova Flysch Zone; Gav, Gavdos; Ka, Karpathos; Rh, Rhodes. The Cyclades Islands are interpreted as underlain by a distinct continental fragment, of unclear provenance, which has Variscan age crust.

mélanges on southern Evia and Syros embedded in a pelagic clastic sequence might be correlated to the subducted Pindos Ocean. Ion-probe study of zircon crystals separated from an eclogitized metagabbro block from the Syros mélange yielded Late Cretaceous ages of *c.* 80 Ma (Tomaschek *et al.* 2003). These were interpreted as the crystallization age of the protolith; however, Late Cretaceous oceanic magmatism has not been recognized anywhere in continental Greece. The residual Cretaceous Pindos basin is a possible source for the Cycladic eclogites, but other oceanic basins that show unequivocal evidence for Late Cretaceous sea-floor spreading, such as those represented by the ophiolites of the Lycian nappes to the east (Fig. 12), should also be considered as potential source terranes for the Syros ophiolitic mélange.

The rocks of the structurally lowest ultramafic horizon on Naxos (MUH) are considered to be mantle flakes that intermingled with an upper continental crustal section while the latter was buried to great depth during the Eocene continental collision. Nevertheless, correlating the underthrust Naxos continental section including ortho- and paragneisses, pelite schists and thick karst-bauxite-bearing marbles with equivalent unmetamorphosed sequences in the Hellenides or elsewhere is not straightforward. Variscan and younger pre-Alpine ages are dominant in zircons separated from various gneisses in the Naxos core (Keay *et al.* 2001). Given its Pan-African basement, correlation with the Menderes Massif of SW Turkey is ruled out (Fig. 12). The provenance of the Naxos section should be sought in areas with Variscan continental basement, such as the Apulian and Pelagonian zones of the external and internal Hellenides, respectively. Pelagonian sequences tectonically overlie the CBU on Evia and are the main source of clasts for the Upper Unit conglomerates on Mykonos (Shaked *et al.* 2000; Sanchez-Gömez *et al.* 2002). It is hard to envision how the Pelagonian section can form both the uppermost and lowermost parts of the present orogenic pile in the Cyclades. Likewise, the occurrence of serpentinized ultramafic rocks within the Upper Series of Naxos makes its correlation with the ophiolite-devoid shallow-water sedimentary cover of Apulia questionable.

The ophiolitic slices of the Tinos Upper Unit were metamorphosed at greenschist- to amphibolite-facies conditions in Late Cretaceous times (Avigad & Garfunkel 1991; Patzak *et al.* 1994; Bröcker & Franz 1998). This period of time was characterized by renewed pelagic sedimentation in the Pindos basin and is regarded as a 'calm' interval between the Eohellenic and Meso-Hellenic events (Robertson *et al.* 1991). The Tinos ophiolite thus either originated in an area outside the Hellenic realm (Avigad & Garfunkel 1989) or indicates remobilization of previously emplaced Eohellenic ophiolites. The Tinos ophiolite slices may be related to a rock assemblage of high-temperature metamorphic rocks, greenschists and granites of Late Cretaceous to Paleocene age that occurs at the eastern edge of the Cyclades (Fig. 12; Dürr *et al.* 1978*b*; Reinecke *et al.* 1982; Altherr *et al.* 1994). Like the Tinos ophiolite they overlie the CBU and possibly record the former existence of a large metamorphic terrane in the central Aegean.

Summing up, the ultramafic rocks and their host sequences in the Cyclades show bimodal provenance. Several ultramafic associations share characteristics with rock sequences in the Hellenides, whereas others have clear non-Hellenic origin. These include ophiolites formed and deformed at *c.* 80–65 Ma, a time of sea-floor spreading, but overall plate convergence, in the Anatolian domain to the east of the Aegean Sea. The mixing of elements from domains of entirely different history located west and east of the Aegean Sea supports the idea that the Cycladic Massif represents a major discontinuity along which these domains were juxtaposed in the Tertiary (Ring *et al.* 1999; Garfunkel 2004).

This paper greatly benefited from thorough reviews by E. Bozkurt and an anonymous reviewer.

References

ALTHERR, R. & SIEBEL, W. 2002. I-type plutonism in a continental back-arc setting: Miocene granitoids and monzonites from the central Aegean Sea, Greece. *Contributions to Mineralogy and Petrology*, **143**, 397–415.

ALTHERR, R., SCHLIESTEDT, M., OKRUSCH, M. *ET AL.* 1979. Geochronolgy of high pressure rocks on Sifnos (Cyclades, Greece). *Contributions to Mineralogy and Petrology*, **70**, 245–255.

ALTHERR, R., KREUZER, H., LENZ, H., WENDT, I., HARRE, W. & DÜRR, S. 1994. Further evidence for a Late Cretaceous low-pressure/high-temperature terrane in the Cyclades, Greece. *Chemie der Erde*, **63**, 319–328.

ANDRIESSEN, P. A. M., BOELRIJK, N. A. I. M., HEBEDA, E. H., PRIEM, H. N. A., VERDURMEN, E. A. TH. & VERSCHURE, R. H. 1979. Dating the events of metamorphism and granitic magmatism in the Alpine orogen of Naxos (Naxos, Greece). *Contributions to Mineralogy and Petrology*, **69**, 215–225.

AVIGAD, D. 1998. High-pressure metamorphism and cooling on SE Naxos (Cyclades, Greece). *European Journal of Mineralogy*, **10**, 1309–1319.

AVIGAD, D. & GARFUNKEL, Z. 1989. Low-angle faults above and below a blueschist belt—Tinos Island, Cyclades, Greece. *Terra Nova*, **1**, 182–187.

AVIGAD, D. & GARFUNKEL, Z. 1991. Uplift and exhumation of high-pressure metamorphic terrains: the example of the Cycladic blueschist belt (Aegaen Sea). *Tectonophysics*, **188**, 357–372.

AVIGAD, D., MATTHEWS, A., EVANS, B. W. & GARFUNKEL, Z. 1992. Cooling during the exhumation of a blueschist terrane: Sifnos (Cyclades), Greece. *European Journal of Mineralogy*, **4**, 619–634.

AVIGAD, D., GARFUNKEL, Z., JOLIVET, L. & AZAÑON, J. M. 1997. Back arc extension and denudation of Mediterranean eclogites. *Tectonics*, **16**, 924–941.

BANNO, S. 1998. Pumpellyite–actinolite facies of the Sanbagawa metamorphism. *Journal of Metamorphic Geology*, **16**, 117–128.

BIJU-DUVAL, B., DERCOURT, J. & LE PICHON, X. 1977. From the Tethys ocean to the Mediterranean seas: a plate tectonic model of the evolution of the Western Alpine system. *In*: BIJU-DUVAL, B. & MONTADERT, L. (eds) *Structural History of the Mediterranean Basins*. Technip, Paris, 143–164.

BLAKE, M. C., BONNEAU, M., GEYSSANT, J., KIENAST, J. R., LEPVRIER, C., MALUSKI, H. & PAPANIKOLAOU, D. 1981. A geologic reconnaissance of the Cycladic blueschist belt, Greece. *Geological Society of America Bulletin*, **92(I)**, 247–254.

BONNEAU, M. 1984. Correlation of the Hellenide nappes in the south-east Aegean and their tectonic reconstruction. *In*: DIXON, J. E. & ROBERTSON, A. H. F. (eds) *The Geological Evolution of the Eastern Mediterranean*. Geological Society, London, Special Publications, **17**, 517–528.

BONNEAU, M. & KIENAST, J. R. 1982. Subduction, collision et schistes bleus: l'exemple de l'Egée (Gréce). *Bulletin de la Société Géologique de France*, **7**, 785–792.

BONNEAU, M., BLAKE, M. C., GEYSSANT, J., KIENAST, J. R., LEPVRIER, C., MALUSKI, H. & PAPANIKOLAOU, D. 1980. Sur la signification des series métamorphique (schistes bleus) des Cyclades (Hellénides, Grèce). L'exemple de l'île de Syros. *Comptes Rendus de l'Académie des Sciences*, **D290**, 1463–1466.

BRÖCKER, M. & ENDERS, M. 1999. U–Pb zircon geochronology of unusual eclogite-facies rocks from Syros and Tinos (Cyclades, Greece). *Geological Magazine*, **136**, 111–118.

BRÖCKER, M. & ENDERS, M. 2001. Unusual bulk-rock compositions in eclogite-facies rocks from Syros and Tinos (Cyclades, Greece): implications for U–Pb zircon geochronology. *Chemical Geology*, **175**, 581–603.

BRÖCKER, M. & FRANZ, L. 1998. Rb–Sr isotope studies on Tinos Island (Cyclades, Greece): additional time constraints for metamorphism, extent of infiltration-controlled overprinting and deformational activity. *Geological Magazine*, **135**, 369–382.

BRÖCKER, M., KREUZER, A., MATTHEWS, A. & OKRUSCH, M. 1993. $^{40}Ar/^{39}Ar$ and oxygen isotope studies of poly-metamorphism from Tinos Island, Cycladic blueschist belt, Greece. *Journal of Metamorphic Geology*, **11**, 223–240.

BRÖCKER, M., BIELING, D., HACKER, B. & GANS, P. 2004. High-Si phengite records the time of greenschist facies overprinting: implications for models suggesting mega-detachments in the Aegean Sea. *Journal of Metamorphic Geology*, **22**, 427–442.

BUICK, I. S. 1988. *The metamorphic and structural evolution of the Barrovian overprint, Naxos, Cyclades, Greece*. PhD thesis, University of Cambridge.

BUICK, I. S. 1991. The late Alpine evolution of an extensional shear zone, Naxos, Greece. *Journal of the Geological Society, London*, **148**, 93–103.

BUICK, I. S. & HOLLAND, T. J. B. 1989. The P–T–t path associated with crustal extension, Naxos, Cyclades, Greece. *In*: DALY, J. S., CLIFF, R. A. & YARDLEY, B. W. D. (eds) *Evolution of Metamorphic Belts*. Geological Society, London, Special Publications, **43**, 365–369.

BUICK, I. S. & HOLLAND, T. J. B. 1991. The nature and distribution of fluids during amphibolite facies metamorphism, Naxos (Greece). *Journal of Metamorphic Geology*, **9**, 301–314.

COLEMAN, R. G. 1984. The diversity of ophiolites. *Geologie en Mijnbouw*, **63**, 1099–1108.

DERMITZAKIS, M. & PAPANIKOLAOU, D. J. 1981. Paleogeography and geodynamics of the Aegean region during the Neogene. *Annales Geologique des Pays Hellénique*, **31**, 245–289.

DIXON, J. E. & RIDLEY, J. 1987. Syros. *In*: HELGESON, H. C. (ed.) *Chemical Transport in Metasomatic Processes*. NATO ASI Series, **218C**, 489–501.

DUBOIS, R. & BIGNOT, G. 1979. Présence d'un 'hardground' nummulitique au de la série Crétacée d'Almyropotamos (Eubée méridionale, Gréce). *Comptes Rendus de l'Académie des Sciences, Série II*, **289**, 993–995.

DUNN, S. R. & MEDARIS, L. D. 1989. Retrograded eclogites in the Western Gneiss region of a portion of the Scandinavian Caledonides. *Lithos*, **22**, 229–245.

DÜRR, S., ALTHERR, R., KELLER, J., OKRUSCH, M. & SEIDEL, E. 1978*a*. The median Aegean crystalline belt: stratigraphy, structure, metamorphism, magmatism. *In*: CLOSS, H., ROEDER, D. H. & SCHMIDT, K. (eds) *Alps, Apennines, Hellenides*. IUGS Report, **38**, 455–477.

DÜRR, S., SEIDEL, E., KREUZER, H. & HARRE, W. 1978*b*. Témoins d'un métamorphisme d'âge Crétacé Supérieur dans l'Egéide: datations radiométrique de minéraux provenant de l'île de Nikourià (Cyclades, Grèce). *Bulletin de la Société Géologique de France*, **2**, 209–213.

ENGLAND, P. C. & RICHARDSON, S. W. 1977. The influence of erosion upon the mineral facies of rocks from different metamorphic environments. *Journal of the Geological Society, London*, **134**, 201–213.

ENGLAND, P. C. & THOMPSON, A. B. 1984. Pressure–temperature–time paths of regional metamorphism I. Heat transfer during the evolution of regions of thickened continental crust. *Journal of Petrology*, **25**, 894–928.

ERNST, W. G. 1988. Tectonic history of subduction zones inferred from retrograde blueschist P–T paths. *Geology*, **16**, 1081–1084.

EVANS, B. W. 1986. Reactions among sodic, calcic and ferromagnesian amphiboles, sodic pyroxene, and deerite in high-pressure metamorphosed ironstone, Siphnos, Greece. *American Mineralogist*, **71**, 1118–1125.

EVANS, B. W. 1990. Phase relations of epidote-blueschists. *Lithos*, **25**, 3–23.

FAURE, M., BONNEAU, M. & PONS, J. 1991. Ductile deformation and syntectonic granite emplacement during the late Miocene extension of the Aegean (Greece). *Bulletin de la Société Géologique de France*, **162**, 3–11.

FEENSTRA, A. 1985. *Metamorphism of bauxites on Naxos*. PhD thesis, Rijks Universiteit, Utrecht.

GARFUNKEL, Z. 2004. Origin of the Eastern Mediterranean basin: a reevaluation. *Tectonophysics*, **391**, 11–34.

GAUTIER, P. & BRUN, J. P. 1994. Crustal-scale geometry and kinematics of late-orogenic extension in the central Aegean (Cyclades and Evvia Island). *Tectonophysics*, **238**, 399–424.

GAUTIER, P., BRUN, J. P. & JOLIVET, L. 1993. Structure and kinematics of Upper Cenozoic extensional detachment on Naxos and Paros (Cyclades Islands, Greece). *Tectonics*, **12**, 1180–1194.

GREGORY, R. T. & TAYLOR, H. P. 1981. An oxygen isotope profile in a section of Cretaceous oceanic crust, Samail Ophiolite, Oman: evidence for $\delta^{18}O$ buffering of the oceans by deep (>5 km) seawater–hydrothermal circulation at mid-ocean ridges. *Journal of Geophysical Research*, **86**, 2737–2755.

HART, S. R. & ZINDLER, A. 1986. In search of bulk-earth composition. *Chemical Geology*, **57**, 242–267.

HEATON, T. H. E. & SHEPPARD, S. M. F. 1977. Hydrogen and oxygen isotope evidence for seawater hydrothermal alteration and ore deposition, Troodos complex, Cyprus. *In*: *Volcanic Processes in Ore Genesis*. Geological Society, London, Special Publications, **7**, 42–57.

HOFMANN, A. W. 1988. Chemical differentiation of the Earth: the relationship between mantle, continental crust and oceanic crust. *Earth and Planetary Science Letters*, **90**, 297–314.

JACOBSHAGEN, V. 1986. *Geologie von Griechenland*. Borntraeger, Berlin.

JANSEN, J. B. H. 1977. *The geology of Naxos. Geological and Geophysical Research, 1*. Institute of Geological and Mining Research, Athens.

JANSEN, J. B. H. & SCHUILING, R. D. 1976. Metamorphism on Naxos: petrology and geothermal gradients. *American Journal of Science*, **276**, 1225–1253.

JOHN, B. E. & HOWARD, K. A. 1995. Rapid extension recorded by cooling-age pattern and brittle deformation, Naxos, Greece. *Journal of Geophysical Research*, **100**, 9969–9979.

JOLIVET, L. & PATRIAT, M. 1999. Ductile extension and the formation of the Aegean Sea. *In*: DURAND, B., JOLIVET, L., HORVÁTH, F. & SÉRANNE, M. (eds) *The Mediterranean Basins: Tertiary Extension within the Alpine Orogen*. Geological Society, London, Special Publications, **156**, 427–456.

JONES, G. & ROBERTSON, A. H. F. 1991. Tectono-stratigraphy and evolution of the Mesozoic Pindos ophiolite and related units. *Journal of the Geological Society, London*, **148**, 267–288.

KATSIKATSOS, G., MIGIROS, G., TRIANTAPHYLLIS, M. & METTOS, A. 1986. Geological structure of the internal Hellenides (E. Thessaly–SW Macedonia, Eboboea–Attica–Northern Cyclades islands and Lesvos). *In*: *Geological and Geophysical Research*. Institute of Geological and Mining Research, Athens, 191–212.

KATZIR, Y., MATTHEWS, A., GARFUNKEL, Z., SCHLIESTEDT, M. & AVIGAD, D. 1996. The tectono-metamorphic evolution of a dismembered ophiolite (Tinos, Cyclades, Greece). *Geological Magazine*, **133**, 237–254.

KATZIR, Y., AVIGAD, D., MATTHEWS, A., GARFUNKEL, Z. & EVANS, B. W. 1999. Origin and metamorphism of ultrabasic rocks associated with a subducted continental margin, Naxos (Cyclades, Greece). *Journal of Metamorphic Geology*, **17**, 301–318.

KATZIR, Y., AVIGAD, D., MATTHEWS, A., GARFUNKEL, Z. & EVANS, B. W. 2000. Origin, HP/LT metamorphism and cooling of ophiolitic mélanges in southern Evia (NW Cyclades), Greece. *Journal of Metamorphic Geology*, **18**, 699–718.

KATZIR, Y., VALLEY, J. W., MATTHEWS, A. & SPICUZZA, M. J. 2002. Tracking fluid flow during deep crustal anatexis: metasomatism of peridotites (Naxos, Greece). *Contributions to Mineralogy and Petrology*, **142**, 700–713.

KEAY, S., LISTER, G. & BUICK, I. 2001. The timing of partial melting, Barrovian metamorphism and granite intrusion in the Naxos metamorphic core complex, Cyclades, Aegean Sea, Greece. *Tectonophysics*, **342**, 275–312.

LEE, J. & LISTER, G. S. 1992. Late Miocene ductile extension and detachment faulting, Mykonos, Greece. *Geology*, **20**, 121–124.

LISTER, G. S., BANGA, G. & FEENSTRA, A. 1984. Metamorphic core complexes of Cordilleran type in the Cyclades, Aegean Sea, Greece. *Geology*, **12**, 221–225.

MALUSKI, H., VERGELY, P., BAVAY, D., BAVAY, P. & KATSIKATSOS, G. 1981. $^{39}Ar/^{40}Ar$ dating of glaucophanes and phengites in Southern Euboa (Greece), geodynamic implications. *Bulletin de la Société Géologique de France*, **7**, 469–476.

MALUSKI, H., BONNEAU, M. & KIENAST, J. R. 1987. Dating metamorphic events in the Cycladic area: $^{39}Ar/^{40}Ar$ data from metamorphic rocks of the island of Syros (Greece). *Bulletin de la Société Géologique de France*, **8**, 833–842.

MARUYAMA, S. & LIOU, J. G. 1985. The stability of Ca–Na pyroxene in low-grade metabasites of high-pressure intermediate facies series. *American Mineralogist*, **70**, 16–29.

MATTHEWS, A. 1994. Oxygen isotope geothermometers for metamorphic rocks. *Journal of Metamorphic Geology*, **12**, 211–219.

MATTHEWS, A. & SCHLIESTEDT, M. 1984. Evolution of the blueschist and greenschist facies rocks of Sifnos, Cyclades, Greece. *Contributions to Mineralogy and Petrology*, **88**, 150–163.

MATTHEWS, A., LIEBERMAN, J. L., AVIGAD, D. & GARFUNKEL, Z. 1999. Fluid–rock interaction and thermal evolution during thrusting in an Alpine metamorphic complex (Tinos Island, Greece). *Contributions to Mineralogy and Petrology*, **135**, 212–224.

MATTHEWS, A., PUTLITZ, B., HAMIEL, Y. & HERVIG, R. L. 2003. Volatile transport during the crystallization of anatectic melts: oxygen, boron and hydrogen stable isotope study on the metamorphic complex of Naxos,

Greece. *Geochimica et Cosmochimica Acta*, **67**, 3145–3163.

MOORES, E. M. 1982. Origin and emplacement of ophiolites. *Reviews in Geophysics and Space Physics*, **20**, 735–760.

MUKHIN, P. 1996. The metamorphosed olistostromes and turbidites of Andros Island, Greece, and their tectonic significance. *Geological Magazine*, **133**, 697–711.

OKRUSCH, M. & BRÖCKER, M. 1990. Eclogites associated with high-grade blueschists in the Cyclades archipelago, Greece: a review. *European Journal of Mineralogy*, **2**, 451–478.

PAPANIKOLAOU, D. 1980. Contribution to the geology of the Aegean Sea: the island of Paros. *Annales Géologiques des Pays Helléniques*, **30**, 65–95.

PAPANIKOLAOU, D. 1987. Tectonic evolution of the Cycladic blueschist belt (Aegean Sea, Greece). *In*: HELGESON, H. C. (ed.) *Chemical Transport in Metasomatic Processes*. NATO ASI Series, **218C**, 429–450.

PATRIAT, M. & JOLIVET, L. 1998. Post-orogenic extension and shallow-dipping shear zones, study of a brecciated décollement horizon in Tinos (Cyclades, Greece). *Comptes Rendus de l'Académie des Sciences*, **326**, 355–362.

PATZAK, M., OKRUSCH, M. & KREUZER, H. 1994. The Akrotiri Unit on the island of Tinos, Cyclades, Greece: witness to a lost terrane of Late Cretaceous age. *Neues Jahrbuch für Geologie und Paläontologie, Abhandlungen*, **194**, 211–252.

PE-PIPER, G., KOTOPOULI, C. N. & PIPER, D. J. W. 1997. Granitoid rocks of Naxos, Greece: regional geology and petrology. *Geological Journal*, **32**, 153–171.

PLATT, J. P. 1993. Exhumation of high-pressure rocks: a review of concepts and processes. *Terra Nova*, **5**, 119–133.

PUTLITZ, B., MATTHEWS, A. & VALLEY, J. W. 2000. Oxygen and hydrogen isotope study of high-pressure metagabbros and metabasalts (Cyclades, Greece): implications for the subduction of oceanic crust. *Contributions to Mineralogy and Petrology*, **138**, 114–126.

PUTLITZ, B., KATZIR, Y., MATTHEWS, A. & VALLEY, J. W. 2001. Oceanic and orogenic fluid–rock interaction in $^{18}O/^{16}O$-enriched metagabbros of an ophiolite (Tinos, Cyclades). *Earth and Planetary Science Letters*, **193**, 99–113.

PUTLITZ, B., COSCA, M. A. & SCHUMACHER, J. C. 2005. Prograde mica $^{40}Ar/^{39}Ar$ growth ages recorded in high pressure rocks (Syros, Cyclades, Greece). *Chemical Geology*, **214**, 79–98.

REINECKE, T. 1986. Phase relationships of sursassite and other Mn-silicates in highly oxidized, high-pressure metamorphic rocks from Evvia and Andros Islands, Greece. *Contributions to Mineralogy and Petrology*, **94**, 110–126.

REINECKE, T., ALTHERR, R., HARTUNG, B. ET AL. 1982. Remnants of a Late Cretaceous high temperature belt on the Island of Anafi (Cyclades, Greece). *Neues Jahrbuch für Geologie und Paläontologie, Abhandlungen*, **145**, 157–182.

RIDLEY, J. 1984*a*. Listric normal faulting and reconstruction of the synmetamorphic structural pile of the Cyclades. *In*: DIXON, J. E. & ROBERTSON, A. H. F. (eds) *The Geological Evolution of the Eastern Mediterranean*. Geological Society, London, Special Publications, **17**, 755–762.

RIDLEY, J. 1984*b*. Evidence of a temperature-dependent 'blueschist' to 'eclogite' transformation in high-pressure metamorphism of metabasic rocks. *Journal of Petrology*, **25**, 852–870.

RING, U., GESSNER, K., GÜNGÖR, T. & PASSCHIER, C. W. 1999. The Menderes Massif of western Turkey and the Cycladic Massif in the Aegean—do they really correlate? *Journal of the Geological Society, London*, **156**, 3–6.

RING, U., THOMSON, S. N. & BRÖCKER, M. 2003. Fast extension but little exhumation: the Vari detachment in the Cyclades, Greece. *Geological Magazine*, **140**, 245–252.

ROBERTSON, A. H. F. & DIXON, J. E. 1984. Aspects of the geological evolution of the Eastern Mediterranean. *In*: DIXON, J. E. & ROBERTSON, A. H. F. (eds) *The Geological Evolution of the Eastern Mediterranean*. Geological Society, London, Special Publications, **17**, 1–74.

ROBERTSON, A. H. F., CLIFT, P. D., DEGNAN, P. J. & JONES, G. 1991. Palaeogeographic and palaeotectonic evolution of the Eastern Mediterranean Neotethys. *Palaeogeography, Palaeoclimatology, Palaeoecology*, **87**, 289–343.

ROSENBAUM, J. M., KYSER, T. K. & WALKER, D. 1994. High temperature oxygen isotope fractionation in the enstatite–olivine–$BaCO_3$ system. *Geochimica et Cosmochimica Acta*, **58**, 2653–2660.

RUBIE, D. C. 1984. A thermal–tectonic model for high-pressure metamorphism and deformation in the Sesia Zone, Western Alps. *Journal of Geology*, **92**, 21–36.

RUPPEL, C. 1995. Extensional processes in continental lithosphere. *Journal of Geophysical Research*, **100**, 24187–24215.

RUPPEL, C., ROYDEN, L. & HODGES, K. 1988. Thermal modeling of extensional tectonics: application to pressure–temperature–time histories of metamorphic rocks. *Tectonics*, **7**, 947–957.

SANCHEZ-GÖMEZ, M., AVIGAD, D. & HEIMANN, A. 2002. Geochronology of clasts in allochthonous Miocene sedimentary sequences on Mykonos and Paros islands: implications for back-arc extension in the Aegean Sea. *Journal of the Geological Society, London*, **159**, 45–60.

SCHLIESTEDT, M. 1986. Eclogite–blueschist relationships as evidenced by mineral equilibria in the high-pressure metabasic rocks of Sifnos (Cyladic Islands), Greece. *Journal of Petrology*, **27**, 1437–1459.

SCHLIESTEDT, M. & MATTHEWS, A. 1987. Transformation of blueschist to greenschist facies rocks as a consequence of fluid infiltration, Sifnos (Cyclades), Greece. *Contributions to Mineralogy and Petrology*, **97**, 237–250.

SCHLIESTEDT, M. & OKRUSCH, M. 1988. Meta-acidites and silicic meta-sediments related to eclogites and glaucophanites in northern Sifnos, Cycladic archipelago, Greece. *In*: SMITH, D. C. (ed.) *Eclogites and Eclogite-Facies Rocks*. Elsevier, Amsterdam, 291–334.

SCHLIESTEDT, M., ALTHERR, R. & MATTHEWS, A. 1987. Evolution of the Cycladic crystalline complex: petrology, isotope geochemistry and geochronology. *In*: HELGESON, H. C. (ed.) *Chemical Transport in Metasomatic Processes*. NATO ASI Series, **218C**, 389–428.

SCHMÄDICKE, E. & WILL, T. M. 2003. Pressure–temperature evolution of blueschist facies rocks from Sifnos, Greece, and implications for the exhumation of high-pressure rocks in the Central Aegean. *Journal of Metamorphic Geology*, **21**, 799–811.

SECK, H. A., KOETZ, J., OKRUSCH, M., SEIDEL, E. & STOSCH, H. G. 1996. Geochemistry of a meta-ophiolite suite: an association of metagabbros, eclogites and glaucophanites on the Island of Syros, Greece. *European Journal of Mineralogy*, **8**, 607–623.

SHAKED, Y., AVIGAD, D. & GARFUNKEL, Z. 2000. Alpine high-pressure metamorphism at the Almyropotamos window (Southern Evia, Greece). *Geological Magazine*, **137**, 367–380.

SMITH, A. G. 1993. Tectonic significance of the Hellenic–Dinaric ophiolites. *In*: PRICHARD, H. M., ALABASTER, T., HARRIS, N. B. W. & NEARY, C. R. (eds) *Magmatic Processes and Plate Tectonics*. Geological Society, London, Special Publications, **76**, 213–243.

STOLZ, J., ENGI, M. & RICKLI, M. 1997. Tectonometamorphic evolution of SE Tinos, Cyclades, Greece. *Schweizerische Mineralogische und Petrographische Mitteilungen*, **77**, 209–231.

SUN, S. S. & MCDONOUGH, W. F. 1989. Chemical and isotopic systematics of oceanic basalts: implications for mantle composition and process. *In*: SAUNDERS, A. D. & NORRY, J. M. (eds) *Magmatism in the Ocean Basins*. Geological Society, London, Special Publications, **42**, 313–345.

THOMPSON, A. B. & RIDLEY, J. 1987. Pressure–time–temperature (P–T–t) histories of orogenic belts. *Philosophical Transactions of the Royal Society of London, Series A*, **321**, 27–45.

TOMASCHEK, F., KENNEDY, A., VILLA, I. M., LAGOS, M. & BALLHAUS, C. 2003. Zircons from Syros, Cyclades, Greece— recrystallization and mobilization of zircon during high-pressure metamorphism. *Journal of Petrology*, **44**, 1977–2002.

TROTET, F., JOLIVET, L. & VIDAL, O. 2001*a*. Tectono-metamorphic evolution of Syros and Sifnos islands (Cyclades, Greece). *Tectonophysics*, **338**, 179–206.

TROTET, F., VIDAL, O. & JOLIVET, L. 2001*b*. Exhumation of Syros and Sifnos metamorphic rocks (Cyclades, Greece): new constraints on the P–T paths. *European Journal of Mineralogy*, **13**, 901–920.

URAI, J. L., SCHUILING, R. D. & JANSEN, J. B. H. 1990. Alpine deformation on Naxos (Greece). *In*: KNIPE, R. J. & RUTTER, E. H. (eds) *Deformation Mechanisms, Rheology and Tectonics*. Geological Society, London, Special Publications, **54**, 509–522.

WAKABAYASHI, J. & DILEK, Y. 2003. What constitutes 'emplacement' of an ophiolite?: Mechanisms and relationship to subduction initiation and formation of metamorphic soles. *In*: DILEK, Y. & ROBINSON, P. T. (eds) *Ophiolites in Earth History*. Geological Society, London, Special Publications, **218**, 427–447.

WIJBRANS, J. R. & MCDOUGALL, I. 1988. Metamorphic evolution of the Attic Cycladic metamorphic belt on Naxos (Cyclades, Greece) utilizing $^{40}Ar/^{39}Ar$ age spectrum measurements. *Journal of Metamorphic Geology*, **6**, 571–594.

WIJBRANS, J. R., SCHLIESTEDT, M. & YORK, D. 1990. Single grain argon laser probe dating of phengites from the blueschist to greenschist transition on Sifnos (Cyclades, Greece). *Contributions to Mineralogy and Petrology*, **104**, 582–593.

ZEFFREN, S., AVIGAD, D., HEIMANN, A. & GVIRTZMAN, Z. 2005. Age resetting of hanging wall rocks above a low-angle detachment fault: Tinos Island (Aegean Sea). *Tectonophysics*, **400**, 1–25.

Structural evolution of Andros (Cyclades, Greece): a key to the behaviour of a (flat) detachment within an extending continental crust

C. MEHL[1], L. JOLIVET[1], O. LACOMBE[1], L. LABROUSSE[1] & G. RIMMELE[2]

[1]*Laboratoire de Tectonique, UMR 7072, Université Pierre et Marie Curie, T 46-00 E2, case 129, 4 place Jussieu, 75252 Paris Cedex 05, France (e-mail: caroline.mehl@lgs.jussieu.fr)*

[2]*Laboratoire de Géologie, UMR 8538, Ecole Normale Supérieure, 24, rue Lhomond, 75005 Paris, France*

Abstract: The continental crust extends in a brittle manner in its upper part and in more distributed (ductile) manner in its lower part. During exhumation of HP metamorphic rocks, brittle features superimpose on earlier ductile ones as a result of the progressive localization of deformation. The islands of Tinos and Andros are part of the numerous metamorphic core complexes exhumed in the Aegean domain. They illustrate two steps of a gradient of finite extension along a transect between Mt. Olympos and Naxos. This study confirms the main role of boudinage as an initial localizing factor at the brittle–ductile transition and emphasizes the continuum of strain from ductile to brittle during exhumation. Early low-angle semi-brittle shear planes superimpose onto precursory ductile shear bands, whereas steeply dipping late brittle planes develop by progressive steepening of structures or sliding across en echelon arrays of veins. The comparison between Tinos and Andros allows us to propose a complete dynamic section of the Aegean extending continental crust and emphasizes that the strain localization process depends on both its rheological stratification and its compositional heterogeneity.

Although post-orogenic extension has been studied for a long time in several regions of the world, such as the Basin and Range Province and the Aegean Sea, numerous questions remain open.

It is commonly admitted that the extending continental crust is characterized by steeply dipping normal faults in its upper part (Jackson 1987; Jackson & White 1989), crustal-scale shear bands at and below the brittle–ductile transition and more distributed ductile deformation in its lower part. Such a model raises two main problems: (1) the initial localizing factor allowing localization of deformation at the brittle–ductile transition zone although, considering the rheological envelopes, a maximum of strength is expected there; (2) the way in which ductile structures evolve towards brittle ones when the rocks pass through the brittle–ductile transition zone.

Metamorphic core complexes (MCC), because they offer the opportunity to observe large portions of the exhumed lower continental crust, are good sites to study the way in which deformation localizes and evolves from ductile to brittle. MCCs were recognized on several islands in the Aegean region during the last 20 years (Lister *et al.* 1984; Avigad & Garfunkel 1989, 1991; Gautier & Brun 1994*a*; Avigad *et al.* 1997). (Jolivet & Patriat (1999) studied a transect starting in the Mt. Olympos region and running through the MCCs of Evia, Andros, Tinos, Mykonos, Paros and Naxos. They concluded that the Aegean metamorphic core complexes are characterized by a gradient of finite extension from continental Greece towards the centre of the Cyclades, maximum extension being observed on the island of Naxos.

Several detailed structural studies were carried out on Tinos. They allowed researchers to emphasize the role of boudinage as an initial efficient localizing factor (Jolivet *et al.* 2004*a*) and to propose a new scenario of evolution of deformation from ductile to brittle (Mehl *et al.* 2005). Despite its key position on the Mt. Olympos–Naxos transect and its situation in the direct vicinity of Tinos, little attention has been paid to the structural framework of Andros. The aim of this study is twofold. First, we present the results of structural fieldwork carried out on Andros, with special emphasis on the progressive evolution of structures from ductile to brittle when rocks in the footwall of the main detachment are exhumed. Field observations allow us to test the mechanism of initiation of ductile deformation first proposed on Tinos and to emphasize the role played by boudinage. Second, we compare the extensional structures of Andros and Tinos, which are situated close to each other. The gradient of extension gives access to different portions of the extending continental crust, from the deeper and more stretched parts in the central

From: TAYMAZ, T., YILMAZ, Y. & DILEK, Y. (eds) *The Geodynamics of the Aegean and Anatolia.* Geological Society, London, Special Publications, **291**, 41–73.
DOI: 10.1144/SP291.3 0305-8719/07/$15.00

Aegean (Naxos and Paros), to the shallower parts near the continent (Tinos and Andros). Studying and comparing Andros with Tinos could lead to the development of a more complete scheme of evolution of a previously thickened continental crust, extending at the brittle–ductile transition.

The localization process: previous studies

Several studies have already been carried out on the localizing factors and the localization process, especially on Tinos. They are summarized below.

Localizing factors

Because localization of deformation occurs at the brittle–ductile transition, where the rheological envelopes predict a maximum of strength, the localization process requires localization factors that induce a local decrease of strength, making the onset of shear bands feasible.

Three localizing factors reducing the deviatoric stress at the brittle–ductile transition are classically described in the literature: increasing temperature (Kirby 1985), dynamic recrystallization and softening reactions, such as, for example, breakdown of strong feldspars to weaker white micas (Mitra 1978; White & Knipe 1978; Dixon & Williams 1983; Marquer *et al.* 1985; Fitz Gerald & Stünitz 1993; Wintsch *et al.* 1995; Wibberly 1999; Gueydan *et al.* 2001, 2003). None of these factors seems to be convenient in the case of the Cyclades: temperature does not play an important role at the brittle–ductile transition, being more efficient at the base of the crust (Kirby 1985). Dynamic recrystallization occurs only after large strains (Weathers *et al.* 1979) and thus cannot be involved in the initiation of shear bands. Finally, replacement of feldspars by micas cannot be advocated here because Cycladic blueschists are initially very rich in phyllosilicates and no significant increase in the concentration of micas can be observed during deformation. An additional localizing factor has been proposed by Jolivet *et al.* (2004*a*): boudinage. Metamorphic core complexes were first interpreted in terms of mega-boudinage in the Basin and Range Province by Davis & Coney (1979) and Davis (1980), but Jolivet *et al.* (2004*a*) pointed out the relation between boudinage and localization of shear bands. Boudinage induces progressive localization of strain in interboudin necks, which finally leads to local stress concentration and higher strain rate. When reaching the brittle–ductile transition, the first localized structures, such as shear bands and faults, will form in the necks between boudins. This mechanism of localization, based on field observations on Tinos, fits observations at metre scale as well as at crustal scale. The efficiency of this localizing factor will be tested from new field observations on Andros.

The localization process: the example of Tinos

Metamorphic core complexes are composed of two tectonic units separated by shallow-dipping detachments. Upper units display brittle steeply dipping extensional structures characteristic of the upper continental crust. Lower units have been exhumed along the detachments and underwent successively ductile and brittle deformation during their way back to the surface (Gautier & Brun 1994*b*; Jolivet & Patriat 1999; Jolivet *et al.* 2004*a*). They are therefore characterized by a superimposition of ductile and brittle structures, as a consequence of progressive localization of deformation during exhumation (Mehl *et al.* 2005). Detachments are considered as the ultimate evolution of shear bands towards more localized deformation (Lister & Davis 1989).

A detailed study of both ductile and brittle features in the footwall of the Tinos detachment allowed Mehl *et al.* (2005) to demonstrate a continuum of strain from ductile to brittle during extensional kinematics and exhumation of HP metamorphic rocks. Brittle extensional structures are characterized, on Tinos, by shallow- and steeply dipping normal faults; both types have formed under a vertical shortening axis, as shown by their association with ubiquitous vertical veins and as confirmed by inversions of fault slip data. The only way to explain the initiation of low-angle brittle extensional structures (including the detachment) is that the main displacement along the detachment was accommodated by ductile deformation and cataclastic flow, only the last increment of deformation being accommodated in a purely brittle manner (Mehl *et al.* 2005).

Tinos and Andros, because they are situated close to each other, have the same kinematic history. Moreover, they are situated on a gradient of finite strain. The comparison of the structures of the two islands could lead us to build a more complete section of the extending continental crust.

Structural setting of Andros

The Aegean domain

Andros is situated in the northern part of the Aegean Sea, which formed in the back-arc of the Hellenic subduction zone (Le Pichon & Angelier 1981) in a region once occupied by the Hellenides–Taurides mountain belt (Aubouin & Dercourt 1965; Brunn *et al.* 1976; Jacobshagen *et al.* 1978). Post-orogenic extension dates back to the late Oligocene–earliest Miocene, as shown by the cooling ages of the metamorphic core complexes and the ages of the basins in the region (Lister *et al.* 1984; Gautier & Brun 1994*a*; Jolivet *et al.* 1994; Jolivet & Faccenna 2000), and affected the whole Aegean domain. It is now localized around the Aegean Sea, in west Turkey, in the Peloponnesus, in the Gulf of Corinth and in Crete (Seyitoglu & Scott 1991, 1996; Taymaz *et al.* 1991; Armijo *et al.* 1992, 1996; Rietbrock *et al.* 1996; Rigo *et al.* 1996; Taymaz *et al.* 2004).

As mentioned above, several MCCs were recognized in the Aegean region during the last 20 years (Lister *et al.* 1984; Avigad & Garfunkel 1989, 1991; Gautier & Brun 1994*a*; Avigad *et al.* 1997). Two types of domes have been described by Jolivet *et al.* (2004*a*) in the basin: 'b-type' domes (Tinos and Andros), having their axis perpendicular to extension, and 'a-type' (Paros, Naxos and Mykonos) domes, elongated parallel to extension. The 'b-type' domes were exhumed *c.* 5 Ma before the 'a-type' domes. The 'a-type' domes correspond to exhumation of deeper and higher-temperature levels of crust and have recorded a constrictional component of deformation shown by north–south-trending fold axes. The main direction of extension is north–south to NE–SW over the entire area.

The Aegean domain is cut by several major NE-dipping normal faults (Taymaz *et al.* 1991, 1994; Jackson 1994) isolating crustal-scale tilted blocks (Papanikolaou *et al.* 1988; Jolivet *et al.* 1994), whose geometry is consistent with crustal-scale boudinage (Fig. 1). Andros belongs to the same block as Evia, Tinos and Mykonos.

Previous studies

A complete morphological study of Andros has already been made by Papanikolaou (1978). The topography of the island shows a structural dome oriented NW–SE (Fig. 1) in continuity with Evia and Tinos. It consists of a succession of NE–SW-trending mountains and valleys. The topography is asymmetric, with sharp slopes on the southern coast of the island and smoother ones on the northern coast, as shown by the topographic profile (Fig. 1, section AA′). The topography is smoother in the northwestern part of the island.

The structural framework of the island has been interpreted as reflecting mega-folds with NE–SW axes (Papanikolaou 1978). The smoother relief in the NW seems to correspond to what has been identified by Papanikolaou (1978) as a separate structural unit, the Makrotantalon Unit; this will be discussed in a later section. This attenuation must be due to differential erosion testifying to a 'weaker' lithology in the western part of the island. In contrast, the sharpness of the southern coast can probably be explained by the presence of an offshore normal fault, dipping to the SW (Fig. 1; map of the Aegean Sea).

Two tectonic units, separated by ophiolites that underline a presumable NE-dipping thrust, were originally described on Andros (see Fig. 2; Papanikolaou 1978; Reinecke *et al.* 1985). The Upper Unit, or Makrotantalon Unit, crops out in the northwestern part of the island. The occurrence of fauna relics in the Makrotantalon Unit supports a Permian age of sedimentation (Papanikolaou 1978). The Lower Unit, or Central Unit, is expected to be of Mesozoic age (Reinecke 1982). Both units are composed of an alternation of metabasites, metapelites and marble horizons, as on Tinos. Manganese-rich minerals have been described in the Central Unit (Reinecke 1982). Serpentinite bodies have been mapped on the northern coast of the island within this unit. Their significance will be discussed below.

The main part of the island has been retromorphosed to greenschist facies. (Reinecke 1982) deduced, from the reaction celsian + water = cymrite, a temperature of 400 °C and a pressure of 5–6 kbar for the greenschist event. Concerning the Makrotantalon Unit, Papanikolaou (1978) described garnet in the lowest horizon of the metapelites and Reinecke (1982) pointed out relics of omphacite and chloromelanite in the metabasites. Garnet and glaucophane are preserved in the metabasites of the Central Unit (Papanikolaou 1978). HP relics are better preserved on the southern coast of the island. Peak P–T conditions are estimated at 450–500 °C and >10 kbar from the reaction of sursassite-bearing to spessartine-bearing assemblages of the manganese-rich layers of the Central Unit (Reinecke 1986).

Andros as a metamorphic core complex

Lister *et al.* (1984) first described metamorphic core complexes in the Aegean Sea on the islands of Naxos and Ios. They suggested that the shallow-dipping faults separating the HP–LT rocks of the Cycladic Blueschist Belt (CBB) from non-metamorphosed ophiolites of Pelagonian affinity could be normal faults, and not thrust faults as previously proposed. They proposed that the exhumation of HP–LT metamorphic rocks could be explained by the presence of a south-dipping low-angle normal fault (or detachment) above the CBB. Faure & Bonneau (1988) further pointed out a top-to-the-NE sense of shear on Mykonos suggesting that the detachment there was not south-dipping, but rather NE-dipping. Following this dynamics, Avigad & Garfunkel (1989) described a NE-dipping detachment on Tinos that separates a

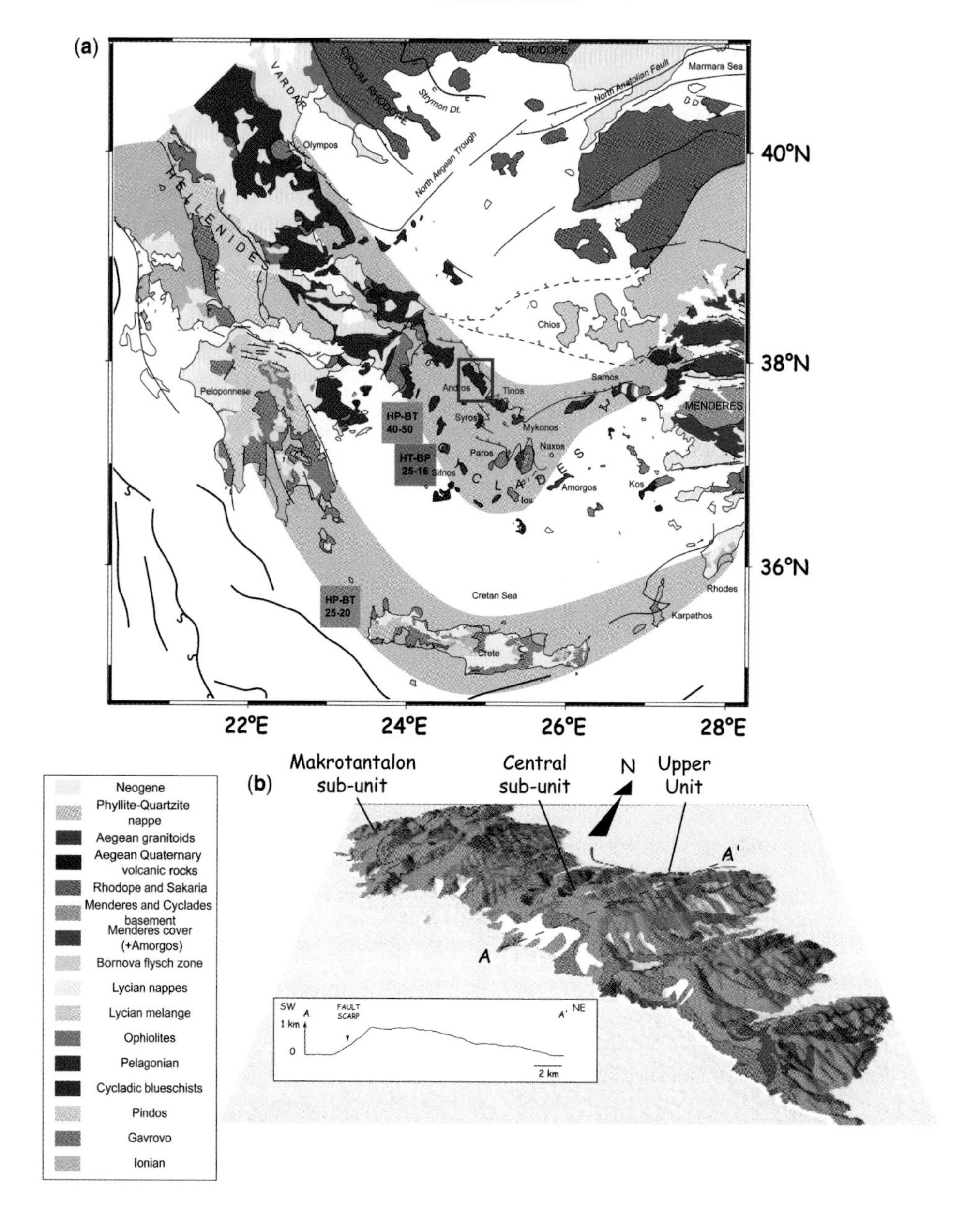

Fig. 1. (**a**) Geological map of the Aegean domain, after Jolivet *et al.* (2004a) The two blueschist belts of the domain are shown in blue–grey. The northern one corresponds to the Cycladic Blueschist Belt (CBB). (**b**) SRTM topographic model of Andros, ×2 vertical exaggeration. The island has been identified as being part of the CBB. The geological map of the island has been overlain on the topography (geological background: same key as in Fig. 2). AA': SW–NE topographic section of the island. The southern coast of the island is affected by a fault scarp. The overall morphology of the island has been related to megafolds with NE–SW axes.

Lower Unit of metamorphic rocks from an upper unit that is affected neither by the Eocene HP event nor by the Oligo-Miocene greenschist overprint. The kinematics of this extensional episode on Tinos was first described by Gautier & Brun (1994*a*, *b*) and Patriat & Jolivet (1998).

Such a vertical succession has been identified on Andros by Patriat (1996). What had been previously

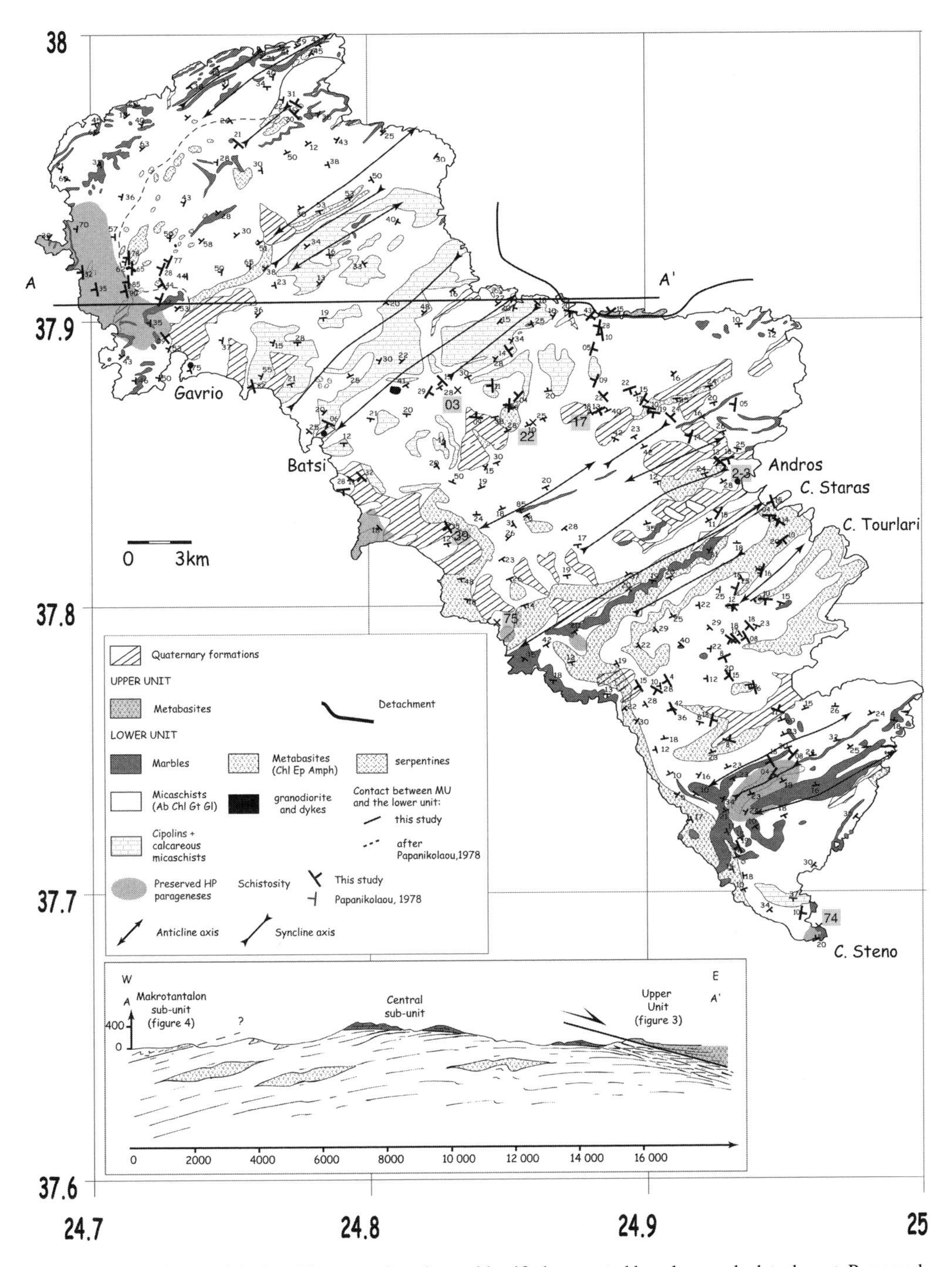

Fig. 2. Geological map of Andros. Two tectonic units are identified, separated by a low-angle detachment. Preserved blueschist parageneses, direction and plunge of schistosity, and ductile megafold NE–SW axes are shown. AA′: conceptual cross-section of the island. Two gradients exist on Andros from SW to NE: a gradient of retrogression and a gradient of finite strain. The closer the detachment, the more retrogressed the rocks and the less coaxial (that is, the more intense) the deformation. Deformation is accommodated, on the northeasternmost part of the island, by localized decametre-scale shear bands.

mapped by Papanikolaou (1978) as a serpentinite body within the Lower Unit and cropping out on the northern coast of the island could be recognized as the upper unit of a metamorphic core complex (Fig. 3). The Upper Unit of Andros is separated by a low-angle normal fault from the Lower Unit (Fig. 3). The contact is underlined by a discontinuous reddish breccia, as on Tinos.

The detachment

The detachment is visible along the NE coast of Andros below two remnants of the Upper Unit on two capes on either side of the wrecked ship *Semiramis*. The Upper Unit is composed of intensely foliated greenschists and serpentinites. A shallow NE-dipping normal fault marks the contact (Fig. 3) and shallow-dipping minor normal faults root in the underlying breccia (Fig. 3d). The basal breccia is stratified, with a 3–5 m thick reddish breccia made of serpentinite clasts resting on top of a 10 m thick greenish serpentinite breccia overlying highly sheared serpentinites mixed with some pelitic schists from the Lower Unit (Fig. 3c). The whole system of breccia rests on top of the sheared schists of the Lower Unit. The direction of slip along the faults in the contact is toward the NE, and semi-ductile features such as sigmoidal schistosity in the cataclasites also indicate top-to-the-NE shear.

In summary, Andros is composed of two structural units separated by a flat-lying detachment, in the sense of metamorphic core complexes. The unmetamorphosed Upper Unit crops out only in a small area on the northern coast of the island. The Lower Unit, metamorphosed under Eocene HP–LT conditions and retromorphosed to greenschist facies during post-orogenic extension, crops out on the major part of the island. Bröcker & Franz (2007) recently performed a Rb–Sr phengite dating on the Lower Unit. The study yielded, as it is common in the Cyclades, an HP–LT event at *c.* 50 Ma and a second retrogression episode at *c.* 20 Ma.

Relics of blueschist facies recorded on Andros are considered to correspond to the Eocene HP event responsible of the formation of the Cycladic Blueschist Belt. Within this framework, the actual significance of the Makrotantalon Unit as a sub-unit thrust onto the main part of the island deserves consideration.

The Makrotantalon Unit

The origin and structural significance and position of the Makrotantalon Unit have been a matter of debate. Assuming the presence of a tectonic contact at the base of the Makrotantalon Unit, two hypotheses can be made on its origin. Some workers (Papanikolaou 1978, 1987; Shaked *et al.* 2000) consider it as part of the Ochi Unit that crops out on the nearby island of Evia. This interpretation implicitly supposes that it is part of the Cycladic Blueschist unit and that it has recorded an HP–LT event of Eocene age. Other workers (Blake *et al.* 1981; Bonneau 1982; Dürr 1986) consider it as part of the Pelagonian domain or as the Upper Unit of the metamorphic core complex (Katzir *et al.* 2000). Assuming a structural definition of the Pelagonian domain, and thus referring it to the late Jurassic ophiolite obduction and associated deformation (Bonneau 1982; Jolivet *et al.* 2004*b*), implicitly supposes that the rocks of the Makrotantalon Unit did not record any Tertiary high-pressure metamorphism.

The Rb–Sr phengite dating by Bröcker & Franz (2007) shows that the Makrotantalon Unit has preserved ages as old as 100 Ma as well as a more recent episode at 20 Ma. This suggests that the Makrotantalon Unit has not recorded the high-pressure event recorded in the Lower Unit at 50 Ma. However, in the field, the contact between the Makrotantalon Unit and the Lower Unit is not clear, except in one outcrop on the northern part of the island. The contact was mapped in a different position by Papanikolaou (1978) and Bröcker & Franz (2007) and we were unable to identify a clear detachment surface similar to those observed on Tinos or on the NE coast of Andros. Some researchers have cast doubt on the existence of the contact (Gautier 1994; Patriat 1996). Two alternative solutions are thus available: (1) the Makrotantalon Unit is part of the Cycladic Blueschists and has escaped re-equilibration in the blueschist facies for some unknown reason; (2) the Makrotantalon Unit is an intermediate unit juxtaposed between the Upper Cycladic Unit and the Lower Unit.

Structures in the footwall of the detachment

Rocks of the Lower Unit underwent first an Eocene HP–LT metamorphic event characteristic of the Hellenides and then an Oligo-Miocene retrogression to greenschist facies during exhumation. During their exhumation, rocks passed through the brittle–ductile transition: brittle features were thus superimposed on ductile ones. We describe below the ductile and brittle extensional structures.

From blueschist- to greenschist-facies deformation

HP relics and syn-HP ductile structures are better preserved on the southern coast of the island west

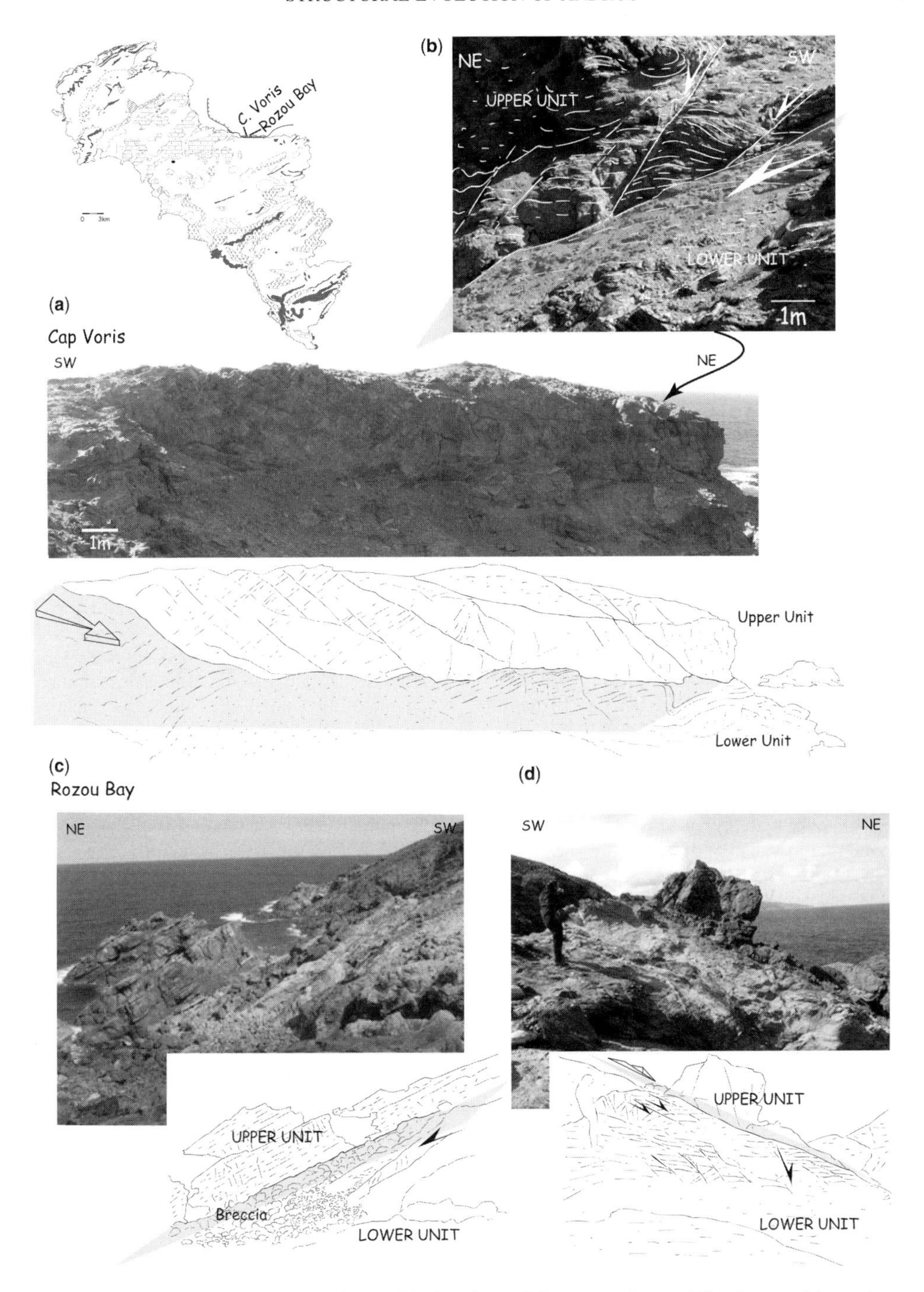

Fig. 3. Hanging wall and the Andros detachment. The hanging wall is composed essentially of greenschists and serpentinites. (**a**) Shallow-dipping normal fault that separates the Upper Unit from the Lower Unit at Cap Voris. The Upper Unit is affected by steeply dipping normal faults. (**b**) Close-up view of the steeply dipping normal faults affecting the Upper Unit at Cap Voris. (**c**) The detachment in Rozou Bay. The contact is sealed by a reddish breccia of serpentinite clasts resting on a greenish breccia of serpentinites. Normal faults have developed in the breccia. The normal sense of motion along the detachment is shown by the sigmoidal schistosity in the cataclasites, which indicates top-to-the-NE shear. (**d**) Both the Upper Unit and the breccia are affected by a dense network of steeply dipping normal faults and veins.

of Gavrio. They are also present sporadically in the Lower Unit especially along the southern coast of the island (Ipsili and Thiaki capes) and locally within metabasite boudins dispersed over the island. HP relics mainly consist of garnet relics or glaucophane-bearing mineral assemblages. Locations of preserved HP parageneses are shown on the geological map of the island (Fig. 2).

A section west of Gavrio shows the progressive evolution of deformation from the blueschist stage to the greenschist retrogression (Fig. 4). Table 1 summarizes this evolution.

The section shows rather well-preserved blueschist-facies metapelites with garnet and glaucophane and glaucophane-rich lenses of metabasites embedded within an alternation of retrograded metapelites and marbles. The best-preserved blueschists are found in the southeastern part of the section on either side of a highly deformed serpentinite lens. Whether this serpentinite represents the trace of a former thrust contact is difficult to ascertain because the lithologies below and above the contact are not very different. The preservation of the high-pressure S_1 foliation below and above the serpentinite lens may be related to the low resistance of the serpentinite that has taken up all the retrograde deformation and thus 'prevented' surrounding rocks deforming, but this hypothesis remains to be ascertained.

This section shows an intense retrograde greenschist-facies deformation that is also ubiquitous in the rest of the island (see below). The HP S_1 foliation is first reworked by P_2 folds and S_2 crenulation cleavage. With increasing shear strain, S_2 and L_2 become more intense and P_2 folds evolve toward sheath folds. This evolution is coeval with greenschist retrogression.

Near the top of the section intensely foliated and folded blue marbles (Fig. 4e) crop out before a west-dipping normal fault. The section ends near Agios Sostis, where dolomitic marbles rest on top of the section above albitic schists. The NW part of the section is cut by several normal faults of various sizes.

Greenschist finite deformation

The S_2 schistosity and the L_2 stretching lineation have been mapped all over the island (Figs 2 and 5). Orientations and dips of schistosity show a succession of kilometre-scale NE–SW folds, already described by Papanikolaou (1978). The stretching lineation shows a remarkable consistency throughout the island with a NE–SW trend with only local distortions, especially west of Gavrio, where it trends more north–south. Because both the stretching lineation and the fold axes show a consistent NE–SW trend (map in Fig. 5), it is difficult to ascertain the chronology of folding, stretching and shearing. Tilting of both schistosity and late normal fault systems nevertheless suggests that folding may have occurred during the latest greenschist evolution. Such folding could correspond to a component of constriction during extension. Such a component of constriction has already been recorded on Tinos (Mehl *et al.* 2005) and in the Menderes Massif, in western Turkey (Bozkurt & Park 1997; Bozkurt 2003). Avigad *et al.* (2001) estimated that folding on Andros accounts for 40–50% of NW–SE crustal shortening.

Greenschist retrogression and associated ductile features

Greenschist deformation occurs in two steps.

The first step consists of the formation of ubiquitous sheath folds, with axes always parallel to the stretching lineation (Fig. 5). Sheath folds result from the evolution of folding of the first schistosity S_1 under intense ductile shearing. They are visible at centimetre to decametre scale and are locally observed in the core of later boudins with parallel axes. Although we have no clear observations that substantiate this conclusion, we suspect that some of the NE–SW trending boudins result from the stretching of the sheath folds.

The second step of greenschist deformation corresponds to the boudinage of the S_2 foliation.

Boudinage is a very common phenomenon on Andros. It extends over the entire area and at every observation scale (Fig. 6; see caption for more details). Although the major part of the boudinaged outcrops consists of boudins of several metres in scale of metabasites in the metapelitic matrix (Figs 6a and 7, outcrops 75, 39, 22, 17 and 2–3), boudinage commonly involves several types of materials, such as quartz or marbles (Fig. 6b and c, respectively).

Some outcrops show evidence of crystallization between boudins (metre scale: Fig. 7, outcrops 75 and 74; centimetre scale: Fig. 6d; millimetre-scale: Fig. 6e). Crystallization of albite and chlorite in the interboudin necks of outcrops 39 and 75 (Fig. 7) shows that boudinage occurred under greenschist-facies conditions. Such crystallization can be explained by the presence of stress gradients established during the development of boudins: the gradients allow the migration of the more mobile mineral elements from the surrounding areas towards the low-pressure zones; that is, the interboudin gaps and/or the ends of boudins (Price & Cosgrove 1990).

Roadcut outcrops between Gavrio and Andros show an evolution in the geometry of boudins and

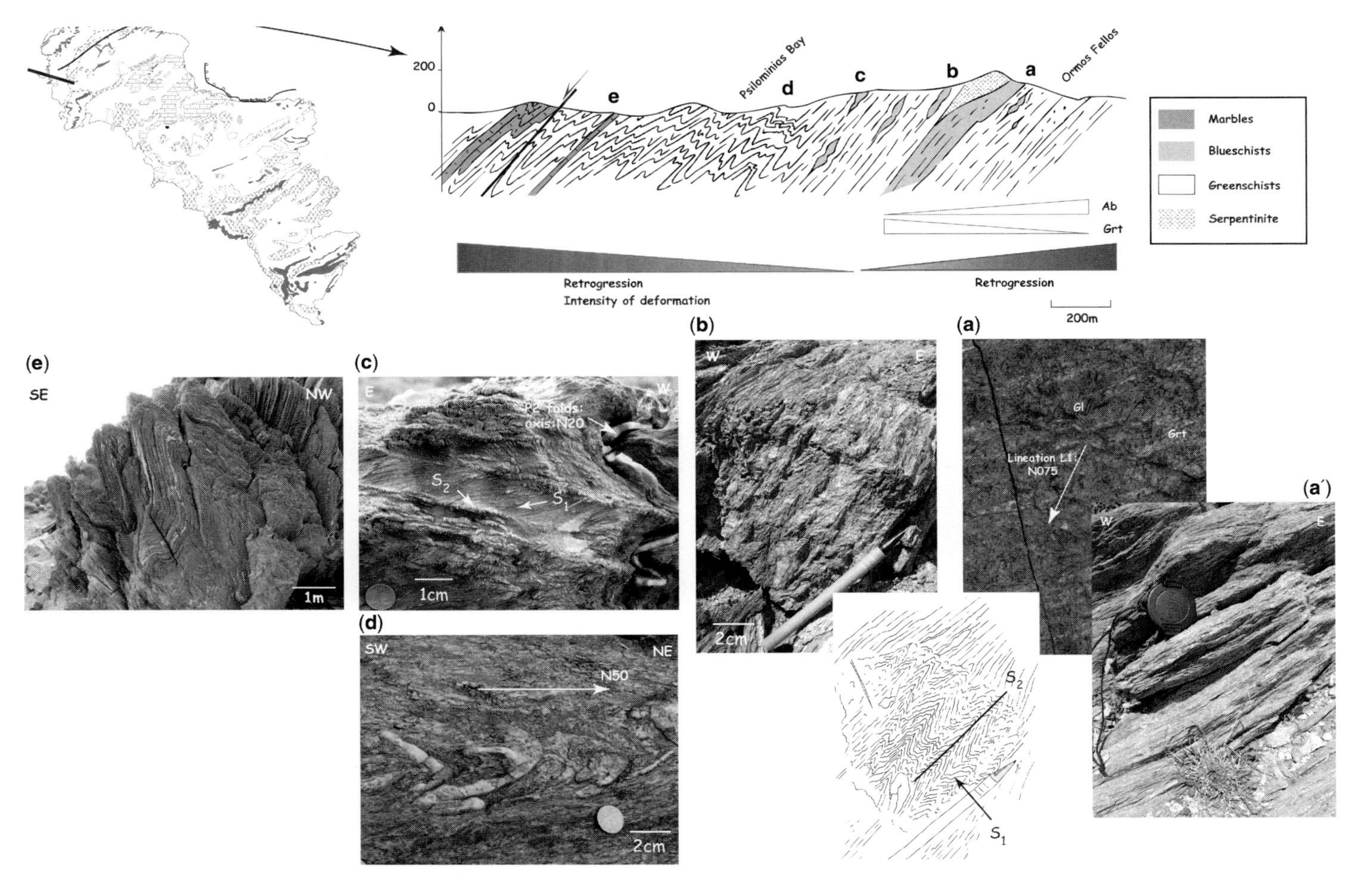

Fig. 4. Cross-section of the Makrotantalon Unit illustrating the evolution from blueschist to greenschist deformation on Andros. A gradient of retrogression is observed below and above the serpentinite body. (**a**) NE–SW-trending blueschist stretching lineation (L_1) recorded by the alignment of blue amphiboles. (**a′**) Pressure shadows on the garnets illustrating the top-to-the-NE sense of shear in blueschist facies. (**b**) Progressive formation of a greenschist S_2 foliation in the west-dipping axial crenulation plane of blueschist S_1 foliation. (**c**, **d**) Progressive curvature of P_2 fold axes (c) towards sheath folds (d) with axes parallel to the stretching lineation L_2. (**e**) Intensely foliated and folded marbles indicating the increase in intensity of deformation towards the NW.

Table 1. *Evolution of deformation from blueschist to greenschist facies along the section west of Gavrio*

Metamorphic facies, Stretching lineation	Foliation	Finite deformation	Complementary remarks
Blueschist facies			
L_1, east–west- for NE–SW-trending (alignment of blue amphibolite needles, Fig. 4a)	S_1	Top-to-the-NE sense of shear (asymmetry of pressure shadows on the garnets and shear bands, Fig. 4a) Open folds with a west-dipping axial-plane crenulation cleavage evolving towards S_2 (Fig. 4b and c)	Blueschists better preserved below and above serpentinite bodies
Greenschist facies			
L_2-NE–SW-trending	S_2	Sense of shear not clear along the section Crenulation Folding (P_2 folds, Fig. 4c) and progressive rotation of folds limbs with increasing shearing Boudinage	Increasing retrogression and intensity of shear strain when moving away from the serpentinite body (cross-section in Fig. 4) Increasing shear strain indicated by evolution of P_2 fold axes towards sheath folds (Fig. 4c and d) and intensively foliated and folded blue marbles at the top of the section (Fig. 4e)

shear bands. In the SW half of the island, boudinage leads to almost symmetrical structures when metabasites are involved (Fig. 7, left side). In the NE the geometry becomes clearly asymmetrical with NE-dipping shear bands and sigmoidal boudins indicating top-to-the-NE shear sense (Fig. 7, right side). Sequences of decametre-scale shear bands are observable in the landscape over all the capes of the northern coast (Fig. 8). They had first been interpreted by Papanikolaou (1978) as normal faults but do not show evidence of significant brittle slip. Foliation boudinage already shows a localization of non-coaxial top-to-the-NE shear in the NE half of the island.

The outcrops of Figure 9a and b illustrate the evolution from ductile to brittle structures for symmetrical (Fig. 9a) and asymmetric (Fig. 9b) boudins of metabasites embedded in a metapelitic matrix. Whatever the shape of boudins, shear bands seem to quasi-systematically localize in the necks or at the end of boudins (Fig. 9a and b; see captions for detailed descriptions), as already described on Tinos (Jolivet *et al.* 2004*a*; Mehl *et al.* 2005). The evolution towards brittle deformation is characterized by the onset of en echelon arrays of veins, whose normal shear movement is consistent with boudinage (Fig. 9a, picture A). Brittle steeply dipping planes seem to have developed on en echelon arrays of quartz veins (Fig. 9a, B; Fig. 9b, B–D), and therefore reflect the progressive and ultimate localization of normal shear. As shown in the diagrams of ductile and brittle data, the dip of shear planes increases while deformation evolves from ductile to brittle.

It is interesting to note that the metapelitic matrix of the outcrop shown in Figure 9b is

Fig. 5. Map of greenschist stretching lineation L_2 with sense of shear recorded on Andros (after Gautier 1994; Patriat 1996; this study). Stretching lineations indicate a consistent NE–SW direction of ductile stretching. Most of the photographs illustrate ubiquitous centimetre- to decimetre-scale sheath folds with NE–SW axes. They correspond to the first increment of ductile deformation. They result from the evolution of folding of S_1 under intense ductile shearing. (**a**, **b**) Sheath folds in a quartz vein included in retromorphosed metabasites of the Makrotantalon Unit. The NE–SW fold axes are parallel to the stretching lineation L_2. (**c**, **d**) Sheath folds with NE–SW axis in a boudin of metabasites. (**e**, **f**) Sheath folding of quartz veins with NE–SW axis in metabasites. Folding is sheared, as indicated by the onset of a top-to-the-NE shear band in the left part of (e). (**g**) Intense folding with NE–SW axis in metabasites of the Halkolimniona Cape. Folding can be here interpreted as resulting from a component of constriction of the deformation. (**h**) Cross-section perpendicular to a sheath fold of quartz vein in the metabasites of the Halkolimniona Cape. (**i**, **j**): Metre-scale sheath folds in metabasites of the Thiaki Cape.

Stretching Lineation L2
This study
Gautier, 1994
Patriat, 1999
Simple arrow: sense of shear
Double arrow: coaxial deformation
(a)
NE
SW
5cm
(b)
NE
5cm
(c)
5cm
NE
(d)
NE
5cm
(e)
ENE
WSW
5cm
(f)
NE
5cm
(g)
NE
15cm
(h)
5cm
NE
(i)
NE
20cm
(j)
1m
NE

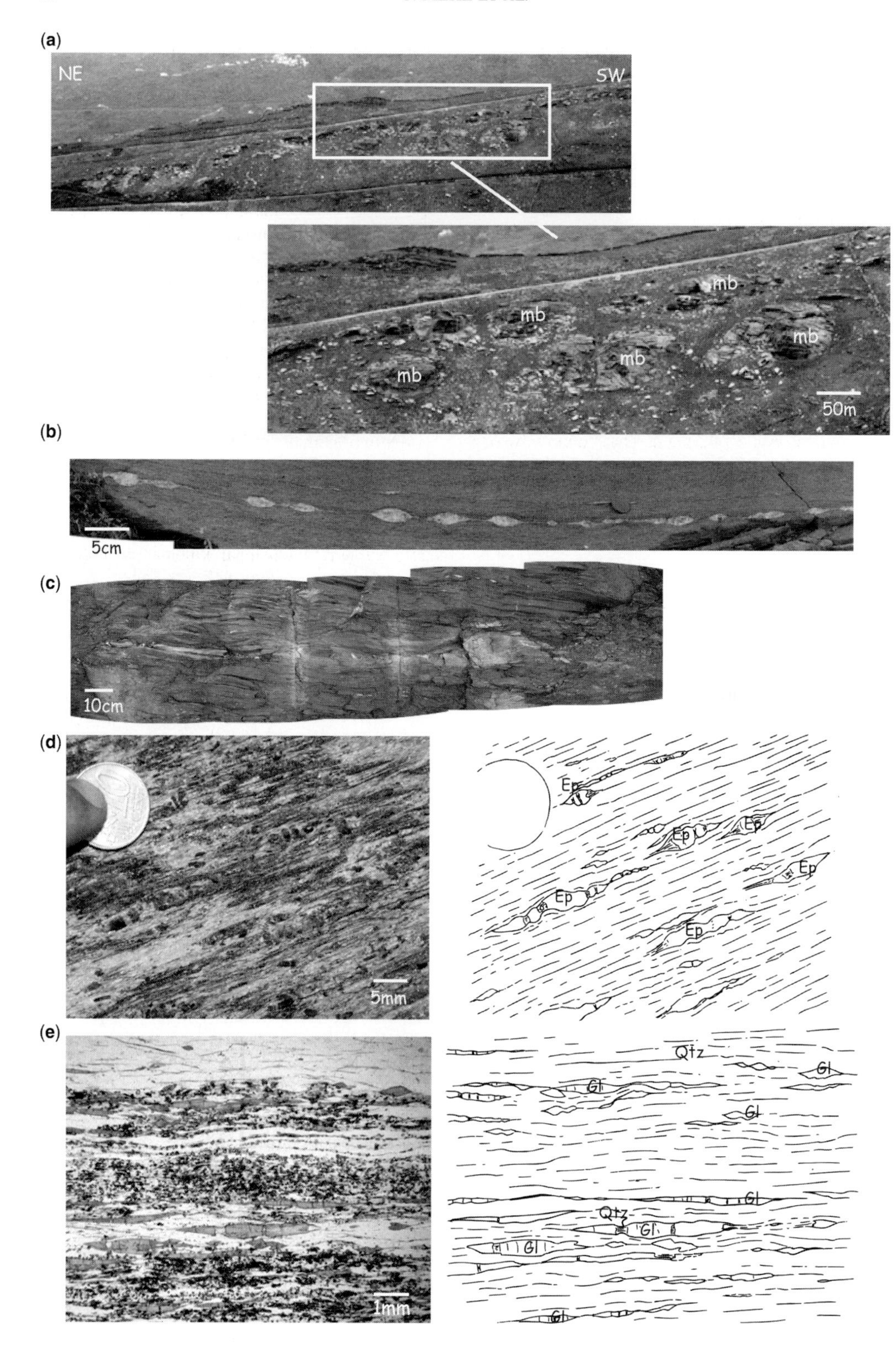
(a)
NE
SW
mb
mb
mb
mb
mb
50m
(b)
5cm
(c)
10cm
(d)
Ep
Ep
Ep
Ep
Ep
Ep
5mm
(e)
Qtz
Gl
Gl
Gl
Gl
Qtz
Gl
Gl
Gl
1mm

apparently not affected by brittle features. Localization of deformation in the metapelites is weak and is marked only by shear bands picture (A), whereas actual brittle deformation concentrates in the metabasites. This emphasizes the role of the lithological contrast in the localization process.

Semi-brittle structures

Semi-brittle shear bands can also be observed on Andros. They correspond to localized shear bands that display a latest brittle increment of extensional kinematics: foliation deviates along the plane but, contrary to a classical ductile shear band, a small offset support a late discontinuous shear movement. Some of these shear structures show slickensides which unambiguously support a brittle and normal sense of motion along the plane. They commonly correspond to planes belonging to a sequence of shear bands with increasing dip. The steeper the plane the more brittle the deformation is (Fig. 10b, A).

Brittle features

Two examples of 'brittle' outcrops are detailed in Figure 10a and b. The two outcrops are made up of an alternation of pelitic and more quartzitic beds. The quartzitic beds, being more competent than pelitic ones, are boudinaged (Fig. 10a and b, views of the entire outcrops).

The pelitic beds show the same straightening sequences as the outcrops of Figure 9a and b (Fig. 10b, picture A). The main brittle features observed here correspond to joints, veins and fault planes. Displacements of beds, striations on fault planes and rotations of en echelon arrays of veins argue for the extensional nature of deformation. Only major faults cut across the entire outcrops (Fig. 10a, picture B).

Conjugate patterns of normal faults, veins and joints are well expressed in the two outcrops. Although one plane cuts across the entire outcrop of Figure 10a, most conjugate sets of faults are concentrated in the quartzitic layers (Fig. 10a, A). When focusing on the northernmost part of the outcrop, we can see that joints concentrate in quartzitic beds (Fig. 10a, B). The same conclusion can be reached concerning the veins of the outcrop shown in Figure 10b: en echelon arrays of veins are concentrated in the light beds (Fig. 10b, panorama and C). This feature is particularly obvious in Figure 10b, picture B: veins clearly stop at the interface of the two beds.

En echelon arrays of veins and joints define rough planes whose orientation, dip and kinematics are comparable and consistent with classical conjugate sets of normal faults. Sometimes, these en echelon structures evolve toward true normal faults. This evolution is statistically more common for NE-dipping planes. However, the brittle planes do not propagate in the pelitic beds: structures flatten in the dark beds (Fig. 10a, C), and they seem to be relayed by shear bands in these pelitic beds (Fig. 10b, B).

Measurements of directions and dips of mesoscale striated faults, postfolial joints and veins were carried out all over the island. The most prominent fault sets trend NW–SE. Whatever the lithology, en echelon veins and normal faults clearly support extensional kinematics. Poles of veins and joints, together with orientation of faults and slip vectors on fault planes, consistently indicate a NE–SW direction of extension (Fig. 11). Reconstruction of stress regimes was carried out from fault slip data. This reconstruction was necessary to determine whether or not stress axes underwent a significant rotation in the latest stage of brittle deformation. Our fault slip data were collected on late, outcrop-scale faults displaying small offsets and a large scatter in attitudes, and cutting generally through the ductile rock fabrics (foliation and shear bands), and away from the detachment zone. Thus palaeostress reconstructions reported in this paper fulfil the assumptions of stress homogeneity and low-finite strain, which can be approximated by nearly coaxial conditions, and therefore probably yield the regional palaeostresses of interest. The *a posteriori* consistency of the stress regimes derived from both the inversion of striated faults and the statistical analysis of vein patterns, from one site to another, in spite of significant lithological variations (metabasites, metapelites, quartzitic beds), supports the reliability of the results. The aim of our stress analysis is, therefore, to

Fig. 6. Boudinage on Andros. The locations of the outcrops are shown on the map of Figure 2. Boudinage occurs at every scale and involves several types of materials over the whole island. (**a**) Outcrop 04; several-metre-scale boudins of metabasites that are visible in the landscape (mb, metabasites; mp, metapelites). (**b**) Outcrop 44; decimetre-scale synfolial quartz veins regularly boudinaged in the metapelitic matrix. (**c**) Outcrop 07; decimetre-scale boudinage also involves layers of marble in the metapelitic matrix. (**d**) Centimetre-scale boudinage of epidotes (Ep) in the matrix of Thiaki Cape. In this case, inter-boudins are filled with quartz precipitated in irregular voids. The schistosity is deflected around boudins: micas follow the shape of boudinaged epidotes. (**e**) Thiaki Cape; thin section illustrating millimetre-scale boudinage. This is an example of syn-greenschist boudinage of glaucophanes (gl) (the foliation is horizontal in this cross-section) and growth of quartz (Qtz) in the inter-boudins or at the end of boudins. Small-scale boudinage is coeval with penetrative stretching and intense ductile shearing.

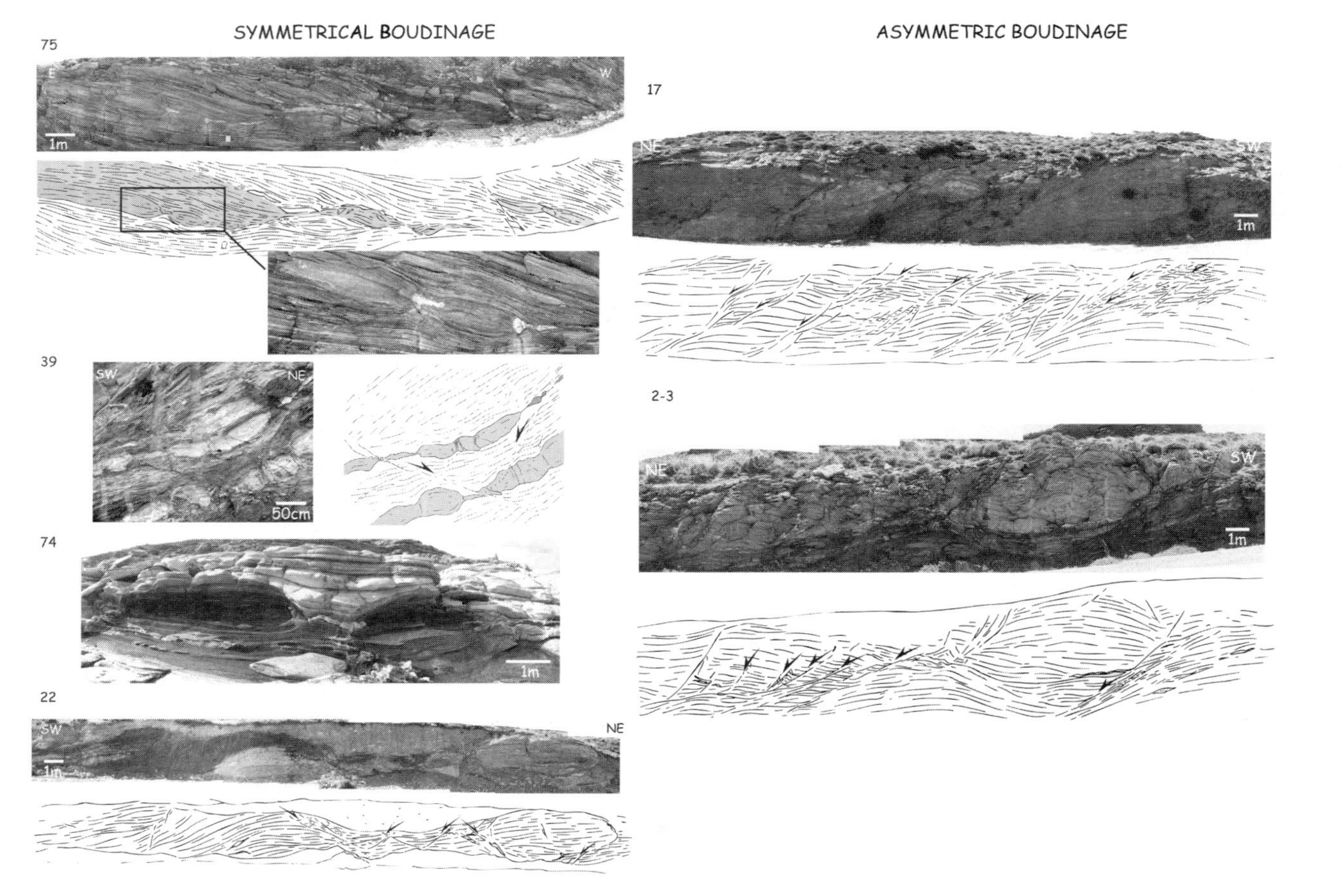

Fig. 7. Boudinage as the initial factor of localization of deformation. The locations of outcrops in this figure are shown in Figure 2. Outcrop 75: symmetrical boudinage in a metabasite matrix. Inter-boudin gaps and the ends of boudins are filled with quartz veins. Outcrop 39: Centimetre-scale symmetrical boudin of metabasite in a metapelitic matrix, Thiaki Cape. Conjugate shear bands develop in the neck between boudins. Outcrop 74: Metre-scale 'shadows' of symmetrical boudins in marbles of the southern cape of the island. Inter-boudin gaps are filled with quartz veins. Outcrop 22: several-metre-scale boudins of metabasite embedded in a metapelitic matrix in the central part of the island. Conjugate shear bands localize in the neck between boudins. Outcrop 17: asymmetric boudins of metapelite in a metapelitic matrix. Shear bands localize at the end of boudins and dip systematically towards NE. Outcrop 2 – 3: several-metre-scale boudins of metabasite in a metapelitic matrix. NE-dipping shear bands localize at the end or in the neck of the boudins. Boudins evolve from a symmetrical shape (left-hand side of the figure) in the southern part of the island towards an asymmetric one in the north-eastern part (right-hand side of the figure). The evolution of shape is interpreted is related to an increase in the intensity of shear deformation from SW to NE.

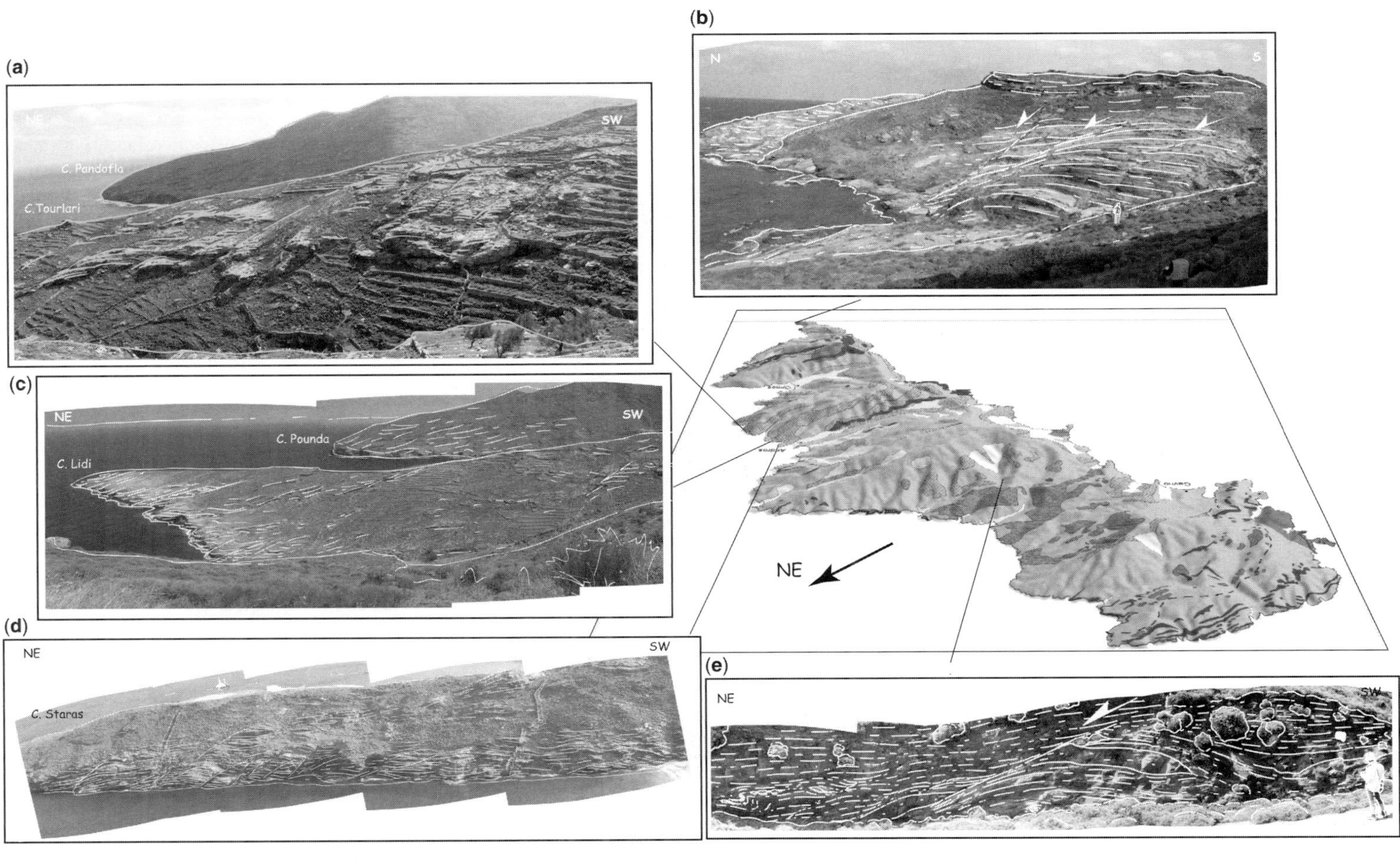

Fig. 8. Structural landscapes illustrating decametre-scale shear bands on the northeastern capes of Andros (for location, see Fig. 2). (**a**) NE-dipping, regularly spaced shear bands on the Tourlari Cape. (**b**) Several-metre-scale, NE-dipping shear band affecting the marbles of the southern cape of the island. (**c**, **d**): Several-metres-scale shear bands on the Lidi and Staras Capes. (**e**) NE-dipping shear band affecting the metapelites of the central part of the island.

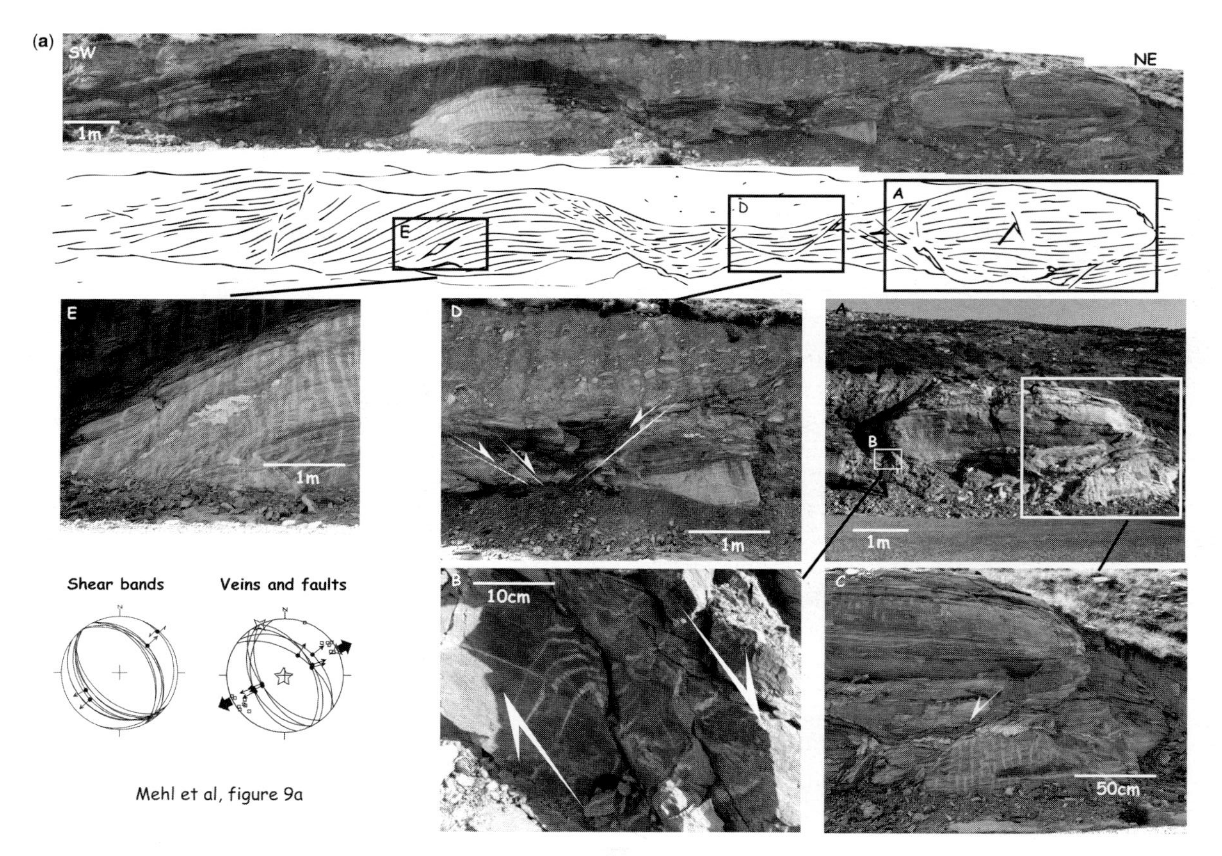

Fig. 9. (**a**) Evolution of deformation from ductile to brittle. The outcrop shows two symmetrical boudins of metabasites embedded in a metapelitic matrix. The ends of boudins show evidence of brittle–ductile to brittle deformation. (**A**) The SW end of the westernmost boudin showing an en echelon array of quartz veins in the metabasite. (**B**) This en echelon pattern testifies to a local normal shear movement that is kinematically consistent with boudinage. The ultimate step of localization consists of the development of an actual normal fault cutting through the en echelon system. (**C**) The NE end of the boudin shows a brittle normal steeply dipping plane. (**D**) Symmetrical patterns of shear bands can be observed in the inter-boudin gaps. As shown in the diagrams of ductile and brittle data, the dip of shear planes increases while deformation evolves from ductile to brittle. (**E**) Focus on a metre-scale boudin of quartz embedded in the pluri-metre-scale boudin of metabasites.

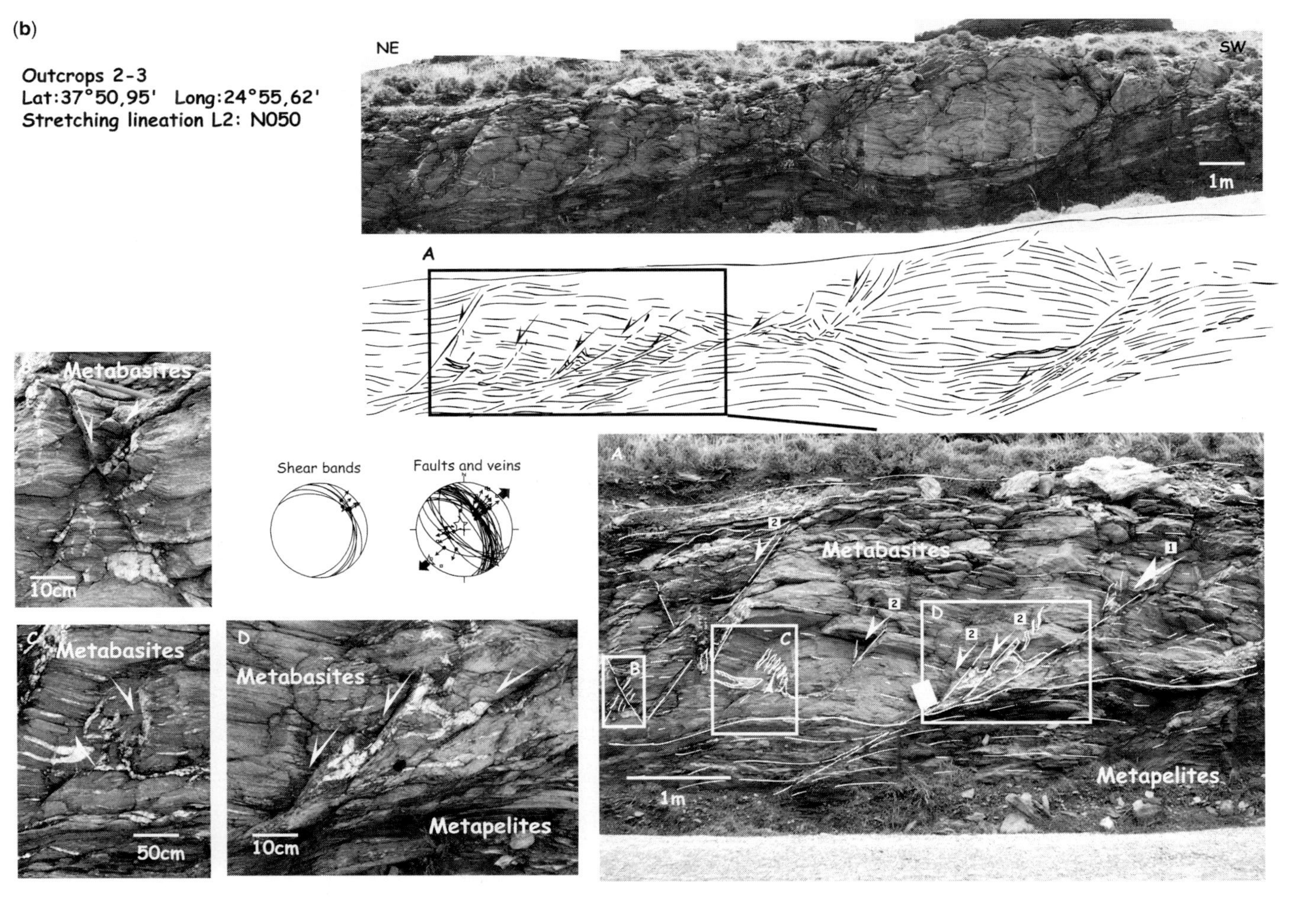

Fig. 9. (*Continued*) (**b**) Evolution of deformation from ductile to brittle around an asymmetric boudin of metabasite embedded in a metapelitic matrix. The metapelitic matrix is apparently not affected by brittle features. Localization of deformation in the metapelites is weak and is marked only by shear bands, whereas actual brittle deformation concentrates in the metabasites. (**A**) NE-dipping shear bands preferentially localize at the end of boudins. (**B**, **C**) En echelon arrays of quartz veins on which brittle planes develop. Brittle planes sometimes display conjugate patterns. (**D**) Steeply dipping brittle planes connected to ductile shear bands. Onset of brittle deformation is shown by onset of en echelon arrays of veins and progressive steepening of ductile structures, as illustrated by the diagram of ductile and brittle data.

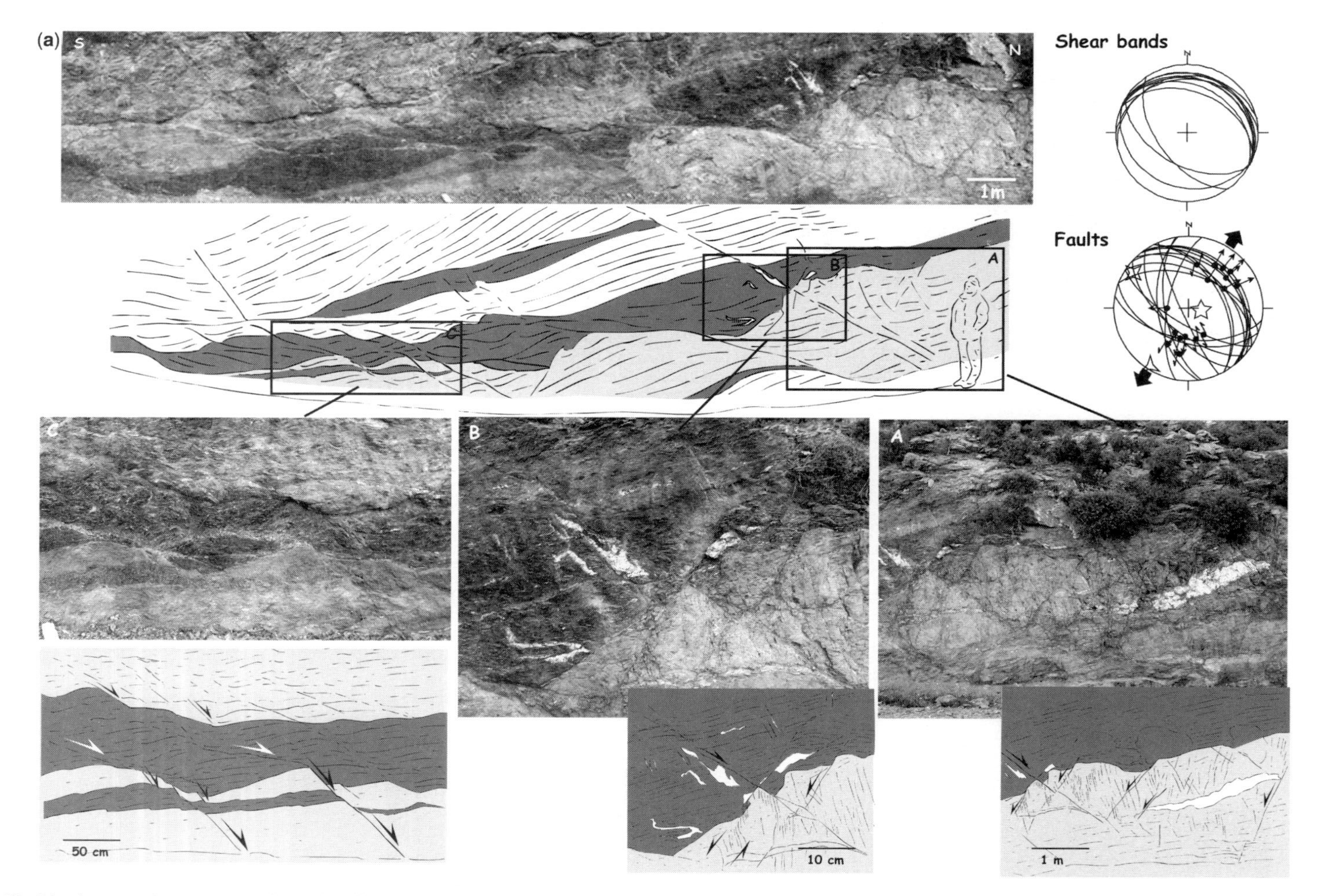

Fig. 10. Rheology as a key parametre in the localization process. Alternation of metapelites (dark grey) and quartzitic (light grey) beds. Quartzitic beds are more competent than metapelites and are boudinaged. Brittle features (en echelon arrays of veins, steeply dipping faults) preferentially localize in quartzitic beds (**a**, **A**–**C**; **b**, **B** and **C**) whereas shear bands are better preserved in metapelitic beds (**a**, **C**; **b**, **B**).

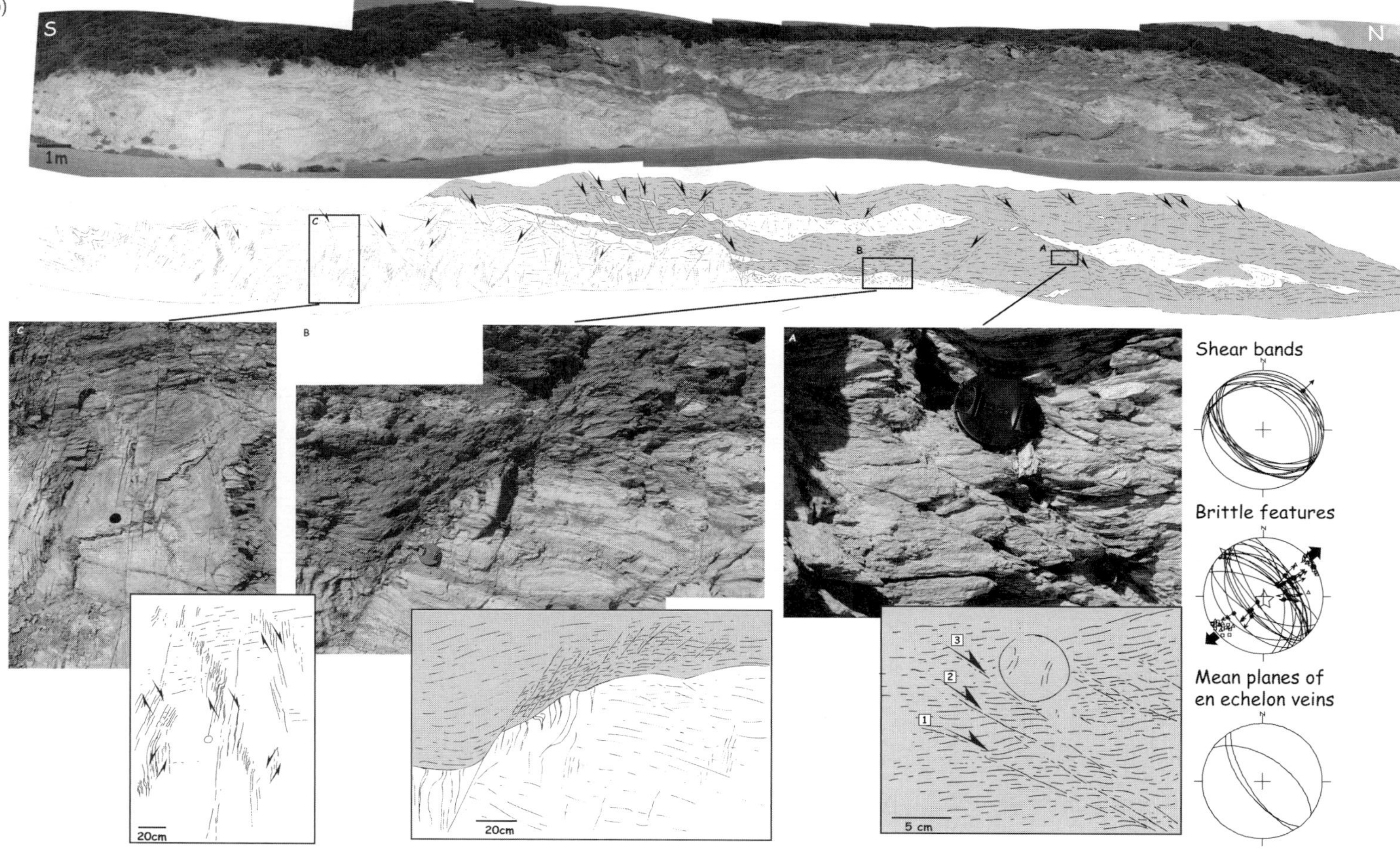

Fig. 10b. (*Continued*)

derive the orientation of the maximum principal stress $\sigma 1$, which will be compared with the attitude of veins. Stress tensors were calculated using a direct analytical inversion method (Angelier 1990), assuming that the slip direction in any given plane is parallel to the direction of maximum resolved shear stress of a large-scale homogeneous stress tensor (Wallace 1951; Bott 1959). For the studied outcrops, all data were retained when computing a single tectonic event. A reduced stress tensor is obtained; that is, the orientation of the principal stress axes σ_1, σ_2 and σ_3 ($\sigma_1 \geq \sigma_2 \geq \sigma_3$, compression positive) and a scalar invariant ϕ characterizing the shape of the stress ellipsoid:

$$\phi = (\sigma_2 - \sigma_3)/(\sigma_1 - \sigma_3), 0 \leq \phi \leq 1.$$

Inversion minimizes the misfit of the predicted shear and observed slip within the fault plane. Where the stress axes are computed from well-defined conjugate fault sets, as at most sites on Andros, the uncertainties in their orientation are lower than 10°. The good agreement between the stress axes computed from striated faults and the orientation of veins measured at the same sites confirms that the results obtained are reliable and accurate.

Results are presented using Schmidt's lower hemisphere projection (Fig. 11); orientations and dips of principal stress axes are reported in Table 2, together with the values of the stress ellipsoid shape ratio ϕ and estimators of the quality of the numerical calculation of the tensor. At most sites where the foliation is subhorizontal or displays a gentle dip, the computed stress axes σ_2 and σ_3 are horizontal or gently dipping and the σ_1 axis is subvertical, despite the variations between lithologies. This subvertical orientation is consistent with the vertical patterns of veins often associated with normal faults and characterizes an extensional tectonic regime. A consistent NE–SW direction of brittle extension is therefore recognized over the whole island. One outcrop is an exception to the rule, with a computed reverse-type tensor (southern coast of the island, Fig. 12a). For all the outcrops, it is noticeable that the maximum stress axis σ_1 is always nearly perpendicular to the foliation whatever its dip.

Some outcrops have been detailed in Figure 12. The brittle deformation of the outcrops shown in Figure 12a and b is characterized by veins and conjugate patterns of brittle faults. In the first outcrop, the foliation is dipping 71° west and indicators on fault planes and displacements of the beds along the faults indicate a reverse sense of motion. In the second outcrop, the foliation is dipping 42° west. Slip indicators and displacement of quartz veins show a normal sense of motion. Veins are perpendicular to the foliation and are bisectors of the acute angle between faults. In the two cases, the structures seem to have been tilted by the value by the foliation dip. When tilting the foliation around its local strike back to horizontal, 'unfolding' (Fig. 12b and c, second diagrams), a consistent NE direction for brittle extension is obtained, the stress axes becoming similar in trend to those determined at sites with subhorizontal foliation. This observation can be made even at sites where the dip of foliation remained gentle (Fig. 13). These results strongly suggest that all brittle structures of the island formed under a vertical maximum stress axis σ_1 and with an almost flat-lying foliation. Both the foliation and the brittle fault systems were therefore tilted later (see below).

Except at two sites where the values are low (0.14), the ϕ values calculated are in the range of 0.21–0.47, suggesting a generally well-defined true triaxial stress regime throughout the island.

Interpretation of field data and discussion: evolution of structures from ductile to brittle

Stretching lineation and stress tensors deduced from the inversion of fault slip data indicate a consistent NE–SW extension during ductile and brittle deformation. This leads us to conclude that there was a continuum of kinematics from ductile to brittle. We now discuss how brittle deformation is superimposed on ductile deformation during extension and the exhumation of the Lower Unit.

At the outcrop scale

Initial localization of ductile shear bands: the role of boudinage. In the presence of boudins, of whatever the scale and type (symmetrical or asymmetric), shear bands often nucleate at the end or in the necks between boudins, as already observed on Tinos (Jolivet *et al.* 2004*a*; Mehl *et al.* 2005): we can thus confirm that boudinage (and more generally lithological heterogeneities) is an efficient localizing factor of ductile and brittle deformation (Fig. 9a and b). Ductile shear bands, observed at decametre to millimetre scale all over the island, evolve, as do boudins, from symmetrical patterns on the southern coast towards asymmetric ones when approaching the detachment.

Onset of brittle deformation. The existence of semi-brittle shear bands shows the way in which brittle structures are superimposed on ductile ones. When brittle slip occurs along previous ductile shear planes in a direction strictly parallel to the stretching lineation, this superimposition

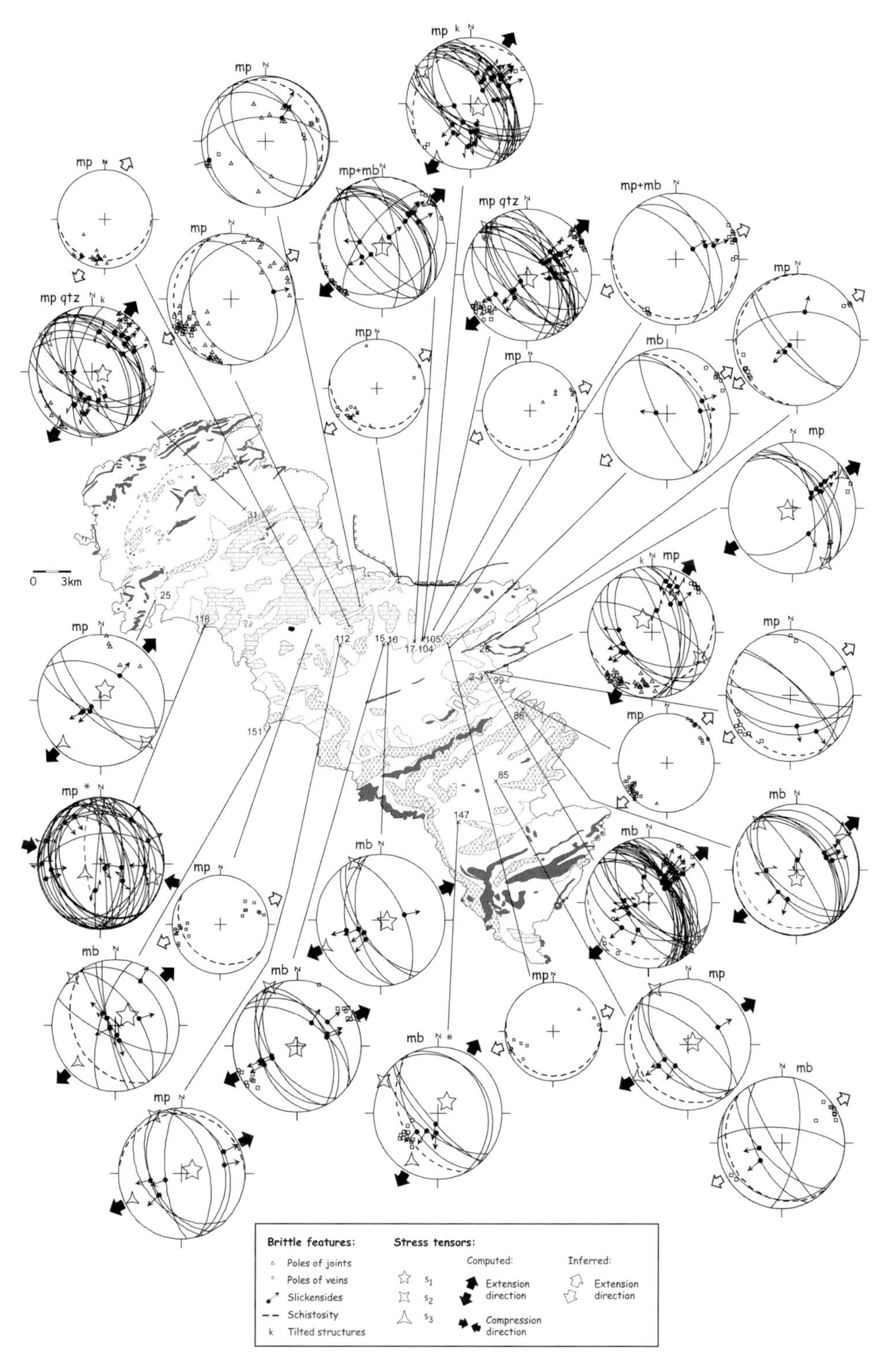

Fig. 11. Schmidt's lower hemisphere equal-area projection of brittle structures of Andros. mp, measurements made in metapelites; mb, measurements made in metabasites. The foliation is indicated by a dashed line. The computed and inferred brittle extension directions are in good agreement with ductile stretching (stretching lineation map of Figure 4).

Table 2. *Trends and plunges of axes of stress tensors deduced from the direct inversion of orientation and striae of faults*

Outcrop	Lithology	Number of fault planes	Strike/dip of foliation	σ_1	σ_2	σ_3	$\phi = (\sigma_2 - \sigma_3)/(\sigma_1 - \sigma_3)$	Quality estimator
15	**Metabasites**	8	–	222/87	334/01	064/03	0.29	A
16	**Metabasites**	5	–	104/82	335/05	244/06	0.32	A–B
17	**Metapelites** + metabasites	9	040/02	186/85	320/04	050/04	0.41	A
25	Metapelites	4	–	021/76	131/05	223/13	0.30	B–C
26		7	–	244/83	148/01	057/07	0.27	B
2–3	**Metabasites**	27	290/16	321/80	138/10	228/01	0.37	A
31	**Quartzitic metapelites**	16	–	097/71	297/18	205/06	0.46	A
85	Metapelites	4	315/17	068/83	328/01	238/07	0.28	B–C
86	**Metabasites**	9	304/14	181/79	320/08	051/07	0.52	A
99	Metapelites	10	254/18	324/71	118/17	210/08	0.30	B
104	Metapelites	25	050/18	116/80	302/10	212/01	0.37	A
105	**Quartzitic metapelites**	20	–	152/82	323/08	053/01	0.31	A
112	Metapelites	5	095/09	076/76	334/03	243/14	0.14	A–B
118*	Metapelites	13	354/71	104/17	009/15	239/67		A–B
				195/56	**347/31**	**085/13**	0.54	
147*	**Metabasites**	8	332/42	038/72	300/03	209/18		A–B
				270/75	**122/13**	**030/08**	0.40	
151	**Metabasites**	7	330/5	051/71	317/02	226/19	0.14	C

A quality estimator (A–C) has been attributed to each numerical result, based on the number and variety of attitudes of faults and on an intra-algorithm estimator accounting for the mean deviation between the computed shear stresses and the actual measured striations. Stress axes are given in their current attitude. Back-tilted stress axes are shown in bold.
Outcrops shown in Figure 14 where structures and foliation have been tilted by a significant amount.

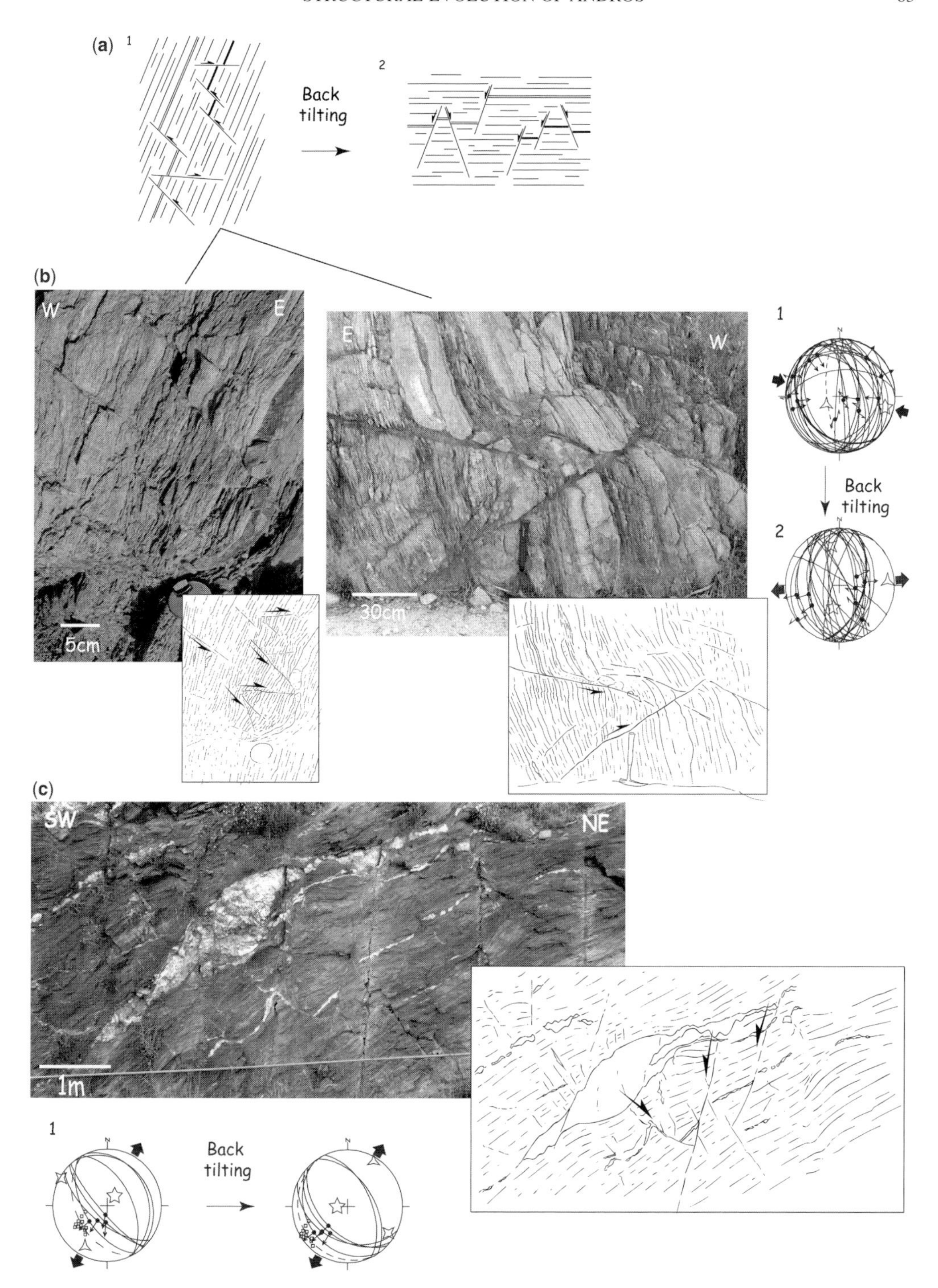

Fig. 12. (**a**) Principle of back-tilting of brittle structures. (**b**, **c**) Examples of tilted structures. (**B**, outcrop 118; **C**, outcrop 147) and associated Schmidt's diagrams showing the present attitude and the attitude of back-tilted fracture sets and foliation. It should be noted that after back-tilting, computed tensors correspond to a consistent NE–SW direction of brittle extension and a nearly vertical position of σ_1, in agreement with the vertical attitude of veins.

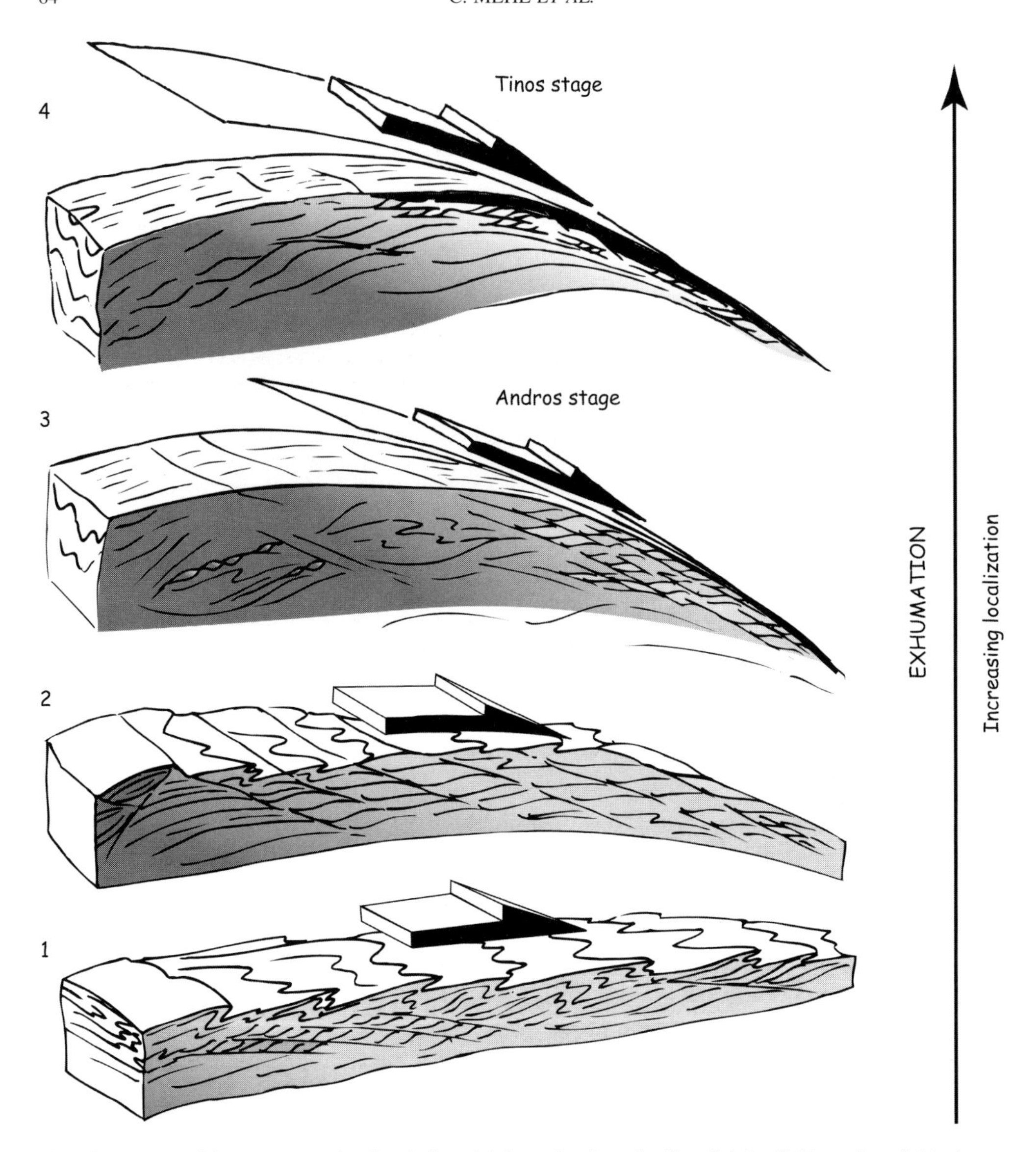

Fig. 13. Conceptual four-step scenario of evolution of deformation from ductile to brittle. (1) Formation of ubiquitous greenschist foliation under high finite strain and non-coaxial flow, as indicated by onset of sheath folds. (2) Boudinage of the foliation and onset of spatial distribution between coaxial and non-coaxial flow. Shear bands localize in interboudin necks. (3) Localization of non-coaxial flow in the northeastern part of the island. (4) Evolution of shear bands from semi-brittle to brittle structures. Deformation localizes in the NE, below the contact with the Upper Unit, to finally form the detachment itself. A new stage in the exhumation process has been proposed for Andros compared with a previous study (Mehl *et al.* 2005) (stage 3), which allows us to propose a complete section of an extending continental crust.

corresponds to a kind of 'reactivation' of a precursory ductile anisotropy. 'Reactivation' commonly refers to sliding along a pre-existing discontinuity; instead, we mean here that the discontinuous brittle slip is 'prepared' by ultimate localization of shear within a precursory shear band under a continuous extensional kinematics. Only the more steeply dipping shear bands show reactivation as brittle faults.

Numerous outcrops of the island show a succession of progressively steepening shear planes. The early shallow planes are almost parallel to the underlying ductile shear zones and we can observe an increasingly brittle behaviour, with first a slight bending of the foliation plane on either side of the semi-brittle shear zone and then a clear offset in a sense compatible with the ductile shear. Predominant brittle normal planes dip to the NE but conjugate planes with slip senses toward the SW are also observed on the same outcrop (Figs 9b and 10b). The ultimate step of evolution corresponds to steeply dipping fault planes showing calcite steps and slickenside lineations. Such an evolution can be explained as follows: during their return to the surface, rocks underwent a decrease in temperature and pressure that induces an evolution towards a more competent rheology (i.e. an increase in the internal friction angle). The increase in the internal friction angle supposes a decrease in the angle between the plane and σ_1, σ_1 remained vertical at least during the late brittle evolution of Andros, as indicated by vertical veins and σ_1 axes computed from faults at sites where foliation is subhorizontal and after back-tilting of vein and fault sets at sites where foliation is steeply dipping. Considering a subvertical σ_1, it is not surprising that the more brittle the regime, the more important is the dip of features. A progressive straightening of structures is a classical evolution from ductile to brittle.

'Reactivation' is not the only way for brittle features to develop. Some of them are newly formed. Joints and veins are often associated in en echelon arrays, already described by Papanikolaou (1978). En echelon arrays of veins and joints define shear zones whose orientation and dip are comparable with classical conjugate sets of normal faults. They occur in the more competent layers of the studied outcrops; that is, in the boudins of metabasites (Fig. 9a) and in quartzitic layers of the pelitic outcrops (Fig. 10a and b). NE-dipping planes are commonly better expressed than SW-dipping ones; that is, en echelon arrays associated with a SW dip are less numerous. En echelon arrays of veins and joints seem to play an important role on Andros in the progressive localization toward brittle deformation. They are interpreted here as the earliest step of localization before the steeply dipping normal patterns of faults initiate.

Scenario of localization process. Field observations allow us to propose a first-order scenario of evolution of deformation from ductile to brittle, under a continuous kinematic evolution. Primary localization of ductile deformation is closely linked to boudinage and the evolution to brittle deformation is marked by progressive straightening of structures and the onset of en echelon arrays of veins or joints. The ultimate step of localization consists in sliding across the en echelon patterns and the onset of actual brittle steeply dipping planes generally displaying conjugate patterns. Exhumation is thus accompanied by an increase in localization of deformation from ductile to brittle, pervasive normal faults reflecting the ultimate step of localization.

Role of the lithological contrast in the localization process. The preferred occurrence of brittle features in the more competent lithologies indicates the control of the rheology on the localization process. The importance of the rheological contrast has already been emphasized in the description of the earliest increments of localization of deformation. By considering boudinage as the initial localizing factor of ductile deformation, we implicitly assume a dominant role of lithological contrast in the first stage of localization. This control is always very important during the last brittle increments of deformation: brittle behaviour is preferentially observed (and presumably appeared earlier) in more competent layers (metabasites and quartzitic layers). Although the first-order scenario we propose is in good agreement with the sequential evolution of structures from ductile to brittle, rheological behaviour of materials appears as a key point in the description and understanding of the localization process. Rheological heterogeneities probably have a dominant affect on the depth at which the structures initially localize during their return to the surface.

At the island scale

The scenario of evolution of deformation from ductile to brittle discussed above at the outcrop scale also applies at the scale of the island, taking into account the distribution of ductile and brittle deformation across Andros. A conceptual scheme of time and space evolution of structures at island scale is proposed in Figure 13, steps 1–3, for Andros. Field observations show a progressive concentration of non-coaxial deformation along the NE coast. Post-HP deformation begins with the formation of a greenschist foliation, present all over the island; that is, associated with evidence for high finite strain rates and non-coaxial flow such as sheath folds. The foliation is further boudinaged. At this stage a spatial distribution of coaxial and non-coaxial deformation is already observed, with symmetrical boudins in the SW and asymmetrical ones in the NE. The sense of shear is consistently toward the NE in the NE part. There is thus localization of non-coaxial flow in the NE at the scale of the island. The formation of boudins is accompanied by the formation of shear bands in interboudin necks. These shear bands are symmetrical in the SW and asymmetrical with a consistent top-to-the-NE shear sense in the NE. Within metapelites a component of non-coaxial strain is always present. During exhumation, when shear bands evolve progressively to semi-brittle then brittle structures, deformation tends to localize in the NE below the contact with the upper plate and finally along the detachment itself. During this evolution, minor additional contacts also concentrate the shear, such as the base of some marble units, as in Paleokastro. The progressive concentration of deformation in a narrow zone explains the better preservation of HP parageneses and of early ductile structures on the southwestern coast of the island. The ultimate step of localization could correspond to the onset of the flat detachment himself; that is, a planar discontinuity that may have experienced cataclastic flow (compare the reddish breccia) before the last increment of brittle sliding. The gentle dome of foliation encompassing the whole island thus can be seen as a crustal-scale boudin with localization of a crustal-scale shear zone and later of a shallow-dipping fault at one extremity.

Doming, interpreted here as crustal-scale boudinage, is thus primarily a syn- to post-greenschist feature. Similar observations were recently made in the Betic Cordillera, where the formation of crustal-scale domes (Sierra Nevada, Sierra de Los Filabres, Sierra Alhamilla) also starts to be recorded by greenschist structures during exhumation (Augier *et al.* 2005). Folding affects greenschist facies on Tinos as well as on Andros and is interpreted to have occurred near (or above?) the brittle–ductile transition (Avigad *et al.* 2001). This means that folding is, like doming, a syn- to post-greenschist feature.

Numerous studied outcrops show palaeostress tensors computed from the measurement of brittle features with stress axes slightly tilted. All the brittle structures of Andros probably formed under a vertical maximum stress axis σ_1, but have been locally tilted in a late stage of deformation. This supposes that the schistosity was nearly flat before the onset of brittle structures. Tilting could be attributed to doming as well as to large-scale open folds described by Papanikolaou (1978) and Avigad *et al.* (2001), probably to a combination of both, but how can we explain a flat schistosity at the time brittle structures developed? Two hypotheses can be made. (1) Early ductile doming developed with a gentle curvature on Andros and ductile folding remained limited before brittle deformation occurred, so the schistosity remained nearly flat at this stage on most of the island. Doming and folding were thus mostly achieved after the onset of the first brittle structures. (2) Despite a first-order continuous evolution from ductile to brittle, local rheological contrasts or strain rate variations could have led to alternation of ductile and brittle behaviour across the transition, leading, for instance, to brittle deformation within metabasites while the pelitic matrix was still deforming more or less ductilely by folding. The two explanations do not contradict each other. Doming and large-scale open folding could have remained limited at the time of occurrence of the first increment of brittle deformation, and have later led to tilting of brittle structures developed mainly in competent material. In addition, folding, which is related to NW–SE shortening perpendicular to extension, certainly initiated in ductile conditions but possibly ended in the brittle field; this could be in good agreement with the component of NW–SE constriction recorded on Tinos and marked by late crenulation and brittle strike-slip faults (Mehl *et al.* 2005).

Comparison with Tinos

Tinos and Andros belong to the same crustal block of the Aegean Sea. They both correspond to b-type metamorphic domes; that is, domes elongated perpendicular to the main stretching direction (Jolivet *et al.* 2004*a*). Two metamorphic units are exposed on the islands and are separated by a reddish brecciated zone and a detachment (Fig. 14). The Upper Unit crops out on the northern coast of the two islands but occurs on Tinos over a larger area. The lithologies of the two Lower Units are comparable, with alternating metapelites, marble horizons and metabasites, the latter two being boudinaged into the less competent matrix of metapelites. Boudins are more numerous on Andros because the finite deformation is less severe, but are also observable in some places on Tinos. Stretching lineation and brittle strain axes indicate a continuum of strain from ductile to brittle on the two islands (Mehl *et al.* 2005). Extension is oriented NE–SW.

Drawing two cross-sections of Tinos and Andros perpendicular to the detachment allows us to further compare the spatial evolution of deformation (Fig. 14). The two islands show the same gradient of retrogression from SW to NE, with better-preserved HP parageneses on the southern coast. Peak P–T conditions, although not well constrained on Andros, seem to be comparable, at 18 or 15 kbar and 500 °C for Tinos (Parra *et al.* 2002) and >10 kbar and 450–500 °C for Andros. The greenschist overprint is estimated at 9 kbar and 400 °C for Tinos (Parra *et al.* 2002) and 5–6 kbar and 400 °C for Andros (Reinecke 1982). Andros seems *a priori* to have undergone lower pressures than Tinos for equivalent temperatures, but, again, we must remain cautious with this conclusion: contrary to Tinos where precise P–T estimates were made, $P-T$ estimates on Andros are based on only one metamorphic reaction.

The same gradient of shear strain exists on the two islands, with a coaxial flow on the southwestern part of the islands, which evolves towards a non-coaxial flow when approaching the detachment, as indicated by the evolution from a symmetrical deformation on the SW coast towards an asymmetric deformation on the NE coast. Symmetrical deformation is expressed in a wide southern zone on Andros whereas it is limited to a narrow band on the southern coast of Tinos. However, the different widths of the two islands, and therefore the difference in area of outcrops away from the detachment, may cause a bias in these observations.

Some structural differences exist between the two islands. Ductile structures, and especially decametre-scale shear bands, are better expressed on Andros than on Tinos, and they are more 'brittle' there. The decametre-scale shear bands of Andros seem to have been frozen during the localization process and to have encompassed brittle deformation, whereas on Tinos the whole NE part of the island below the main detachment is a large-scale shear zone with a large concentration of strain. The progressive evolution with increasingly numerous shear bands seen on Andros is not as clearly visible on Tinos, where the spatial transition from coaxial to non-coaxial is more abrupt. We interpret this observation as the result of a greater strain localization and a larger finite strain for Tinos. This interpretation is in agreement with the fact that Tinos is closer to the centre of the Cyclades, where extension has its maximum rate, and closer to Mykonos and Naxos, where the units that have experienced the highest temperatures have been exhumed (migmatites) (Avigad & Garfunkel 1989; Jolivet & Patriat 1999).

A study was made on the boudins of the two islands, because they testify for the amount to finite strain the rocks encompassed from the beginning of the greenschist overprint to the end of the localization process. We analysed 39 photographs of trains of boudins of different scales and in different material, taken in the maximum stretching plane. Twenty-one photographs were taken on Andros, and 18 on Tinos. Assuming a conservation of surface, we transform them into trains of rectangles whose heights correspond to the maximum height of the initial trains of boudins (Fig. 15, 1–3). We used the strain reversal program (Lloyd & Condliffe 2003) to constrain the elongation coefficient responsible for the boudinage (Fig. 15, 4). Results are presented in Figure 15, 5. The peaks of frequency of elongation coefficient are between 300 and 400% and 200 and 300% and the means of the elongation coefficients are 412% and 245% for Tinos and Andros, respectively. Despite the roughness of these estimates, it can be proposed that Tinos recorded, at least until boudinage started, a greater finite strain than Andros, thus supporting conclusions derived solely from field observations. The peak of greenschist overprint is dated to 21–23 Ma on Tinos (Avigad & Garfunkel 1989; Stolz *et al.* 1997). Assuming boudinage to be coeval with greenschist deformation and the peaks being nearly coeval on Andros and Tinos, it is possible to calculate the mean strain rate rocks sustained: it is roughly evaluated $4.5e-15\ s^{-1}$ for Tinos and $2.1e-15\ s^{-1}$ for Andros.

The fact that the shear bands of Andros were 'frozen' could perhaps have been favoured by an earlier arrival of Andros in the brittle domain. A gradient of the P/T ratios from Mt. Olympus to Naxos has been proposed by Jolivet & Patriat (1999) by comparison of the $P-T$ paths of the islands. A first-order estimation of the gradients of temperature in the final part of the $P-T$ paths gives 20 °C km^{-1}, 30 °C/km and 40 °C km^{-1} for Olympus, Tinos and Naxos, respectively. These gradients seem very large, but a systematic increase of the P/T ratios in the shallower part of the crust (i.e. the depths around the brittle–ductile transition zone) can be deduced from the $P-T$ paths. As Andros is situated between Olympus and Tinos, it seems logical to consider its P/T ratio to be between 20 and 30 °C km^{-1}. Because no $P-T$ paths have been precisely calculated on Andros, it is difficult to confirm this hypothesis without doubt. We further assume that P/T ratios seen on the $P-T$ grid give an idea of geothermal gradients.

As strain rate estimates on the two islands are not significantly different and the P/T ratios of Andros are probably less important than those of Tinos, we can hypothesize that the brittle–ductile transition zone of Andros develops at deeper levels of the crust than that of Tinos: this could explain why the shear bands of Andros seem more brittle than those of Tinos.

Furthermore, the brittle evolution is furthermore different on Tinos and Andros. Tinos seems to have recorded an episode of semi-brittle to brittle deformation that Andros has not: small-scale shallow-dipping normal faults were found in the footwall of Tinos (Mehl *et al.* 2005) that never appear on Andros below the detachment. In contrary, Andros has preserved one early stage of brittle deformation that Tinos does not: en echelon arrays of veins are well preserved in metabasitic and quartzitic lithologies of Andros, whereas actual faults are the rule on Tinos. These differences tend to prove that the localization process is more mature on Tinos than on Andros.

No major tilting of structures has been recorded on Tinos, as pointed out by the inversion of stress

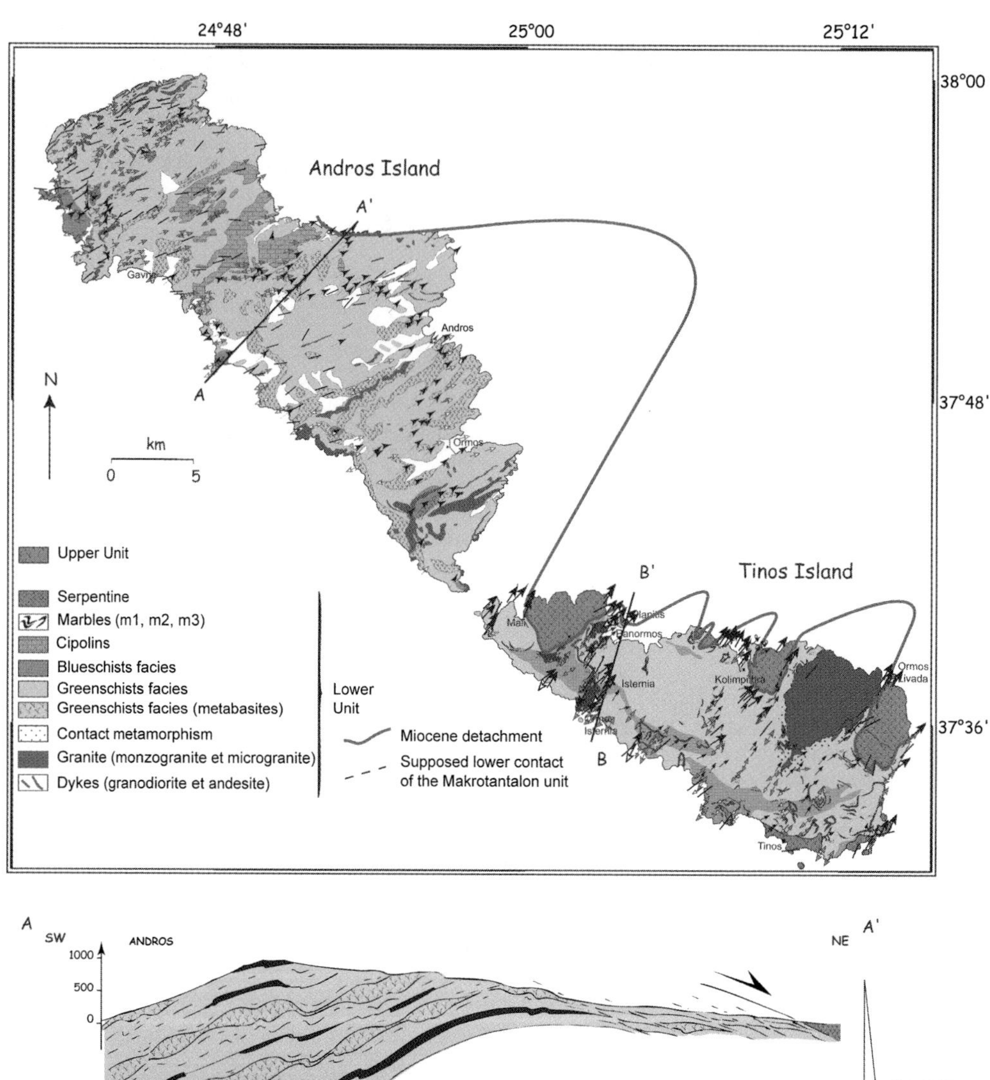

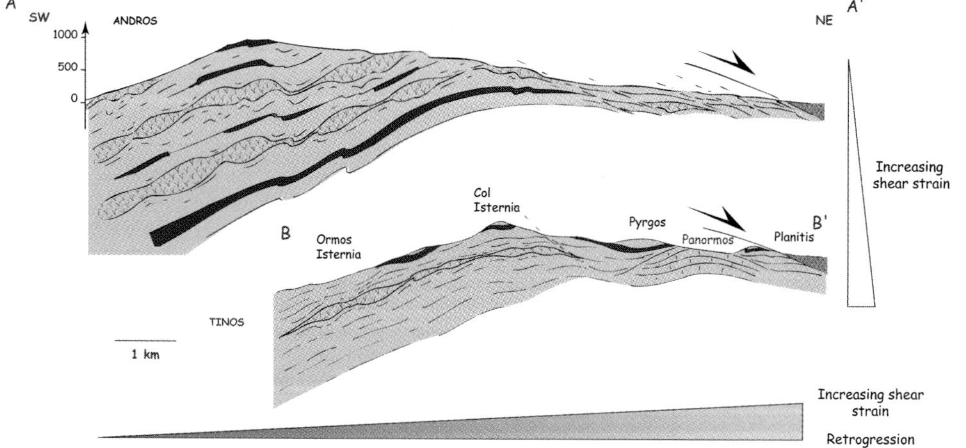

Fig. 14. Comparison in map and cross-section of Tinos and Andros. The two islands display the same gradients of retrogression and increasing shear strain from SW to NE. Decametre-scale shear bands are better preserved on Andros. We interpret this as the result of greater strain localization and larger finite strain for Tinos.

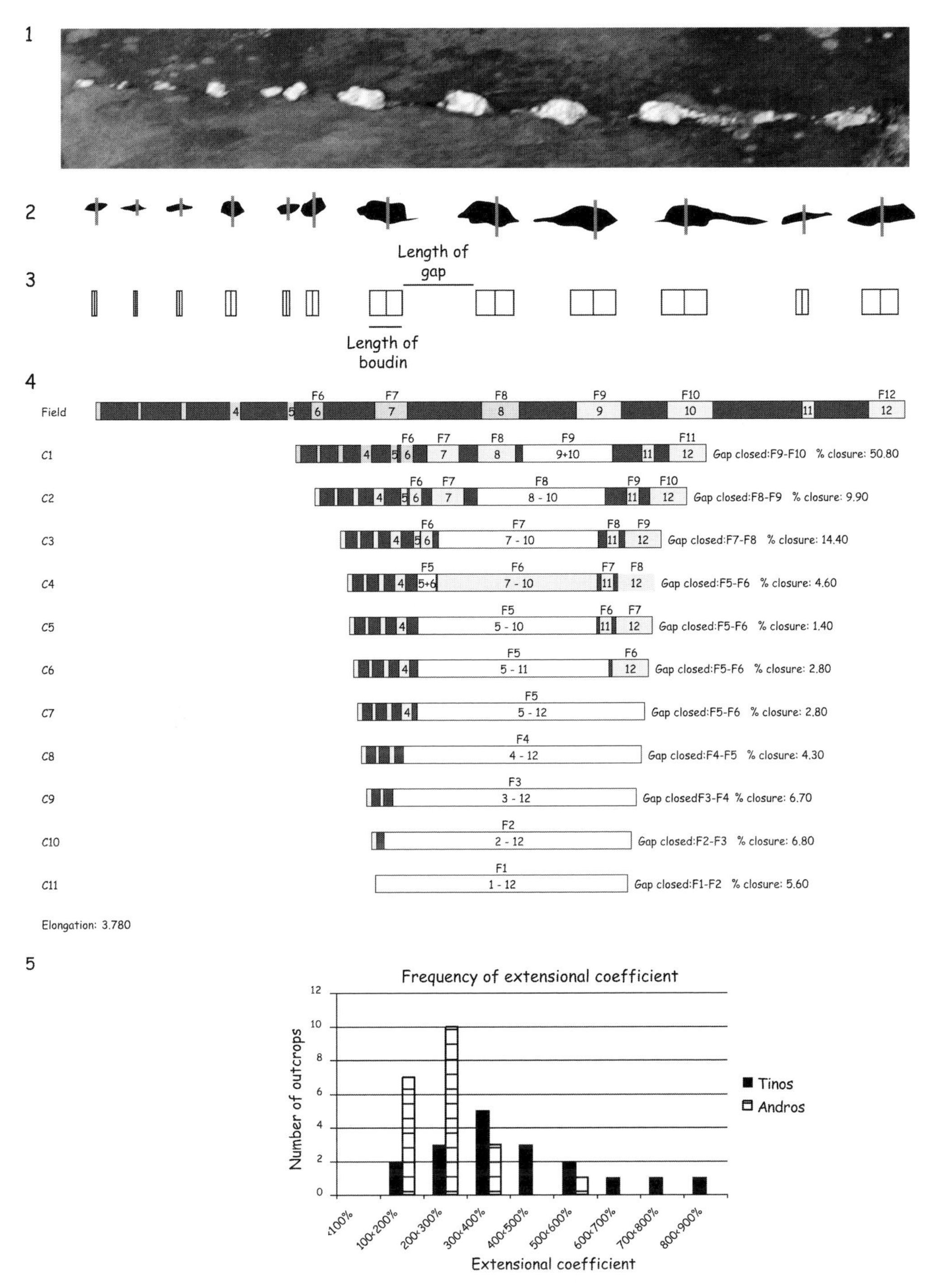

Fig. 15. Principle of calculation of coefficients of extension based on boudinage. Several photographs have been analysed for Tinos and Andros. The first step consists of calculating the surface of the trains of boudins. Assuming a conservation of surface, the boudins were replaced by rectangles whose position on a horizontal was deduced from the position of the maximum high of each boudin. The coefficient of elongation was computed from the Strain Reversal Program® (Lloyd & Condliffe 2003). The extensional coefficients of the two islands are shown in the lower part of the figure.

tensor. σ_1 is vertical or close to vertical for the major part of the studied outcrops, in good agreement with the ubiquitous vertical attitude of veins all over the island (Mehl *et al.* 2005). This absence of tilting is confirmed by the palaeomagnetic data on the granodiorite and on post-folial dykes of Tinos (Morris & Anderson 1996; Avigad *et al.* 1998). The major part of extension was accommodated before 19 Ma on Tinos, which corresponds to the age of the granite intrusion, and the post-ductile extension was also achieved without tilting. All brittle structures of Andros have similarly formed under a vertical maximum stress axis σ_1, but contrary to the major part of the outcrops of Tinos, they have been locally tilted in a late stage of deformation, probably in response to large-scale open NE–SW folds (Papanikolaou 1978; Avigad *et al.* 2001).

Conclusion: toward a complete section of an extending continental crust

Our observations on Andros and Tinos show two different stages of a continuous process that exhumed metamorphic rocks below a crustal-scale detachment. Earlier stages can be seen near Mt. Olympus and on the island of Evia, and a later and extreme stage on Mykonos and Naxos (Jolivet & Patriat 1999). As proposed by Avigad & Garfunkel (1989), deeper units are exhumed from Evia to Mykonos. This observation led those workers to postulate a NW–SE direction of extension, before the deformation was first described by Gautier (1994) and the top-to-the-NE shear sense ascertained. We assume that the same extension process has caused Oligo-Miocene post-HP exhumation from Mt. Olympus to Mykonos–Naxos during the formation of the Aegean Sea and that the only difference lies in the finite extension, which is greater in the centre of the Cyclades. An active equivalent of this deformation process can be found along the NE coast of continental Greece (Laigle *et al.* 2000) and in the Gulf of Corinth (Jolivet *et al.* 1994; Jolivet 2001), where brittle faults roots on shallow north- or NE-dipping shear zones within the brittle–ductile transition (Rigo *et al.* 1996; Sorel 2000). Following this assumption we can propose a scheme of vertical stratification of deformation regimes from the extending upper brittle crust to the lower crust (Fig. 16).

The following rheological stratification is proposed. (1) The upper crust is brittle and shallow- and steeply dipping normal faults control the deposition of synrift basins. (2) Cataclastic deformation along the main detachment allows it to continue with a shallow dip for most of its life (Mehl *et al.* 2005). The cataclastic shear zone becomes wider downward and a progressive change to ductile conditions is observed depending on the nature of the material involved and the thermal and fluid conditions. (3) At depth this shear zone becomes shallow dipping and merges with a shallow NE-dipping shear zone below the brittle–ductile transition. (4) The lower crust is weak because of partial melting as recorded on Mykonos or Naxos and the deformation is thus less localized. The deformation there is thus partitioned between bulk coaxial thinning and simple shear induced by the motion of the hanging wall of the detachment.

At the brittle–ductile transition, the deformation is progressively localized in the footwall, first in the necks between boudins and along shear zones. An evolution toward more non-coaxial conditions is observed toward the detachment. The overall structure corresponds to a megaboudinage of the crust with the localization of a shear zone and then a fault at the extremity of this crustal-scale boudin.

This evolution thus emphasizes that the rheological stratification and the intrinsic compositional heterogeneity of the continental crust (leading to boudinage) both control strain localization processes. This process is, furthermore, under the control of the behaviour of fluids. Famin *et al.* (2004) have shown that surface-derived fluids invade the brittle–ductile transition and favour strain localization at this level of the crust. A connected vein network ensures the channelization of these fluids from the surface, along the uppermost faults and within the brittle–ductile transition. Further down, veins do not make a connected network and fluid accumulate at the brittle–ductile transition. The presence of these fluids in active extensional context is well illustrated by the Corinth Rift case (Pham *et al.* 2000). As mentioned above, fluids can lower the resistance of the rocks and favour strain localization but they need the formation of conduits to reach the brittle–ductile transition, they thus need some strain localization to have already happened. This early strain localization, as shown on Andros, can be explained by boudinage and the formation of shear bands and faults in the necks between boudins. Andros illustrates this stage, whereas Tinos shows the later evolution toward the formation of a narrow shear zone when an intense shearing is recorded. The latest stage of extension shows a more coaxial deformation pattern and is marked by widespread conjugate sets of steeply dipping mesoscale normal faults all over the islands. These faults cut across the cataclastic zone itself, indicating that this zone was no longer active at that time and that the cataclasites have become progressively stiffer and

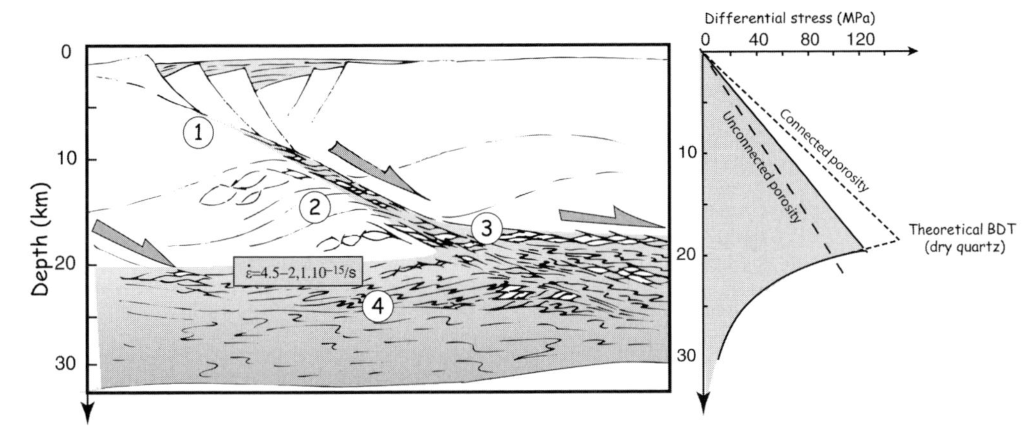

Fig. 16. Synthetic cross-section of an extending continental crust. (1) Brittle upper crust: shallow- and steeply dipping normal faults control the deposition of synrift basins. (2) The cataclastic shear zone broadens with depth, evolving progressively towards ductile deformation. (3) Decrease in the dip of the shear zone, which merges below the brittle–ductile transition (BDT). (4) Weak lower crust, as a result of partial melting. We show on the cross-section the strain rates calculated from boudinage on the islands of Tinos and Andros. A rheological envelope, calculated from the rheological parametres of dry quartz, is shown in the right part of the picture to better calibrate the depth scale of the cross-section.

more brittle during exhumation, while the shear movement localizes along the flat brittle detachment, which accommodates the last increments of extension.

References

ANGELIER, J. 1990. Inversion of field data in fault tectonics to obtain the regional stress—III. A new rapid direct inversion method by analytical means. *Geophysical Journal International*, **103**, 363–376.

ARMIJO, R., LYON-CAEN, H. & PAPANIKOLAOU, D. 1992. East–West extension and Holocene normal fault scarps in the Hellenic arc. *Geology*, **20**, 491–494.

ARMIJO, R., MEYER, B., KING, G. C. P., RIGO, A. & PAPANASTASSIOU, D. 1996. Quaternary evolution of the Corinth Rift and its implications for the Late Cenozoic evolution of the Aegean. *Geophysical Journal International*, **126**, 11–53.

AUBOUIN, J. & DERCOURT, J. 1965. Sur la géologie de l'Egée: regard sur la Crète (Grèce). *Bulletin de la Société Géologique de France*, **7**, 787–821.

AUGIER, R., JOLIVET, L. & ROBIN, C. 2005. Late Orogenic doming in the eastern Betic Cordilleras: Final exhumation of the Nevado-Filabride complex and its relation to basin genesis. *Tectonics*, **24**, doi: 10.1029/2004TC001687.

AVIGAD, D. & GARFUNKEL, Z. 1989. Low-angle faults above and below a blueschist belt: Tinos Island, Cyclades, Greece. *Terra Nova*, **1**, 182–187.

AVIGAD, D. & GARFUNKEL, Z. 1991. Uplift and exhumation of high-pressure metamorphic terranes: the example of the Cycladic blueschists belt (Aegean Sea). *Tectonophysics*, **188**, 357–372.

AVIGAD, A., GARFUNKEL, Z., JOLIVET, L. & AZAÑÓN, J. M. 1997. Back-arc extension and denudation of Mediterranean eclogites. *Tectonics*, **16**, 924–941.

AVIGAD, D., BAER, G. & HEIMANN, A. 1998. Block rotations and continental extension in the Central Aegean Sea: paleomagnetic and structural evidence from Tinos and Mykonos. *Earth and Planetary Science Letters*, **157**, 23–40.

AVIGAD, D., ZIV, A. & GARFUNKEL, Z. 2001. Ductile and brittle shortening, extension-parallel folds and maintenance of crustal thickness in the Central Aegean. *Tectonics*, **20**, 277–287.

BLAKE, M. C. J., BONNEAU, M., GEYSSANT, J., KIENAST, J. R., LEPVRIER, C., MALUSKI, H. & PAPANIKOLAOU, D. 1981. A geologial reconnaissance of the Cycladic blueschist belt, Greece. *Geological Society of America Bulletin*, **92**, 247–254.

BONNEAU, M. 1982. Evolution géodynamique de l'arc égéen depuis le Jurassique Supérieur jusqu'au Miocène. *Bulletin de la Société Géologique de France*, **7**, 229–242.

BOTT, M. H. P. 1959. The mechanisms of oblique slip faulting. *Geological Magazine*, **96**, 109–117.

BOZKURT, E. 2003. Origin of the NNE-trending basins in Westren Turkey. *Geodinamica Acta*, **16**, 61–81.

BOZKURT, E. & PARK, R. G. 1997. Evolution of a mid-Tertiary extensional shear zone in the southern Menderes massif, western Turkey. *Bulletin de la Société Géologique de France*, **168**, 3–14.

BRÖCKER, M. & FRANZ, L. 2006. Dating metamorphism and tectonic stacking on Andros Island (Cyclades, Greece): results of a Rb–Sr study. *Geological Magazine*, **143**, 609–620.

BRUNN, J. H., ARGYRIADIS, I., RICOU, L. E., POISSON, A., MARCOUX, J. & DE GRACIANSKY, P. C. 1976. Eléments majeurs de liaison entre Taurides et Hellénides. *Bulletin de la Société Géologique de France*, **18**, 481–497.

DAVIS, G. H. 1980. Structural characteristics of metamorphic core complexes, southern Arizona. *In*: CRITTENDEN, M. D. J., CONEY, P. J. & DAVIS, G. H.

(eds) *Cordilleran Metamorphic Core Complexes. Geological Society of America, Memoirs*, **153**, 35–77.

DAVIS, G. H. & CONEY, P. J. 1979. Geologic development of the Cordilleran metamorphic core complexes. *Geology*, **7**, 120–124.

DIXON, J. & WILLIAMS, G. 1983. Reaction softening in mylonites from Arnaboll thrust, Sutherland. *Scottish Journal of Geology*, **19**, 157–168.

DÜRR, S. 1986. Das Attisch-kykladische Kristallin. *In*: JACOBSHAGEN, V. (ed.) *Geologie von Grienchenland* Gebruder Borntraeger, Berlin, 116–148.

FAMIN, V., PHILIPPOT, P., JOLIVET, L. & AGARD, P. 2004. Evolution of hydrothermal regime along a crustal shear zone, Tinos Island, Greece. *Tectonics*, **23**, doi:10.1029/2003TC001509.

FAURE, M. & BONNEAU, M. 1988. Données nouvelles sur l'extension néogène de l'Egée: la déformation ductile du granite miocène de Mykonos (Cyclades, Grèce). *Comptes Rendus de l' Academie des Sciences*, **307**, 1553–1559.

FITZ GERALD, J. D. & STÜNITZ, H. 1993. Deformaton of granitoids at low metamorphic grade. Reactions and grain size reduction. *Tectonophysics*, **221**, 269–297.

GAUTIER, P. 1994. Géométrie crustale et cinématique de l'extension tardi-orogénique dans le domaine centre-égéen (iles des Cyclades et d'Eubée, Grèce). Unpublished Thesis, Université de Rennes.

GAUTIER, P. & BRUN, J. P. 1994*a*. Ductile crust exhumation and extensional detachments in the central Aegean (Cyclades and Evvia islands). *Geodinamica Acta*, **7**, 57–85.

GAUTIER, P. & BRUN, J. P. 1994*b*. Crustal-scale geometry and kinematics of late-orogenic extension in the central Aegean (Cyclades and Evvia island). *Tectonophysics*, **238**, 399–424.

GUEYDAN, F., LEROY, Y. & JOLIVET, L. 2001. Grain-size sensitive flow and shear stress enhancement at the brittle to ductile transition of the continental crust. *International Journal of Earth Sciences*, **90**, 181–196.

GUEYDAN, F., LEROY, Y., JOLIVET, L. & AGARD, P. 2003. Analysis of continental midcrustal strain localization induced by microfracturing and reaction-softening. *Journal of Geophysical Research*, **108**, 2064, doi:10.1029/2001JB000611.

JACOBSHAGEN, V., DÜRR, S., KOCKEL, F., KOPP, K. O., KOWALCZYK, G., BERCKHEMER, H. & BÜTTNER, D. 1978. Structure and geodynamic evolution of the Aegean region. *In*: CLOOS, H., ROEDER, D. & SCHMIDT, K. (eds) *Alps, Apennines, Hellenides*. IUGG, **report 38** Stuttgart, 537–564.

JACKSON, J. A. 1987. Active normal faulting and continental extension. *In*: COWARD, M. P., DEWEY, J. F. & HANCOCK, P. L. (eds) *Continental Extensional Tectonics*. Geological Society, London, Special Publications, **28**, 3–18.

JACKSON, J. 1994. Active tectonics of the Aegean region. *Annual Review of Earth and Planetary Sciences*, **22**, 239–271.

JACKSON, J. A. & WHITE, N. J. 1989. Normal faulting in the upper continental crust: observations from regions of active extension. *Journal of Structural Geology*, **11**, 15–36.

JOLIVET, L. 2001. A comparison of geodetic and finite strain pattern in the Aegean, geodynamic implications. *Earth and Planetary Science Letters*, **187**, 95–104.

JOLIVET, L. & FACCENNA, C. 2000. Mediterranean extension and the Africa–Eurasia collision. *Tectonics*, **19**, 1095–1106.

JOLIVET, L. & PATRIAT, M. 1999. Ductile extension and the formation of the Aegean Sea. *In*: DURAND, B., JOLIVET, L., HORVÀTH, F. & SÉRANNE, M. (eds) *The Mediterranean Basins: Tertiary Extension within the Alpine Orogen*. Geological Society, London, Special Publications, **156**, 427–456.

JOLIVET, L., BRUN, J. P., GAUTIER, P., LALLEMANT, S. & PATRIAT, M. 1994. 3-D kinematics of extension in the Aegean from the Early Miocene to the Present, insight from the ductile crust. *Bulletin de la Société Géologique de France*, **165**, 195–209.

JOLIVET, L., FAMIN, V., MEHL, C., PARRA, T., AVIGAD, D. & AUBOURG, C. 2004*a*. Progressive strain localisation, crustal-scale boudinage and extensional metamorphic domes in the Aegean Sea. *In*: WHITNEY, D. L., TEYSSIER, C. & SIDDOWAY, C. S. (eds) *Gneiss Domes in Orogens*. American Geological Society, Special Papers, **380**, 185–210.

JOLIVET, L., RIMMELÉ, G., OBERHÄNSLI, R., GOFFÉ, B. & CANDAN, O. 2004*b*. Correlation of syn-orogenic tectonic and metamorphic events in the Cyclades, the Lycian Nappes and the Menderes massif, geodynamic implications. *Bulletin de la Société Géologique de France*, **175**, 217–238.

KATZIR, Y., AVIGAD, D., MATTHEWS, A., GARFUNKEL, Z. & EVANS, B. 2000. Origin, HP/LT metamorphism and cooling of ophiolitic mélanges in southern Evia (NW Cyclades), Greece. *Journal of Metamorphic Geology*, **18**, 699–718.

KIRBY, S. H. 1985. Rock mechanics observations pertinent to the rheology of the continental lithosphere and the localization of strain along shear zones. *Tectonophysics*, **119**, 1–27.

LAIGLE, M., HIRN, A., SACHPAZI, M. & ROUSSOS, N. 2000. North Aegean crustal deformation: an active fault imaged to 10 km depth by reflection seismic data. *Geology*, **28**, 71–74.

LE PICHON, X. & ANGELIER, J. 1981. The Aegean Sea. *Philosophical Transaction of the Royal Society of London*, **300**, 357–372.

LISTER, G. S. & DAVIS, G. A. 1989. The origin of metamorphic core complexes and detachment faults formed during Tertiary continental extension in the northern Colorado River region, U.S.A. *Journal of Structural Geology*, **11**, 65–94.

LISTER, G. S., BANGA, G. & FEENSTRA, A. 1984. Metamorphic core complexes of cordilleran type in the Cyclades, Aegean Sea, Greece. *Geology*, **12**, 221–225.

LLOYD, G. E. & CONDLIFFE, E. 2003. 'Strain Reversal': a Windows™ program to determine extensional strain from rigid–brittle layers or inclusions. *Journal of Structural Geology*, **25**, 1141–1145.

MARQUER, D., GAPAIS, D. & CAPDEVILA, R. 1985. Chemical changes and mylonitisation of a granodiorite within low-grade metamorphism (Aar Massif, Central Alps). *Bulletin de Minéralogie*, **108**, 209–221.

MEHL, C., JOLIVET, L. & LACOMBE, O. 2005. From ductile to brittle: evolution and localization of deformation below a crustal detachment (Tinos, Cyclades, Greece). *Tectonics*, **24**, TC4017, doi: 10.1029/2004TC001767.

MITRA, G. 1978. Ductile deformation zones and mylonites: the mechanical processes involved in the deformation of crystalline basement rocks. *American Journal of Science*, **278**, 1057–1084.

MORRIS, A. & ANDERSON, A. 1996. First paleaomagnetic results from the Cycladic Massif, Greece, and their implications for Miocene extension directions and tectonic models in the Aegean. *Earth and Planetary Science Letters*, **142**, 397–408.

PAPANIKOLAOU, D. J. 1978. Contribution to the geology of the Aegean Sea: the island of Andros. *Annales Géologiques des Pays Helléniques*, **29**, 477–553.

PAPANIKOLAOU, D. 1987. Tectonic evolution of the Cycladic blueschist belt (Aegean Sea, Greece). *In*: HELGESON, H. C. (ed.) *Chemical Transport in Metasomatic Processes*. D. Reidel Publishing Company, Dordrecht, 429–450.

PAPANIKOLAOU, D., LYKOUSIS, V., CHRONIS, G. & PAVLAKIS, P. 1988. A comparative study of neotectonic basins across the Hellenic arc: the Messiniakos, Argolikos, Saronikos and Southern Evoikos gulfs. *Basin Research*, **1**, 167–176.

PARRA, T., VIDAL, O. & JOLIVET, L. 2002. Relation between deformation and retrogression in blueschist metapelites of Tinos island (Greece) evidenced by chlorite–mica local equilibria. *Lithos*, **63**, 41–66.

PATRIAT, M. 1996. Etude de la transition cassant-ductile en extension, application au transect Olympe-Naxos, Grèce. Unpublished Thèse de Doctorat, Université Pierre et Marie Curie, Paris.

PATRIAT, M. & JOLIVET, L. 1998. Post-orogenic extension and shallow-dipping shear zones, study of a brecciated décollement horizon in Tinos (Cyclades, Greece). *Comptes Rendus de l'Académie des Sciences*, **326**, 355–362.

PHAM, V. N., BERNARD, P., BOYER, D., CHOULIARAS, G., LE MOUEL, J. L. & STAVRAKAKIS, G. N. 2000. Electrical conductivity and crustal structure beneath the central Hellenides around the Gulf of Corinth (Greece) and their relationship with the seismotectonics. *Geophysical Journal International*, **142**, 948–969.

PRICE, N. J. & COSGROVE, J. W. 1990. Boundinage and pinch-and-swell structures. *In*: PRICE, N. J. & COSGROVE, J. W. *Analysis of Geological Structures*. Cambridge University Press, Cambridge, 405–443.

REINECKE, T. 1982. Cymrite and celsian in manganese-rich metamorphic rocks from Andros island, Greece. *Contributions to Mineralogy and Petrology*, **79**, 333–336.

REINECKE, T. 1986. Phase relationships of sursassite and other Mn-silicates in highly oxidized low-grade, high-pressure metamorphic rocks from Evvia and Andros Islands, Greece. *Contributions to Mineralogy and Petrology*, **84**, 110–126.

REINECKE, T., OKRUSCH, M. & RICHTER, P. 1985. Geochemistry of ferromanganoan metasediments from the island of Andros, Cycladic Blueschist Belt, Greece. *Chemical Geology*, **53**, 249–278.

RIETBROCK, A., TIBÉRI, C., SCHERBAUM, F. & LYON-CAEN, H. 1996. Seismic slip on a low angle normal fault in the Gulf of Corinth: evidence from high resolution cluster analysis of microearthquakes. *Geophysical Research Letters*, **23**, 1817–1820.

RIGO, A., LYON-CAEN, H., ARMIJO, R. *ET AL*. 1996. A microseismicity study in the western part of the Gulf of Corinth (Greece): implications for large-scale normal faulting mechanisms. *Geophysical Journal International*, **126**, 663–688.

SEYITOGLU, G. & SCOTT, B. 1991. Late Cenozoic crustal extension and basin formation in West Turkey. *Geological Magazine*, **128**, 155–166.

SEYITOGLU, G. & SCOTT, B. C. 1996. The cause of N–S extensional tectonics in western Turkey: tectonic escape vs back-arc spreading vs orogenic collapse. *Journal of Geodynamics*, **22**, 145–153.

SHAKED, Y., AVIGAD, D. & GARFUNKEL, Z. 2000. Alpine high-pressure metamorphism at the Almyropotamos window (southern Evia, Greece). *Geological Magazine*, **137**, 367–380.

SOREL, D. 2000. A Pleistocene and still-active detachment fault and the origin of the Corinth–Patras rift, Greece. *Geology*, **28**, 83–86.

STOLZ, J., ENGI, M. & RICKLI, M. 1997. Tectonometamorphic evolution of SE Tinos, Cyclades, Greece. *Swiss Bulletin of Mineralogy and Petrology*, **77**, 209–231.

TAYMAZ, T., JACKSON, J. & MCKENZIE, D. 1991. Active tectonics of the north and central Aegean Sea. *Geophysical Journal International*, **106**, 433–490.

TAYMAZ, T., WESTAWAY, R. & REILINGER, R. (eds) 2004. Active faulting and crustal deformation in the Eastern Mediterranean Region. *Tectonophysics* (special issue), **391**.

WALLACE, R. E. 1951. Geometry of shearing stress andrelation to faulting. *Journal of Geology*, **59**, 118–130.

WEATHERS, M. S., BIRD, J. M., COOPER, R. F. & KOHLSTEDT, D. L. 1979. Differential stress determined from deformation-induced microstructures of the Moine Thrust Zone. *Journal of Geophysical Research*, **84**, 7495–7509.

WHITE, S. H. & KNIPE, R. J. 1978. Transformation- and reaction-enhanced ductility in rocks. *Journal of the Geological Society, London*, **135**, 513–516.

WIBBERLY, C. 1999. Are feldspar-to-micas reactions necessarily reaction-softening process in fault zones? *Journal of Structural Geology*, **21**, 1219–1227.

WINTSCH, R. P., CHRISTOFFERSEN, R. & KRONENBERG, A. K. 1995. Fluid–rock reaction weakening of fault zones. *Journal of Geophysical Research*, **100**, 13021–13032.

Late Miocene igneous rocks of Samos: the role of tectonism in petrogenesis in the southeastern Aegean

G. PE-PIPER[1] & D. J. W. PIPER[2]

[1]*Department of Geology, Saint Mary's University, Halifax, N.S. B3H 3C3, Canada (e-mail: gpiper@smu.ca)*

[2]*Geological Survey of Canada (Atlantic), Bedford Institute of Oceanography, PO Box 1006, Dartmouth, N.S. B2Y 4A2, Canada*

Abstract: Late Miocene igneous rocks of Samos, in the southeastern Aegean Sea, comprise monzodiorite and minor granite of the Katavasis complex, trachyte and rhyolite of the Ambelos volcanic centre, and bimodal basalt–rhyolite at basin margins. Six new K–Ar ages, together with existing geochronology and biostratigraphy, show that the Katavasis complex and Ambelos centre date from 10–11 Ma and basalt–rhyolite from 8 Ma, correlating with cooling ages for the Katavasis complex and an unconformity in the basin fill. Monzodiorite, granite, trachyte and basalt all have similar radiogenic isotopes. Monzodiorite and basalt have similar trace element compositions and could result from 5–10% partial melting of enriched garnet lherzolite in the subcontinental lithosphere. Variations in trace elements suggest that trachyte and monzodiorite evolved by fractional crystallization from a parental magma similar to the younger basalt. The Katavasis and Ambelos rocks were synchronous with regional extension and listric faulting, which created opportunities for mid-crustal magma chambers and magma fractionation. Basalt extrusion was synchronous with the onset of north–south strike-slip faulting, which permitted more rapid transfer of magma to the surface. Late Miocene strike-slip faulting propagated from north to south in western Anatolia and the southeastern Aegean Sea, providing pathways for different types of mantle melts.

Miocene igneous rocks of the island of Samos in the southeastern Aegean Sea (Fig. 1) are part of a series of late Miocene–Quaternary extension-related rocks in a continental back-arc setting in the Aegean Sea (Pe-Piper & Piper 2006). In northwestern Anatolia and the central Aegean Sea, to the north of Samos, is an extensive area of shoshonites of early Miocene age (Pe-Piper & Piper 1992; Aldanmaz *et al.* 2000). I-type late Miocene granitoids in the Cyclades (Altherr *et al.* 1982) were interpreted by Altherr & Siebel (2002) as a consequence of extension and upwelling asthenospheric melts. The Pliocene–Quaternary south Aegean arc includes subduction-related andesite–dacite. In western Anatolia (and the island of Patmos), alkaline basalts of late Miocene age and younger are widespread (Aldanmaz *et al.* 2000).

Three groups of Late Miocene igneous rocks are present in Samos (Fig. 2). An intrusive complex of monzodiorite and granitoid dykes crops out at Katavasis in western Samos (Mezger *et al.* 1985). Potassic trachytes and minor rhyolite form a volcanic complex near Ambelos (Theodoropoulos 1979). Rhyolites and basalts crop out along the margins of two Miocene basins, in the eastern and western parts of the island (Robert & Cantagrel 1979). This wide range of rock types provides an opportunity to determine the petrogenetic relationships between these rock types. This interpretation is placed within the general context of the Late Miocene of the southeastern Aegean Sea, in the light of the tectonic evolution of the Aegean back-arc area.

Geological setting

Regional setting

The Miocene igneous rocks of Samos closely resemble those of the nearby islands of Patmos and Kos to the south, and the Bodrum peninsula of Turkey just east of Kos. Rather similar rocks are also present at Urla, east of the Karaburun peninsula, to the north of Samos.

During the Miocene, palaeomagnetic data and stretching lineations show that the west Aegean block (Fig. 1), bounded by the Scutari–Pec line and the Mid-Cycladic lineament, rotated clockwise as a coherent block with respect to the southeastern Aegean (Walcott & White 1998). The southeastern Aegean appears to have rotated counterclockwise together with western Anatolia, although local clockwise rotation (e.g. at Karaburun) points to complex local tectonics (Kissel *et al.* 1987; Kondopoulou 2000). Yilmaz *et al.* (2000), from detailed mapping in western Anatolia, showed that north–south-trending grabens developed in

From: TAYMAZ, T., YILMAZ, Y. & DILEK, Y. (eds) *The Geodynamics of the Aegean and Anatolia.* Geological Society, London, Special Publications, **291**, 75–97.
DOI: 10.1144/SP291.4 0305-8719/07/$15.00

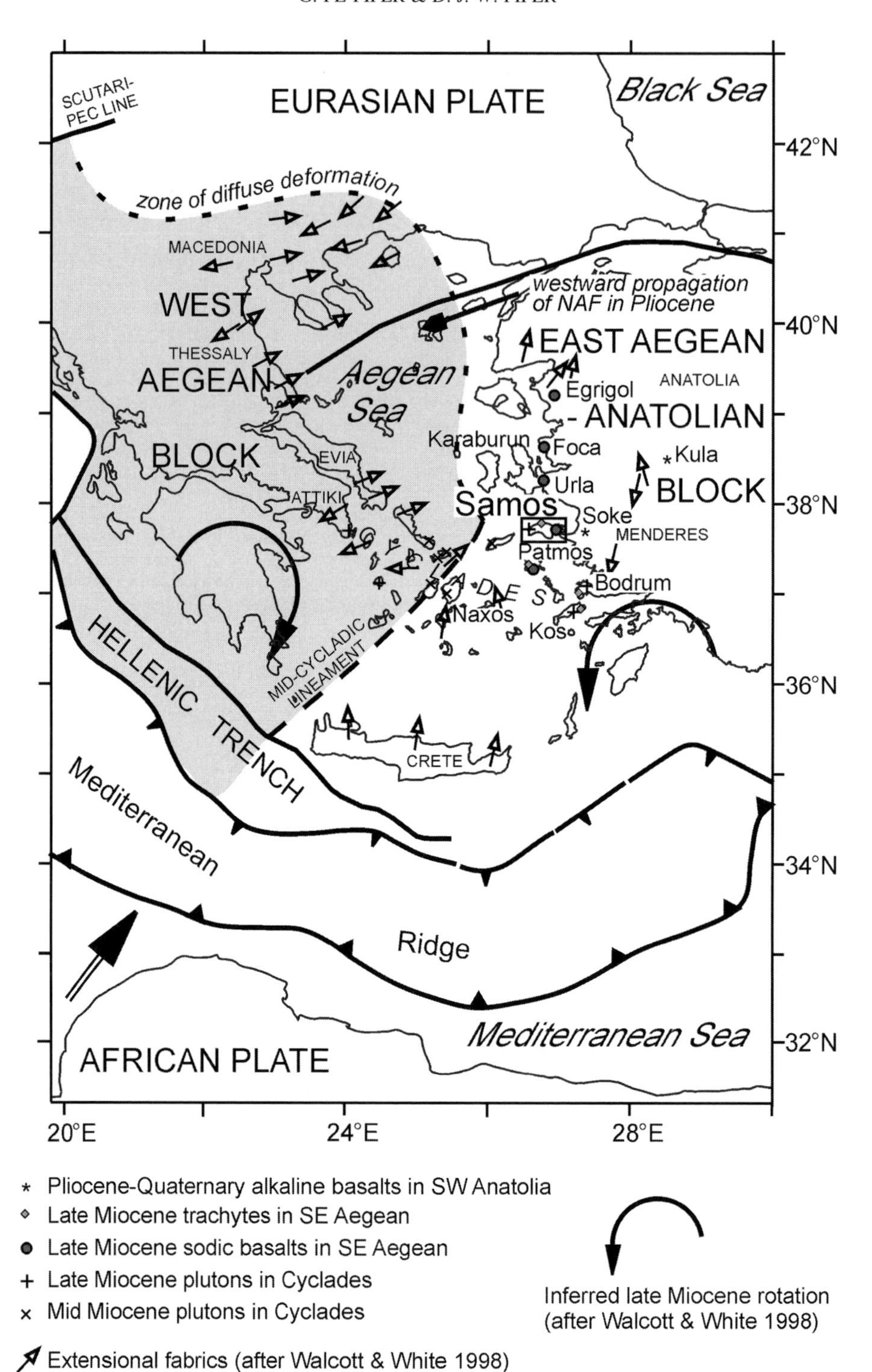

Fig. 1. Map showing the location of Samos (box), the present major tectonic features of the Aegean Sea region and late Miocene igneous rocks, and the interpretation by Walcott & White (1998) of major late Miocene tectonic blocks and post-Miocene rotation.

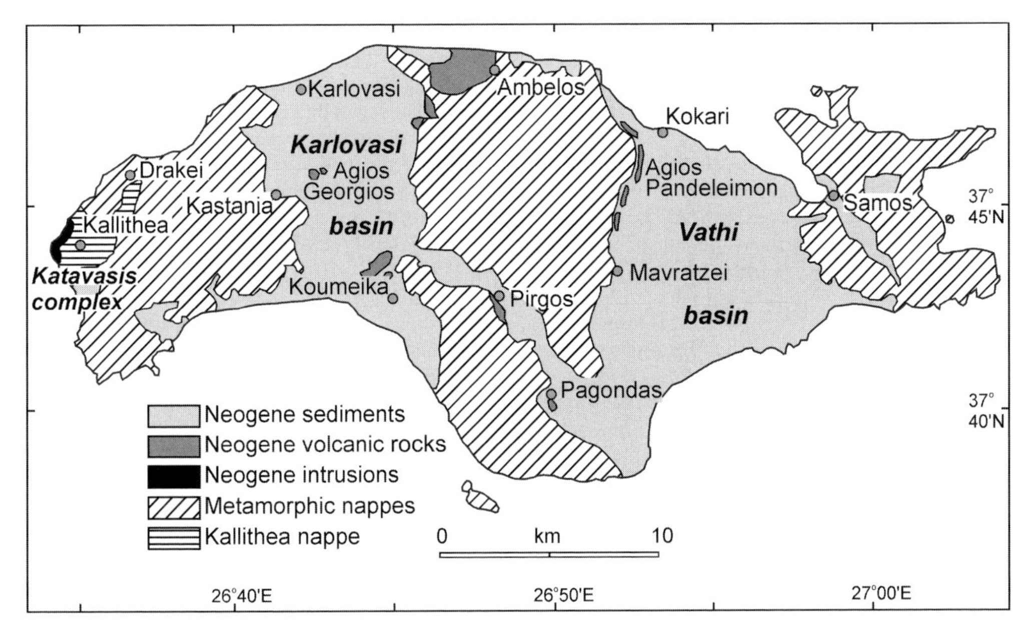

Fig. 2. Geological map of the island of Samos, showing location of Cenozoic igneous rocks (simplified from Theodoropoulos 1979).

the Early Miocene. North–south extension with detachment faulting began in the Late Miocene, leading by early Pliocene to the formation of east–west-trending grabens.

South of Samos, on the island of Patmos (Fig. 1), trachyte and rhyolite, with minor basalt of the Old Volcanic Series (OVS) date from 7.4 Ma. Trachyte flows and minor phonolite of the Intermediate Volcanic Series (IVS) date from 6.1 to 7.2 Ma. The potassic Main Volcanic Series (MVS) of Patmos consists of lavas of trachyandesite (= shoshonite *sensu stricto*) and trachyte and includes hawaiite bombs (nepheline trachybasalt of Wyers & Barton 1986), with an age of about 5.6 Ma. Younger basaltic dykes and lavas have been dated at about 4 Ma (Robert & Cantagrel 1979) and are termed the Young Volcanic Series (YVS). Their Nd and Sr isotopic composition suggests that they are related to the alkaline basalts of western Anatolia (e.g. Altunkaynak & Dilek 2006).

On the island of Kos, the late Miocene Dikeos pluton of monzonite and quartz monzonite and dykes of lamprophyre (kersantite) to monzonitic porphyry are found in the central part of the island (Altherr *et al.* 1976, 1982; Altherr & Siebel 2002). Andesite lava flows and ignimbrites have yielded ages of 10.0–10.6 Ma, similar to the age of the monzonite intrusion. Minor trachytic volcanic rocks and dykes are found in the NE part of the island and appear to be of Tortonian age (Besang *et al.* 1977; Bellon *et al.* 1979; Bellon & Jarrige 1979).

On the west coast of the nearby peninsula of Bodrum (Turkey), the Bodrum Volcanic Complex represents the remnants of a stratovolcano with basalt to trachyandesite flows and trachytic domes (Ulusoy *et al.* 2004). Several small monzodiorite bodies were emplaced as ring dykes. Radiometric ages from the volcanic complex range from 8.6 to 12 Ma, but mostly cluster around 10 Ma (Robert & Montigny 2001). Dykes of alkaline trachyte and ultrapotassic basalt, trending NW–SE, cut the central part of the volcanic complex and have yielded ages of 7.7–7.9 Ma (Robert & Cantagrel 1979). Primitive mafic rocks represented in the complex are ultrapotassic basalt dykes and sodic basalt flows.

In western Anatolia, rocks similar to those of Samos include 12.7 Ma K-rich basalt at Foça (Innocenti *et al.* 1982) and 11.3 Ma quartz-normative basalt and hawaiite and 11.9 Ma trachyte and rhyolite SE of Urla (Borsi *et al.* 1972; Innocenti & Mazzuoli 1972). In western Anatolia alkaline rocks mostly of late Miocene to Pliocene age (Yılmaz 1990; Aldanmaz *et al.* 2000; Altunkaynak & Dilek 2006) include the late Miocene Eğrigöl basalts, a 7 Ma volcano at Soke near Samos, and the Quaternary volcanic centre of Kula (Westaway *et al.* 2004; Fig. 1).

Geology of Samos

The island of Samos (Fig. 2) exposes stacked metamorphic nappes of the Cycladic blueschist unit that were emplaced in the Eocene to early Oligocene (Ring *et al.* 1999). In the late Oligocene and Miocene, the area experienced crustal extension that emplaced the Kallithea nappe over the blueschists. The Katavasis complex is in fault contact with the Kallithea nappe, but it was regarded as the same palaeogeographical entity by Ring *et al.* (1999). At the same time, extensional sedimentary basins formed, with the oldest sediments being of Serravallian (late mid-Miocene) age. Volcanism was synchronous with basin formation. A contractional event within the basins (Boronkay & Doutsos 1994) is dated by a mid-Tortonian unconformity within the basins (9–8.6 Ma: Weidmann *et al.* 1984). Basin sedimentation continued until the early Pliocene. Basin-margin faults trend north–south. Apparently correlative faults in the marine areas both north and south of the Karaburun peninsula, imaged in seismic reflection profiles, show predominant strike-slip motion (Ocakoğlu *et al.* 2004, 2005).

The Katavasis (or Kallithea) igneous complex has been interpreted as part of the younger group of plutonic rocks of the Cyclades (Pe-Piper *et al.* 2002). Mezger *et al.* (1985), in their detailed petrographic and geochemical study of the Katavasis dykes, focused on the problem of the origin of coexisting felsic and mafic magmas. They concluded that at least some of the different magma pulses were genetically unrelated, and that net-veined parts of the composite dykes were formed by multiple injections of felsic melt into mafic magma. They obtained a K–Ar age of 10.2 ± 0.15 Ma on hornblende from monzodiorite. Emplacement of the Katavasis complex post-dated this intrusion age, but predated the mid-Tortonian contraction event (Ring *et al.* 1999).

The trachyte and minor rhyolite in the volcanic complex near Ambelos on the northern coast of the island (Theodoropoulos 1979) overlie basement rocks and have not been previously studied in detail.

Rhyolite and sodic basalt crop out along the margins of the Vathy and Karlovasi basins, in the eastern and western parts of the island, respectively (Robert & Cantagrel 1979). Three whole-rock K–Ar ages on basalts from the basin margins range from 7.8 ± 0.5 to 8.3 ± 0.4 Ma (Robert & Cantagrel 1979; Table 1). At the eastern margin of the Karlovasi basin, volcanic rocks are interbedded with Tortonian lacustrine sediments, ash-fall tuffs and tuffites. The volcanic rocks are overlain and underlain by calcareous marlstone in the main part of the basin (Meissner 1976; Stamatakis 1989*a*, *b*).

On the western margin of the Vathi basin, in eastern Samos, an extensive mafic sheet crops out (Robert & Cantagrel 1979). Where best exposed at Agios Pandeleimon (Fig. 2), the basalt is 12 m thick, with a weathered top, and is overlain by 40 m of felsic pyroclastic deposits including clasts of basalt. The jointed top of the felsic pyroclastic deposits contains overlying marl up to 50 cm below the regional top surface. At Pagondas and Pirgos (Fig. 2), there is a similar section of basalt with a weathered top overlain by a few tens of metres of felsic pyroclastic deposits. This volcanic unit appears to correlate with the Mytilene Formation in the centre of the basin.

Within the Vathi basin, the occurrence of tuffs is known precisely, as a result of stratigraphic studies of the setting of the famous mammalian faunas (Weidmann *et al.* 1984; Sen & Valet 1986; Fig. 3). The Basal Conglomerate Formation is overlain by the 200 m thick Pythagorion Formation, consisting principally of lacustrine limestone. At the top of

Table 1. *Radiometric ages from the Late Miocene igneous rocks of Samos*

Rock type (sample number)	Component	%K (ppm)	Rad. ^{40}Ar	Age (Ma)	Location	Source
Monzodiorite (38)	Hornblende	0.9	0.000671	10.7 ± 0.9	Katavasis	This study
Monzodiorite (42)	Biotite	6.655	0.003776	8.2 ± 0.4	Katavasis	This study
Quartz monzodiorite (50)	Biotite	7.082	0.003966	8.1 ± 0.4	Katavasis	This study
Micro-Diorite	Hornblende			10.19 ± 0.15	Katavasis	Mezger *et al.* (1985)
Rhyolite (11)	Whole rock	3.596	0.002553	10.2 ± 0.3	Ambelos	This study
Trachyte (35)	Whole rock	5.405	0.003730	9.9 ± 0.3	Ambelos	This study
Rhyolite (77)	Whole rock	4.961	0.003006	8.7 ± 0.4	Koumeika	This study
Basalt (43)	Whole rock	0.697	0.00380	7.8 ± 0.5	Pagondas	Robert & Cantagrel (1979)
Basalt (29)	Whole rock	1.32	0.00745	7.9 ± 0.3	Koumeika	Robert & Cantagrel (1979)
Basalt (9)	Whole rock	1.76	0.01035	8.3 ± 0.4	Mavratsei	Robert & Cantagrel (1979)

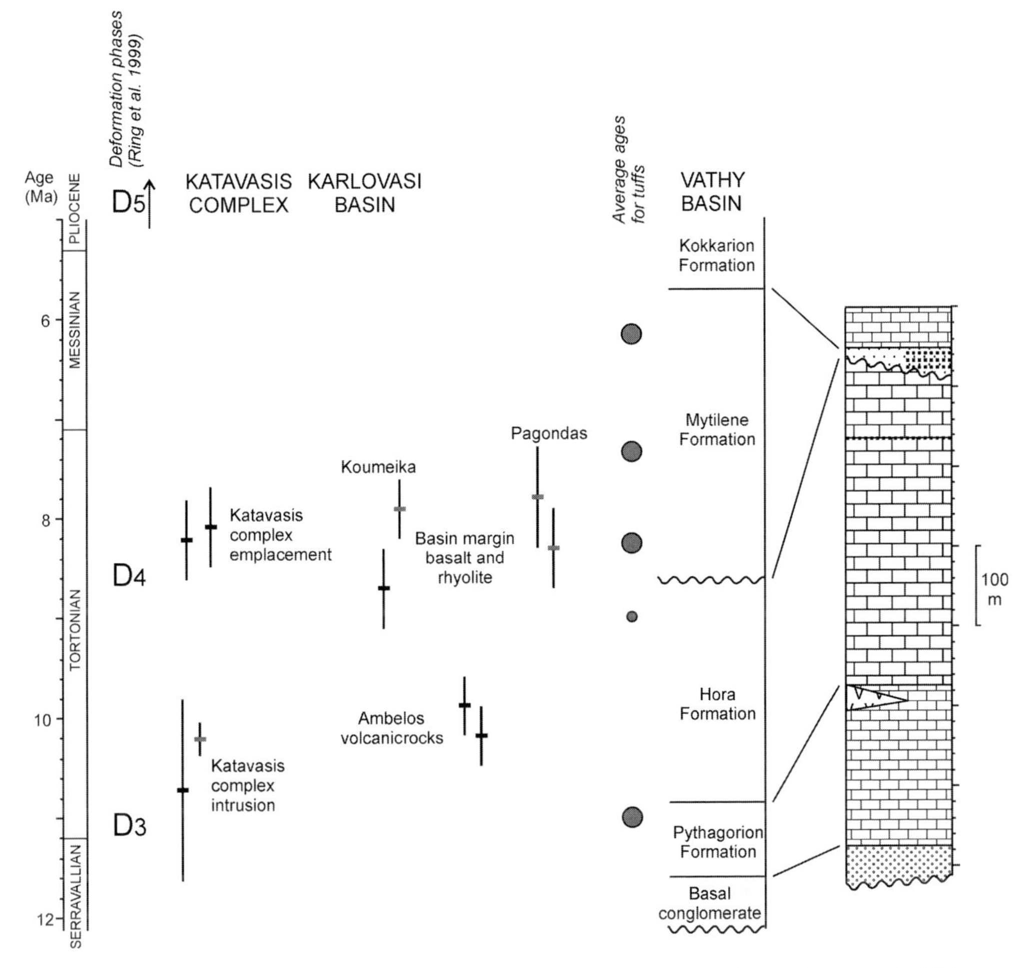

Fig. 3. Stratigraphy and geochronology of the late Miocene of Samos (based on data from Meissner 1976; Weidmann *et al.* 1984; Mezger *et al.* 1985; Sen & Valet 1986; Stamatakis 1989*a*, *b*; Ring *et al.* 1999).

the Pythagorion Formation in the western part of the basin is a 3–8 m thick subaqueous basalt flow overlain by felsic lahar tuffs with thin basalt flows. These volcanic rocks have yielded ages of 10.8–11.2 Ma. A thin tuffaceous turbidite in the overlying 400 m thick Hora Formation (lacustrine limestone) yielded an age of 9.0 ± 0.3 Ma. The Hora Formation includes evaporites such as nitre, halite and sylvite (Stamatakis & Zagouroglou 1984). It is overlain unconformably by the 20 m thick clastic Mytilini Formation, which contains groups of tuffs with mean ages of 8.26 Ma, 7.35 Ma and 6.18 Ma. The base of the overlying Kokkarion Formation (50 m, principally limestone), which includes tuffaceous silts, was dated at 5.7 Ma by magnetostratigraphy (Sen & Valet 1986).

In the Karlovasi basin (Stamatakis 1989b), tuffs 250 m thick overlie 50 m of limestone in the Pythagorion Formation. Tuffs also occur near the base of the Hora Formation. These two formations were deposited in a saline alkaline lake. Authigenic zeolites, feldspars and silica polymorphs are abundant in the tuffs (Stamatakis 1989*a*, *b*; Pe-Piper & Tsoli-Katagas 1991).

Methods

Field geology and sampling

The section at Katavasis (Fig. 2; Mezger *et al.* 1985) consists of schists cut by irregular monzodiorite dykes up to several metres thick striking SSW (160°E), orthogonal to sparse stretching lineations in the underlying nappes (Ring *et al.* 1999, Fig. 8). Granitoid rocks occur in composite dykes, some contain oriented mafic inclusions, and in places the rocks form net veins within the monzodiorite. Diabase sills also occur. Sampling was directed towards avoiding rocks where interaction of mafic and felsic magmas was obvious, and thus focused on granite and monzodiorite, at the expense of monzonite or granodiorite.

The rocks at Ambelos (Fig. 2; Theodoropoulos 1979) consist principally of trachytes. Minor rhyolites occur at the apparently intrusive contact of the trachytes and schists. The rocks are interpreted as a subvolcanic intrusive complex.

On the eastern margin of the Karlovasi basin, in western Samos, the best outcrops are NW of Koumeika (Fig. 2), where several tens of metres of rhyolitic tuffs and flow-banded rhyolite are exposed in road cuts. In the same area, small basalt cones and marls with tufa (indicating hot spring activity) are exposed. The rhyolitic rocks are intensely altered, mostly along veins, which are associated with either zeolitization or silicification. Volcaniclastic conglomerate immediately overlying the rhyolites contain clasts of both basalt and fresh (dark grey) flow-banded rhyolite. Basalt from the western margin of the Karlovasi basin was sampled at Moni Agios Georgios (Fig. 2). Samples from the western margin of the Vathi basin were collected from Agios Pandeleimon, Pagondas and Pirgos (Fig. 2).

Laboratory analyses

All mineral analyses were made with a JEOL-733 electron microprobe with four wavelength spectrometers and a Tracor Northern 145 eV energy-dispersive detector. Operating conditions were 15 kV at 5 nA beam current. Geological standards were used. Data were reduced using a Tracor Northern ZAF matrix correction program.

A total of 39 representative igneous rock samples were analysed for 10 major and minor element oxides and 14 trace elements on a Philips PW 1400 sequential X-ray fluorescence spectrometer using a Rh-anode X-ray tube. Rare earth elements (REE) were determined on selected samples by instrumental neutron activation analysis. Analytical precision is as given by Pe-Piper & Piper (2002, appendix 1). Lead and Sm–Nd isotopic ratios were determined by Geospec Consultants Limited, Edmonton, using methods summarized by Pe-Piper & Piper (2001). Geochronology was carried out on whole-rock samples, or mineral separates, as noted in Table 1, by the K–Ar method at Kruger Laboratories Inc.

A full set of whole-rock geochemical analyses, radiogenic isotope analyses, mineral analyses and details of the petrography of analysed samples is available online at http://www.geolsoc.org.uk/SUP 18299. A hard copy can be obtained from the Society Library.

Geochronology

Six new K–Ar ages have been obtained (Table 1; Fig. 3). One age on hornblende in monzodiorite from Katavasis was 10.7 ± 0.9 Ma, confirming the early Tortonian age obtained by Mezger *et al.* (1985) for the Katavasis complex. Two ages from biotite in monzodiorite and quartz monzodiorite at Katavasis were a little younger, 8.2 and 8.1 ± 0.4 Ma, and probably represent cooling ages associated with the juxtaposition of the Katavasis complex and the Kallithea nappe. Samples of rhyolite and trachyte at Ambelos yielded consistent ages of 10.2 and 9.9 ± 0.3 Ma, respectively, which, within the range of analytical error, is synchronous with or slightly younger than the Katavasis intrusive complex.

One new age has been obtained from the basin margin rhyolites and basalts: a rhyolite from Koumeika yielded a whole-rock age of 8.7 ± 0.4 Ma. The adjacent and overlying basalts have yielded ages between 8.3 and 7.8 Ma (Robert & Cantagrel 1979). These mid-Tortonian ages are synchronous with, or a little younger than, the unconformity between the Hora and Mytilene formations (Fig. 3).

Petrography and mineral chemistry

Petrography of the major rock types

The lithologies identified in the Katavasis intrusive complex are monzodiorite, quartz monzodiorite, tonalite and leucogranite. The main ferromagnesian minerals in the monzodiorite, quartz monzodiorite and tonalite are hornblende and biotite, with rare clinopyroxene; titanite is also common; and the principal opaque mineral is magnetite, although hematite and pyrite are also present. Some rocks are very heterogeneous, with 'pools' of quartz, plagioclase and titanite. All crystals in these pools contain abundant apatite inclusions. The only ferromagnesian mineral present in the leucogranites is biotite, with accessory minerals titanite, allanite and rarely actinolite and actinolitic hornblende.

The phenocrysts commonly present in the Ambelos trachyte are plagioclase, K-feldspar, clinopyroxene, biotite and quartz, generally forming glomeroporphyries. The groundmass ranges from 72% to 82% of the whole rock and consists generally of the same minerals. The trachytes in places contain many xenoliths, including granite, chloritized gabbro or diorite, very fine-grained igneous lithologies, garnetiferous pelites and quartz–feldspathic schists.

The basin margin basalt contains principally plagioclase and clinopyroxene. The associated rhyolite shows a variety of textures: granophyric, spherulitic, vitrophyric, and vesicular, and the glassy groundmass may make up to 90% of the rock. Common phenocrysts are K-feldspar, quartz and plagioclase.

Pyroxene

Pyroxene is rare in monzodiorite, reported by Mezger *et al.* (1985) only from their leucodiorites. Monzodiorite sample SV46 has resorbed crystals of diopside (Fig. 4) with Mg/(Mg + Fe) 0.63 and Al_2O_3 0.7%, partially altered to amphibole, dusty opaque minerals and clays. All pyroxenes from monzodiorite (both our analyses and those of Mezger *et al.* (1985)) have low TiO_2 content.

Clinopyroxene is the only ferromagnesian mineral present in the basalt, showing complex zoning and ranging in composition from diopside to augite (Fig. 4). Three early clinopyroxene populations are present in basalt, forming cores to phenocrysts: green augite (Ts component 13–30%), colourless diopside (Di component 80–86%) and brown augite. All of these cores may be resorbed and are mantled with intermediate augite compositions, with variable Cr_2O_3 (from below detection limit to 1%). These augites are similar to some normally zoned phenocrysts and to the microphenocryts in these rocks.

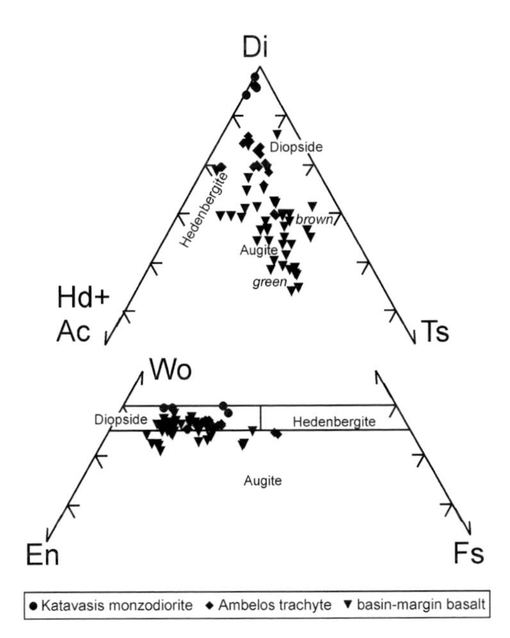

Fig. 4. Composition of clinopyroxene from mafic and intermediate rocks of Samos.

Clinopyroxene is also the main ferromagnesian mineral in the trachyte. Phenocrysts are of light green diopside (Mg/(Mg + Fe) 0.62–0.75), in some cases with augite cores with a high Ts component. Some samples contain dark green Fe-rich augite or hedenbergite crystals with opaque rims, partly replaced by dusty opaques and clays. Reverse zoning is common.

Amphibole

The granite contains <2% amphibole, either actinolite or actinolitic hornblende. The monzodiorite contains <60% amphibole, including edenitic hornblende, ferroan pargasitic hornblende and magnesiohornblende, some of which shows sub-solidus alteration to actinolite. Al^{iv} is as high as 1.8 in some samples. Estimates of the temperature of crystallization, using the Blundy & Holland (1990) geothermometer, are about 900 °C, which yields a pressure estimate of 1 kbar for the most aluminous cores, using the Anderson & Smith (1995) geobarometer. Anderson & Smith noted that the geobarometer is less reliable for high temperatures, low oxygen fugacity and plagioclase outside the range of An_{25-35}, all conditions that may apply to the Katavasis monzodiorite. Amphibole is absent from the basalts, trachytes and rhyolites.

Biotite

Biotite is present in monzodiorite, granite and trachytes, belonging to the phlogopite–annite series (Fig. 5). Some biotite in granite has Al-rich rims. Biotite in granite is much less magnesian than biotite in monzodiorite for the same FeO^T content. The rare biotite from trachyte has lower FeO^T/ FeO^T + MgO) compared with those from monzodiorite, indicating that the trachyte crystallized under high fO_2, and has high TiO_2 content. All biotites fall in the calc-alkali field in the ternary discrimination diagram of Abdel-Rahmen (1994).

Feldspars

The monzodiorite contains normally zoned andesine to oligoclase (An_{50-26}). Rarely, K-feldspar is found as small patches within amphibole or plagioclase crystals or as isolated crystals. In granite, both K-feldspar and plagioclase are present, with oligoclase (An_{42-21}) the dominant plagioclase. Leucocratic vein SV37 contains 85% K-feldspar and about 10% K_2O. Trachyte contains reversely zoned andesine (An_{42-33}) and sanidine phenocrysts. Basalt generally contains bytownite to labradorite (An_{82-63}), but sample SV102 has

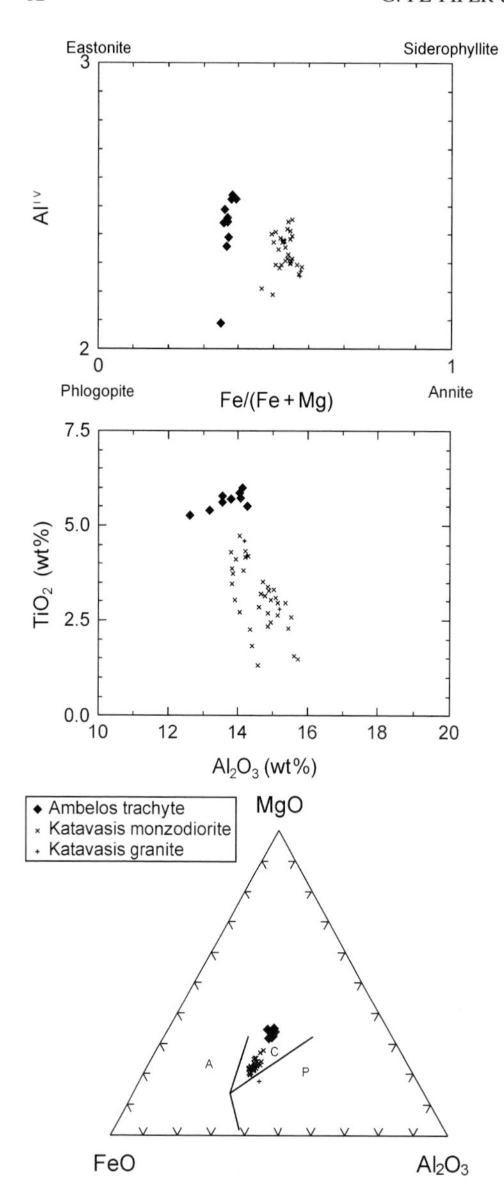

Fig. 5. Composition of biotite from Katavasis complex and Ambelos trachyte. Fields from Abdel-Rahmen (1994): A, alkaline; C, calc-alkaline; P, peraluminous.

microphenocrysts of andesine (An_{54-40}) and phenocrysts with sieve texture have cores of andesine (An_{37}) rimmed with labradorite (An_{67}).

Sulphides and Fe–Ti oxides

The monzodiorite contains equant crystals of magnetite, some with chalcopyrite inclusions that contain globular inclusions of pyrite. Rutile occurs along fractures or as corona rims on magnetite. Some magnetite is developed in fractures. Rutile crystals with hematite and chalcopyrite inclusions, the latter containing both hematite and pyrite inclusions, occur in some samples (e.g. SV47).

The granite contains magnetite and rutile either as coronas upon magnetite or as scattered hypidiomorphic crystals. Some magnetite occurs in fractures and some is closely associated with brown biotite. Late leucogranitic veins (SV37) contain very few opaque crystals of ilmenite. The mineral assemblage titanite + magnetite + quartz found in the granites indicates high oxygen fugacity (Wones 1989).

Titano-magnetite is the most common Fe–Ti oxide mineral in the trachyte and may contain globular inclusions of chalcopyrite. Idiomorphic crystals of pyrite are partially replaced by rutile. Small inclusions of pyrite are present in clinopyroxene phenocrysts. Ambelos rhyolite contains only magnetite, which may be intergrown with rutile.

The basalts generally contain hypidiomorphic and equant crystals of titano-magnetite. Rutile is the dominant opaque mineral in altered samples (e.g. SV71b). The basalt from Agios Georgios with unusual feldspar mineralogy contains composite crystals of magnetite–ilmenite throughout the groundmass and magnetite crystals as overgrowths on chromian spinel.

The most common opaque mineral in the altered rhyolite from Koumeika is pyrite, as hypidiomorphic crystals, large idiomorphic crystals, small equant anhedral grains dispersed throughout the groundmass, or intergrown with rutile. Chalcopyrite as hypidiomorphic crystals dispersed throughout the groundmass is less common. Hematite in some samples forms xenomorphic grains intergrown with rutile, mostly confined to spherulitic grain boundaries.

Titanite and allanite

In the monzodiorite, pleochroic titanite is abundant (up to 5%), as hypidiomorphic to xenomorphic crystals with magnetite and plagioclase inclusions. Granite contains minor titanite, with higher Al content than in monzodiorite. Some titanite from monzodiorite has much higher $\sum$REE than those from granite. In titanite from monzodiorite, $\sum$REE generally decreases from core to rim, but increases in the titanite from granite. The size, shape and $\sum$REE content of titanite in monzodiorite suggest early crystallization. The lower $\sum$REE in titanite from granite may be a consequence of allanite crystallization. Titanite is rare in trachytes, but chemically similar to titanite from monzodiorite.

Mezger *et al.* (1985) reported allanite in both the mafic and felsic rocks of the Katavasis complex. The common pleochroic allanite in the granite shows a wide range of compositions and no systematic zoning. Apatite is common in monzodiorite and leucogranite.

Rock geochemistry

Nomenclature

We follow Altherr & Siebel (2002) in naming the mafic plutonic rocks as monzodiorite. The analysed granitoid rocks classify either as granite or as quartz syenite in the chemical classification system of Streckeisen & Le Maitre (1979). In the IUGS chemical classification system (Fig. 6) all the analysed felsic volcanic rocks classify as rhyolites, the intermediate rocks as trachytes and the mafic rocks as basalts (Pagondas, Agios Pandeleimon), trachybasalts (Koumeika, Pirgos) and basaltic trachyandesites (Agios Georgios and one sample from Koumeika). We use the general term basalt for all mafic rocks.

Katavasis monzodiorite and granite

The analysed monzodiorites are nepheline normative, except for the most felsic rocks (57% SiO_2), which are hypersthene normative (see Table 2). Trace element variation, normalized to primitive mantle (Fig. 7) is very uniform, with a progressive decrease in normative abundance from the large ion lithophile elements (LILE) to the high field strength elements (HFSE), prominent troughs at Nb and Ta, and minor troughs at P, Zr, Hf and Ti for some samples. REE show progressive decrease in normative light (LREE) and middle REE (MREE), but lesser fractionation of the heavy REE (HREE; Fig. 8).

All analysed granites are slightly peraluminous (mol Al_2O_3 > mol ($CaO + Na_2O + K_2O$)), with a small amount of normative corundum. Using the tectonic environment discriminant diagrams of Pearce *et al.* (1984) they classify either as post-collision or volcanic arc granites. Compared with the monzodiorite, the granites show enrichment in LILE and overall rather lower HFSE, including strong depletion in Nb, P, Ti and Y. The granites have lower contents of most REE than the monzodiorites (Fig. 8), but similar La content (as discussed in detail by Mezger *et al.* (1985)).

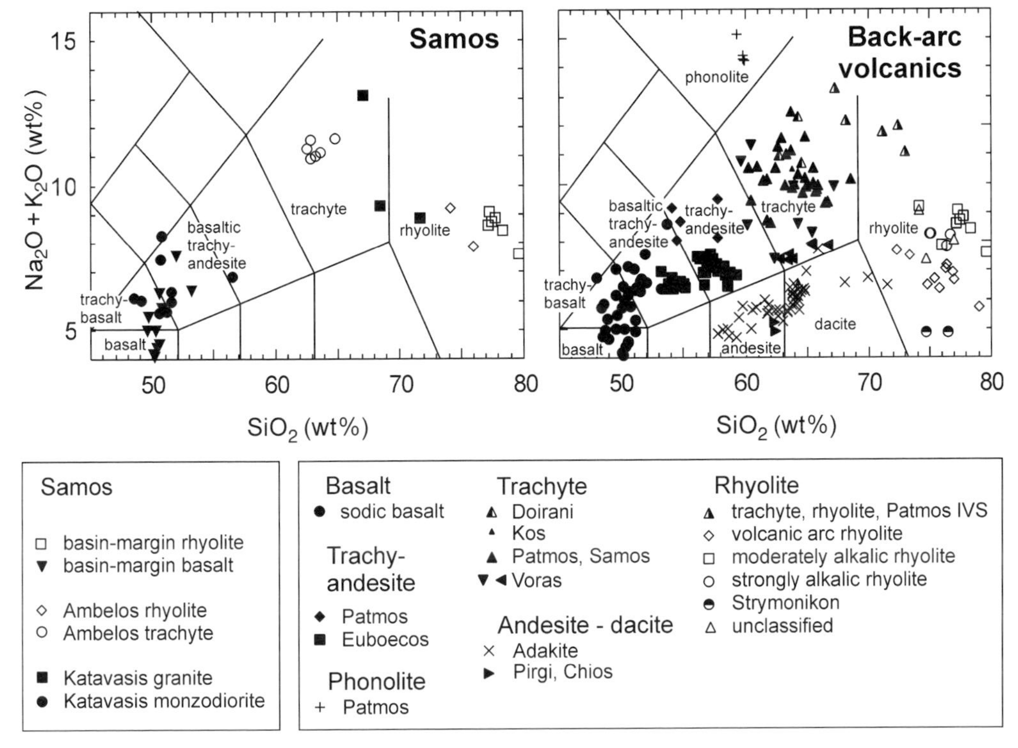

Fig. 6. Variation in total alkalis and silica (IUGS nomenclature) for (**a**) the Miocene igneous rocks of Samos and (**b**) mid-Miocene to Pliocene back-arc volcanic rocks of Greece. Data sources for (b) have been given by Pe-Piper & Piper (2002).

Table 2. *Representative chemical analyses of Katavasis plutonic rocks*

Lithology:	Monzodiorite		Micro-Diorite sill	Monzodiorite			Granite vein	Granite
Sample:	SV36	SV46	SV42	SV38	SV50	SV47	SV37	SV40
Major elements (wt %)								
SiO_2	47.18	48.11	49.2	49.91	50.26	55.4	66.82	70.86
TiO_2	1.99	1.79	1.48	1.54	1.32	0.76	0.12	0.23
Al_2O_3	17.51	17.52	17.47	18.31	17.28	13.59	16.65	15.21
$Fe_2O_3^T$	10.79	11.05	9.80	9.57	9.06	6.87	0.50	1.07
MnO	0.17	0.21	0.16	0.17	0.18	0.15	0.02	0.03
MgO	5.23	4.85	5.33	3.70	5.28	6.68	1.12	1.18
CaO	7.92	7.94	8.02	6.17	7.99	7.57	1.21	1.47
Na_2O	3.49	3.86	3.55	5.09	4.22	2.94	2.80	3.23
K_2O	2.45	2.04	1.89	3.03	1.62	3.76	10.26	5.54
P_2O_5	0.51	0.61	0.46	0.85	0.39	0.27	0.04	0.07
LOI	1.6	1.20	1.40	0.90	1.30	1.20	0.40	0.60
Total	98.84	99.18	98.76	99.24	98.90	99.19	99.94	99.49
Trace elements (ppm) by XRF								
Ba	828	786	572	1361	282	817	2432	870
Rb	73	54	73	115	69	109	284	130
Sr	697	687	783	796	681	400	398	263
Y	40	47	37	38	33	40	<5	17
Zr	188	166	195	114	184	236	141	171
Nb	21	23	17	20	13	22	<5	11
Th	<10	<10	<10	<10	11	14	15	26
Pb	16	20	15	16	16	16	51	43
Ga	20	20	19	20	19	14	13	14
Zn	94	120	107	102	109	85	11	14
Cu	30	23	14	23	25	9	10	<5
Ni	35	15	10	<5	12	89	5	<5
V	302	246	255	161	240	147	10	11
Cr	<5	<5	<5	<5	5	222	<5	8
REE and other trace elements (ppm) by INAA								
La	42	51	46	55	45	53	36	50
Ce	90	109	94	115	89	104	55	82
Nd	44	52	41	55	42	44	12.2	24
Sm	8.8	9.4	8.6	9.4	7.7	8.8	2.3	4.2
Eu	2.6	2.5	2.2	2.5	2.1	1.52	1.10	0.88
Tb	0.88	1.02	1.09	0.95	0.89	1.10	1.01	1.05
Yb	2.9	3.3	2.5	2.7	2.3	3.2	0.78	2.0
Lu	0.42	0.49	0.40	0.40	0.39	0.51	0.16	0.37
Cs	1.71	1.53	3.4	2.7	2.6	1.56	5.0	1.91
Hf	4.7	4.2	4.4	2.9	4.3	6.2	4.7	5.0
Sc	19.7	17.3	17.7	10.6	17.8	23	1.28	1.58
Ta	1.13	1.28	1.09	0.85	0.90	1.78	0.94	1.77
Th	9.6	5.1	12.6	8.8	14.2	19.5	24	31
U	1.88	0.99	2.4	1.28	2.9	2.5	5.5	3.5

Ambelos trachyte and rhyolite

The Ambelos trachytes have a rather narrow range of geochemistry, with 62–65% SiO_2 (Table 3). They are hypersthene normative. Trace element patterns differ from those for monzodiorite in that most incompatible elements are more abundant and troughs for Sr, P and Ti are more pronounced (Fig. 7). REE patterns are similar to those of the monzodiorite, except that the LREE are a little more abundant (Fig. 8). Unlike the granites, the associated rhyolites show strong Eu depletion and enrichment in HREE compared with trachyte. Compared with the trachytes, the rhyolites show strong depletion in Ba, Sr, P, Eu and Ti, but some enrichment in LILE, Nb and Ta (Fig. 7).

Basin margin basalt and rhyolite

Most basalt is nepheline-normative and has similar major element composition to the monzodiorite,

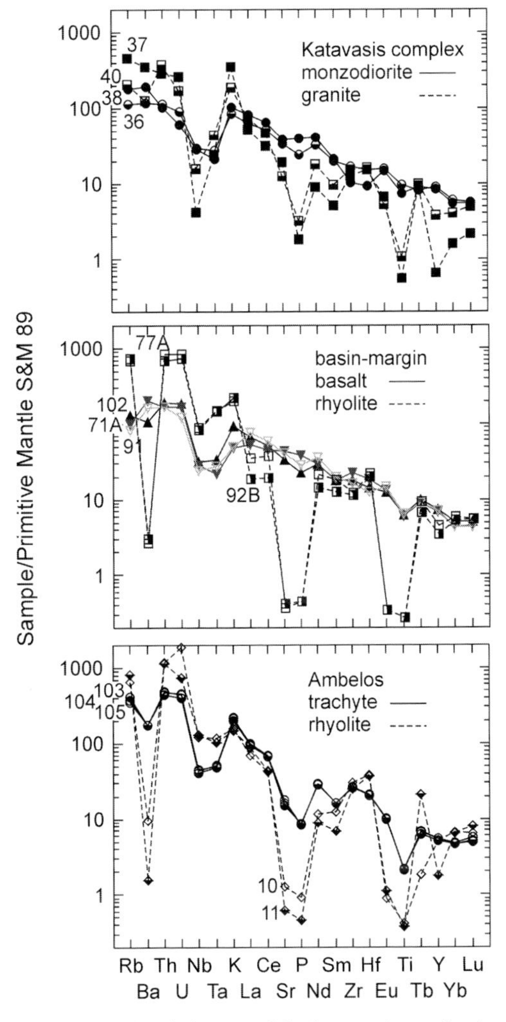

Fig. 7. Variations in incompatible elements (normalized to primitive mantle: Sun & McDonough 1989 (S&M 89)) for selected Late Miocene igneous rocks of Samos. Sample numbers are as in Tables 2–4.

except that MgO and CaO are a little higher and Na_2O a little lower. Least-altered basalt samples show incompatible trace element variation remarkably similar to that of the monzodiorite, with the trough at Ti a little more pronounced (Fig. 7). The basalt has higher Ni and Cr and lower Y than the monzodiorite, reflecting the presence of clinopyroxene and possible pseudomorphs after olivine in the former and amphibole in the latter. REE abundances are also very similar to those for monzodiorite (Fig. 8).

Rhyolite associated with the basalt has similar abundances of most HFSE, but is enriched in LILE, Nb and Ta, and strongly depleted in Ba, Sr, P, Eu and Ti (Fig. 7). The LREE abundance in the rhyolite is slightly lower than in the basalt, but HREE abundance is similar (Fig. 8).

Alteration in basalts was evaluated from two samples from the same outcrop, SV71A (fresh) and SV71A (altered) (Table 4). These samples show that TiO_2, CaO, Y, Zr, Cr, Ba, Rb and Sr decrease during alteration, whereas Al_2O_3, MgO, Na_2O, K_2O, Nb, Cu and V increase. Hydrothermal alteration of the rhyolites at Koumeika results in a decrease in alkalis and an increase in FeO^T, MgO, CaO, loss on ignition (LOI), Rb and Sr. Basalt from Agios Pandeleimon, which is hypersthene-normative, has low K_2O and abnormally high Rb compared with other LILE, and appears altered. A set of four basalt samples (SV4B, 71A, 91 and 102, Table 4) is regarded as showing the least alteration.

Lead isotopes

The Samos rocks exhibit a narrow range of Pb isotopic ratios, varying between 18.86 and 19.10 in $^{206}Pb/^{204}Pb$. Most data points form a well-defined line in a $^{207}Pb/^{204}Pb$ v. $^{206}Pb/^{204}Pb$ diagram (Fig. 9) with two outlying values for one basalt and Ambelos rhyolite, which have high $^{206}Pb/^{204}Pb$. The outlying basalt (SV71A) shows no unusual characteristics in its trace element composition, except that Cs is particularly high. The μ value of the source rock reservoir appears to be around 10.0, indicating involvement of upper crustal material, either as subducted sediment or by assimilation and fractional crystallization in the upper crust. Data distribution in $^{208}Pb/^{204}Pb$ v. $^{206}Pb/^{204}Pb$ space (not illustrated) implies an average present-day Th/U ratio of 2.3, lower than the average crustal value of 3.78 (Stacey & Kramers 1975). Pb isotopic values show no systematic variation with SiO_2 content.

Nd–Sm isotopes

The monzodiorite, granite, Ambelos trachyte and basin margin basalt all show very similar, ε_{Nd} of -1 to -2 (Fig. 10) and model ages of 0.7–0.9 Ga. These values are similar to those shown by rocks of similar age in Bodrum in western Turkey (Robert *et al.* 1992), some basalts from Patmos (Wyers & Barton 1987), monzodiorite from Kos (Altherr & Siebel 2002), and some mildly alkaline volcanic rocks from western Anatolia (Altunkaynak & Dilek 2006). The rhyolites from both Ambelos and the basin margin have more negative ε_{Nd} (-4.4, -6.1) and higher model ages (0.9, 1.9 Ga), together with high $^{207}Pb/^{204}Pb$.

Table 3. *Representative chemical analyses of Ambelos volcanic rocks*

Lithology:	Trachyte			Rhyolite	
Sample:	SV105	SV103	SV104	SV10	SV11
Major elements (wt %)					
SiO_2	61.41	61.59	61.84	73.86	74.88
TiO_2	0.46	0.45	0.46	0.09	0.08
Al_2O_3	17.51	17.51	17.51	13.41	13.21
$Fe_2O_3^T$	3.49	3.44	3.57	1.39	1.36
MnO	0.10	0.10	0.10	0.04	0.01
MgO	1.67	1.57	1.48	1.02	1.22
CaO	1.51	1.81	1.80	0.66	0.05
Na_2O	4.25	5.21	4.94	4.43	3.45
K_2O	6.42	6.15	5.80	4.75	4.32
P_2O_5	0.18	0.18	0.18	0.02	0.01
LOI	2.20	1.10	1.30	1.00	2.30
Total	99.20	99.11	98.98	100.67	100.89
Trace elements (ppm) by XRF					
Ba	1218	1211	1205	66	11
Rb	260	232	220	416	524
Sr	321	346	380	27	13
Y	23	25	25	23	8
Zr	296	291	295	334	275
Nb	29	31	32	86	92
Th	40	40	40	112	99
Pb	30	28	37	38	7
Ga	17	18	19	23	21
Zn	50	45	48	39	30
Cu	5	5	6	<5	<5
Ni	5	7	6	<5	<5
V	27	21	26	6	<5
Cr	<5	7	9	6	9
REE and other trace elements (ppm) by INAA					
La	65	68	66	48	59
Ce	118	122	121	76	78
Nd	38	40	39	15.6	12.1
Sm	7.1	7.2	7.1	5.5	3.1
Eu	1.69	1.70	1.68	0.15	0.19
Tb	0.67	0.74	0.72	0.20	2.3
Yb	2.3	2.2	2.3	3.2	3.2
Lu	0.38	0.37	0.41	0.48	0.59
Cs	3.4	3.5	3.9	4.6	8.3
Hf	6.3	6.4	6.5	11.9	11.5
Sc	2.6	2.6	2.6	0.63	bd
Ta	2.0	2.1	2.0	4.8	4.2
Th	37	40	40	100	98
U	8.4	9.6	9.3	39	15.4

bd, below detection.

Petrogenesis

A common source for monzodiorite, basalt and trachyte

A plot of ε_{Nd} v. SiO_2 for Late Miocene igneous rocks of the southeastern Aegean (Fig. 7a) shows no systematic change in ε_{Nd} v. SiO_2 for the igneous rocks of Samos, with the exception of rhyolites. This indicates that crustal assimilation with fractional crystallization or mixing with crustal melts is unlikely to account for most of the isotope variation. Altherr & Siebel (2002) argued that the difference between a kersantite (lamprophyric) dyke ($\varepsilon_{Nd} = +0.55$) and various monzonite and more evolved lamprophyric dykes in Kos implied some mixing process between kersantitic melts and crustal materials, and the Bodrum rocks show a similar trend of decreasing ε_{Nd} with increasing SiO_2 (Robert *et al.* 1992). However, in the plutonic rocks of Samos, as in the intermediate rocks of Patmos and Bodrum, there is no systematic change in ε_{Nd} with SiO_2. Such isotopic variation thus supports interpretations based on geochemistry (Robert *et al.* 1992) that the late Miocene basalts and trachytes of the southeastern Aegean were derived principally from the mantle and that there was no significant crustal assimilation involved in the fractionation of the trachytes. Conversely, the more negative ε_{Nd} and higher radiogenic Pb in the rhyolites suggests a crustal contribution to the felsic magma.

In addition to the isotopic evidence, the uniformity of incompatible trace element distribution for basalt and monzodiorite (Fig. 7) and their similarity in REE distribution (Fig. 8) argue for a common source. The REE distribution (Figs 8 and 11) suggests partial melting of enriched mantle in the presence of garnet. The trachyte also has similar incompatible trace element composition to the basalt and monzodiorite, except that elements fractionated by feldspars (Ba, Sr, Eu), clinopyroxene (Cr, Ni, Eu), amphibole (Y), apatite (P), and titanite (Ti) are relatively depleted and other elements are relatively enriched (Figs 7 and 8). It is concluded that they too are derived from the same mantle source.

The evidence for magma mixing

Petrographic and mineral chemical data, in particular, the complex zoning patterns shown by the clinopyroxenes and plagioclase phenocrysts, provide compelling evidence that mixing of small magma batches has occurred in the mafic volcanic rocks. At least three types of mafic magma were mixed in the basalts: one of these was relatively primitive and carried phenocrysts of diopsidic clinopyroxene and anorthitic plagioclase, whereas the other two were more evolved and carried phenocrysts of either augitic or salitic clinopyroxene and more albitic plagioclase. After mixing, pyroxene and plagioclase of intermediate composition precipitated from the hybrid magma as microphenocrysts or mantles around the xenocrysts. The green augite cores with high Al_2O_3 content represent crystallization at elevated pressures, whereas the diopside overgrowths with high Cr_2O_3 represent crystallisation from a less evolved magma batch at lower pressures (see Pe-Piper 1984). Reverse zoning in clinopyroxene from trachyte may be related to the appearance of magnetite at the liquidus.

In the Katavasis complex, Mezger *et al.* (1985) showed that chemical variations could be best explained by multiple intrusions of small magma batches and that Sr isotope ratios suggested that some magma batches were genetically unrelated.

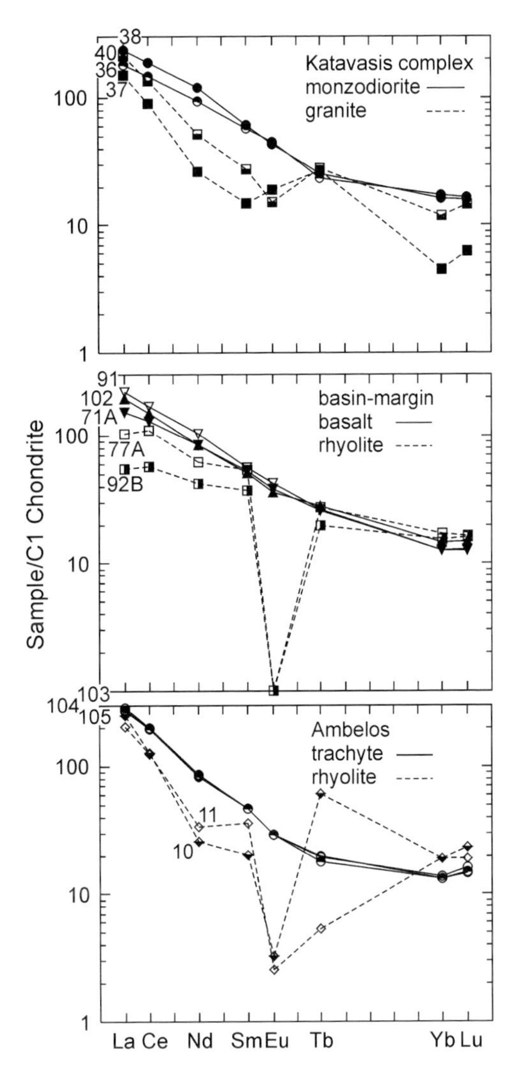

Fig. 8. Variations in REE (normalized to C1 chondrite, after Sun & McDonough 1989) for selected Late Miocene igneous rocks of Samos. Sample numbers are as in Tables 2–4.

New Nd isotope data from this study and from Altherr & Siebel (2002), however, show a remarkable overall uniformity in isotopic composition (Fig. 10). The minor variations in isotope composition could be a result of different source points in a compositionally inhomogeneous enriched mantle (Pe-Piper & Piper 2001) and/or minor interaction with crustal melts (Altherr & Siebel 2002).

The role of fractionation

Using geochemical characteristic such as $MgO/(MgO + FeO^T)$ ratio (Fig. 12d) and Ni and Cr

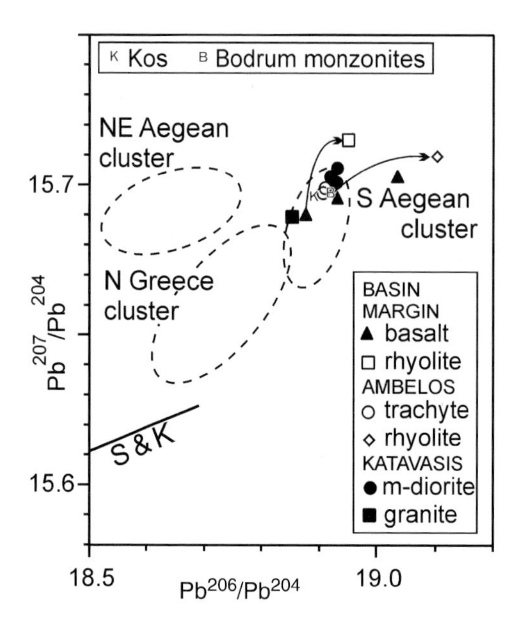

Fig. 9. Pb isotopic compositions of Late Miocene igneous rocks from the southeastern Aegean (Bodrum and Kos intrusions from Juteau *et al.* (1986); other rocks from Pe-Piper (1994)). S&K is Stacey & Kramers' (1975) reference growth curve. Geographical clusters of Pb isotope values are from Pe-Piper & Piper (2001).

abundance (Fig. 13b), the basalts appear more primitive than the monzodiorite. Element ratios including La/Sm, Nd/Sm and Nb/Zr show no systematic differences between basalt and monzodiorite, although Sm/Yb tends to be higher in basalt (Fig. 11). Ba/Rb ratio shows more scatter in unaltered basalt than in monzodiorite, whereas V/Zr and Y/Zr show more scatter in the monzodiorite (Fig. 12).

Geochemical relationships between monzodiorite, basalt and trachyte have been examined using binary plots of trace elements that are influenced by fractionation of particular minerals. Comparisons are made between the set of little altered basalt (outlined by the circle in Fig. 13) and the monzodiorite and trachyte samples to examine the nature of fractionation assuming that the basalt most closely represents parental magma. Low Cr and Ni abundances in both monzodiorite and trachyte imply predominant clinopyroxene (or perhaps olivine) fractionation (Fig. 13b). Values of Zr and Nb in monzodiorite are scattered around values for basalt (Fig. 13c), consistent with the variable amounts of titanite, zircon and various ferromagnesian minerals in the monzodiorite. Trachyte compositions could be derived from parental basalt magma by fractionation of amphibole (Fig. 12c). Fractionation of clinopyroxene or feldspar is suggested by some element plots (Fig. 13c–e), and some plagioclase fractionation is confirmed by the small Eu anomaly in the trachyte (Fig. 8) and the fractionation of Sr against Ba (Fig. 13f). Ti content is high in biotite from the trachyte (Fig. 5), but whole-rock TiO_2 is lower in the trachyte (Fig. 13a), probably as a result of magnetite or titanite fractionation. The trachyte is depleted in P relative to basalt and monzodiorite (Fig. 7), probably as a result of apatite fractionation. Variations in U and Th suggest that zircon fractionation may be involved in petrogenesis of monzodiorite.

The fractionation relationships discussed above suggest that the monzodiorites and trachytes (of similar age) were derived from parental magma similar to that of the basin margin basalts. The trachyte resulted from fractionation of hornblende, plagioclase and magnetite, together with lesser apatite and titanite fractionation. These minerals are the principal mineral phases in the monzodiorite, suggesting that much of the monzodiorite represents residual fractionating material retained in a magma chamber.

This genetic link between monzodiorite and trachyte is further supported by the opaque mineralogy of the two rock types. In the monzodiorite, ilmenite is altered to rutile (with sulfide inclusions) and there are two generations of magnetite: equant crystals and in fractures. The presence of pyrite suggests low fO_2 during the early stages of crystallization of the dioritic magma, but during later stages the presence of magnetite and titanite suggests high fO_2. Pyrite and chalcopyrite inclusions in titanomagnetite in the trachytes suggest evolution of fO_2 similar to that in the monzodiorite.

Where was the source?

Bulk chemical composition and specific trace element ratios in the mafic rocks of Samos are consistent with partial melting of enriched hydrous mantle peridotite within the stability field of phlogopite, amphibole and garnet (e.g. Turner *et al.* 1996). The ratios of REE (Fig. 11) suggest 5–10% partial melting. The Sr and Nd isotope data (Fig. 10) do not support significant involvement of asthenospheric melts, which are important in the more alkalic magmas in the YVS of Patmos and in western Anatolia (Güleç 1991; Seyitoğlu *et al.* 1997; Aldanmaz *et al.* 2000). Rather, the source in enriched subcontinental lithosphere is similar to that of the lamprophyres of Kos (Altherr & Seibel 2002) and Bodrum (Robert *et al.* 1992). The isotopic data for the Katavasis granites indicate a common source with the monzodiorite, although the details of their evolution is uncertain, as noted by Mezger *et al.* (1985).

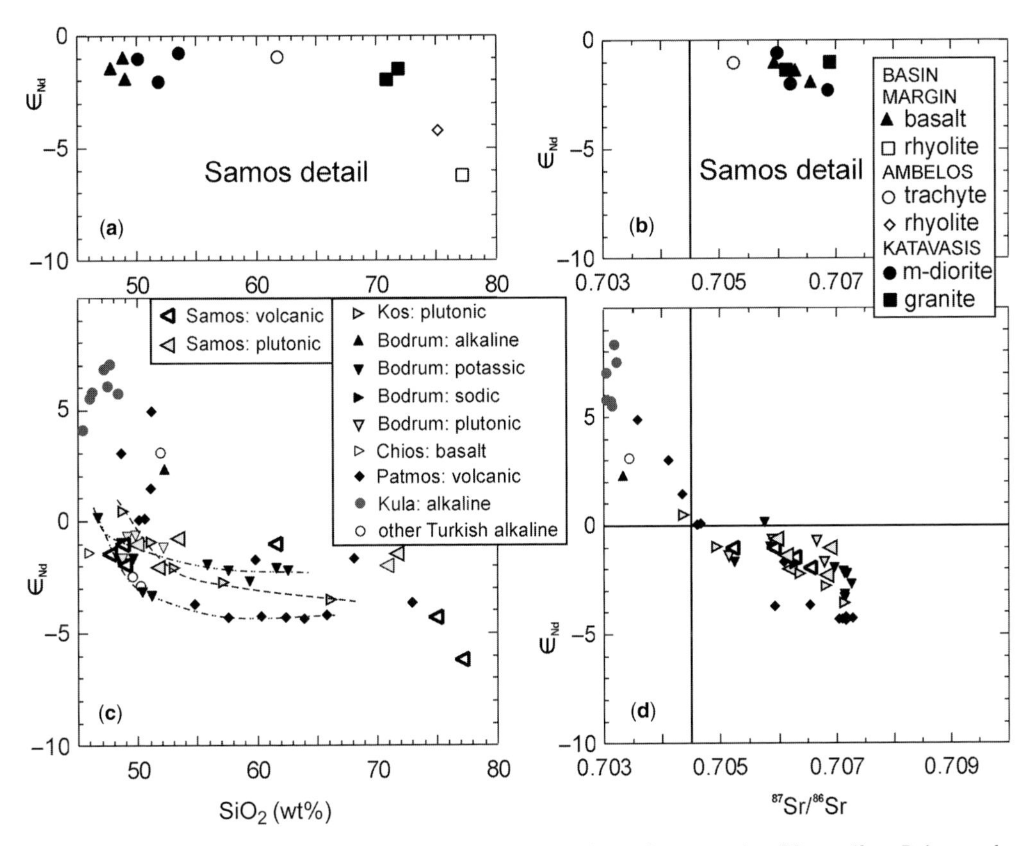

Fig. 10. Nd and Sr isotope compositions for (**a**) and (**b**) the Late Miocene igneous rocks of Samos (from Robert *et al.* 1992; Pe-Piper & Piper 2001; Altherr & Siebel 2002) compared with (**c**) and (**d**) other Cenozoic igneous rocks from the southeastern Aegean (for which data sources have been given by Pe-Piper & Piper (2002)).

Evolution of the rhyolites

The Nd and Pb isotope composition of the rhyolites, compared with that of the associated trachyte and basalt, suggests that both have evolved through crustal assimilation. Their relatively high ε_{Nd} (-4 to -6.5) indicates that any old crustal component is likely to be minor; for example, the Hercynian paragneisses of the central Aegean have ε_{Nd} of near -11 (Tarney *et al.* 1998) and granites of Naxos have ε_{Nd} of -7 to -10 (Pe-Piper 2000).

The basin margin rhyolite shows simple REE behaviour with a strong Eu anomaly and contains 3–8 ppm Cr, suggesting mixing with 1–2% basaltic magma. This may indicate that rhyolite eruption was triggered by injection of hot mafic magma into a fractionating magma chamber, as argued, for example, by Druitt *et al.* (1999) at Santorini. Strong depletion in Ba, Sr, Eu and Y points to significant plagioclase and clinopyroxene and/or amphibole fractionation, implying that fractionation took place within the crust, but at what depth is uncertain.

Complex REE patterns in the Ambelos rhyolite (Fig. 8) and some binary element plots (Fig. 13c and d) suggest the importance of accessory mineral fractionation. This rhyolite is much less voluminous than the basin margin rhyolite, and Ni and Cr are below detection.

Tectonic implications

The change from trachyte to basalt

As in Samos, many back-arc volcanic rocks of the Aegean area correlate in time with basin subsidence (Pe-Piper *et al.* 1994), suggesting a relationship between faulting and extrusion. Ring *et al.* (1999) showed that Neogene sediments in the Pirgos area of Samos occupy a half-graben bounded by a WNW–ESE striking master fault, corresponding to their D_3 phase, which they interpreted as correlative with the Kallithea detachment. This is then cut by D_4 north–south- or NE–SW-striking faults, which appear to localize the basin margin basalts

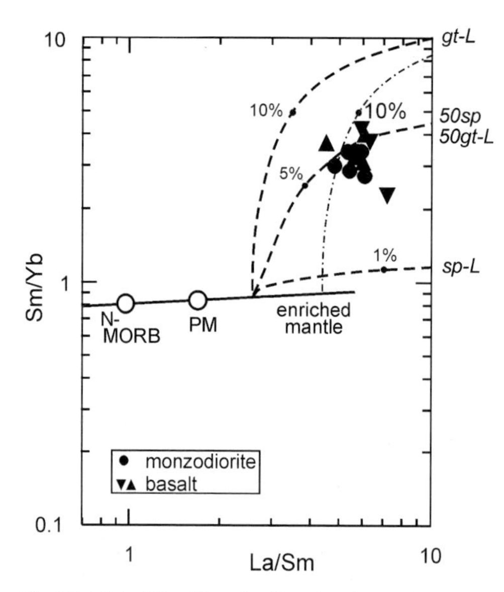

Fig. 11. Plot of Sm/Yb v. La/Sm showing melt curves obtained using non-modal batch melting equations for spinel lherzolite (sp-L) and garnet lherzolite (gt-L) of various mantle compositions (after Aldanmaz *et al.* 2000). PM, primitive mantle.

and correspond in time to the major unconformity between the Hora and Mytilene formations. These faults show transpressional kinematics (Boronkay & Doutsos 1994) and are correlated with a 'sinistral wrench corridor' extending east of Karaburun proposed by Ring *et al.* (1999; Fig. 14). The D_4 strike-slip faulting was mapped on either side of the Karaburun peninsula by Ocakoğlu *et al.* (2004, 2005). The faulting appears to continue northeastward into the late Miocene Zeytindağ, Örenli-Eğiller and Altınova basins mapped by Yılmaz *et al.* (2000), which cross-cut earlier structures and contain late Miocene lacustrine sediments. These sediments interbed with the Eğrigöl basalt, approximately dated by Rb–Sr at 9–6 Ma (Borsi *et al.* 1972; Ercan *et al.* 1985) and erupted from NE–SW-trending fissures (Yılmaz *et al.* 2000). Also on this trend are the 10–12 Ma bimodal basalt and rhyolite at Urla and 4–5.6 Ma basalts of the Young Volcanic Series at Patmos, perhaps indicating that this sinistral strike-slip system propagated southward through time. This 'wrench corridor' may also have been responsible for the localization of north–south lamprophyre dykes in Kos of uncertain age that cut the 10 Ma Dikeos monzonite and the faulted margin of the monzonite against country rock. Older, middle Miocene strike-slip faulting is suggested by the north–south-trending dykes of the Mytilene Formation in southeastern Lesbos (Pe-Piper & Piper 2007) and the parallel orientation of rhyolite domes and fissure eruptions in Chios (Pe-Piper *et al.* 1994; Fig. 14).

The change from extensional listric faulting (D_3) to strike-slip faulting (D_4) in Samos is analogous to the pattern seen in the Pliocene–Quaternary of the south Aegean arc, where evolved andesites and dacites are associated with listric faulting in the western arc, with a much higher proportion of basalt and rhyolite associated with strike-slip faulting in the eastern part of the arc (Pe-Piper & Piper 2005). In Samos, evolved Ambelos trachytes were extruded during listric extensional faulting and bimodal basalts and rhyolites during strike-slip faulting. Large-scale strike-slip faulting provides crust-penetrating pathways for basaltic magma. In contrast, mid-crustal extensional detachment faulting associated with upper crustal listric faulting creates barriers to the upward migration of magma and promotes fractional crystallization in the resulting mid-crustal magma chambers, with the production of monzodiorite and trachyte.

Relationship to regional tectonics

The changes in fault patterns and volcanism in the late Miocene of the southeastern Aegean are a result of the progressive convergence of Gondwanan continental crust with the Anatolian–Aegean microplates (Fig. 15), as reviewed for example by Aksu *et al.* (2005). By the Miocene, there was already collision of Arabia with eastern Anatolia. However, oceanic subduction was active south of Cyprus at least in the early Miocene (Robertson 2000). Widespread north–south extension in the Aegean Sea region began in the early Miocene (Gautier & Brun 1994; Dinter 1998), but less extension appears to have taken place in western Anatolia (Yılmaz *et al.* 2000). The widespread north–south graben faulting in western Anatolia mapped by Yılmaz *et al.* (2000) may therefore have had a component of wrench faulting to accommodate the Aegean extension, as suggested for the late Miocene by Ring *et al.* (1999). In western Anatolia, north–south extension with mid-crustal detachment developed in the late Oligocene (Çemen *et al.* 2006). Roll-back continued along the subduction zone south of the Aegean Sea (Royden 1993) across oceanic crust, whereas collision had already taken place south of Cyprus (Robertson 2000).

Palaeomagnetic data from Anatolia show no evidence for significant rotation from Eocene to Miocene, but with rotation beginning probably in the late Miocene (Platzman *et al.* 1998) or Pliocene (Kissel *et al.* 2003). Arabia started moving northward faster than Africa at about 12 Ma with the development of the Dead Sea fault zone (Garfunkel 1981), and many researchers have recognized this as the time of final collision and the onset of orogen-parallel movement of Anatolia (Şengör

Table 4. *Representative chemical analyses of basin margin basalt and rhyolite*

Lithology:	Basalt					Rhyolite	
	Altered	Fresh	Altered	Fresh	Fresh		
Location:	Ag. Pand	Koumeika	Koumeika	Pirgos	Ag.Geo.	Koumeika	Koumeika
Sample	SV29	SV71A	SV71B	SV91	SV102	SV77A	SV92B
Major elements (wt %)							
SiO_2	47.54	47.58	48.80	46.99	50.25	77.09	78.28
TiO_2	1.42	1.26	0.76	1.29	1.27	0.06	0.06
Al_2O_3	16.47	15.56	18.86	15.75	16.13	12.08	11.89
$Fe_2O_3^T$	8.27	7.82	7.41	8.78	5.46	0.72	0.57
MnO	0.14	0.13	0.13	0.13	0.27	0.01	0.01
MgO	5.90	6.45	7.54	5.29	3.76	0.79	0.69
CaO	10.74	11.56	3.03	10.10	10.84	0.07	0.04
Na_2O	2.89	3.44	4.15	3.25	3.37	2.44	1.70
K_2O	1.07	1.37	2.97	1.41	2.64	6.24	6.71
P_2O_5	0.58	0.79	0.31	0.57	0.47	0.01	0.01
LOI	4.80	3.80	6.10	6.90	4.90	1.10	0.70
Total	99.82	99.76	100.06	100.46	99.36	100.61	100.66
Trace elements (ppm) by XRF							
Ba	1136	1449	295	1136	747	19	22
Rb	309	62	40	51	83	439	478
Sr	896	906	233	863	710	8	9
Y	22	31	22	32	33	21	16
Zr	192	248	49	183	197	138	131
Nb	16	20	21	17	22	63	59
Th	11	<10	<10	<10	16	73	64
Pb	20	23	<10	18	21	52	51
Ga	17	19	14	18	15	16	15
Zn	66	69	62	74	68	65	17
Cu	47	44	91	31	33	<5	<5
Ni	59	55	32	67	48	<5	<5
V	209	189	334	203	144	<5	<5
Cr	179	144	41	149	137	<5	5
REE and other trace elements (ppm) by INAA							
La	50	36	28	52	46	24	13.1
Ce	95	79	49	104	91	68	35
Nd	44	39	19	49	39	29	19.5
Sm	8.5	8.0	4.0	8.6	7.8	8.2	5.7
Eu	2.2	2.2	1.50	2.4	2.1	0.06	0.06
Tb	0.84	0.97	0.52	0.98	1.05	1.03	0.74
Yb	2.0	2.2	1.76	2.2	2.5	2.9	2.7
Lu	0.32	0.33	0.28	0.32	0.38	0.42	0.41
Cs	8.8	29	1.81	8.9	3.3	5.0	11.6
Hf	3.7	5.8	1.20	3.9	4.4	6.9	6.4
Sc	22	24	35	21	18.8	1.07	0.95
Ta	1.09	0.89	0.96	1.12	1.38	6.2	6.0
Th	12.5	14.4	2.7	14.4	16.5	71	59
U	3.7	3.5	0.76	2.5	3.9	17.7	15.4

et al. 1985). The initiation of the North Anatolian Fault is generally interpreted to date from about this time (Barka 1992), but slip along the East Anatolian Fault did not begin until the latest Miocene (Hempton 1985).

The rotation of the west Aegean block relative to the SE Aegean–Anatolian block throughout the Miocene (Walcott & White 1998) would have resulted in sinistral strike-slip motion along the block boundary, most clearly seen along the mid-Cycladic lineament in the central Cyclades (Pe-Piper *et al.* 2002). Walcott & White (1998) suggested that elsewhere the shear across the boundary between the two blocks was more distributed. This distributed shear is manifested in the mid-Miocene faulting in Lesbos and Chios, and the better documented late Miocene (–?early Pliocene) sinistral shear extending from the Örenli–Eğiller

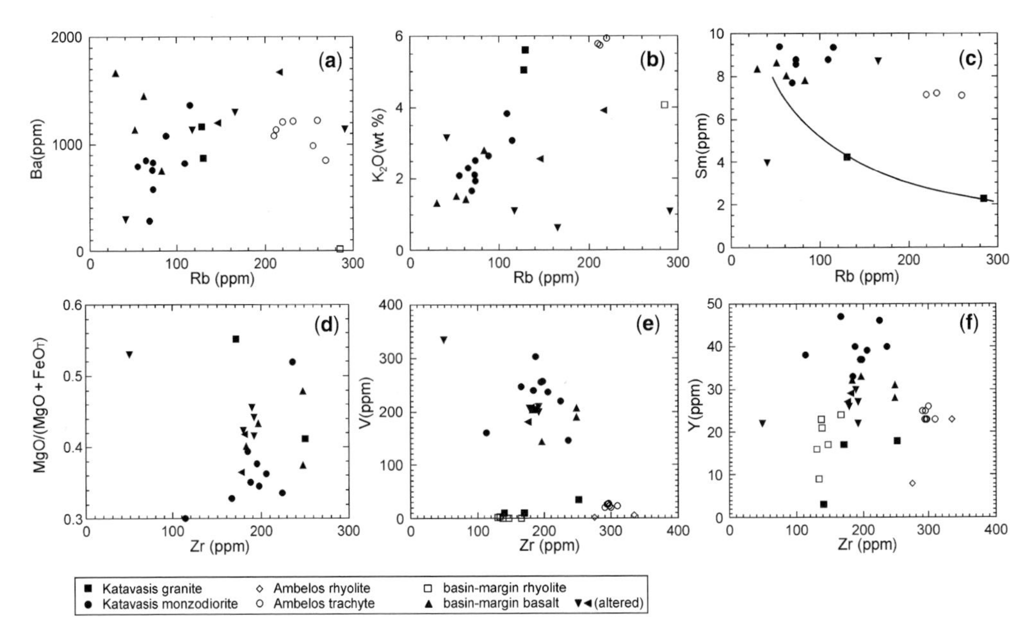

Fig. 12. Binary trace element plots showing source-related trends. Trend in (**c**) parallel amphibole fractionation vector for intermediate rocks (Aldanmaz *et al.* 2000).

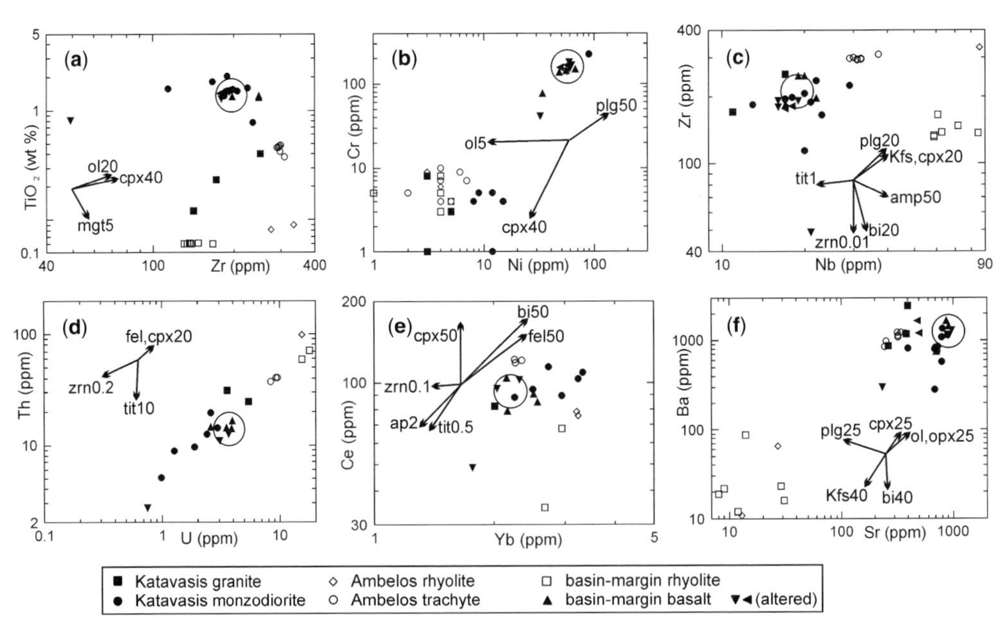

Fig. 13. Binary trace element plots showing composition of Samos rocks and Rayleigh fractionation vectors for particular minerals. Large open circles show assumed primitive composition corresponding to unaltered basalts. (**a**), (**c**) from Macdonald *et al.* (1981); (**b**), (**d**) from Thompson & Fowler (1986); (**e**), (**f**) from Tindle & Pearce (1981). amp, amphibole (hornblende); ap, apatite; bi, biotite; cpx, clinopyroxene; fel, feldspar; Kfs, K-feldspar; mgt, magnetite; ol, olivine; opx, orthopyroxene; plg, plagioclase; tit, titanite; zrn, zircon.

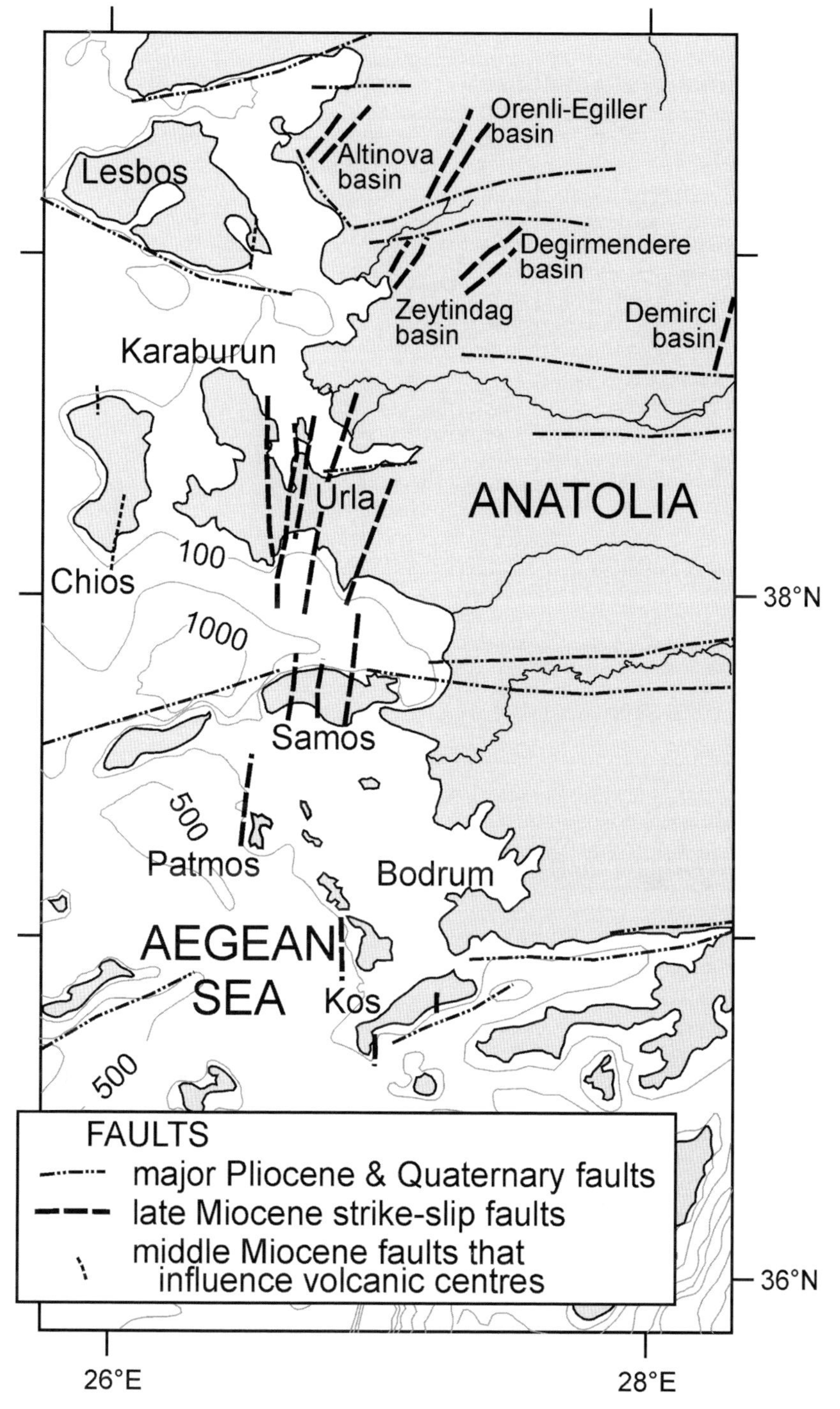

Fig. 14. Map showing prominent faults of various late Cenozoic ages in the southeastern Aegean and adjacent western Anatolia. Data for Turkey from Yılmaz *et al.* (2000) and Ocakoğlu *et al.* (2004, 2005); data for Greece from our own work (principally Pe-Piper *et al.* 2005).

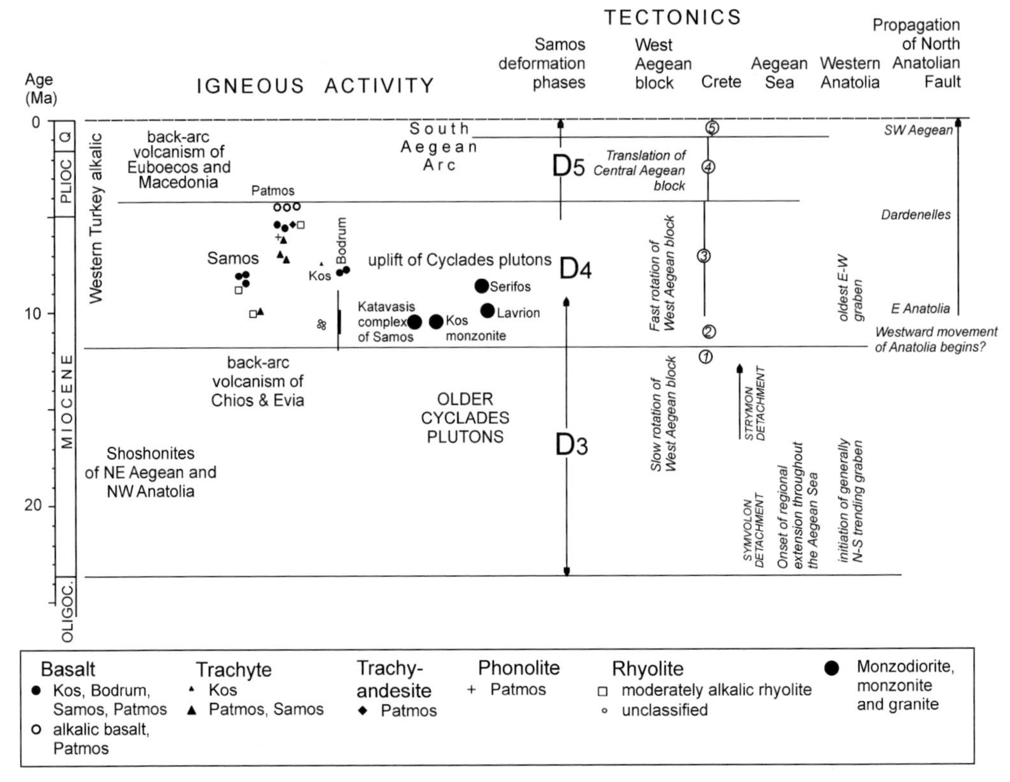

Fig. 15. Chronology of volcanism and deformation in the southeastern Aegean region. Igneous activity modified from Pe-Piper & Piper (2002). Samos deformation from Ring *et al.* (1999). West Aegean block from Walcott & White (1998). Crete from ten Veen & Meijer (1998). Aegean Sea from Dinter (1998). Western Anatolia from Yılmaz *et al.* (2000). North Anatolian Fault from Armijo *et al.* (1999).

basin in the north through Samos to the Dikeos monzonite of Kos in the south.

The initial development of the Samos basins dates from the late Serravallian to early Tortonian (12–11 Ma), and the Katavasis complex and oldest tuffs date from this time. The mid-Tortonian, at about 10 Ma, was a time of change in tectonic style throughout the Aegean (Le Pichon & Angelier 1979), including NE–SW compression in Crete that affected sediments as young as latest Serravallian–early Tortonian (12–11 Ma) (Meulenkamp & Hilgen 1986; phase 1 of ten Veen & Meijer 1998). Palaeomagnetic data suggest that the rotation rate of the west Aegean block may have increased at this time. The Ambelos volcanic rocks and the mid-Tortonian volcanic rocks of Kos date from this time. The deformation at the late Tortonian unconformity in the Vathy basin and the extrusion of the basin margin basalts of Samos is a little younger than this mid-Tortonian event, but correlative volcanic rocks are present in Kos. Volcanism in Patmos is a little younger and ended when subduction-related volcanism began in the south Aegean arc.

Implications

Late Miocene igneous rocks of Samos are part of a suite of rocks in the southeastern Aegean Sea derived from partial melting of enriched subcontinental lithospheric mantle during regional extension. Monzodiorite, granite, trachyte and basalt magmas all had a common source and experienced no significant crustal assimilation. Older monzodiorite, granite and trachyte were emplaced during listric faulting, which created conditions for mid-crustal magma chambers to form, within which trachyte fractionated. Basalts and associated rhyolites were extruded during later strike-slip faulting, which provided efficient pathways for magma to rise through the crust. This sinistral north–south faulting marked the diffuse eastern margin of the west Anatolian block and appears to have propagated southward from the Karaburun peninsula through Samos to Patmos and Kos in the late Miocene and (?)early Pliocene.

This work was supported by a Natural Sciences and Engineering Research Concil of Canada (NSERC) discovery grant to G.P.-P. The geochemical work was carried out at the Saint Mary's University Regional Geochemical Gentre and the Dalhousie University Regional Electron Microprobe Centre. We thank U. Robert, N. Güleç and T. Taymaz for their constructive reviews, which greatly improved this paper.

References

ABDEL-RAHMEN, A.-F. M. 1994. Nature of biotites from alkaline, calc-alkaline and peraluminous magmas. *Journal of Petrology*, **35**, 525–541.

AKSU, A. E., HALL, J. & YALTIRAK, C. 2005. Miocene to Recent tectonic evolution of the eastern Mediterranean: new pieces of the old Mediterranean puzzle. *Marine Geology*, **221**, 1–13.

ALDANMAZ, E., PEARCE, J. A., THIRLWALL, M. F. & MITCHELL, J. G. 2000. Petrogenetic evolution of late Cenozoic post-collisional volcanism in western Anatolia, Turkey. *Journal of Volcanology and Geothermal Research*, **102**, 67–95.

ALTHERR, R. & SIEBEL, W. 2002. I-type plutonism in a continental back-arc setting: Miocene granitoids and monzonites from the central Aegean Sea, Greece. *Contributions to Mineralogy and Petrology*, **143**, 397–415.

ALTHERR, R., KELLER, J. & KOTT, K. 1976. Der jungtertiäre Monzonit von Kos und sein Kontakthof (Ägäis, Griechenland). *Bulletin de la Société Géologique de France*, **18**, 403–412.

ALTHERR, R., KREUZER, H., WENDT, I., ET AL. 1982. A late Oligocene/early Miocene high temperature belt in the Attic–Cycladic crystalline complex (S.E. Pelagonian, Greece). *Geologische Jahrbuch*, **E23**, 97–164.

ALTUNKAYNAK, Ş. & DILEK, Y. 2006. Timing and nature of postcollisional volcanism in western Anatolia and geodynamic implications. *In*: DILEK, Y. & PAVLIDES, S. (eds) *Post-collisional Tectonics and Magmatisum in the Mediterranean Region and Asia.* Geological Society of America, Special Papers, **409**, 321–351.

ANDERSON, J. L. & SMITH, D. R. 1995. The effects of temperature and fO_2 on the Al-in-hornblende barometer. *American Mineralogist*, **80**, 549–559.

ARMIJO, R., MEYER, B., HUBERT, A. & BARKA, A. 1999. Westward propagation of the North Anatolian fault into the northern Aegean: timing and kinematics. *Geology*, **27**, 267–270.

BARKA, A. A. 1992. The North Anatolian fault zone. *Annales Tectonicae*, **6**, 164–195.

BELLON, H. & JARRIGE, J.-J. 1979. Le magmatisme néogène et quaternaire de l'île de Kos (Grèce): données géochronologique. *Comptes Rendus de l'Academie des Sciences, série D* **228**, 1359–1362.

BELLON, H., JARRIGE, J.-J. & SOREL, D. 1979. Les activités magmatiques égéennes de l'Oligocène à nos jours et leurs cadres géodynamiques. Données nouvelles et synthèse. *Revue de Géologie Dynamique et Géographie Physique*, **21**, 41–55.

BESANG, C., ECKHARDT, F. J., HARRE, W., KREUZER, G. & MULLER, P. 1977. Radiometrische Altersbestimmungen an neogenen Eruptivgesteinen der Turkei. *Geologische Jahrbuch*, **25**, 3–36.

BLUNDY, J. D. & HOLLAND, T. J. B. 1990. Calcic amphibole equilibria and a new amphibole–plagioclase geothermometer. *Contributions to Mineralogy and Petrology*, **104**, 208–224.

BORONKAY, K. & DOUTSOS, T. 1994. Transpression and transtension within different structural levels in the central Aegean region. *Journal of Structural Geology*, **11**, 1555–1573.

BORSI, S., FERRARA, G., INNOCENTI, F. & MAZZUOLI, R. 1972. Geochronology and petrology of recent volcanics in the eastern Aegean Sea (West Anatolia and Lesvos Island). *Bulletin Volcanologique*, **36**, 473–496.

ÇEMEN, I., CATLOS, E. J., GÖĞÜS, O. & ÖZERDEM, C. 2006. Postcollisional extensional tectonics and exhumation of Menderes massif in the Western Anatolia extended terrane, Turkey. *In*: DILEK, Y. & PAVLIDES, S. (eds) *Post-collisional Tectonics and Magmatisum in the Mediterranean Region and Asia.* Geological Society of America, Special Papers, **409**, 353–379.

DINTER, D. A. 1998. Late Cenozoic extension of the Alpine collisional orogen, northeastern Greece: origin of the north Aegean basin. *Geological Society of America Bulletin*, **110**, 1208–1230.

DRUITT, T. H., EDWARDS, L., MELLORS, R. M., ET AL. (eds) 1999. *Santorini Volcano*. Geological Society, London, Memoirs, **19**, 165.

ERCAN, T., SATIR, M., KREUZER, H., ET AL. 1985. Interpretation of new chemical, isotopic and radiometric data on Cenozoic volcanics of western Anatolia. *Bulletin of the Geological Society of Turkey*, **28**, 121–136 [in Turkish, English abstract].

GARFUNKEL, Z. 1981. Internal structure of the Dead Sea leaky transform (rift) in relation to plate kinematics. *Tectonophysics*, **80**, 81–108.

GAUTIER, P. & BRUN, J.-P. 1994. Crustal-scale geometry and kinematics of late-orogenic extension in the central Aegean (Cyclades and Evvia Island). *Tectonophysics*, **238**, 399–424.

GÜLEÇ, N. 1991. Crust–mantle interaction in western Turkey; implications from Sr and Nd isotope geochemistry of Tertiary and Quaternary volcanics. *Geological Magazine*, **128**, 417–435.

HEMPTON, M. R. 1985. Structure and deformation history of the Bitlis suture zone near Lake Hazar, southeastern Turkey. *Geological Society of America Bulletin*, **96**, 233–243.

INNOCENTI, F. & MAZZUOLI, R. 1972. Petrology of the Izmir–Karaburun volcanic area (West Turkey). *Bulletin Volcanologique*, **36**, 83–104.

INNOCENTI, F., MANETTI, P., MAZZUOLI, R., PASQUARÉ, G. & VILLARI, L. 1982. Anatolia and north-west Iran. *In*: THORPE, R. S. (ed.) *Andesites*. Wiley, New York, 327–349.

JUTEAU, M., MICHARD, A. & ALBARÈDE, F. 1986. The Pb–Sr–Nd isotope geochemistry of some recent circum-Mediterranean granites. *Contributions to Mineralogy and Petrology*, **92**, 331–340.

KISSEL, C., LAJ, C., ŞENGÖR, A. M. C. & POISSON, A. 1987. Palaeomagnetic evidence for rotation in opposite senses of adjacent blocks in the northeastern Aegean and western Anatolia. *Geophysical Research Letters*, **14**, 907–910.

KISSEL, C., LAJ, C., POISSON, A. & GÖRÜR, N. 2003. Palaeomagnetic reconstruction of the Cenozoic evolution of the Eastern Mediterranean. *Tectonophysics*, **362**, 199–217.

KONDOPOULOU, D. 2000. Palaeomagnetism in Greece: Cenozoic and Mesozoic components and their geodynamic implications. *Tectonophysics*, **326**, 131–151.

LE PICHON, X. & ANGELIER, J. 1979. The Hellenic arc and trench system: a key to the evolution of the Eastern Mediterranean. *Tectonophysics*, **60**, 1–42.

MACDONALD, R., GOTTFRIED, D., FARRINGTON, M. J., BROWN, F. W. & SKINNER, N. G. 1981. Geochemistry of a continental tholeiite suite: late Palaeozoic quartz dolerite dykes of Scotland. *Transactions of the Royal Society of Edinburgh, Earth Sciences*, **72**, 57–74.

MEISSNER, B. 1976. Das Neogen von Ost-Samos, Sedimentationsgeschichte und Korrelation. *Neues Jahrbuch für Geologie und Palaeontologie, Abhandlungen*, **152**, 161–176.

MEULENKAMP, J. E. & HILGEN, F. 1986. Event stratigraphy, basin evolution and tectonics of the Hellenic and Calabro-Silician arcs. *In*: WEZEL, F.-C. (ed.) *The Origin of Arcs*. Elsevier, Amsterdam, 327–350.

MEZGER, K., ALTHERR, R., OKRUSCH, M., HENJES-KUNST, F. & KREUZER, H. 1985. Genesis of acid/basic rock associations: a case study of the Kallithea intrusive complex, Samos, Greece. *Contributions to Mineralogy and Petrology*, **90**, 353–366.

OCAKOĞLU, N., DEMIRBAĞ, E. & KUŞÇU, İ., 2004. Neotectonic structures in the area offshore of Alaçatı, Doğanbey and Kuşadası (western Turkey): evidence of strike-slip faulting in the Aegean extensional province. *Tectonophysics*, **391**, 67–83.

OCAKOĞLU, N., DEMIRBAĞ, E. & KUŞÇU, İ. 2005. Neotectonic structures in İzmir Gulf and surrounding regions (western Turkey): evidences of strike-slip faulting with compression in the Aegean extensional regime. *Marine Geology*, **219**, 155–171.

PEARCE, J. A., HARRIS, N. B. W. & TINDLE, A. G. 1984. Trace element discrimination diagrams for tectonic interpretation of granitic rocks. *Journal of Petrology*, **25**, 952–983.

PE-PIPER, G. 1984. Zoned pyroxenes from shoshonite lavas of Lesbos, Greece: inferences concerning shoshonite petrogenesis. *Journal of Petrology*, **25**, 453–472.

PE-PIPER, G. 1994. Lead isotopic compositions of Neogene volcanic rocks from the Aegean extensional area. *Chemical Geology*, **118**, 27–41.

PE-PIPER, G. 2000. Origin of S-type granites coeval with I-type granites in the Hellenic subduction system, Miocene of Naxos, Greece. *European Journal of Mineralogy*, **12**, 859–875.

PE-PIPER, G. & PIPER, D. J. W. 1992. Geochemical variation with time in the Cenozoic high-K volcanic rocks of the island of Lesbos, Greece: significance for shoshonite petrogenesis. *Journal of Volcanology and Geothermal Research*, **53**, 371–387.

PE-PIPER, G. & PIPER, D. J. W. 2001. Late Cenozoic, post-collisional Aegean igneous rocks: Nd, Pb and Sr isotopic constraints on petrogenetic and tectonic models. *Geological Magazine*, **138**, 653–668.

PE-PIPER, G. & PIPER, D. J. W. 2002. *The Igneous Rocks of Greece*. Borntraeger, Stuttgart.

PE-PIPER, G. & PIPER, D. J. W. 2005. The South Aegean active volcanic arc: relationships between magmatism and tectonics. *Developments in Volcanology*, **7**, 113–133.

PE-PIPER, G. & PIPER, D. J. W. 2006. Unique features of the Cenozoic igneous rocks of Greece. *In*: DILEK, Y. & PAVLIDES, S. (eds) *Post-collisional Tectonics and Magmatisum in the Mediterranean Region and Asia*. Geological Society of America, Special Papers, **409**, 259–282.

PE-PIPER, G. & PIPER, D. J. W. 2007. Neogene back-arc volcanism of the Aegean: new insights into the relationship between magmatism and tectonics. *In*: BECCALUVA, L., BIANCHINI, G. & WILSON, M. *Cenozoic Volcanism in the Mediterranean Area*. Geological Society of America, Special Papers, **418**, 17–31.

PE-PIPER, G. & TSOLI-KATAGAS, P. 1991. K-rich mordenite from late Miocene rhyolitic tuffs, island of Samos, Greece. *Clays and Clay Minerals*, **39**, 239–247.

PE-PIPER, G., PIPER, D. J. W., KOTOPOULI, C. N. & PANAGOS, A. G. 1994. Neogene volcanoes of Chios, Greece: the relative importance of subduction and back-arc extension. *In*: SMELLIE, J. L. (ed.) *Volcanism Associated with Extension at Consuming Plate Margins*. Geological Society, London, Special Publications, **81**, 213–232.

PE-PIPER, G., PIPER, D. J. W. & MATARANGAS, D. 2002. Regional implications of geochemistry and style of emplacement of Miocene I-type diorite and granite, Delos, Cyclades, Greece. *Lithos*, **60**, 47–66.

PE-PIPER, G., PIPER, D. J. W. & PERISSORATIS, C. 2005. Neotectonics and the Kos Plateau Tuff eruption of 161 ka, South Aegean arc. *Journal of Volcanology and Geothermal Research*, **139**, 315–338.

PLATZMAN, E. S., TAPIRDAMAZ, C. & SANVER, M. 1998. Neogene anticlockwise rotation of central Anatolia (Turkey): preliminary palaeomagnetic and geochronological results. *Tectonophysics*, **299**, 175–189.

RING, U., LAWS, S. & BERNET, M. 1999. Structural analysis of a complex nappe sequence and late-orogenic basins from the Aegean island of Samos, Greece. *Journal of Structural Geology*, **21**, 1575–1601.

ROBERT, U. & CANTAGREL, J. M. 1979. Le volcanisme basaltique dans le Sud-Est de la Mer Egée. Données géochronologiques et rélations avec la tectonique. *In*: KALLERGIS, G. (ed.) *Proceedings of the VI Colloquium on the Geology of the Aegean Region, Athens. IGME*, Athens, 961–967.

ROBERT, U. & MONTIGNY, R. 2001. A new age data set for the Bodrum volcanic complex (SW Anatolia). *In*: *Fourth International Turkish Geology Symposium, Adana, Turkey*, 24–28 September 2001, Curova University, 303.

ROBERT, U., FODEN, J. & VARNE, R. 1992. The Dodecanese Province, SE Aegean: a model for tectonic control on potassic magmatism. *Lithos*, **28**, 241–260.

ROBERTSON, A. H. F. 2000. Mesozoic–Tertiary tectonic–sedimentary evolution of a south Tethyan oceanic basin and its margins in southern Turkey. *In*: BOZKURT, E., WINCHESTER, J. A. & PIPER, J. D. A. (eds) *Tectonics and Magmatism in Turkey and the Surrounding Area*. Geological Society, London, Special Publications, **173**, 353–384.

ROYDEN, L. H. 1993. Evolution of retreating subduction boundaries formed during continental collision. *Tectonics*, **12**, 629–638.

SEN, S. & VALET, J.-P. 1986. Magnetostratigraphy of late Miocene continental deposits in Samos, Greece. *Earth and Planetary Science Letters*, **80**, 167–174.

ŞENGÖR, A. M. C., GÖRÜR, N. & ŞAROĞLU, F. 1985. Strike-slip faulting and related basin formation in zones of tectonic escape: Turkey as a case study. *In*: BIDDLE, K. T. & CHRISTIE-BLICK, N. (eds) *Strike-slip Deformation, Basin Formation and Sedimentation*. Society of Economic Paleontologists and Mineralogists, Special Publications, **37**, 227–264.

SEYITOĞLU, G., ANDERSON, D., NOWELL, G. & SCOTT, B. 1997. The evolution from Miocene potassic to Quaternary sodic magmatism in western Turkey: implications for enrichment processes in the lithospheric mantle. *Journal of Volcanology and Geothermal Research*, **76**, 127–147.

STACEY, J. S. & KRAMERS, J. D. 1975. Approximation of terrestrial lead evolution by a two-stage model. *Earth and Planetary Science Letters*, **26**, 207–221.

STAMATAKIS, M. 1989*a*. Authigenic silicates and silica polymorphs in the Miocene saline–alkaline deposits of the Karlovassi Basin, Samos, Greece. *Economic Geology*, **84**, 788–798.

STAMATAKIS, M 1989*b*. A boron-bearing potassium feldspar in volcanic ash and tuffaceous rocks from Miocene lake deposits, Samos Island, Greece. *American Mineralogist*, **74**, 230–235.

STAMATAKIS, M. & ZAGOUROGLOU, K. 1984. On the occurrence of niter in Samos island *Orychtos Plutos*, **33**, 17–26 [in Greek with English abstract].

STRECKEISEN, A. & LE MAITRE, R. W. 1979. A chemical approximation to the Modal QAPF classification of the igneous rocks. *Neues Jahrbuch fur Mineralogie, Abhandlungen*, **136**, 169–206.

SUN, S.-S. & MCDONOUGH, W. F. 1989. Chemical and isotopic systematics of oceanic basalts: implications for mantle composition and processes. *In*: SAUNDERS, A. D. & NORRY, M. J. (eds) *Magmatism in the Ocean Basins*. Geological Society, London, Special Publications, **42**, 313–345.

TARNEY, J., BARR, S. R., MITROPOULOS, P., SIDERIS, K., KATERINOPOULOS, A. & STOURAITI, C. 1998. Santorini: geochemical constraints on magma sources and eruption mechanisms. *In*: CASALE, R., FYTIKAS, M., SIGVALDASSON, G. & VOUGIOUKALAKIS, G. (eds) *The European Laboratory Volcanoes*. EUR **18161**, 89–111.

TEN VEEN, J. H. & MEIJER, P. TH. 1998. Late Miocene to Recent tectonic evolution of Crete (Greece): geological observations and model analysis. *Tectonophysics*, **298**, 191–208.

THEODOROPOULOS, D. 1979. *Geological map of Greece, 1:50 000, Island of Samos*. IGME, Athens.

THOMPSON, R. N. & FOWLER, M. B. 1986. Subduction-related shoshonitic and ultrapotassic magmatism: a study of Siluro-Ordovician syenites from the Scottish Caledonides. *Contributions to Mineralogy and Petrology*, **94**, 507–522.

TINDLE, A. & PEARCE, J. A. 1981. Petrogenetic modelling of *in situ* fractional crystallization in the zoned Loch Doon pluton, Scotland. *Contributions to Mineralogy and Petrology*, **78**, 196–207.

TURNER, S., ARNAUD, N., LIU, J., ET AL. 1996. Post-collision shoshonitic volcanism on the Tibetan Plateau: implications for convective thinning of the lithosphere and the source of ocean island basalts. *Journal of Petrology*, **37**, 45–71.

ULUSOY, I., ÇUBUKÇU, E., AYDAR, E., LABAZUY, P., GOURGAUD, A. & VINCENT, P. M. 2004. Volcanic and deformation history of the Bodrum resurgent caldera system (southwestern Turkey). *Journal of Volcanology and Geothermal Research*, **136**, 71–96.

WALCOTT, C. R. & WHITE, S. H. 1998. Constraints on the kinematics of post-orogenic extension imposed by stretching lineations in the Aegean region. *Tectonophysics*, **298**, 155–175.

WEIDMANN, M., SALOUNIAS, N., DRAKE, R. E. & CURTIS, G. H. 1984. Neogene stratigraphy in the eastern basin, Samos island, Greece. *Geobios*, **17**, 477–490.

WESTAWAY, R., PRINGLE, M., YURTMEN, S., DEMIR, T., BRIDGLAND, D., ROWBOTHAM, G. & MADDY, D. 2004. Pliocene and Quaternary regional uplift in western Turkey: the Gediz River terrace staircase and the volcanism at Kula. *Tectonophysics*, **391**, 121–169.

WONES, D. R. 1989. Significance of the assemblage titanite + magnetite + quartz in granitic rocks. *American Mineralogist*, **74**, 744–749.

WYERS, G. P. & BARTON, M. 1986. Petrology and evolution of transitional alkaline–subalkaline lavas from Patmos, Dodecanesos, Greece: evidence for fractional crystallisation, magma mixing and assimilation. *Contributions to Mineralogy and Petrology*, **93**, 297–311.

WYERS, G. P. & BARTON, M. 1987. Geochemistry of a transitional ne-trachybasalt–Q-trachyte lava series from Patmos (Dodecanesos), Greece: further evidence for fractionation, mixing and assimilation. *Contributions to Mineralogy and Petrology*, **97**, 279–291.

YILMAZ, Y. 1990. Comparison of young volcanic associations of western and eastern Anatolia formed under a compressional regime: a review. *Journal of Volcanology and Geothermal Research*, **44**, 69–87.

YILMAZ, Y., GENÇ, C., GÜRER, F., ET AL. 2000. When did the western Anatolian grabens begin to develop? *In*: BOZKURT, E., WINCHESTER, J. A. & PIPER, J. D. A. (eds) *Tectonics and Magmatism in Turkey and the Surrounding Area*. Geological Society, London, Special Publications, **173**, 353–384.

Distribution and chronology of submarine volcanic rocks around Santorini and their relationship to faulting

D. J. W. PIPER[1], G. PE-PIPER[2], C. PERISSORATIS[3] & G. ANASTASAKIS[4]

[1]*Geological Survey of Canada (Atlantic), Bedford Institute of Oceanography, PO Box 1006, Dartmouth, N.S., B2Y 4A2, Canada (e-mail: dpiper@nrcan.gc.ca)*

[2]*Department of Geology, Saint Mary's University, Halifax, N.S., B3H 3C3, Canada*

[3]*IGME, 70 Mesogion Street, 11527 Athens, Greece*

[4]*University of Athens, Panepistimiopolis, Athens 15784, Greece*

Abstract: Seismic reflection profiles from the marine areas around Santorini in the south Aegean arc show the distribution of active faults and the occurrence of submarine volcanic rocks interfingering with stratified basinal sediment. Santorini is located at the intersection of fault sets of different ages. To the west, active faults trend east–west, whereas to the east, active faults trend ENE–WSW and a slightly older set of faults trends NE–SW. Subsurface submarine volcanic rocks can be dated using ages estimated for the stratified basin sediments elsewhere in the Milos–Christiani Basin. Volcanic horizons off Santorini correlate with the main (young) Akrotiri (0.65–0.55 Ma) and old Akrotiri (1.6 ka) volcanic episodes, respectively. Off Christiani, the upper unit of volcanic rocks is of similar age to the young Akrotiri episode and the lower volcanic unit is likely to be of latest Pliocene age. Late Neogene basin subsidence and volcanism are a consequence of changing patterns of faulting resulting from the collision of the African and Aegean–Anatolian plates.

The south Aegean volcanic arc, resulting from the subduction of the African plate beneath the Aegean–Anatolian plate, is developed in an area of thinned continental crust in the southern Aegean Sea. The area has a complex tectonic history in the Neogene and Quaternary, recently reviewed by Taymaz *et al.* (2004), Aksu *et al.* (2005) and Kokkalas *et al.* (2006), and some details of active tectonic deformation are known from geodetic (e.g. Nyst & Thatcher 2004) and seismological (e.g. Bohnhoff *et al.* 2006) studies. Volcanic history is well known on the volcanic islands (e.g. Pe-Piper & Piper 2002), but recent studies of seismic reflection profiles have shown that large volumes of late Cenozoic volcanic rocks occur in submarine areas of the south Aegean arc; for example, near Methana (Papanikolaou *et al.* 1988) and Milos (Anastasakis & Piper 2005). Thus, where volcanic rocks interfinger with stratified marine sediment, there is the potential to estimate the age of the deeper parts of volcanic edifices.

The purpose of this study is two-fold. First, we use high-resolution sparker profiles collected by the Institute of Geology and Mineral Exploration (IGME) (Perissoratis 1995) and seismic reflection profiles from the 1974 *Meteor* cruise (Jongsma *et al.* 1977) to determine the distribution and age of submarine volcanism around Santorini and Christiani. Second, we explore the consequences of the fault patterns seen in seismic reflection profiles to evaluate the role of faulting in the evolution of the volcanic centres, building on the regional interpretation of Pe-Piper & Piper (2005).

Methods and materials

Approximately 800 line-km of sparker profiles from the Santorini area (Fig. 1) were acquired by IGME with a 0.7–1 kJ SIG sparker source and a single-channel EG&G hydrophone array, recorded at a 0.25 s sweep. Sub-bottom penetration is typically 0.1–0.2 s. Navigation was by global positioning system (GPS) with precision better than 10 m. Data quality is variable as a result of variable weather conditions and the top 10 ms of data is partially obscured by the bubble pulse.

We have used microfilm archive copies of seismic profiles reported by Jongsma *et al.* (1977), which used 1000 cm^3 and 5000 cm^3 air guns and an Aquatronix streamer. Positioning was by Loran-C and transit satellite, estimated good to 0.5 km. The interpretation of these lower-resolution records was aided by comparison with the higher-resolution IGME sparker profiles. In addition, we have used one multichannel seismic reflection profile reported by Martin (1987) (see also Mascle & Martin 1990).

From: TAYMAZ, T., YILMAZ, Y. & DILEK, Y. (eds) *The Geodynamics of the Aegean and Anatolia.* Geological Society, London, Special Publications, **291**, 99–111.
DOI: 10.1144/SP291.5 0305-8719/07/$15.00

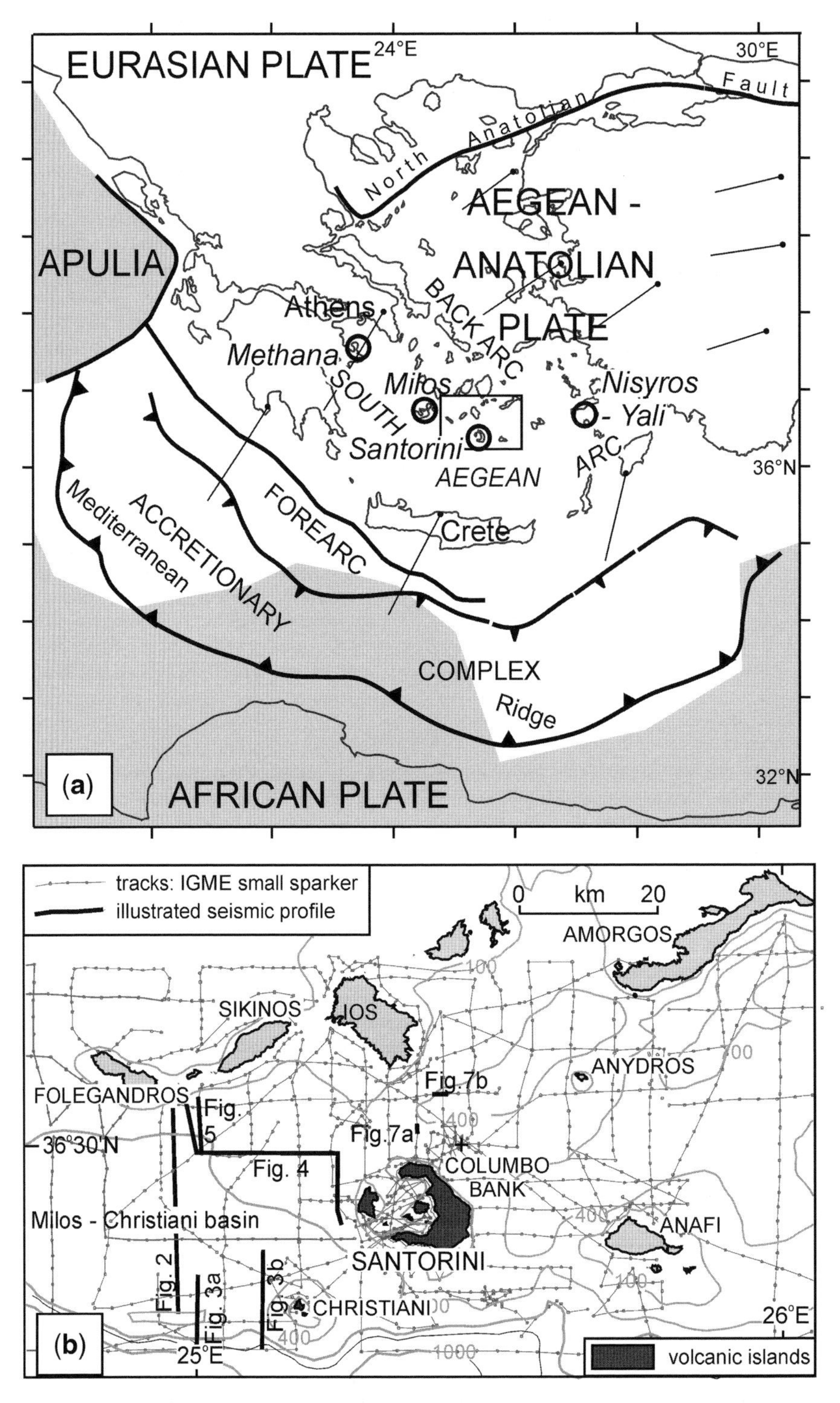

Fig. 1. (**a**) Regional map showing location of the south Aegean arc. Circles indicate dormant volcanoes; lines are motion vectors relative to Eurasia (from Reilinger *et al.* 1997). (**b**) Map of the area around Santorini, showing location of seismic profiles illustrated in the text. Generalized bathymetry from Piper & Perissoratis (2003).

Small, active, high-angle faults can be readily recognized in seismic reflection profiles from offsets in sub-sea-bed reflectors, with displacement increasing with depth. Larger faults are more difficult to define with certainty, because they are characterized by steep slopes that return hyperbolic diffractions and may be difficult to distinguish from steep constructional volcanic lava slopes. Evidence of recent movement on such faults may be provided by progressive tilting of adjacent reflectors, whereas inactive faults show only a drape of sediment.

Geological setting and previous work

Basement rocks on Santorini and adjacent islands consist of metamorphic rocks of the Attico-Cycladic complex that were deformed in a compressional orogeny in the early Cenozoic and then experienced major Miocene and younger extension. The Cretan Basin south of Santorini and the Milos–Christiani Basin to the west probably developed as a result of this Miocene extension (Fig. 1). Both basins have a prominent Messinian marker in seismic profiles (Bartole *et al.* 1983), which was drilled by Deep Sea Drilling Project (DSDP), Hole 378, in the Cretan Basin (Hsü *et al.* 1978). This marker consists of evaporites in the centre of the basins and of an erosion surface nearer basin margins (Fig. 2).

Pliocene basins in the southern Aegean Sea and adjacent areas were defined principally by east–west-trending listric faults, terminating against north–south-trending transfer faults (e.g. Papanikolaou *et al.* 1988; Collier & Dart 1991; Taymaz *et al.* 1991; ten Veen & Meijer 1998; Piper & Perissoratis 2003; Purvis & Robertson 2004). Volcanism of the south Aegean arc began in the Pliocene in the western part of the arc (Pe-Piper & Piper 2005).

In the eastern part of the arc, prominent NE–SW-trending strike-slip faults developed in the late Pliocene (ten Veen & Kleinspehn 2002) to early Quaternary (Fytikas & Vougioukalakis 1993; Pe-Piper *et al.* 2005; Uluğ *et al.* 2005). These faults probably represent the first consequence of collision of African continental crust with the Aegean–Anatolian plate (Taymaz *et al.* 1990; Bohnhoff *et al.* 2001; ten Veen & Kleinspehn 2002), although other interpretations are possible (Kokkalas & Doutsos 2001).

In the mid- to late Quaternary, continuing collision resulted in new active faults in the south Aegean arc and adjacent areas (Piper & Perissoratis 2003). The faults parallel modern seismically active faults (Taymaz 1996; Hatzfeld 1999): north–south- or NNW–SSW-trending in the western part of the arc and the Peloponnese (Armijo *et al.* 1992) and ENE–WSW-trending in the eastern part of the arc (Pe-Piper *et al.* 2005) and eastern Crete (Taymaz *et al.* 1990; ten Veen & Kleinspehn 2003). The various changes in dominant fault direction in the Pliocene–Quaternary are marked by regional deep-water unconformities on the flanks of basins, the age of which were estimated (with substantial uncertainty) by Piper & Perissoratis (2003) from extrapolation of sedimentation rates and correlation with coastal progradation units controlled by eustatic sea-level change. Near Santorini, those workers recognized two regional unconformities, one at about 0.8 ± 0.2 Ma, and the other at about 0.2 Ma.

The oldest known volcanism on Santorini, the older Akrotiri volcanic unit, is submarine and dates from the beginning of the Pleistocene age. The younger Akrotiri unit dates from around 0.6 Ma and is overlain by flows of the Peristeria stratovolcano. In the last 0.25 Ma, a series of major eruptions deposited the Thera Pyroclastic Formation (Druitt *et al.* 1999; Pe-Piper & Piper 2002, pp. 408–409; Vespa *et al.* 2006). Columbo Bank is a young submarine volcanic edifice NE of Santorini (Fig. 1b). The Christiani islets SW of Santorini consist entirely of volcanic materials of uncertain age.

The Milos–Christiani Basin to the west of Santorini has a regional Messinian marker (Bartole *et al.* 1983; Martin 1987), which is represented by an erosion surface near the margins of the basin (Fig. 2). The Messinian marker is overlain by stratified Pliocene to Quaternary basinal sediments in the centre of the basin and shallow-water Pliocene facies at the basin margins, passing up into Quaternary basinal sediments (Fig. 2). Anastasakis & Piper (2005) recognized three correlatable key reflections (A–C, of latest Pliocene to Quaternary age) in *Meteor* 1974 seismic profiles (Fig. 3a) and noted that the seismic character of these reflections was remarkably constant in different basins near the central part of the south Aegean arc (Anastasakis *et al.* 2006). They assigned a chronology to these reflectors (1) by making ties to dated volcanic horizons in Milos; and (2) by correlating stacked coastal progradation sequences on subsiding shelves near Milos and assigning chronology on the basis of eustatic sea-level changes estimated from the global oxygen isotope curve (Skene *et al.* 1998). Using these techniques, they estimated A to date from marine oxygen isotope stage (MIS) 12 (0.45 Ma), B from about 1.0 Ma, and C from about 2.0 Ma.

Basins to the east of Santorini show a complex late Neogene history. (?) Pliocene to Early Pleistocene sediments (unit 'B' of Piper & Perissoratis 2003, not to be confused with reflection B of this paper) show both basinal and shelf facies, as in the Milos–Christiani Basin, but the Messinian marker has not been recognized. These sediments were cut by a regional unconformity prior to

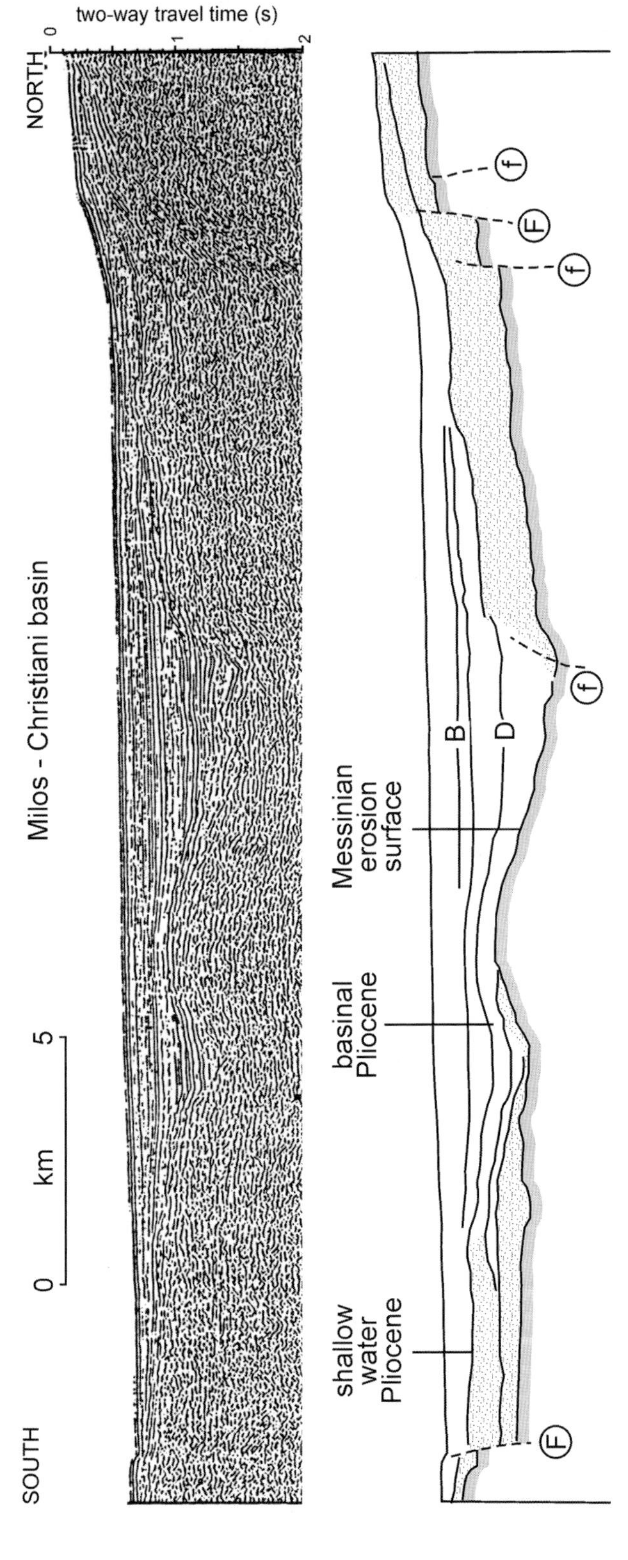

Fig. 2. Regional cross-section of the Milos–Christiani Basin just west of Santorini. Multichannel seismic profile from Martin (1987). Messinian erosion surface correlated from Bartole *et al.* (1983); reflections B and D from Anastasakis & Piper (2005). f, inactive fault; F, active fault.

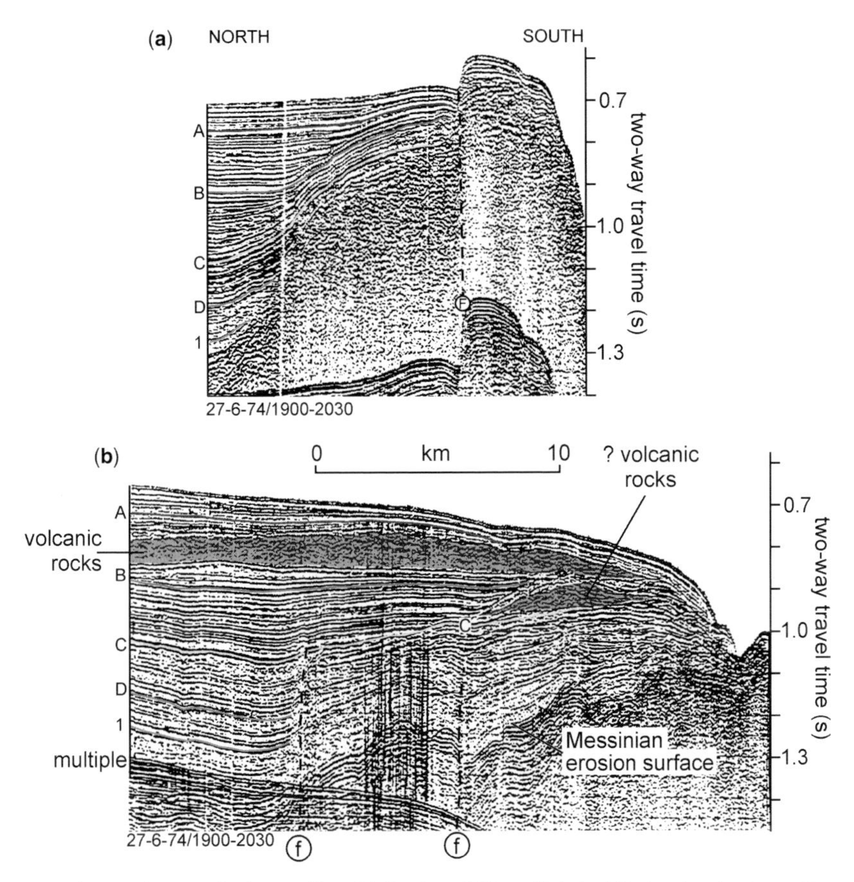

Fig. 3. *Meteor* airgun seismic reflection profiles. (**a**) Southern Milos–Christiani Basin, showing key reflectors of Anastasakis & Piper (2005). (**b**) West of Christiani, showing stratigraphic position of volcanic units.

deposition of mid- to late Pleistocene sediments (units 'A2' and 'A1' of Piper & Perissoratis 2003). Widespread basin inversion has taken place since the early Pleistocene and the modern basins are bounded principally by ENE–WSW trending faults (Piper & Perissoratis 2003). Seismicity suggests modern NW–SE extension on these faults, within a dextral transtensional zone (Bohnhoff *et al.* 2005, 2006).

Interpretation of seismic reflection profiles

Volcanic horizons in the Meteor *airgun profiles*

Further correlation to the central part of the profile in Figure 4, off NW Santorini, is also possible. This shows a volcanic unit between reflections A and B and a second horizon about one-third of the way down from B to C, but any deeper occurrences are obscured by the sea-bed multiple. A thick development of volcanic rocks west of Santorini, on the basis of overlying sediment thickness, corresponds to the unit between A and B. West of Santorini, ponded incoherent deposits stratigraphically above the volcanic rocks, but mostly below A, are either pyroclastic flows or mass-transport deposits, interbedded with stratified marine sediment.

Basin subsidence

The style of subsidence is seen clearly at the northern margin of the Milos–Christiani Basin south of Folegandros, which is imaged in three almost coincident profiles collected with different systems: a *Meteor* profile (Fig. 4), an IGME sparker profile (Fig. 5), and the multichannel seismic profile (Fig. 2). Beneath the modern shelf are stacked prograded shelf sediment units and the shelf edge is marked by one or more major active faults (Figs 4 and 5). Correlation to adjacent seismic reflection profiles (e.g. Piper & Perissoratis 2003, fig. 10;

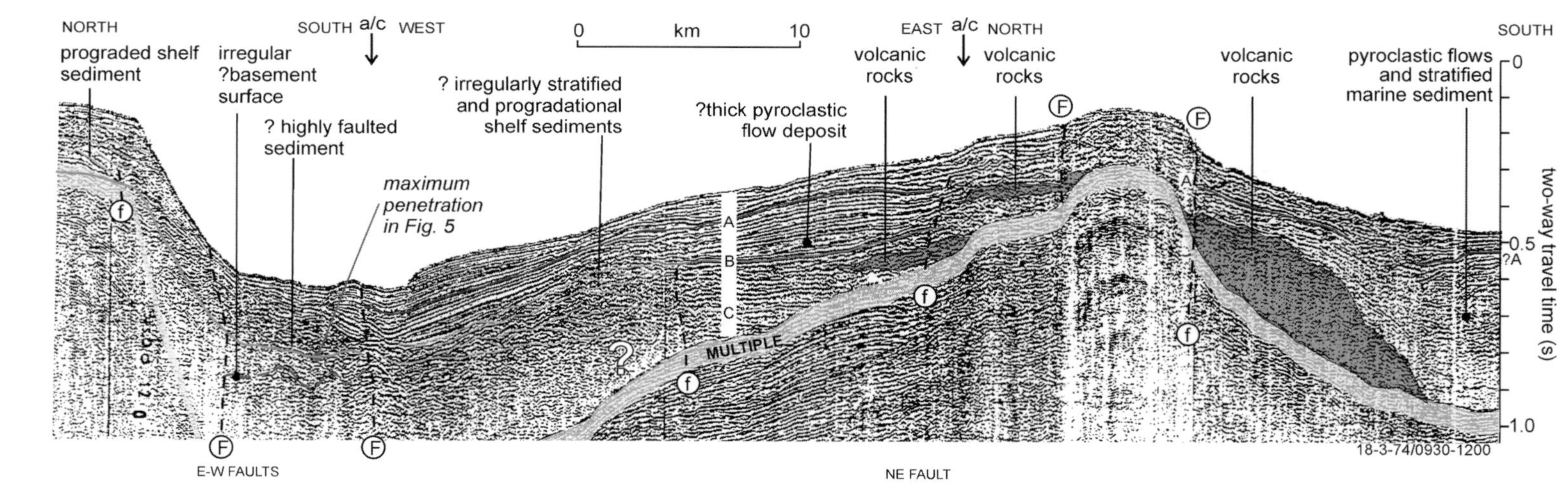

Fig. 4. *Meteor* airgun seismic reflection profile from NW of Santorini. ? indicates uncertain interpretation. f, inactive fault; F, active fault; a/c, alter course. A–C are key reflectors of Anastasakis & Piper (2005).

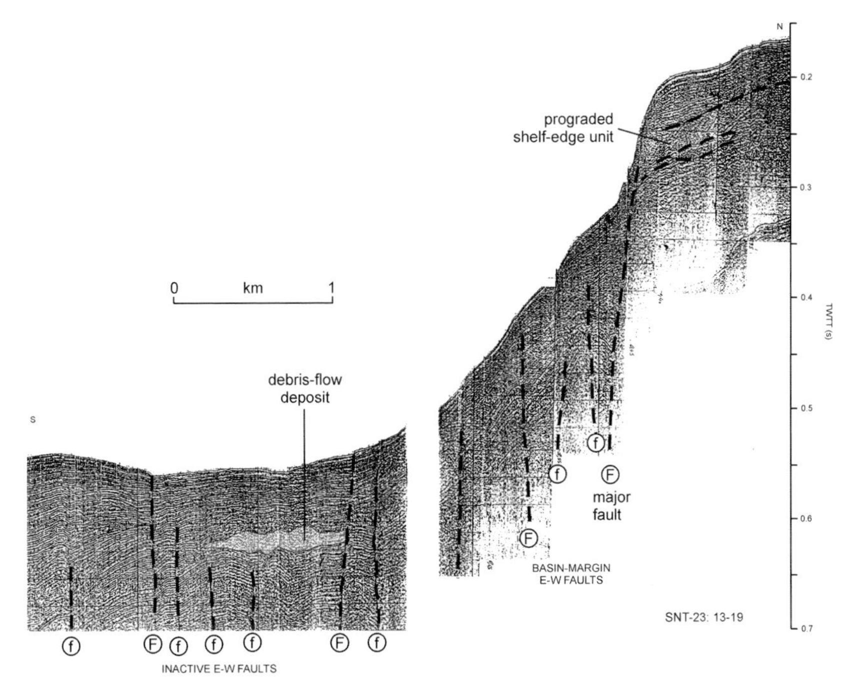

Fig. 5. IGME sparker profile corresponding to the northwestern end of the profile in Figure 4. TWTT, two-way travel time.

and other unillustrated profiles located in Fig. 1b) and bathymetric trends clearly show that this active fault trends east–west. To the south, well-stratified basinal marine sediments are cut by numerous minor faults, many of which do not affect the uppermost strata (also illustrated by Piper & Perissoratis 2003, fig. 10A and B) and these faults also appear to have an east–west trend. Farther to the SE, unfaulted well-stratified basinal marine sediments above regional reflection B overlie a unit that consists of a stacked succession of planar erosion surfaces and eastward-prograding clinoforms, all tilted gently westward (Fig. 4). Following Piper & Perissoratis (2003), the inflection point of the clinoforms is interpreted as being close to sea level at lowstands. Thus using the age estimate of 1 Ma for reflection B implies a mean subsidence rate of 0.5 mm a^{-1} since the youngest progradational package.

Fault patterns

The dense grid of seismic reflection profiles in the Milos–Christiani Basin (Piper & Perissoratis 2003; Anastasakis & Piper 2005) shows that many faults trend east–west. These include some that affect only older sediment and now appear inactive (e.g. Figs 3b and 5), but a major active fault defines the northern margin of the Milos–Christiani Basin (Figs 4–6).

In contrast, to the east of Santorini, at least two prominent fault sets are present. Faults trending NE–SW (N40°E to N55°E) are widespread to the north and east of Santorini, defining many of the major bathymetric features (Fig. 6a). Figure 6b illustrates how sea-floor offsets resulting from active faulting can be measured on closely spaced seismic reflection profiles to correlate large faults from profile to profile and hence determine their orientation. This technique assumes that faults showing sea-floor offsets of tens to hundreds of metres are correlatable over horizontal distances of 10–20 km, which is a reasonable assumption based on mapping on land. For example, the prominent fault immediately SE of Ios is defined from sea-floor scarps in six seismic reflection profiles and has a sea-floor offset greater than 100 m in its central part. The same trend continues to the NE, where fault offsets are only 25 m. To the SW, the surface expression of the fault disappears, but the trend continues into the buried fault, as illustrated in Figure 4 (labelled NE fault), to the NW of Santorini. At Santorini, linear trends of volcanic vents have been termed the Kameni and Columbos lines (Heiken & McCoy 1984) and they are also on the NE–SW trend (Fig. 6b).

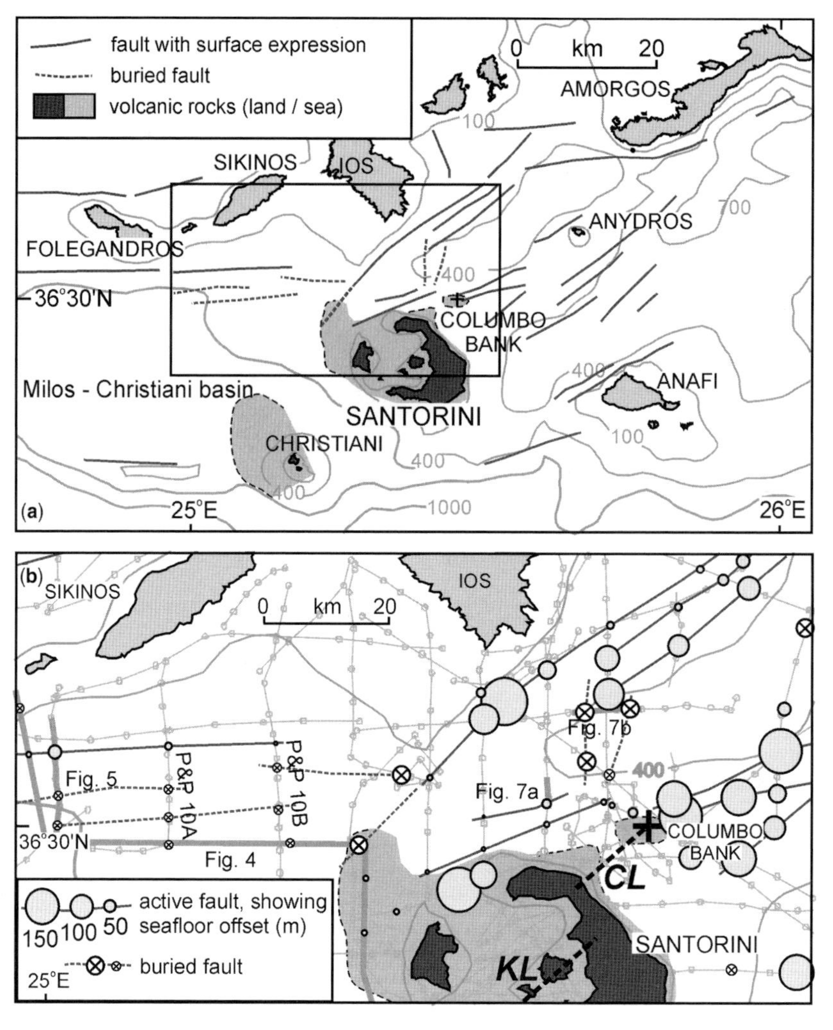

Fig. 6. (**a**) Interpretative map of major faults and submarine volcanic rocks (based on seismic reflection profiles located in Fig. 1 and bathymetry). (**b**) Detail of area north of Santorini showing how fault orientations were determined from closely spaced seismic reflection profiles. CL, Columbos line; KL, Kammeni line. P&P refers to Piper & Perissoratis (2003).

A second set of faults trends ENE–WSW (N65°E to N75°E). Piper & Perissoratis (2003) demonstrated that faults on this trend define the shelf south of Amorgos and north of Anafi, the former being seismically active (Hatzfeld 1999). Faults on this trend are mapped from closely spaced seismic lines north of Santorini and Columbo Bank (Figs 6b and 7a) and SE of Santorini (Fig. 6a). Faults with a similar orientation are common farther east, between Amorgos, Astypalea and Kos (Piper & Perissoratis 2003; Pe-Piper *et al.* 2005; Bohnhoff *et al.* 2006).

Locally, older faults on a north–south trend can be mapped from closely spaced seismic lines north of Santorini (Figs 6a and 7b). Piper & Perissoratis (2003, figs 13A and 15) identified common buried inactive faults trending east–west in the area between Santorini and Kos.

Interpretation

Using the Milos chronology of Anastasakis & Piper (2005), the volcanic horizons off Santorini correlate with the younger Akrotiri (0.65–0.55 Ma) and older Akrotiri (1.6 ka) volcanic episodes respectively (Fig. 8). The upper unit of volcanic rocks off Christiani is of similar age to the younger Akrotiri

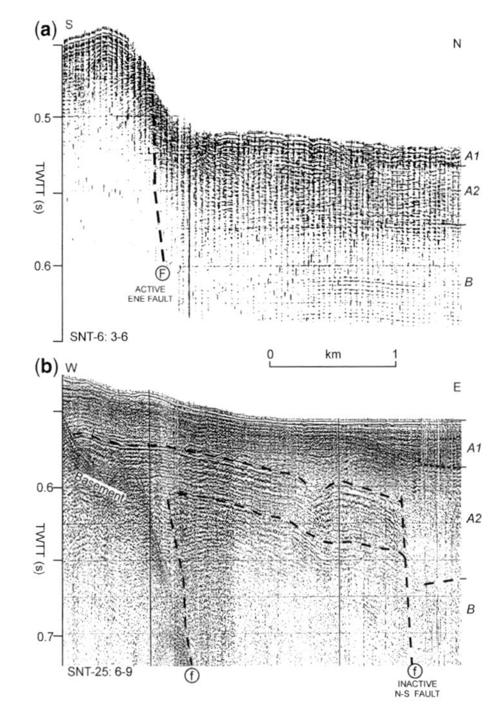

Fig. 7. IGME sparker profiles showing character of active and inactive faults north of Santorini. Stratigraphic units A1, A2 and B from Piper & Perissoratis (2003) and Fig. 1b are shown in Figure 8.

episode, whereas the lower unit appears deeper in the stratigraphy than the lower volcanic rocks at Santorini, close to reflection C, and is thus likely to be of latest Pliocene age. Volcanic rocks of similar age are found in the western part of the Milos–Christiani Basin. The volcanic rocks identified in the seismic reflection profiles were deposited in an entirely submarine environment and are onlapped by well-stratified basinal marine sediment. Their extent shown in Figure 6 is based on interpretations of both *Meteor* and IGME seismic profiles.

West of Santorini, there is no evidence in our seismic reflection profiles for thick pyroclastic deposits derived from the major felsic eruptions of the Thera Pyroclastic Formation over the past 250 ka. The clearest pyroclastic units are below regional reflection A and thus predate the Thera Pyroclastic Formation: they are 20–30 ms thick, close to the limits of resolution of the *Meteor* seismic profiles. Thick marine pyroclastic deposits of the Thera Pyroclastic Formation lie to the south of Santorini (Sparks *et al.* 1983; Anastasakis 2007).

This chronology allows us to propose a tectonic model for the setting of Santorini. Observations on sedimentary basins from high-resolution seismic reflection profiles provide evidence for when particular faults were active, but do not allow the role of inherited basement structures to be assessed. In the late Pliocene, the east–west-trending

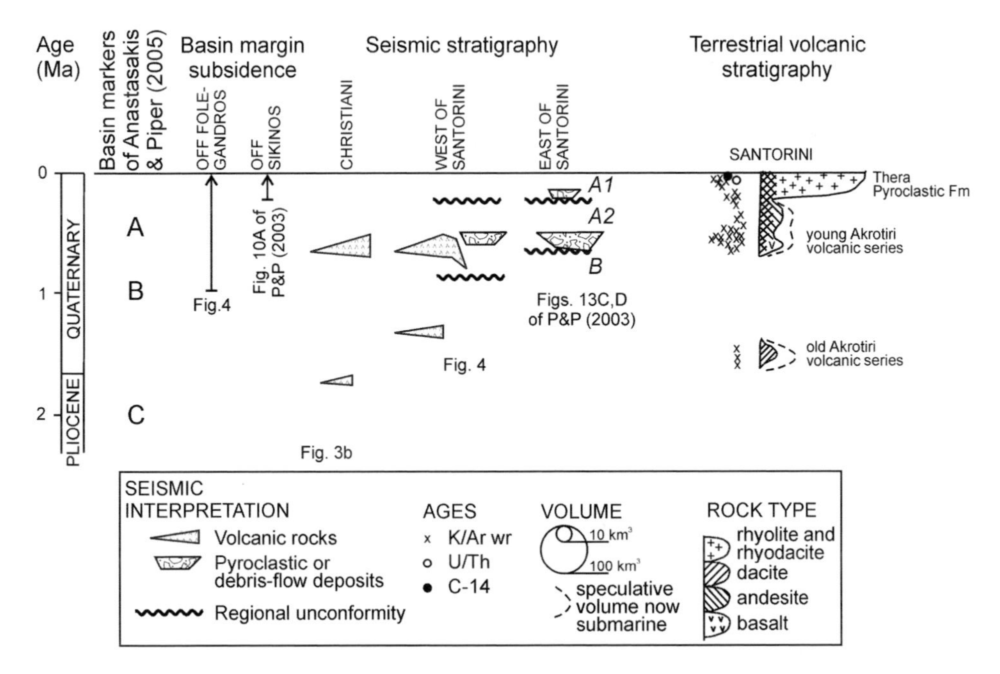

Fig. 8. Chronology of features around Santorini discussed in text. Terrestrial volcanic stratigraphy from compilation by Pe-Piper & Piper (2002, fig. 262). P&P (2003) refers to Piper & Perissoratis (2003).

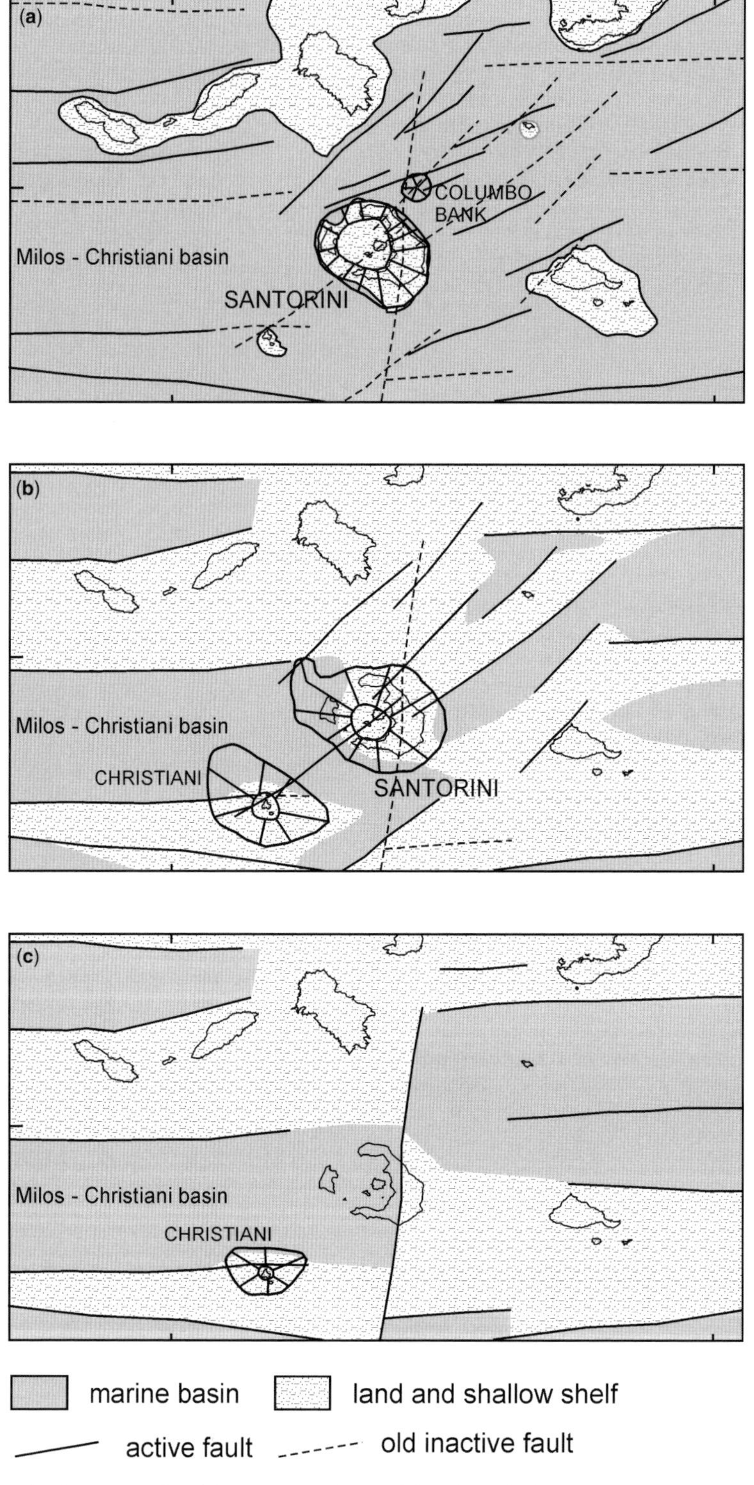

Fig. 9. Interpreted palaeogeography of the area around Santorini. (**a**) Late Pleistocene; (**b**) Mid-Pleistocene; (**c**) Late Pliocene. Faults and sediment facies in part from Piper & Perissoratis (2003).

Milos–Christiani Basin terminated eastward against an Anafi–Santorini horst (Fig. 4) and to the west, basinal sediments terminated against an Ios horst along a north–south fault (Piper & Perissoratis 2003) (Fig. 7b). Major north–south faults are widespread in basement rocks of the eastern Cyclades and controlled the emplacement of Miocene plutonic rocks (Pe-Piper *et al.* 2002). The distribution of basins east of Santorini is poorly constrained, but Piper & Perissoratis (2003) showed evidence that east–west faults were important at this time, so we tentatively show basin margins as trending east–west in Figure 9c. The Mid-Pleistocene (Fig. 9b) and Late Pleistocene reconstructions are based on data presented in this paper together with the detailed interpretation of the Anafi–Amorgos area presented by Piper & Perissoratis (2003, figs 13 and 14). There was widespread basin inversion in the Mid-Pleistocene and rapid subsidence of several horsts in the Late Pleistocene.

It is beyond the scope of this paper to provide a geodynamic explanation for the sequence of fault patterns seen in the Santorini area. Rather, our intent is to relate the observed fault pattern to the distribution of volcanic rocks. Farther east in the volcanic arc, the explosive volcanism of the Kos Plateau Tuff eruption has been argued to be the result of sinistral ENE–WSW-trending strike-slip faults causing extension on older north–south- or NE–SW-trending faults, thus allowing upward percolation of magma and downward percolation of water (Pe-Piper *et al.* 2005). In contrast, east of Santorini, ENE–WSW-trending faults are interpreted as dextral strike-slip on the basis of seismicity (Bohnhoff *et al.* 2006). They show major sea-floor offsets near the young volcanic rocks of Columbo Bank. Volcanism itself appears localized along NE–SW-trending faults (Fig. 9), orthogonal to the regional stress direction of SE–NW-directed extension inferred from geodetic measurements (Nyst & Thatcher 2004) and earthquake focal mechanisms (Bohnhoff *et al.* 2005). These faults appear to have been initiated in the early Quaternary as regional crustal-scale strike-slip faults (Piper & Perissoratis 2003) and thus protentially provide pathways for base-of-crust magma. The older volcanism at Santorini and Christiani predates the ENE–WSW fault system (Fig. 9a), which was initiated at about 0.2 Ma (Piper & Perissoratis 2003). This older volcanism may have been localized by the intersection of north–south-trending transfer faults with early Quaternary NE–SW-trending sinistral strike-slip faults (Fig. 9b and c).

On a wider scale in the South Aegean arc, faulting appears to be important in localizing the distribution of volcanoes. Magma may be generated along the length of the subduction zone, but volcanic activity is localized (Fig. 1). As discussed above, volcanism at Kos and Nisyros is strongly influenced by intersecting faults (Pe-Piper *et al.* 2005). To the west of Santorini, Milos is the most voluminous volcanic centre of the arc, although a large part of the volcanic products has now subsided beneath the sea (Anastasakis & Piper 2005). It is located at the intersection of a north–south fault that bounds the late Miocene Myrtoon basin (Anastasakis *et al.* 2006) with the major NE–SW-trending mid-Cycladic lineament, which was the strike-slip boundary between two crustal blocks in the middle Miocene (Walcott & White 1998) and was reactivated near Milos in the Early Pleistocene (Fytikas & Vougioukalakis 1993). Renewed fault activity on a north–south trend has taken place in the Mid- to Late Quaternary (Piper & Perissoratis 2003, fig. 9). Fault control of the volcanic centre of Methana is the subject of current study by the authors.

Implications

The main phase of lava eruption at Santorini (namely, the younger Akrotiri volcanic unit and the Peristeria stratovolcano) is represented by submarine lava flows over a large area NW of Santorini. Minor flows are also recognized equivalent to the older Akrotiri volcanic unit. The main volcanism at Christiani was synchronous with the younger Akrotiri volcanic episode, with minor earlier volcanism in the latest Pliocene.

Volcanism at Santorini is a result of changing patterns of active faulting. NE–SW-trending strike-slip faulting was initiated in the latest Pliocene or early Quaternary and triggered the Akrotiri volcanism of Santorini and the main volcanism at Christiani. These faults then acted as extensional pathways for magma and downward percolation of water with further change to the modern stress regime in the mid-Quaternary. Substantial parts of the volcanic edifices of the central south Aegean arc lie beneath the sea.

We thank the staff of the Marine Geology Department of IGME for their help during the collection of the IGME sparker data. These field cruises were partly funded by the research programmes Marine Science and Technology (MAST I and II) and Environment of the XII Directorate of Science Research and Technology of the European Commission. G.P.P. is funded by an NSERC Discovery Grant. Thorough reviews by N. Güleç, J. ten Veen, J. Woodside, I. Koukouvelas and T. Taymaz resulted in substantial improvement to the documentation and interpretation.

References

AKSU, A. E., HALL, J. & YALTIRAK, C. 2005. Miocene to Recent tectonic evolution of the eastern

Mediterranean: new pieces of the old Mediterranean puzzle. *Marine Geology*, **221**, 1–13.

ANASTASAKIS, G. 2007. The anatomy and provenance of thick volcaniclastic flows in the Cretan Basin, South Aegean Sea. *Marine Geology* **240**, 113–135.

ANASTASAKIS, G. & PIPER, D. J. W. 2005. Late Neogene evolution of the western South Aegean volcanic arc: sedimentary imprint of volcanicity around Milos. *Marine Geology*, **215**, 135–158.

ANASTASAKIS, G., PIPER, D. J. W., DERMITZAKIS, M. D. & KARAKITSIOS, V. 2006. Upper Cenozoic stratigraphy and paleogeographic evolution of Myrtoon and adjacent basins, Aegean Sea, Greece. *Marine and Petroleum Geology*, **23**, 353–369.

ARMIJO, R., LYON-CAEN, H. & PAPANASTASSIOU, D. 1992. East–west extension and Holocene normal-fault scarps in the Hellenic arc. *Geology*, **20**, 491–494.

BARTOLE, R., CATANI, G., LENARDON, G. & VINCI, A. 1983. Tectonics and sedimentation of the southern Aegean Sea. *Bollettino Oceanologia Teorica ed Applicata*, **1**, 319–340.

BOHNHOFF, M., MAKRIS, J., PAPANIKOLAOU, D. & STAVRAKAKIS, G. 2001. Crustal investigation of the Hellenic subduction zone using wide-aperture seismic data. *Tectonophysics*, **343**, 239–262.

BOHNHOFF, M., HARJES, H.-P. & MEIER, T. 2005. Deformation and stress regimes in the Hellenic subduction zone from focal mechanisms. *Journal of Seismology*, **9**, 341–366.

BOHNHOFF, M., RISCHE, M., MEIER, T., BECKER, D., STAVRAKAKIS, G. & HARJES, H.-P. 2006. Microseismic activity in the Hellenic Volcanic Arc, Greece, with emphasis on the seismotectonic setting of the Santorini–Amorgos zone. *Tectonophysics*, **423**, 17–33.

COLLIER, R. E. LL. & DART, C. J. 1991. Neogene to Quaternary rifting, sedimentation and uplift in the Corinth Basin, Greece. *Journal of the Geological Society, London*, **148**, 1049–1065.

DRUITT, T. H., EDWARDS, L., MELLORS, R. M. *ET AL.* (eds) 1999. *Santorini Volcano*. Geological Society, London, Memoirs, **19**.

FYTIKAS, M. & VOUGIOUKALAKIS, G. 1993. Volcanic structure and evolution of Kimolos and Polyegos (Milos island group). *Bulletin of the Geological Society of Greece*, **28**, 221–237 [in Greek with English abstract].

HATZFELD, D. 1999. The present-day tectonics of the Aegean as deduced from seismicity. *In*: DURAND, B., JOLIVET, L., HORVÁTH, F. & SERANNE, M. (eds) *The Mediterranean Basins: Tertiary Extension within the Alpine Orogen*. Geological Society, London, Special Publications, **156**, 416–426.

HEIKEN, G. & MCCOY, F., JR. 1984. Caldera development during the Minoan eruption, Thira, Cyclades, Greece. *Journal of Geophysical Research*, **89**, 8441–8462.

HSÜ, K. J., MONTADERT, L. *ET AL.* 1978. Site 378, Cretan Basin. *In*: *Initial Reports of the Deep Sea Drilling Project 42(1)*. US Government Printing Office, Washington, DC, 321–357.

JONGSMA, D., WISSMAN, G., HINZ, K. & GARDÉ, S. 1977. Seismic studies in the Cretan Sea. *Meteor Forschungs-Ergebnisse*, **C27**, 3–30.

KOKKALAS, S. & DOUTSOS, T. 2001. Strain dependent stress field and plate motions in the south-east Aegean region. *Journal of Geodynamics*, **32**, 311–332.

KOKKALAS, S., XYPOLIAS, P., KOUKOUVELAS, I. & DOUTSOS, T. 2006. Postcollisional contractional and extensional deformation in the Aegean region. *In*: DILEK, Y. & PAVLIDES, S. (eds) *Post-collisional Tectonism and Magmatism in the Mediterranean Region and Asia*. Geological Society of America, Special Papers, **409**, 97–123.

MARTIN, L. 1987. *Structure et évolution récent de la mer Egée: rapports d'une étude par sismique réflexion*. PhD thesis, Université Paris 6.

MASCLE, J. & MARTIN, L. 1990. Shallow structure and recent evolution of the Aegean Sea: a synthesis based on continuous seismic reflection profiles. *Marine Geology*, **94**, 271–299.

NYST, M. & THATCHER, W. 2004. New constraints on the active tectonic deformation of the Aegean. *Journal of Geophysical Research*, **109**, B11406, doi:10.1029/2003JB002830.

PAPANIKOLAOU, D., LYKOUSIS, V., CHRONIS, G. & PAVLAKIS, P. 1988. A comparative study of neotectonic basins across the Hellenic arc: the Messiniakos, Argolikos, Saronikos and southern Evoikos gulfs. *Basin Research*, **1**, 167–176.

PE-PIPER, G. & PIPER, D. J. W. 2002. *The Igneous Rocks of Greece*. Borntraeger, Stuttgart.

PE-PIPER, G. & PIPER, D. J. W. 2005. The South Aegean active volcanic arc: relationships between magmatism and tectonics. *Developments in Volcanology*, **7**, 113–133.

PE-PIPER, G., PIPER, D. J. W. & MATARANGAS, D. 2002. Regional implications of geochemistry and style of emplacement of Miocene I-type diorite and granite, Delos, Cyclades, Greece. *Lithos*, **60**, 47–66.

PE-PIPER, G., PIPER, D. J. W. & PERISSORATIS, C. 2005. Neotectonics and the Kos Plateau Tuff eruption of 161 ka, South Aegean arc. *Journal of Volcanology and Geothermal Research*, **139**, 315–338.

PERISSORATIS, C. 1995. The Santorini volcanic complex and its relationship to the stratigraphy and structure of the Aegean arc, Greece. *Marine Geology*, **128**, 37–58.

PIPER, D. J. W. & PERISSORATIS, C. 2003. Quaternary neotectonics of the South Aegean arc. *Marine Geology*, **198**, 259–288.

PURVIS, M. & ROBERTSON, A. 2004. A pulsed extension model for the Neogene–Recent E–W-trending Alaşehir Graben and the NE–SW-trending Seledi and Gördes Basins, western Turkey. *Tectonophysics*, **391**, 171–201.

REILINGER, R. E., MCCLUSKY, S. C., ORAL, M. B. *ET AL.* 1997. Global Position System measurements of present-day crustal movements in Arabia–Africa–Eurasia plate collision zone. *Journal of Geophysical Research*, **102**, 9983–9999.

SKENE, K. I., PIPER, D. J. W., AKSU, A. E. & SYVITSKI, J. P. 1998. Evaluation of the global oxygen isotope curve as a proxy for Quaternary sea level by modelling of delta progradation. *Journal of Sedimentary Research*, **68**, 1077–1092.

SPARKS, R. S. J., BRAZIER, S., HUANG, T. C. & MUERDTER, D. 1983. Sedimentology of the Minoan deep-sea tephra layer in the Aegean and eastern Mediterranean. *Marine Geology*, **54**, 131–167.

TAYMAZ, T. 1996. S–P-wave traveltime residuals from earthquakes and lateral inhomogeneity in the upper mantle beneath the Aegean and the Hellenic Trench near Crete. *Geophysical Journal International*, **127**, 545–558.

TAYMAZ, T., JACKSON, J. & WESTAWAY, R. 1990. Earthquake mechanisms in the Hellenic Trench near Crete. *Geophysical Journal International*, **102**, 695–731.

TAYMAZ, T., JACKSON, J. & MCKENZIE, D. 1991. Active tectonics of the north central Aegean Sea. *Geophysical Journal International*, **106**, 433–490.

TAYMAZ, T., WESTAWAY, R. & REILINGER, R. 2004. Active faulting and crustal deformation in the Eastern Mediterranean region. *Tectonophysics*, **391**, 1–9.

TEN VEEN, J. H. & KLEINSPEHN, K. L. 2002. Geodynamics along an increasingly curved convergent plate boundary: late Miocene–Pleistocene Rhodes, Greece. *Tectonics*, **21**, doi: 10.1029/2001TC001287.

TEN VEEN, J. H. & KLEINSPEHN, K. L. 2003. Incipient continental collision and plate-boundary curvature: Late Pliocene–Holocene transtensional Hellenic forearc, Crete, Greece. *Journal of the Geological Society, London*, **160**, 161–181.

TEN VEEN, J. H. & MEIJER, P. TH. 1998. Late Miocene to Recent tectonic evolution of Crete (Greece): geological observations and model analysis. *Tectonophysics*, **298**, 191–208.

ULUĞ, A., DURMAN, M., ERSOY, S., ÖZEL, E. & AVCI, M. 2005. Late Quaternary sea-level change, sedimentation and neotectonics of the Gulf of Gökova, southeastern Aegean Sea. *Marine Geology*, **221**, 381–395.

VESPA, M., KELLER, J. & GERTISSER, R. 2006. Interplinian explosive activity of Santorini volcano (Greece) during the past 150,000 years. *Journal of Volcanology and Geothermal Research*, **153**, 262–286.

WALCOTT, C. R. & WHITE, S. H. 1998. Constraints on the kinematics of post-orogenic extension imposed by stretching lineations in the Aegean region. *Tectonophysics*, **298**, 155–175.

From syn- to post-orogenic Tertiary extension in the north Aegean region: constraints on the kinematics in the eastern Rhodope–Thrace, Bulgaria–Greece and the Biga Peninsula, NW Turkey

N. BONEV[1] & L. BECCALETTO[2]

[1]*Department of Geology and Palaeontology, Sofia University 'St. Kliment Ohridski', 1504 Sofia, Bulgaria (e-mail: niki@gea.uni-sofia.bg)*

[2]*Institute of Geology and Palaeontology, University of Lausanne, Humense Building, CH-1015 Lausanne, Switzerland*

Abstract: The Aegean region experienced back-arc extension related to the Hellenic subduction system at least from the latest Oligocene to the present. We document Tertiary extension-related kinematics in the north Aegean, in the eastern Rhodope–Thrace of Bulgaria–Greece and the Biga Peninsula of NW Turkey. A regionally consistent NNE–SSW- to NE–SW-oriented kinematic direction, delineated in both areas by stretching lineations and associated ductile–brittle shear fabrics in exhumed metamorphic domes beneath detachments, suggests that they were kinematically coupled during the Tertiary extension. This kinematic framework, combined with regional geochronological data and the stratigraphic record in hanging-wall supradetachment basins, defines an extensional history that includes syn- and post-orogenic episodes from Paleocene to Miocene times. Paleocene–early Eocene synorogenic extension in the Kemer micaschists of the northern Biga Peninsula and in the Kesebir–Kardamos dome in Rhodope–Thrace accommodated gravitationally induced hinterland-directed exhumation of the orogenic stack, coeval with the closure of the Vardar Ocean. Then, following collision within the region, it was succeeded by latest Oligocene–Early Miocene extension as recorded in the Kazdağ Massif in the southern Biga Peninsula, which overlaps the Aegean back-arc post-orogenic extension, widely recognized in the central Aegean and southern Greek Rhodope. The protracted record of extension is interpreted to reflect progressive exhumation of the orogenic wedge along the Eurasian plate margin. Southward migration of extension and magmatism across the study areas accounts for sequential shift and roll-back of the subduction boundary at that margin, from the latest Cretaceous in the Rhodope to its present position at the Hellenic trench. The results allow recognition of the investigated areas as an important extensional domain in the north Aegean region, which underwent Tertiary syn- and post-orogenic extension.

The Aegean region has experienced back-arc extension related to the Hellenic subduction system since the latest Oligocene to the present (e.g. McKenzie 1978; Le Pichon & Angelier 1979; Meulenkamp *et al.* 1988). Seismic and geodetic data indicate that this active tectonic regime developed above the northward subducting East Mediterranean slab, plunging under the Eurasian plate (Taymaz *et al.* 1991; Jackson 1994; Le Pichon *et al.* 1995; McClusky *et al.* 2000).

Since the first recognition of metamorphic core complexes of cordilleran-type in the Cyclades (Lister *et al.* 1984), the Aegean region has been considered as a natural laboratory for studying processes of crustal extension and exhumation of metamorphic terranes (Jolivet & Patriat 1999; Jolivet & Faccenna 2000). Many studies of metamorphic core complexes in the Aegean region and its surroundings (e.g. south Rhodope, Cyclades and Menderes Massif) have documented pronounced latest Oligocene–Miocene extension (e.g. Dinter & Royden 1993; Sokoutis *et al.* 1993; Bozkurt & Park 1994; Gautier & Brun 1994; Hetzel *et al.* 1995). The onset of the Aegean extension in the Balkan area and within the Rhodope complex was originally set in the Late Eocene–Oligocene from stratigraphic constraints (Burchfiel *et al.* 2003), or in the latest Early Oligocene from radiometric dating (Lips *et al.* 2000). However, the extensional regime in the Rhodope complex obviously started earlier, in pre-Eocene times (Bonev *et al.* 2006*a*), implying a protracted Cenozoic extensional history of the north Aegean domain.

Ductile shear fabrics associated with stretching lineations in the footwall mylonites beneath detachments were used in the central Aegean region both to establish the kinematic directions of extension (e.g. Gautier & Brun 1994; Vandenberg & Lister 1996;

From: Taymaz, T., Yilmaz, Y. & Dilek, Y. (eds) *The Geodynamics of the Aegean and Anatolia.* Geological Society, London, Special Publications, **291**, 113–142.
DOI: 10.1144/SP291.6 0305-8719/07/$15.00

Bozkurt & Park 1997; Ring *et al.* 1999), and in comparison with the present deformation (Jolivet *et al.* 1994; Walcott & White 1998; Jolivet 2001). Generally, a dominant north–south to NE–SW direction of extension characterizes the central–southern Aegean region (Fig. 1). In contrast, only few kinematic data on the direction of ductile extension are available for the northern Aegean region, when compared with extensive studies on fault kinematics and basin patterns (e.g. Dinter & Royden 1993; Bozkurt 2001; Koukouvelas & Aydin 2002).

In this paper, we therefore present a regional kinematic study of exhumed basement rocks in the northernmost Aegean region. The key investigated areas are the eastern Rhodope–Thrace region of the Rhodope Massif of southern Bulgaria and northeastern Greece, and the Biga Peninsula of NW Turkey (Fig. 2). In the eastern Rhodope–Thrace region, the investigation combines kinematics of large-scale extensional domes in Bulgaria with additional data from Greece used to support the kinematic frame. In the Biga Peninsula, the study encompasses a northern domain along the Marmara Sea coast with previously undocumented kinematics, and the southern domain of the Kazdağ Massif, where further details on the kinematics are provided. We summarize the Tertiary kinematics in both areas, focusing on ductile then brittle extensional structures, to constrain the poorly known extension-related kinematic framework in this part of the Aegean domain. Finally, based on the available temporal constraints on the distinct processes involved, we discuss the kinematic frame and geodynamic context of the extensional tectonics, which collectively delineate episodes of syn- and post-orogenic extension.

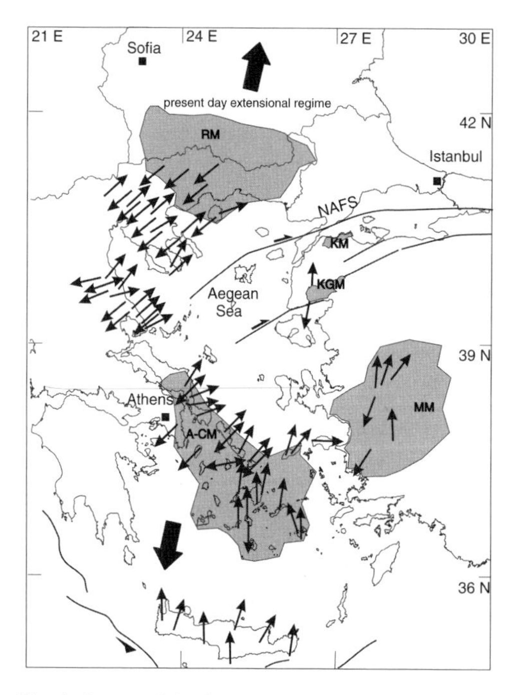

Fig. 1. Senses of ductile shear of Oligocene–Miocene age within the exhumed metamorphic basement in the Aegean region. Data sources: Dinter & Royden (1993); Schermer (1993); Gautier & Brun (1994); Jolivet *et al.* (1994); Hetzel *et al.* (1995); Bozkurt & Park (1997); Walcott & White (1998); Wawrzenitz & Krohe (1998); Kilias *et al.* (1999); Ring *et al.* (1999); Lips *et al.* (2000); Okay & Satır (2000*b*); Beccaletto & Steiner (2005). RM, Rhodope Massif; A-CM, Attica–Cycladic Massif; MM, Menderes Massif; KGM, Kazdağ Massif; KM, Kemer micaschists; NAFS, North Anatolian Fault System.

Large-scale tectonic setting and geological background of the north Aegean region

Belonging to the Alpine–Himalayan mountain chains, the Rhodope Massif and the Biga Peninsula lie in the northernmost sector of the Aegean region, in the Eastern Mediterranean domain (Fig. 2). The orogenic belt exposed in this region results from the Mesozoic–Cenozoic closure of several oceanic basins belonging to the Tethyan realm, and the subsequent collision between the bordering microcontinents (e.g. Robertson & Dixon 1984; Dercourt *et al.* 1986; Robertson *et al.* 1996; Stampfli & Borel 2004). Both areas occur immediately north of the main regional suture zones, namely, the Vardar and the Izmir–Ankara sutures, and the Intra-Pontide suture propagates through the Biga Peninsula (e.g. Okay & Tüysüz 1999; Stampfli 2000).

The Vardar suture zone separates the Rhodope Massif and Serbo-Macedonian massifs in the north, from the inner Hellenides in the south. The Rhodope Massif is limited in the north by the Late Alpine Maritza dextral strike-slip fault, against the Late Cretaceous volcanic arc of the Sredna Gora Zone. It consists of a metamorphic basement comprising pre-Alpine and Alpine units of continental and oceanic affinities, derived from magmatic and sedimentary protoliths. This metamorphic basement is intruded by Late Cretaceous to Early Miocene granitoids (e.g. Meyer 1968; Soldatos & Christofides 1986; Del Moro *et al.* 1988; Dinter *et al.* 1995; Peytcheva *et al.* 1998; Pe-Piper & Piper 2002), and covered by Late Cretaceous to Neogene sedimentary and volcanic sequences (e.g. Ivanov & Kopp 1969; Boyanov & Goranov 2001), including Late Eocene–Oligocene volcanic and volcano-sedimentary successions (e.g. Innocenti *et al.* 1984; Harkovska *et al.* 1989; Yanev & Bardintzeff 1997). The Rhodope metamorphic complex was formed as a stack of crustal-scale ductile nappes during the Late

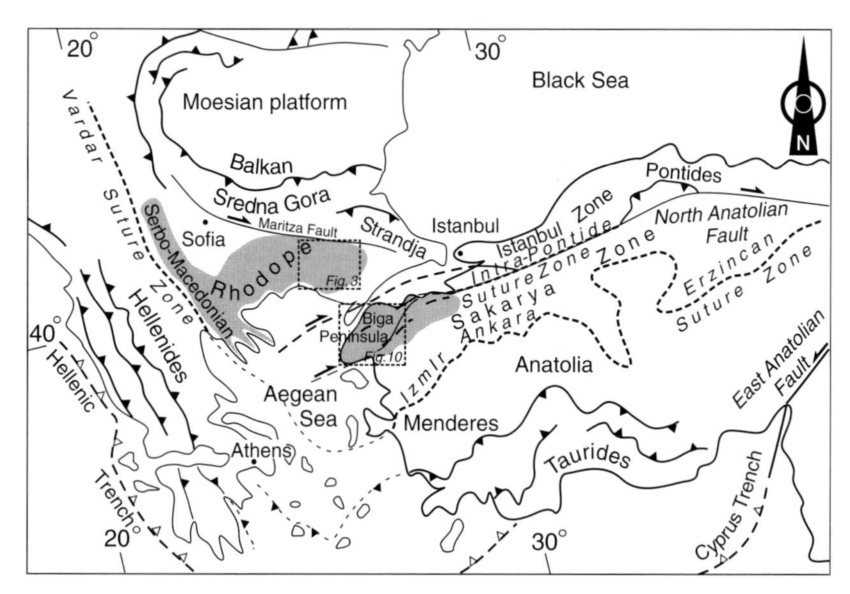

Fig. 2. Synthetic sketch map showing the tectonic elements of the Alpine collisional system in the Eastern Mediterranean, and location of the Rhodope Massif and the Biga Peninsula in the framework of the Aegean region. Data sources: NAFS after Lyberis (1984), Taymaz *et al.* (1991) and Şengör *et al.* (1995); Tethyan sutures and tectonic units in Turkey after Okay and Tüysüs (1999); Hellenic–Balkan area after Bonev (2006).

Cretaceous–Early Tertiary subduction and closure of the Vardar Ocean (Koukouvelas & Doutsos 1990; Burg *et al.* 1996). The nappes were assembled by southward synmetamorphic thrusting in the hanging wall of the north-dipping subduction of the Vardar slab, rooted in the present-day Vardar suture zone (Koukouvelas & Doutsos 1990; Ricou *et al.* 1998). Regionally consistent NNE–SSW-trending stretching lineations, associated with top-to-the-SSW ductile shearing coeval with amphibolite-facies metamorphism, were interpreted as the transport direction during synmetamorphic nappe stacking (Burg *et al.* 1996). Opposite directions of tectonic transport (i.e. towards ENE and NNE), also documented in some thrust sheets, were probably related to coeval or subsequent extensional deformation (Burg *et al.* 1996; Kilias *et al.* 1999). However, Krohe and Mposkos (2002) later argued, mainly on petrological grounds, that the predominantly SSW-directed shearing and kinematic patterns do not reflect progressive thrusting, but were probably associated with extensional deformations.

In NW Turkey, the Biga Peninsula is bounded on its eastern side by the westernmost end of the Sakarya Zone (e.g. Şengör & Yılmaz 1981; Okay *et al.* 1991; Bozkurt & Mittwede 2001); (Fig. 2). The latter is bordered northwestward by the Istanbul and Rhodope–Strandja zones along the Intra-Pontide suture (Okay & Tüysüz 1999). To the south, the Izmir–Ankara–Erzincan suture zone separates the Sakarya Zone from the composite Anatolide–Tauride block. The Sakarya Zone represents a continental fragment consisting in its western part of a Variscan basement of granitic and metamorphic rocks, and widespread occurrences of the Permo-Triassic Karakaya Complex (Bingöl *et al.* 1975). The Karakaya Complex is unconformably overlain by various sedimentary and volcano-sedimentary units ranging in age from Jurassic to Paleocene (Altiner *et al.* 1991; Royaj & Altiner 1998). Early Eocene to Pliocene–Quaternary sediments and volcanic rocks (Sıyako *et al.* 1989) represent younger cover sequences. Late Eocene–Miocene volcanic activity was accompanied by the intrusion of numerous granitoids, mainly of Late Oligocene–Miocene age (Ercan *et al.* 1995; Delaloye & Bingöl 2000; Okay & Satır 2000*b*). From a geodynamic point of view, the Sakarya Zone contains a record of a complex Palaeo- and Neo-Tethyan evolution (Okay *et al.* 1996). Indeed, the Palaeotethyan subduction setting is still controversial, because of the chosen interpretation for the Karakaya Complex evolution (Pickett & Robertson 1996; Okay & Göncüoglu 2004). The final stage of the Neo-Tethyan evolution is characterized by the Paleocene to Miocene south-directed emplacement of the distinct units along the Izmir–Ankara–Erzincan suture zone (Okay & Tüysüz 1999).

Comparative studies have recently shown that Late Palaeozoic–Mesozoic units of the Biga

Peninsula may be correlated with similar units in the Rhodope Massif (Beccaletto & Jenny 2004; Beccaletto *et al.* 2005). Moreover, a large-scale correlation of Cretaceous–Tertiary metamorphic terranes in the northernmost Aegean region suggests continuity between the Serbo-Macedonian, Rhodope and Strandja massifs and the metamorphic basement of the Biga Peninsula (Okay *et al.* 2001). However, there are only few structural and kinematic studies that constrain the deformation pattern in the metamorphic basement of the Biga Peninsula, and no real attempts at correlation with westerly terranes in the north Aegean have been made yet. Kinematic data are available only for the Kazdağ Massif, where NNE–SSW-trending stretching lineations were related to NNE–SSW-directed extension (Walcott & White 1998; Okay & Satır 2000*b*), involving also SSW-directed ductile shear and brittle displacement on low-angle detachments (e.g. Şelale detachment, Beccaletto & Steiner 2005).

The eastern Rhodope–Thrace region

Geological context

The geology of the eastern Rhodope–Thrace region, which constitutes the eastern part of the Rhodope Massif, is dominated by a high- to medium-grade metamorphic basement and Tertiary supracrustal rocks (Fig. 3). The tectono-stratigraphic pile comprises four main units, subdivided according to their structural position, tectono-metamorphic history, and the radiometric ages of coherent lithologies in their footwalls and hanging walls (Fig. 3). These units are predominantly bounded by extensional tectonic contacts of at present inactive early Eocene low-angle detachments. In a structurally ascending order, they are: (1) a lower high-grade unit, (2) an upper high-grade unit, both constituting the high-grade metamorphic basement, (3) an overlying low-grade Mesozoic subduction–accretion unit; and (4) a Maastrichtian–Paleocene to Pliocene sedimentary and volcanic unit representing the cover sequences, including voluminous Late Eocene–Oligocene volcanic rocks and volcano-sedimentary successions (Fig. 3). Moreover, the high-grade basement rocks are intruded by late Cretaceous–Oligocene granitoids (Del Moro *et al.* 1988; Koukouvelas & Pe-Piper 1991; Ovtcharova *et al.* 2003; Marchev *et al.* 2004*a*).

The lower high-grade unit, of continental origin, is mainly composed of various types of orthogneisses and migmatites, with subordinate paragneisses and amphibolite intercalations. The upper high-grade unit is lithologically heterogeneous, consisting of meta-sedimentary, meta-igneous rocks and meta-ophiolite slices with mixed continental and oceanic affinity. The lower high-grade basement unit broadly corresponds to the Kardamos and Kechros Complexes, and the upper high-grade unit to the Kimi Complex (these names refer to the original subdivision made in Greece, by Krohe & Mposkos 2002; see Bonev *et al.* 2006*b* for discussion).

The low-grade Mesozoic rocks (Boyanov & Russeva 1989), which are regarded as the extension of the Circum-Rhodope Belt (e.g. Kaufmann *et al.* 1976; Kockel *et al.* 1977; Papanikolaou 1997), consist of greenschists and phyllites at the base, overlain by island-arc tholeiitic basalt–andesite lava flows and meta-pyroclastic rocks. Stratigraphically, they are in turn overlain by clastic sediments, carbonaceous and phyllitic shales, and reworked clastic deposits of Late Permian and Middle–Late Triassic shallow-water carbonates. Based on the petrology and geochemistry of the arc magmatic suite and the bulk lithological context, the low-grade unit is interpreted as a Jurassic–Early Cretaceous island-arc–accretionary assemblage, originally located along to the northern active margin of the Vardar Ocean (Bonev & Stampfli 2003).

The supracrustal sedimentary and volcanic unit of cover sequences, representing syn- to post-tectonic deposits, ranges in age from the Maastrichtian–Paleocene to the Pliocene (Boyanov & Goranov 2001). At the base of the sedimentary unit, the Maastrichtian–Paleocene to Lower Eocene (Ypresian) coarse clastic deposits of the Krumovgrad Group (Goranov & Atanasov 1992) derived only from the upper high-grade unit. They form part of a syntectonic hanging-wall suite of supradetachment half-grabens, in fault contact with the detachment (Bonev *et al.* 2006*a*), and limited by graben-bounding faults or lying unconformably over the high-grade units. Up section, the coarse clastic deposits are unconformably overlain by Upper Eocene–Oligocene clastic deposits, coal-bearing and carbonate sediments, which mark a renewed cycle of continental freshwater to marine sedimentation. They were accompanied by Late Eocene–Oligocene volcanic and sedimentary–volcanogenic successions (Harkovska *et al.* 1989). The Miocene–Pliocene alluvial–proluvial top sediments represent a new transgressive cycle covering unconformably the metamorphic basement and Palaeogene successions.

The regional-scale tectonic pattern is dominated by two late Alpine metamorphic core complexes, namely the Kesebir–Kardamos and the Byala reka–Kechros domes (Figs 3 and 4; Bonev *et al.* 2006*a*, *b*). Both middle Eocene large-scale structures expose a lower high-grade unit in their cores, structurally flanked by lithologies belonging to the upper high-grade unit and the overlying low-grade Mesozoic unit. Basement units are bounded by contractional, synmetamorphic thrust contacts,

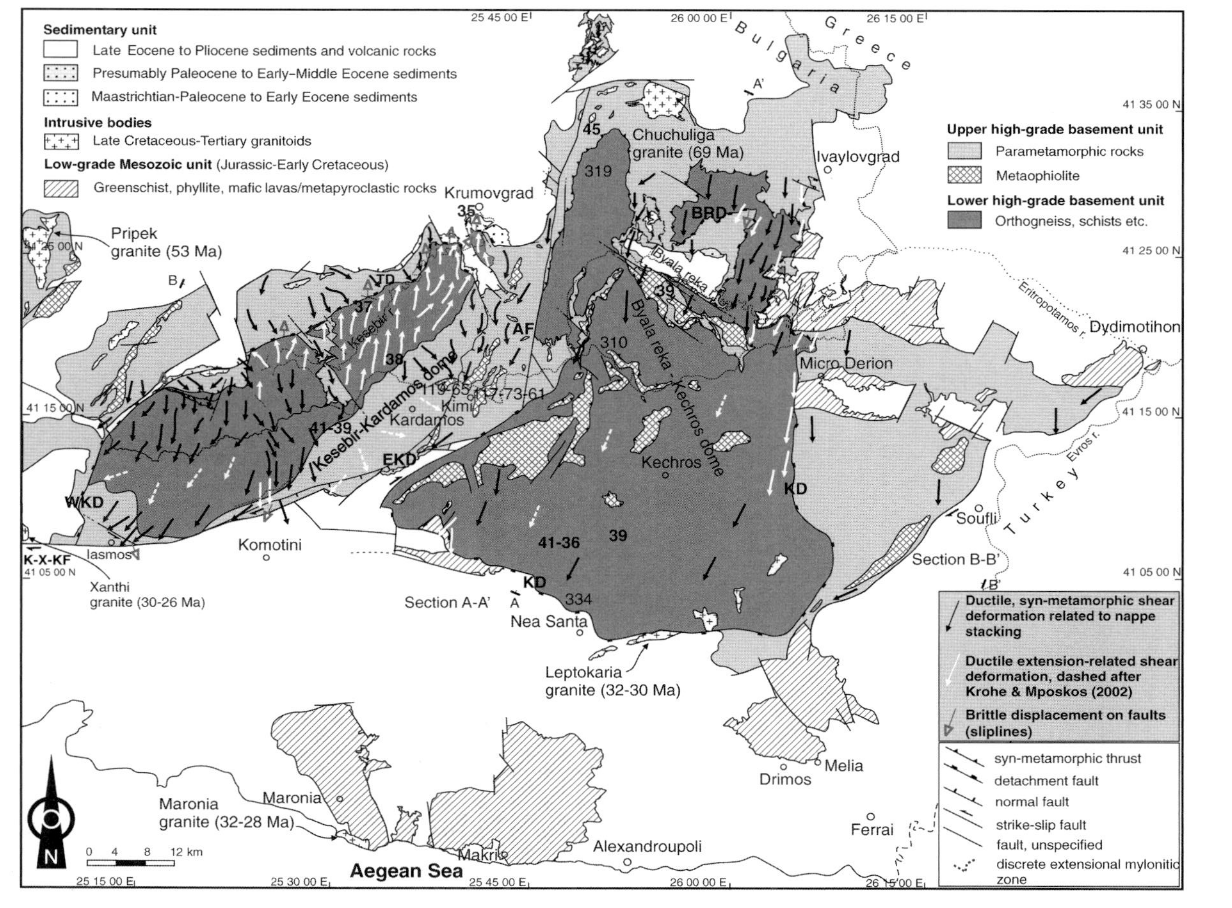

Fig. 3. Structural map showing the kinematic pattern of the high-grade basement units in the eastern Rhodope–Thrace region. Each symbol summarizes numerous measurements of stretching lineations, and macro- and microscopic shear-sense determinations. Kinematic data for metamorphic domes of Bulgaria are adapted from Bonev (2006). (See also Burg *et al.* (1996) and Krohe & Mposkos (2002)). Numbers represent locations and ages of published geochronological data quoted in the text and Table 1. K–Ar and $^{40}Ar/^{39}Ar$ ages are indicated by bold numbers. Isotopic ages of plutons are after Del Moro *et al.* (1988). BRD, Byala Reka detachment; WKD, West Kardamos detachment; EKD, East Kardamos detachment; KD, Kechros detachment; TD, Tokachka detachment; K-X-KF, Kavala–Xanthi–Komotini fault; AF, Avren fault.

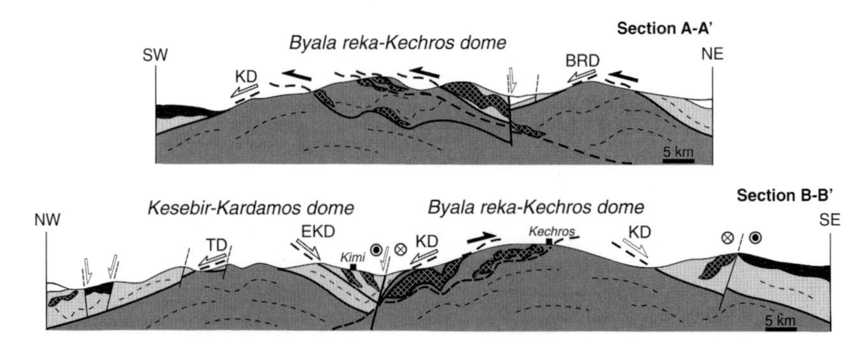

Fig. 4. Cross-sections of the extensional domes in the eastern Rhodope located in Figure 3. Symbols and abbreviations as in Figure 3.

or low-angle extensional detachments, respectively, related to pre-latest Cretaceous crustal thickening and Tertiary extension (Bonev 1996, 2001, 2006; Burg *et al.* 1996; Krohe & Mposkos 2002; Bonev *et al.* 2006*a*). The Kesebir–Kardamos dome constitutes the footwall of the NNE-directed Tokachka detachment, bounding the dome in the north. The west and east Kardamos detachments, with inferred SW sense of movement, limit the dome in the south. The Byala reka–Kechros dome is bounded by the Kechros detachment in the south, whose extension and counterpart in the north is the Byala reka detachment, showing southward transport direction (Fig. 3).

P–T–t evolution and geochronological data

High-pressure eclogite facies, then medium-pressure amphibolite and greenschist facies are the successive metamorphic phases of the Alpine metamorphic cycle in both high-grade basement units (e.g. Mposkos & Liati 1993; see Table 1).

In the upper high-grade unit, eclogite- or granulite-facies conditions at *P c.* 12–17 kbar and *T c.* 750–811 °C were preceded by an earliest ultrahigh-pressure event at >26 kbar and 900 °C. The metamorphic conditions decreased to *c.* 8–10 kbar and 520–650 °C during a medium-pressure event, subsequently retrogressed into the greenschist facies (Table 1). A Sm–Nd isochron age of 119.6 ± 3.5 Ma from garnet–spinel pyroxenite was interpreted as the time of the high-pressure metamorphism (Wawrzenitz & Mposkos 1997), whereas a U–Pb sensitive high-resolution ion microprobe (SHRIMP) age of 117.4 ± 1.9 Ma obtained from the core of zircons in another garnet-rich mafic rock was interpreted as the age of crystallization of the protolith (Liati *et al.* 2002). The zircon rims also yielded an age of 73.5 ± 3.5 Ma for the high-pressure metamorphism and 61.9 ± 1.9 Ma for the late retrogressive stage (Liati *et al.* 2002). The Rb–Sr age of 65.4 ± 0.7 Ma from an undeformed pegmatite is interpreted to date the amphibolite-facies metamorphism (Mposkos & Wawrzenitz 1995), which occurred during cooling between 45 and 39 Ma (Mukasa *et al.* 2003).

In the lower high-grade unit, eclogite-facies metamorphism at 13–15 kbar and 450–600 °C, was followed by nearly isothermal decompression in amphibolite-facies conditions at *P c.* 3–9 kbar and *T* 550–620 °C, and a greenschist-facies overprint at *P c.* 2–3 kbar and *T c.* 400 °C (Table 1). Variscan protolith ages ranging between 295 and 319 Ma were obtained from orthogneisses of the lower high-grade unit by conventional U–Pb zircon and SHRIMP dating (Peytcheva & Quadt 1995; Liati 2005), and 334.6 ± 3.5–Ma by the Rb–Sr method (Mposkos & Wawrzenitz 1995; see fig. 3). Ages for the cooling history that accompanied the extensional exhumation of the unit range between 36 and 42 Ma (Lips *et al.* 2000; Krohe & Mposkos 2002; Bonev *et al.* 2006*b*), thus implying that the high- and medium-pressure metamorphic events occurred prior to the Middle Eocene. Moreover, $^{40}Ar/^{39}Ar$ adularia ages of 35–36.5 Ma (Marchev *et al.* 2003, 2004*b*) document brittle extension in the hanging wall of the detachments.

Latest Cretaceous and Early Eocene late- to post-tectonic granitoids (*c.* 69 Ma and 53 Ma, U–Pb zircon method, Ovtcharova *et al.* 2003; Marchev *et al.* 2004*a*) intrude the upper high-grade unit, and Oligocene granitoids (32–28 Ma, Rb–Sr method, Del Moro *et al.* 1988) intrude both high-grade units (Fig. 3).

Kinematics of the Bulgarian eastern Rhodope

Details on the kinematic pattern of the metamorphic domes in the Bulgarian eastern Rhodope have been given by Bonev (2006). A brief summary is given below (see also Fig. 3). The domes display bulk NNW–SSE to NNE–SSW stretching lineations pattern that are equated with the kinematic

Table 1. *Summary of the metamorphic and geochronological data for the high-grade basement in the eastern Rhodope (Bulgaria–Greece)*

Metamorphism	Units, rock type, regional structure	Method and minerals	Age (Ma)	References
Ultrahigh-pressure $P > 26$ kbar and $T > 900$ °C, based on mineralogical criteria	Upper high-grade unit (Kimi Complex, Greece) metapelite, grt–bt–ky gneiss, SE flank of the Kesebir–Kardamos dome			Mposkos & Kostopoulos (2001)
	Upper high-grade unit (Kimi Complex, Greece) garnet-rich mafic rock, magmatic crystallization of mafic protolith	U–Pb SHRIMP, zr	117.4 ± 1.9	Liati *et al.* (2002)
High-pressure/high-temperature (eclogite/granulite facies) $P > 13.5–16$ kbar, $T = 750–775$ °C	Upper high-grade unit (Kimi Complex, Greece) garnet pyroxenite with assemblage hbl–sp–ol–grt–cpx, SE flank of the Kesebir–Kardamos dome	Sm–Nd, grt–cpx–w.r.	119.6 ± 3.5	Wawrzenitz & Mposkos (1997)
$P > 12–17$ kbar, $T = 750–811$ °C	Upper higit-grade unit (Bulgaria), amphibolized eclogite, SE flank of the Kesebir–Kardamos dome			Kozhoukharova (1998)
	Upper high-grade unit (Kimi Complex, Greece) garnet-rich mafic rock	U–Pb SHRIMP, zr	73.5 ± 3.57 1.5 ± 3.5	Liati *et al.* (2002) Liati (2005)
P 13–16 kbar, T *c.* 600 °C	Lower high-grade unit (Kardamos Complex, Greece), metapelite, gneiss			Mposkos & Liati (1993)
P 13 kbar, $T = 450$ °C	Lower high-grade unit, Byala reka dome (Bulgaria), metapelite			Macheva (1998)
Medium-pressure amphibolite facies $P > 10$ kbar, $T = 600–650$ °C	Upper high-grade unit (Kimi Complex, Greece), metapelite			Krohe & Mposkos (2002)
	Upper high-grade unit (Kimi Complex, Greece), undeformed pegmatite	Rb–Sr, ms	65.4 ± 0.7	Mposkos & Wawrzenitz (1995)
	Upper high-grade unit (Kimi Complex, Greece), undeformed pegmatite	U–Pb SHRIMP, zr	61.9 ± 1.9	Liati *et al.* (2002)
P *c.* 10–12 kbar, $T = 440–520$ °C	Upper high-grade unit (Bulgaria), retrogressed eclogite, Byala reka dome	$^{40}Ar–^{39}Ar$, hbl, ms	45–39 ± 1/2	Mukasa *et al.* (2003)
$P < 8$ kbar, $T = 560–620$ °C	Lower high-grade unit, metapelite, Byala reka dome (Greece)			Mposkos & Liati (1993)
P 3–9 kbar, T *c.* 550 °C	Lower high-grade unit, metapelite, Byala reka dome (Bulgaria)			Macheva (1998)
Amphibolite to greenschist facies P *c.* 3 kbar, T *c.* 400 °C	Lower high-grade unit, metapelite, Byala reka dome (Bulgaria)			Macheva (1998)

(*Continued*)

Table 1. Continued

Metamorphism	Units, rock type, regional structure	Method and minerals	Age (Ma)	References
Cooling of high-grade basement rocks after amphibolite facies in the footwall of the detachments	Lower high-grade unit, Byala reka–Kechros dome (Greece)	^{40}Ar–^{39}Ar, ms	41–36 ± 1/6	Lips *et al.* (2000)
	Upper high-grade unit, Kesebir–Kardamos dome (Greece)	K–Ar, bt, ms	42–39 ± 1	Krohe & Mposkos (2002)
	Lower high-grade unit, Kesebir–Kardamos dome (Bulgaria)	^{40}Ar–^{39}Ar, bt, ms	38–37 ± 0.3	Bonev *et al.* (2006*b*)
Hydrothermal alteration along faults in supradetachment sedimentary deposits $T < 200$ °C	Byala reka–Kechros dome (Bulgaria)	^{40}Ar–^{39}Ar adularia	36.5 ± 0.3	Marchev *et al.* (2003)
	Kesebir–Kardamos dome (Bulgaria)	^{40}Ar–^{39}Ar adularia	35 ± 0.2	Marchev *et al.* (2004*b*)

ol, olivine; hbl, hornblende; grt, garnet; cpx, clinopyroxene; bt, biotite; ms, muscovite; zr, zircon; w.r., whole-rock.

direction. Extensive record of asymmetric ductile shear fabrics and metamorphic crystallization–deformation relationships indicate that the basement rocks experienced two distinct events of Alpine deformation: a NNW–SSE- to NNE–SSW-oriented contraction related to the nappe stacking, and then a top-to-the-SSW and/or -NNE extension. Top-to-the-SSE–SSW ductile fabric elements are coeval with the main metamorphism in the amphibolite facies, and are associated with synmetamorphic thrust imbrications of the high-grade basement units. This contractional event, time-constrained by radiometric ages of the metamorphism (see above), occurred before the intrusion of latest Cretaceous–Eocene granitoids (69–53 Ma). The south-directed kinematics of the contractional event continued in the lower metamorphic grade (greenschist-facies retrogression), and temperature conditions of deformation with top-to-the-SSW ductile to brittle extension in the Byala reka–Kechros dome (Bonev *et al.* 2006*b*), and top-to-the-NNE ductile then brittle extension in the Kesebir–Kardamos dome (Bonev *et al.* 2006*a*). Extension was partly coeval and developed concurrently with the previous stacking event. This led to the tectonic denudation and exhumation of the footwall lower high-grade unit in the cores of large-scale metamorphic domes, through the activity of ductile to semi-ductile shear zones, located under low-angle brittle detachments. These relationships are particularly evident in the western Kesebir–Kardamos dome, where the interplay between contraction and extension-related fabrics is best portrayed (see Fig. 3). The extensional exhumation was accompanied by widespread cooling of the footwall rocks between 42 and 37 Ma, followed by late brittle faulting at 36–35 Ma (Fig. 3). The kinematic pattern in the high-grade basement units is therefore interpreted to reflect the spatially and vertically partitioned shear sense and kinematic direction, defined by stretching lineations in the metamorphic pile, and formed in response to the transition from crustal thickening to late-orogenic extension. The kinematic continuity of shear direction of the contractional and extensional events suggests a lower metamorphic grade reactivation of the pre-extension ductile thrusts as extensional detachments; this is particularly evident in the Byala reka–Kechros dome (Figs 3 and 4, and next section). This possibility has also been invoked by Bonev *et al.* (2006*a*) for the southeastern flank of the Kesebir–Kardamos dome.

Kinematics of the Greek eastern Rhodope

In Greek eastern Rhodope and Thrace (Fig. 3), low-angle detachments bounding the large-scale domes have been found by Krohe & Mposkos (2002), although those workers did not provide details on

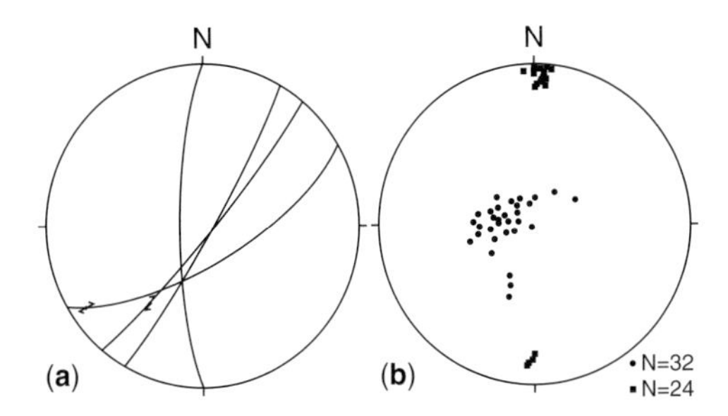

Fig. 5. Equal area lower hemisphere stereographic projection of structural data for the extensional shear zone beneath the Kechros detachment west of Micro Derion and north of Nea Santa. (**a**) Fault planes in a fault zone reworking the detachment, and bounding the western side of the hanging-wall sedimentary basin. (**b**) Poles to foliation (•), and stretching lineation (▪).

the kinematic pattern, in particular for the mylonites associated with the Kechros detachment. In Greece, our study focused on structures and kinematics beneath the detachments in the southern flanks of the Kesebir–Kardamos and the Byala reka–Kechros domes, respectively in the poorly known areas north of Komotini and west of Micro Derion (see Fig. 3).

At Micro Derion in the eastern flank of the Byala reka–Kechros dome, the Kechros detachment (Krohe & Mposkos 2002) separates the lower high-grade unit in the footwall (e.g. Kechros Complex), from the hanging-wall upper high-grade unit (e.g. Kimi Complex), and the low-grade Mesozoic unit (Fig. 3). The detachment is truncated by set of late normal and dextral strike-slip faults (Fig. 5a), confined in a zone that bounds the hanging-wall east–west-trending fault-bounded graben. This fractured zone is underlain by an ENE-dipping decametre-thick mylonitic shear zone in the gneisses of the lower high-grade unit, showing pervasive ductile fabric elements. The mylonitic foliation parallels the north–south to NNW–SSE-trending shear zone and dips moderately to the ENE. Strongly oriented mica flakes and quartz–feldspar aggregates define the stretching lineation, with shallow plunges to the north (Fig. 5b). Kinematic indicators associated with the mylonitic fabrics indicate top-to-the-south tectonic transport and shearing under ductile to semi-ductile deformation conditions (Figs 3 and 6). Mineral constituents in the mylonitic

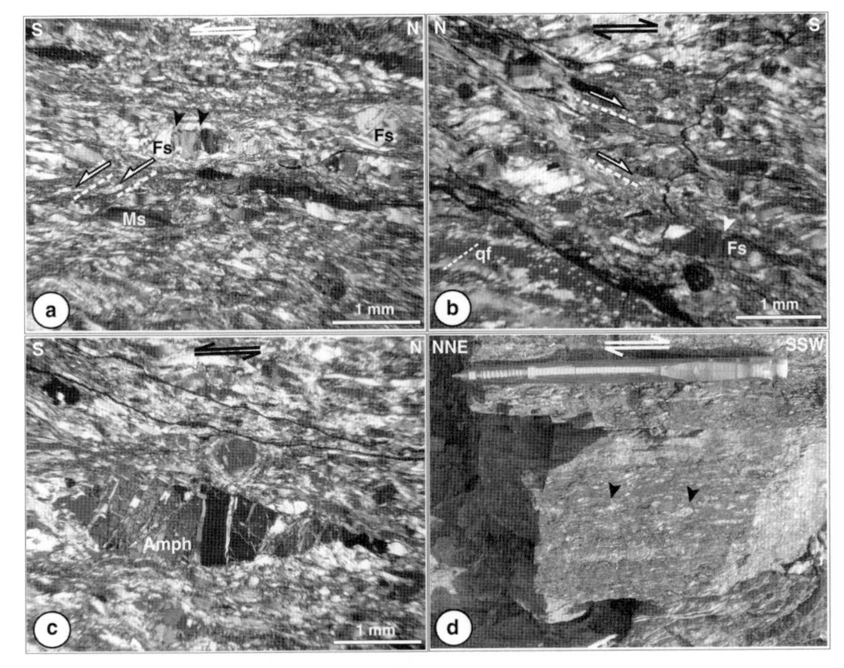

Fig. 6. Ductile to brittle structures and shear criteria in gneiss mylonites below the Kechros detachment. (**a**) Extensional-type shear bands (half-arrows) and associated mica fish (Ms). The cataclastic deformation of feldspar (Fs) along syn- and antithetic fractures (black arrowheads) should be noted. (**b**) Asymmetric extensional shear bands (half-arrows), cataclasis of feldspar (Fs; white arrowheads), and oblique grain-shape fabric (qf) in quartz ribbon. Shear band overprinted by steep micro-fault in the upper edge of the quartz ribbon should be noted. (**c**) Brittle deformation of amphibole (Amph) along syn- and antithetic fractures with respect to bulk flow direction. (**d**) Asymmetric sigma-type feldspar porphyroclasts in mylonites (black arrowheads).

gneisses show strong internal deformations. Micas are bent, forming 'fish' separated by tightly spaced asymmetric shear bands, and quartz exhibits common undulose extinction in recrystallized quartz ribbons, which display a strong grain-shape preferred orientation (Fig. 6a and b). The feldspar and amphibole are deformed in a brittle fashion, showing cataclastic deformation mechanisms that involve fracturing along syn- and antithetic fractures with respect to the bulk flow direction (Fig. 6a and c). The shear zone underlying the Kechros detachment can be traced further south. There, the flat-lying and/or NE–ENE-dipping mylonitic foliation contains shallow NNE-plunging stretching lineations, and is associated with top-to-the-SSW ductile to semi-ductile shearing (Fig. 6d). Microscopic observations on the relationships between metamorphic crystallization and shear fabrics revealed little (i.e. chlorite after biotite and garnet) or no retrogression to the greenschist facies.

In the west and north of Nea Santa, the low-angle (<25°) detachment juxtaposes the footwall lower high-grade unit with the low-grade Mesozoic unit in the hanging wall along a zone of strong alteration, involving chlorite in a cataclastic ledge. The omission of the upper high-grade unit in the tectonostratigraphic pile is interpreted to result from extensional excisement at the contact. The SW-dipping mylonitic foliation in the footwall rocks contains a south-plunging stretching lineation that is associated with a bulk south-directed shear sense. The latter is similar to the direction of tectonic transport associated with the contractional nappe stacking-related fabrics in the structurally deeper levels of the footwall to the north, as well as in the hanging wall to the east (Fig. 3). We interpret the observed mylonitic fabrics and kinematics as related to a south-directed low-temperature ductile to semi-ductile extensional shear zone below the Kechros detachment.

North of Komotini, the east Kardamos detachment (Krohe & Mposkos 2002) limits the southern flank of the Kesebir–Kardamos dome (Fig. 3). Northward in the dome core, structurally deeper gneiss levels of the footwall lower high-grade unit exhibit a moderately NE–ESE-dipping regional foliation and a NNW–SSE- to north–south-trending stretching lineation, with moderate to shallow plunges to the SSE (Fig. 7). Asymmetric minor folds have axes oblique or parallel to the lineation (Figs 7b and 8a). These structural elements are associated with a top-to-the-SSE ductile shear in amphibolite-facies conditions that pertains to the synmetamorphic nappe stacking (Fig. 3). To the south, towards structurally upper levels against the upper high-grade unit, a brittle fault contact, which can be equated with the eastern Kardamos detachment, cuts through the metamorphic pile. It occurs immediately above the mylonitic contact separating both high-grade units all along the southeastern flank of the dome (Fig. 3; Bonev *et al.* 2006*a*, fig. 3). The tectonic contact is a fault surface dipping moderately toward the SE (*c.* 50°), marked by intense cataclasis of the hanging-wall marble horizon, and synthetic steep faults in a brecciated zone more than 20 m thick. Slip lines on slickensides and fault-surface criteria indicate down-dip or dextral oblique southwestward brittle movement (Fig. 8c). In the immediate footwall, the regional foliation is transposed into the hinges of SE-vergent tight folds that developed an axial planar NW-dipping crenulation cleavage. Field and microscopic observations on the asymmetry of the shear structures down section in the footwall indicate top-to-the-south ductile to semi-ductile tectonic transport (Fig. 8b), parallel to the southward-plunging stretching lineation. Strongly recrystallized and newly formed quartz grains in the pressure shadows of feldspars, mica fish between shear bands and fractured feldspar clasts imply low-temperature deformation conditions (<450 °C), involving ductile to brittle mechanisms. These structures are confined in a discrete decametre-thick mylonitic shear zone structurally below the brittle fault surface. The ductile to semi-ductile then brittle deformation conditions, the SSW-directed displacement and the overprinting relationships of planar structures at the east Kardamos detachment are all consistent with the late-stage extension and exhumation of the metamorphic pile in the southern flank of the Kesebir–Kardamos dome. However, the dip angle of the detachment, as well as its similarly expressed counterpart fault surface in Bulgaria (see Figs 3 and 4), suggests that this tectonic contact is mostly a high-angle normal fault. This possible incipient detachment cut into the upper high-grade unit to assist the exhumation of the dome along its southern flank (see Bonev *et al.* 2006*a*, for discussion).

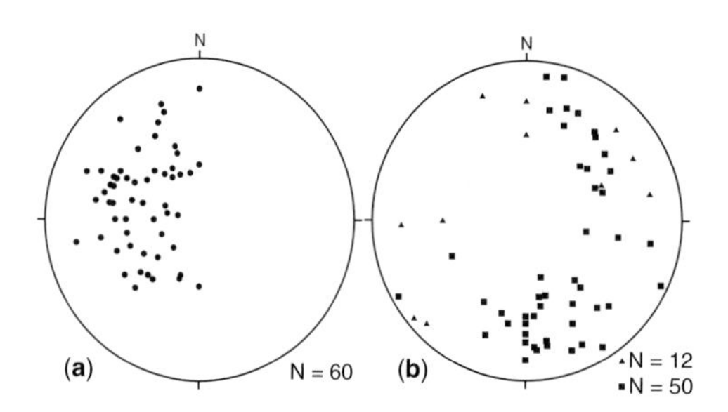

Fig. 7. Equal area lower hemisphere projection of structural data for basement rocks north of Komotini. (**a**) Poles to foliation. (**b**) Stretching lineation (■) and minor fold axes (▲).

Fig. 8. Ductile and brittle structures at the eastern Kardamos detachment. (**a**) Tight intrafolial fold in a banded amphibolite. (**b**) Extensional-type shear bands (sb; half-arrows) and associated mica fish (Ms), and asymmetric sigma-type feldspar porphyroclasts (Fs) in mylonitic gneiss. The cataclastic deformation of feldspars (black arrowheads) should be noted. (**c**) Exposure of a fault surface equated with the eastern Kardamos detachment north of Komotini, looking NNW. The wide arrow indicates the sense of displacement of the hanging wall. The stereoplot depicts fault surfaces and associated slip lines. Inset photographs show aspects of fault zone breccias and striated surfaces.

NE of Iasmos, the upper high-grade unit displays an ENE–ESE-dipping regional tectonic foliation bearing a moderately NE-plunging mineral and/or stretching lineation. Associated asymmetric east-vergent folds with axes nearly parallel to the lineation also occur (Fig. 7). A top-to-the-SW shear is associated with the amphibolite-facies ductile fabrics (Fig. 3). In the same area, slip lines of a NW–SE-striking high-angle fault (60°) cross-cutting the ductile structures indicate down-dip or sinistral oblique SE displacement (Fig. 9a). This fault probably represents an excisement splay of the discrete west Kardamos detachment (Krohe & Mposkos 2002), which bounds the southwestern closure of the Kesebir–Kardamos dome.

NW–SE-striking steep faults overprint the ductile fabrics in the metamorphic pile north of Komotini (Fig. 9b). These normal and strike-slip faults developed nearly orthogonal to the ductile kinematic direction (i.e. stretching lineation), indicating shallow-level NE–SW-oriented brittle extension. The latest brittle deformation is represented by the normal or dextral strike-slip Xanthi–Komotini fault (Lyberis 1984), the extensive trace of which

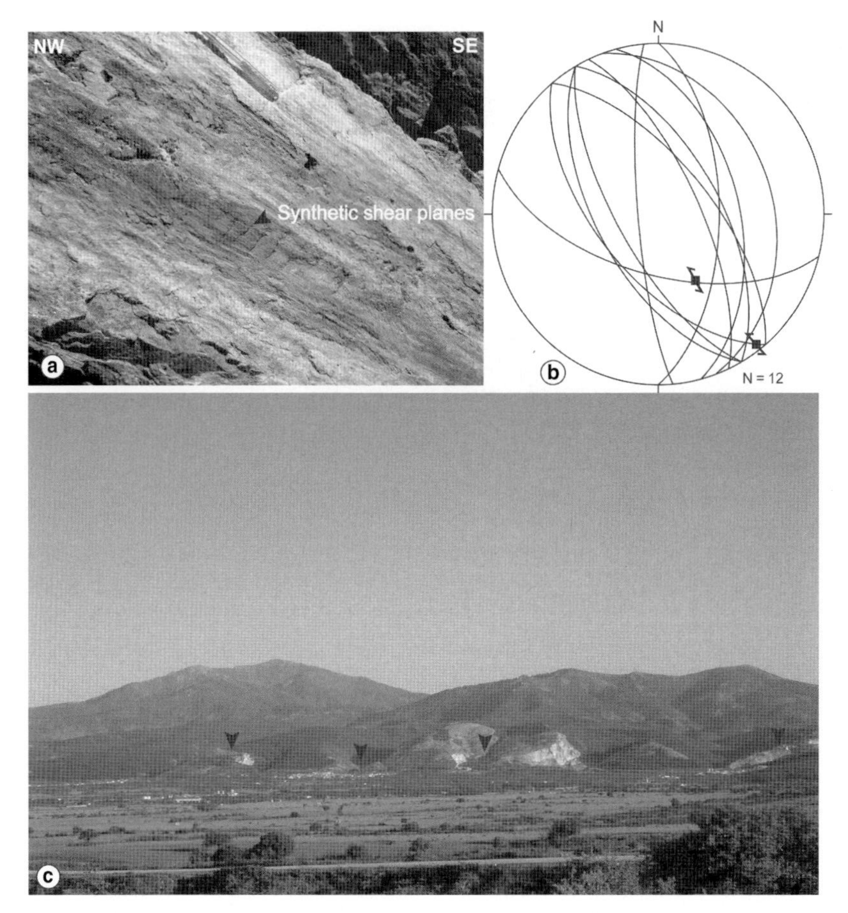

Fig. 9. Brittle structures NE of Iasmos at the southwestern closure of the Kesebir–Kardamos dome. (**a**) Slickensides and associated slip lines and small-scale shears on a fault surface probably representing splay of the west Kardamos detachment. (**b**) Equal area lower hemisphere stereographic projection of upper crustal brittle fault planes and slip data. (**c**) Panoramic view, looking NW, of the southern slope of the East Rhodope Mountains north of Komotini. The facets in the foreground are traces of the Xanthi–Komotini fault (black arrow heads).

dominates the regional topography in the southern slope of the East Rhodope Mountains (Fig. 9c). This steeply (60–80°) SSE-dipping fault abruptly separates the metamorphic basement from the Pliocene–Quaternary sedimentary fill of the Xanthi–Komotini basin, suggesting continuing extension in the late Neogene.

In summary, structural and kinematic data in northeastern Greece provide new supporting information on the kinematic framework of the large-scale metamorphic domes in the eastern Rhodope–Thrace region. Observations show predominantly north–south- to NE–SW- trending stretching lineations associated with a consistent south- to SW-directed tectonic transport. The sense and direction of shear remained in kinematic continuity from the ductile nappe stacking to the ductile or semi-ductile then brittle extension in the late retrogressive stage following the main amphibolite-facies metamorphism. The progressive extensional exhumation of the metamorphic pile was assisted by ductile–brittle mylonitic shear zones beneath detachments and/or high-angle faults. Finally, the latest strike-slip faulting, combining extension and transtension, obviously relates to the propagation of the southern Rhodope basin system after latest Eocene–Oligocene to Miocene times (Karfakis & Doutsos 1990; Koukouvelas & Aydın 2002).

The Biga Peninsula

Geological context

In NW Turkey, the Biga Peninsula exposes the main following units (Fig. 10): (1) high- to medium-grade

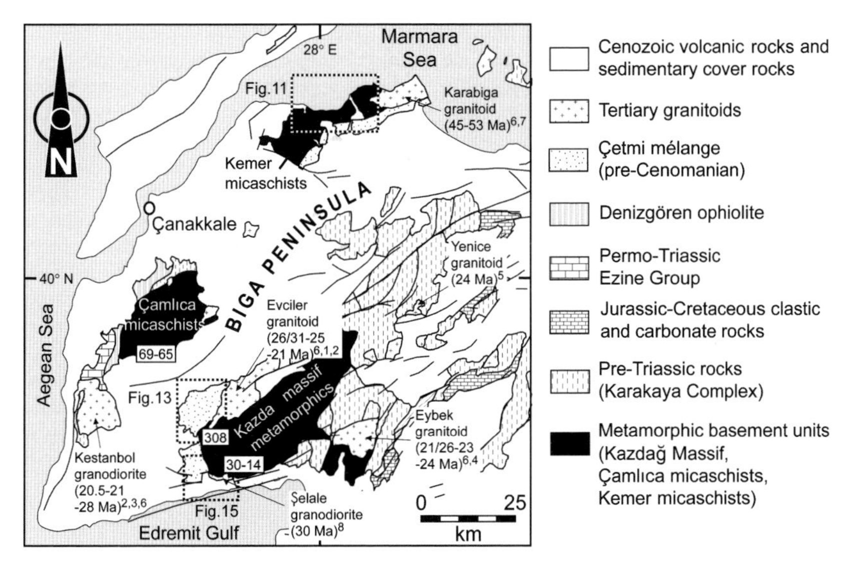

Fig. 10. Synthetic geological map of the Biga Peninsula, adapted from Siyako *et al.* (1989). Boxes represent ages of published geochronological data for the Çamlıca and Kazdağ metamorphic rocks referred to in the text. Sources of the isotopic ages of plutons: 1, Okay & Satır (2000*b*); 2, Birkle & Satır (1995); 3, Fytikas *et al.* (1976); 4, Krushensky (1976); 5, Anıl *et al.* (1989); 6, Delaloye & Bingöl (2000); 7, Beccaletto *et al.* (2007); 8, Beccaletto & Steiner (2005).

basement rocks in several occurrences, including the Kazdağ Massif, the Çamlıca metamorphic series and the Kemer micaschists, (2) the various units of the Karakaya Complex, which crop out only along its eastern border; (3) the accretion-related mid-Cretaceous Çetmi mélange; (4) the Ezine Zone, where the Permo-Triassic sedimentary Ezine Group is tectonically overlain by the Lower Cretaceous Denizgören ophiolite; (5) widespread occurrence of Tertiary volcanic and sedimentary cover sequences and plutonic rocks. Distinct units are tectonically bounded by shallow-dipping mylonitic shear zones, extensional detachments and late faults belonging to the North Anatolian Fault System (NAFS) (Okay *et al.* 1991; Okay & Satır 2000*a*, *b*; Beccaletto & Steiner 2005).

The Kazdağ Massif exposes a high-grade basement consisting of intercalated felsic gneisses, amphibolites and marbles, with meta-ophiolites in the basal–middle part of the section, and sillimanite-bearing gneisses and migmatites at the top (e.g. Kazdağ Group, Okay *et al.* 1991; Duru *et al.* 2004). The subduction–accretionary Karakaya Complex, including intra-oceanic arc and forearc units (Okay *et al.* 1996), tectonically overlies the Kazdağ Massif metamorphic units. The Çamlıca metamorphic rocks (Okay *et al.* 1991) comprise medium-grade metamorphic rocks cropping out west of the peninsula, which occur in fault contact with the sedimentary and ultramafic rocks of the Ezine Zone. They predominantly consist of quartz-micaschists, intercalated with calc-schist, marble, amphibolite and quartzite. Similar rocks occur in the northern part of the peninsula (Okay & Satır 2000*a*). There, the metamorphic rocks, namely the Kemer micaschists (Bonev & Beccaletto 2005; Beccaletto *et al.* 2007), exhibit a similar medium-grade metasedimentary sequence. The Ezine Zone (Okay *et al.* 1991) consists of the weakly metamorphosed Permo-Triassic synrift sequence of the Ezine Group (Beccaletto & Jenny 2004), tectonically overlain by the Denizgören ophiolite. Its amphibolitic metamorphic sole yielded 125–117 Ma $^{40}Ar/^{39}Ar$ ages (Okay *et al.* 1996; Beccaletto & Jenny 2004). The Çetmi mélange (Okay *et al.* 1991; Beccaletto *et al.* 2005) occurs in fault contact with the metamorphic basement units. It is a tectonic mélange amalgamating slices and/or blocks of mafic lavas and pyroclastic rocks, Triassic shallow- and deep-water carbonates, Jurassic–Cretaceous radiolarites, eclogite and ultramafic rocks in a greywacke–shale matrix. The geodynamic evolution of the Çetmi mélange ended in mid-Cretaceous (Albian–Cenomanian) time (Beccaletto *et al.* 2005). Paleocene–early Eocene fluvio-deltaic clastic and volcanic units are unconformably overlain by Middle Eocene–Oligocene carbonate sediments and volcanic, rocks forming the basal part of the cover sequences (Siyako *et al.* 1989). Then, the erosion of the underlying lithologies and the subsequent deposition of lacustrine shale-dominated Early–Middle Miocene sediments in fault-bounded grabens was contemporaneous with widespread calc-alkaline lavas, pyroclastic rocks and intusive bodies. The sedimentation persisted into the Late Miocene, whereas magmatic activity apparently ceased (Yılmaz & Karacık 2001). Pliocene–Quaternary fluvial sediments and lacustrine carbonates mark the renewed depositional cycle, with an internal erosional phase, graben formation, and extensional to strike-slip faulting related to the propagation of the NAFS into the Biga Peninsula (Şengör *et al.* 1985; Yılmaz *et al.* 2000; Bozkurt 2001). Overall, the sedimentary, volcano-sedimentary, volcanic and associated plutonic rocks are related to the transition from a collisional to an extensional tectonic regime during the Cenozoic (Ercan *et al.* 1995; Aldanmaz *et al.* 2000; Yılmaz *et al.* 2001).

Metamorphism and geochronological data

High- to medium-pressure metamorphic events are recognized within the metamorphic basement rocks in the Biga Peninsula. A metabasite lens belonging to the Çamlıca metamorphic series records an early event of eclogite-facies metamorphism at a pressure of 11 kbar and a temperature of 510 °C. This event, dated between 65 and 69 Ma by the Rb–Sr method on phengites (Okay & Satır 2000*a*), was subsequently retrogressed to the greenschist facies. The peak conditions of the amphibolite-facies metamorphism in the Kazdağ Massif at 5 kbar and 650 °C were reached around 24 Ma (Rb–Sr mica ages, Okay & Satır 2000*b*). An earlier HP–LT event is recognized within the Çetmi mélange, where eclogite lenses have been dated at *c.* 100 Ma (Rb–Sr mica ages, Okay & Satır 2000*b*). Carboniferous inherited zircon ages of 308 $\pm$ 16 Ma in the metamorphic rocks (Pb evaporation method, Okay *et al.* 1996) are thought to reflect a presumably Variscan metamorphic event, which has been completely overprinted by a late Alpine metamorphism (Okay & Satır 2000*b*). Tertiary granitoids, systematically intruding most of the units in the Biga Peninsula have K–Ar, Rb–Sr or U–Pb ages ranging from 53 to 21 Ma (Fig. 10 and references in the figure caption).

Kinematics of the northern domain: the Kemer micaschists

In the northern Biga Peninsula, the Kemer micaschists form a NE–SW-trending strip of medium-grade metamorphic rocks along the southern coast of the Marmara Sea (Figs 10 and 11). The micaschists are tectonically overlain by the Çetmi mélange, and both units are intruded by the Karabiga granitic pluton (Güçtekin *et al.* 2004), with a K–Ar

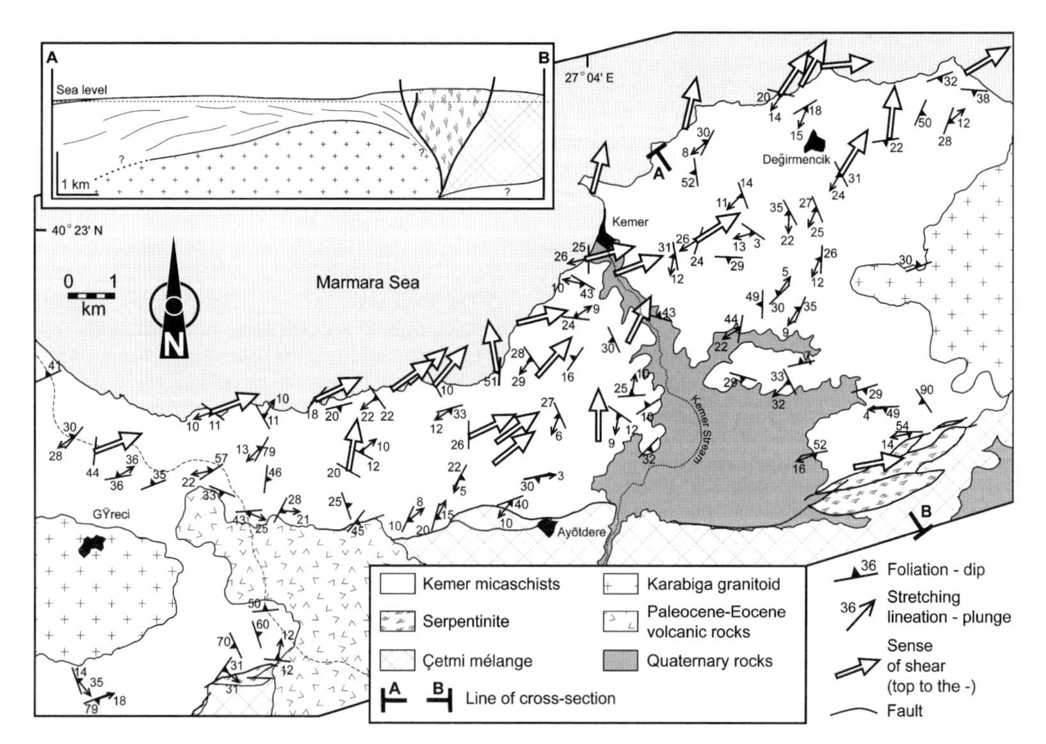

Fig. 11. Geological map and kinematic pattern of the Kemer micaschists.

biotite age of 45.3 ± 0.9 Ma (Delaloye & Bingöl 2000). Paleocene–Eocene volcanic rocks and fluvio-deltaic Lower Eocene sediments (Siyako *et al.* 1989) rest unconformably on the micaschists and the Çetmi mélange. The predominantly marly–carbonaceous and clastic metasedimentary sequence mainly consists of garnet-bearing quartz–white micaschists, intercalated with quartz–chlorite schists, and subordinate quartz–chlorite–albite schists, phyllites, calc-schists, rare quartzites and metabasites. The mineral assemblage, consisting of Qtz + Ms + Chl + Ab ± Grt ± Bt ± Spn ± Ap, is common in metapelites. Structurally, the Kemer micaschists display an overall flat-lying to moderately south-dipping succession, defined by a regional foliation representing schistosity and/or metamorphic layering that parallels the lithological contacts. The foliation contains a stretching lineation defined by strongly oriented micas, streaky quartz–chlorite aggregates and pressure fringes around garnets. The lineation is dominantly oriented NE–SW with shallow plunges in both directions, but predominantly south (Fig. 11). These structural elements are associated with a regionally penetrative shear deformation that has imprinted the pervasive ductile fabrics in the micaschists. The most prominent shear structures are decametre to metre-sized asymmetric extensional shear bands, gently dipping NE. Shear bands noticeably offset the metamorphic layering and/or foliation, which progressively become mylonitic adjacent to them, leading to an omnipresent asymmetric foliation boudinage between them (Fig. 12a). Kinematic indicators such as porphyroclast systems and sigmoidal-shaped objects (Fig. 12b), micro-scale shear bands and drag folds are abundant at various scales. Sense-of-shear criteria indicate a regionally consistent top-to-the-NE ductile tectonic transport direction, parallel to the stretching lineation (Figs 11 and 12). Moreover, the shear deformation continued with the same direction from ductile to brittle–ductile conditions, as demonstrated by tension gashes (Fig. 12c), suggesting that the shearing took place during decreasing temperatures. Finite strain determination using quartz boudins of stretched layers that localized the deformation yielded axial ratios of 1.30–1.80 on the xy-plane and 3–4 on the xz-plane of finite strain, indicating close to plane strain conditions (Flinn parameter, $k \approx 0.9$). Generally, a structurally upward increase in the finite strain is observed. It is mostly defined by the decrease of the angle between the shear bands and the foliation, as displayed along the Marmara Sea coast. This corresponds to a shear gradient toward the structurally upper levels of the metamorphic pile. Deformation–crystallization relationships, such as crystallization of syntectonic albite and titanite porphyroblasts and concomitant garnet growth, indicate that the ductile shear structures formed coevally with the greenschist-facies metamorphism (Fig. 12d and e). Late brittle deformation encompasses a predominant set of north- to NE-dipping high-angle faults (Fig. 12f), which indicate relatively pronounced subsequent brittle faulting at higher structural level.

Overall, the ductile structures and kinematics depict an intense non-coaxial shear deformation characterized by bulk NE-directed ductile flow, which then continued under semi-ductile to brittle deformation conditions with the same kinematic direction. The nature of dominant shear structures (e.g. extensional-type shear bands), which have attenuated the metamorphic layering, and the ductile then brittle character and the kinematic continuity of the deformation are consistent with a NE–SW-oriented extension. This extensional regime has contributed to the stretching and ductile thinning of the metamorphic pile, accommodating its exhumation. We interpret the shear fabrics as related to a NE–SW-directed ductile–brittle extensional shear zone. Because of the pervasive character of shearing, we have not been able to recognize fabrics and/or contact linked to the pre-extension contraction-related crustal stacking. Moreover, the possible pre-extensional tectonic contact between the Kemer micaschists and the Çetmi mélange has been later reworked by splays of the NAFS. Our U–Pb dating on xenotimes of the post-kinematic Karabiga pluton has yielded a 52.7 ± 1.9 Ma crystallization age (Beccaletto *et al.* 2006), which gives a pre-Eocene age for the ductile extensional shear deformation in the Kemer micaschists.

Kinematics of the southern domain: the northwestern flank of the Kazdağ Massif

In the southern domain, the NE–SW-oriented Kazdağ Massif exposes a structural dome, which is regarded as a latest Oligocene metamorphic core complex (Okay *et al.* 1991; Okay & Satır 2000*b*). According to Okay & Satır, the massif is flanked to the NW by a mylonitic shear zone, the Alakeçi shear zone (ASZ), which separates high-grade metamorphic rocks in the footwall from the unmetamorphosed Çetmi mélange in its hanging wall. Our structural mapping and kinematic analysis were focused on the ASZ, to collect shear structures, establish the deformation mechanism and determine the kinematics associated with the shear zone activity. Our purpose is therefore to better constrain the role of the ASZ in the extensional exhumation history of the Kazdağ Massif.

Field and microscopic observations reveal that mylonites of >2 km thickness were derived at

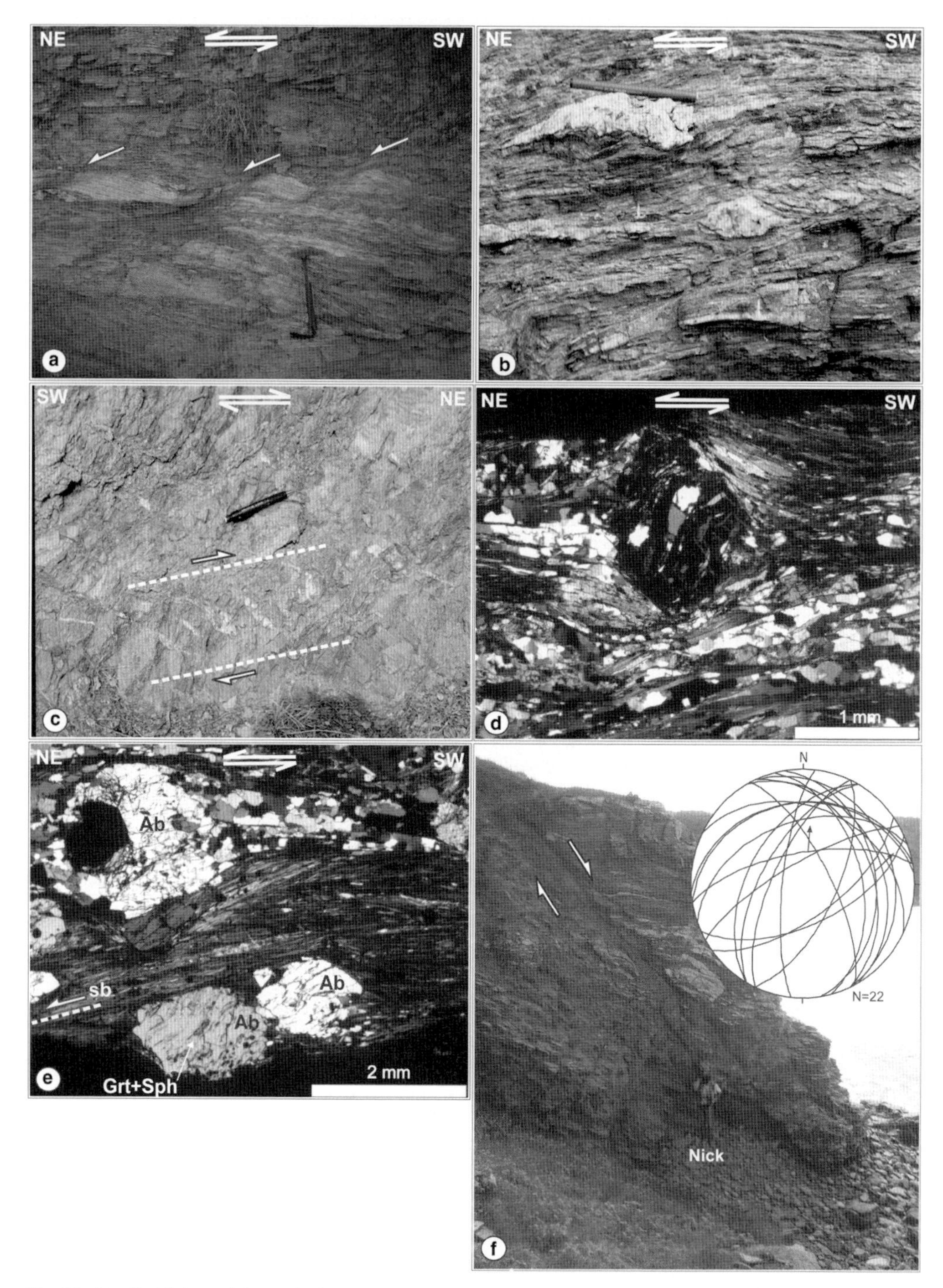

Fig. 12. Ductile fabrics and shear criteria in the Kemer micaschists. (**a**) Extensional-type shear bands (half-arrows) and asymmetric boudinage of metamorphic layering between them. (**b**) Asymmetric sigmoidal-shaped quartz lenses. (**c**) Calcite-filled tension gashes in calc-schists. (**d**) Garnet porphyroblast with helical inclusion trails. (**e**) Syntectonic albite porphyroblasts (Ab) with inclusion trails of graphite, titanite, apatite and elongated garnet nuclei associated with shear bands (half-arrows). (**f**) NNE-dipping fault (half-arrows) crosscutting ductile fabrics in the micaschists. Stereoplot shows equal area lower hemisphere projection of late faults in the micaschists; arrows indicate slip lines.

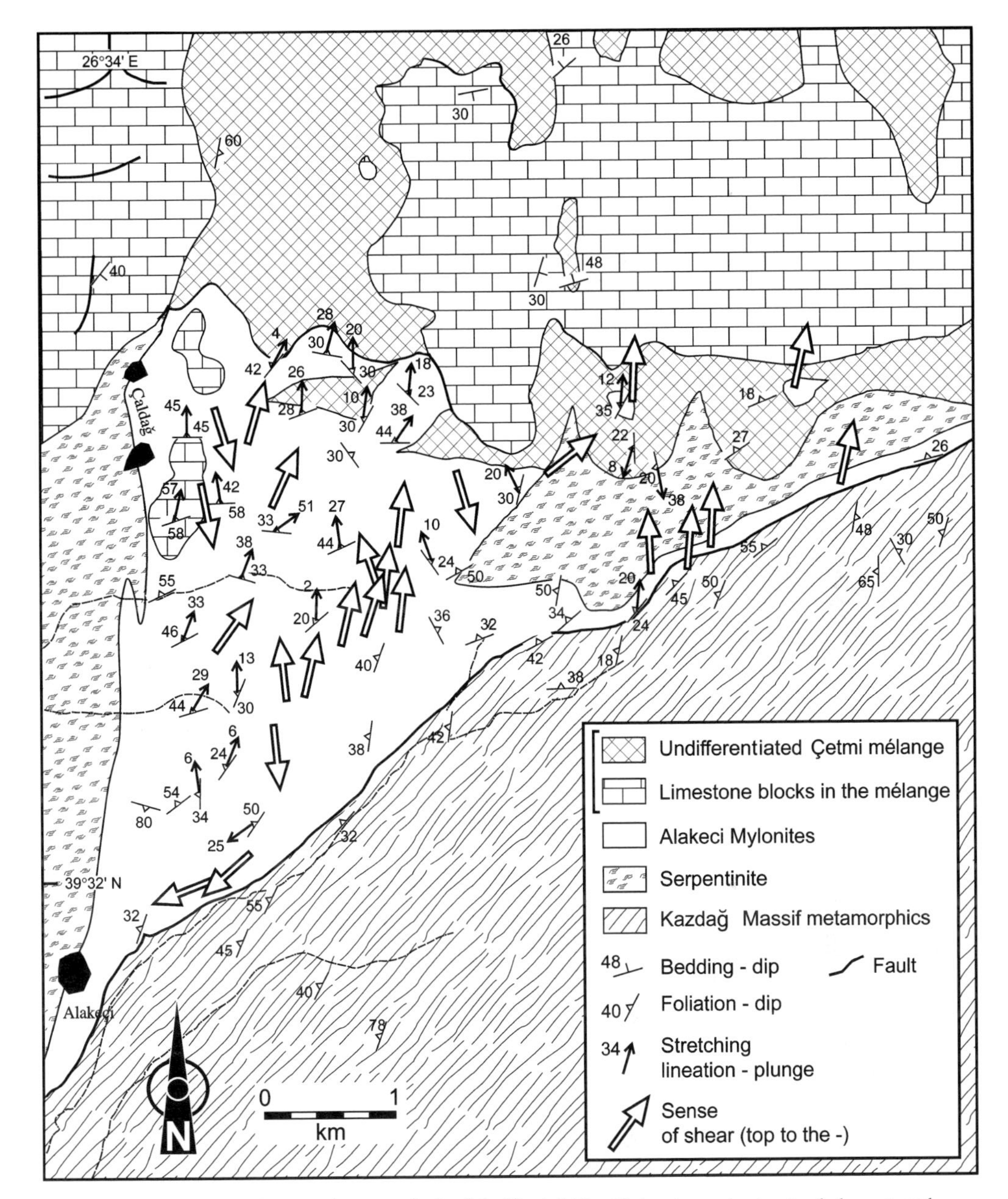

Fig. 13. Geological map of the northwestern flank of the Kazdağ Massif showing main structural elements and associated kinematics in the Alakeçi shear zone.

the expense of both the high-grade metamorphic rocks and the Çetmi mélange. The whole shear zone exhibits a well-developed mylonitic foliation that dips moderately to the NNW (Fig. 13). The foliation contains moderately north- to NNE-plunging stretching lineation, defined by strongly oriented quartz–feldspar aggregates and elongated micas in gneiss, or streaking quartz and phyllosilicates in mélange lithologies. Penetrative mylonitic fabrics and strain gradient from protomylonites to mylonites in the ASZ indicate intense non-coaxial deformation. Sense-of-shear criteria such as tightly spaced C′ type shear bands, porphyroclast systems, mica fish and flanking structures, are abundant (Fig. 14). Kinematic indicators consistently demonstrate a top-to-the-NNE shearing (i.e. north-side), down-dip tectonic displacement, parallel to the stretching lineation (Figs 13 and 14). Scarce

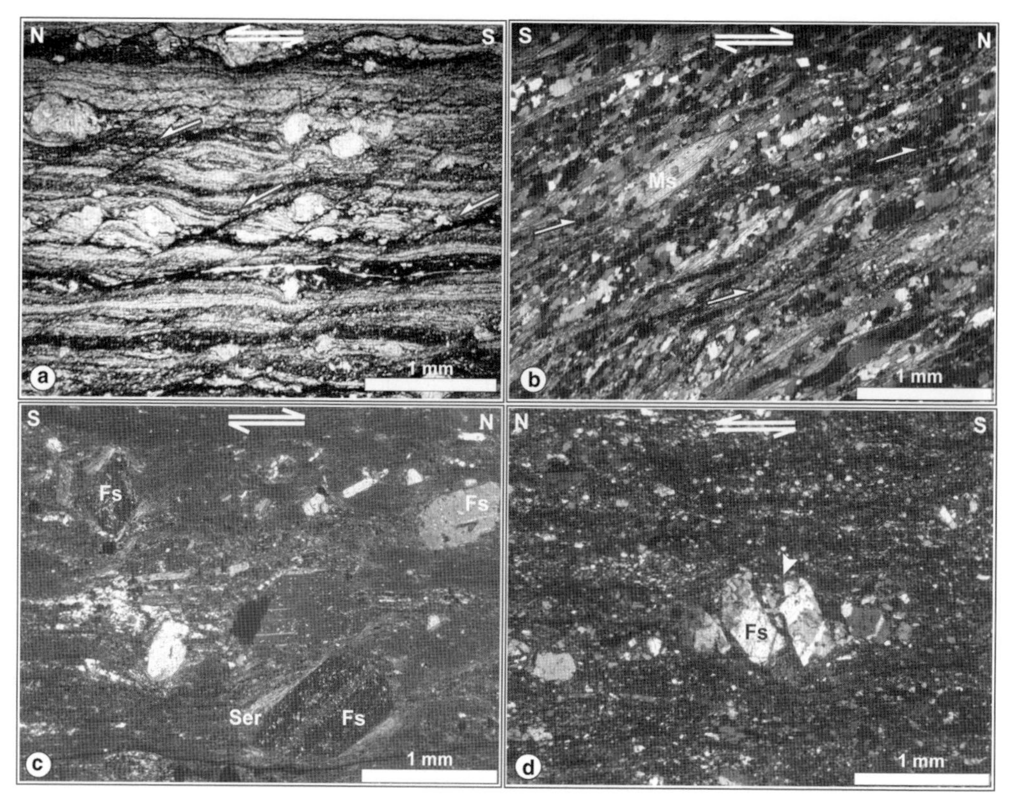

Fig. 14. Ductile to brittle structures and kinematic indicators in the mylonites of the Alakeçi shear zone. (**a**) Asymmetric shear bands (half-arrows) and associated feldspar porphyroclasts (**b**) Tightly spaced shear bands (half-arrows) and associated mica fish (Ms) in a gneiss mylonite. (**c**) Sigmoidal feldspar porphyroclasts rotated into bulk flow direction (Fs), recrystallized into sericite (Ser) in pressure shadows. (**d**) Brittle behaviour of feldspar (Fs) showing cataclastic deformation mechanism along antithetic fractures (arrowhead) with respect to the shear flow direction.

top-to-the-south or -SW indicators in the mélange lithologies probably relate to heterogeneous backward flows induced by the rheological contrast between its distinct components (e.g. matrix or blocks), leading to local opposite shear senses. However, top-to-the-NNE shear sense unequivocally prevails. Deformation–crystallization relationships indicate that the shearing took place from amphibolite- to greenschist-facies conditions. Micaceous mylonites, confined to gneiss levels at the base of the shear zone, display high-temperature ductile fabrics with no retrogression (Fig. 14b). In contrast, structurally higher fine-grained mélange-derived mylonites show retrogression of feldspar to sericite in pressure shadows of the feldspar clasts, and shear deformation involves mechanisms from ductile flow to brittle fracture (Fig. 14c and d). The microstructural characteristics together imply that the deformation occurred at decreasing temperatures and metamorphic grade from the ductile to the brittle shear regime. All these data demonstrate that the ASZ exhibits extensional characteristics. This pattern is deduced not only from its structural geometry and kinematics, which indicate a major normal sense of movement throughout the shear zone, but also from a well-defined metamorphic transition from sillimanite-bearing gneisses at the lowest structural levels to chlorite schists at shallow structural levels, towards the top of mylonite zone. We interpret the ASZ as a major extensional ductile–brittle displacement zone, responsible for the exhumation of the NW flank of the Kazdağ Massif. Moreover, the metamorphic break between the high-grade rocks and the mélange is consistent with an extensional interpretation of the ASZ (e.g. Wheeler & Butler 1994). The timing of the shear zone activity remains problematic. It commenced prior to 24 Ma as shown by Okay and Satır (2000*b*), or before 27 Ma as indicated by radiometric data of Delaloye and Bingöl (2000) for the Evciler pluton that intrudes the ASZ. However, it might have initiated even earlier in Paleocene–Eocene times as deduced from a single

$^{40}Ar/^{39}Ar$ mica age from a carbonate mylonite north of the ASZ (55.05 ± 6.2 Ma; Lips 1998).

Kinematics of the southern domain: the southern flank of the Kazdağ Massif

In the opposite southern flank of the Kazdağ Massif, the Şelale detachment (Beccaletto & Steiner 2005) separates the high-grade Kazdağ core rocks in the footwall from the Çetmi mélange and the Lower Miocene sedimentary Küçükkuyu Formation in the hanging wall (Fig. 15). The Küçükkuyu Formation is interpreted as a supradetachment basin, whose filling was coeval with the detachment activity. The Şelale detachment is a low-angle (<20°) south-dipping fault surface, whose regionally extensive trace still dominates the topography of the southern flank of the Kazdağ Massif. The detachment fault plane is marked by brecciation and cataclasis of the footwall marbles and gneisses. Slip lineations on the detachment surface trend NNE–SSW and plunge SSW, and associated slickenside striations reveal a ubiquitous top-to-the-SSW, sense of brittle movement (Fig. 15). Footwall gneisses exhibit mylonitic fabrics with top-to-the south shear sense. The bulk of the shearing took place during retrogression from the amphibolite to the greenschist facies of the footwall rock, which is more pronounced towards the detachment (chlorite is common in shear bands and replaces earlier mineral phases). A remarkable late deformation feature is represented by abundant east–west-trending and south-dipping step-like normal faults, Pliocene–Quaternary in age, that cross-cut the detachment surface at depth and all the above-mentioned units. The Küçükkuyu Formation records clastic fluviatile–lacustrine syntectonic sedimentation at the base, which progressed under tectonic control to shale-dominated rhythmic turbiditic basin fill toward the top of the sequence, accompanied by volcanic activity. The stratigraphic age of the Küçükkuyu Formation (Early Miocene), and the 29.94 ± 0.37 Ma U–Pb zircon crystallization age of granitoids cut by the detachment itself, provides a latest Oligocene lower age limit for the onset of the Şelale detachment activity (Beccaletto & Steiner 2005). The Şelale detachment is interpreted as contributing to the initial exhumation of the Kazdağ Massif along its southern flank. The exhumation process was subsequently assisted by high-angle step-like normal faulting from the Pliocene–Quaternary onwards, delineating the southerly neotectonic Edremit graben. The regional tectonic regime was possibly influenced by escape tectonics and associated compressive stresses along the NAFS. Data from the southern flank of the Kazdağ Massif imply therefore a two-stage

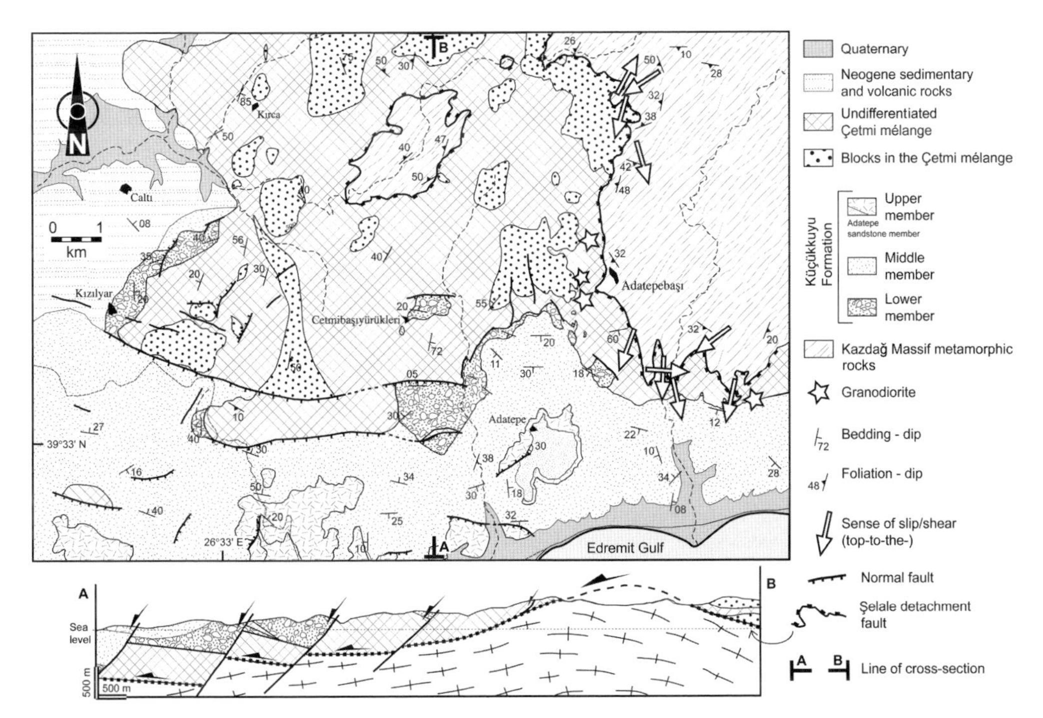

Fig. 15. Geological map of the southern flank of the Kazdağ Massif showing structural relationships and kinematics around the Şelale detachment (simplified after Beccaletto & Steiner 2005).

extensional exhumation, respectively in latest Oligocene and Pliocene–Quaternary times, as also proposed for the northern flank (Okay & Satır 2000*b*).

In summary, the metamorphic basement of the Biga Peninsula documents a NNE–SSW to NE–SW kinematic direction, delineated by stretching lineations associated with ductile–brittle fabrics. The latter show pervasive NNE-directed extension-related shear in the Kemer micaschists and northwestern flank of the Kazdağ Massif, and SW-directed shear in the southwestern flank of the Kazdağ Massif.

Discussion

Kinematic frame

The kinematic framework in the exhumed basement rocks of the eastern Rhodope–Thrace and the Biga Peninsula shows a regionally consistent NNE–SSW to NE–SW orientation of the stretching lineations. It represents the kinematic direction during the Tertiary ductile extension, thus defining an overall NNE–SSW trend of extension direction (Figs 3, 11, 13 and 15). Palaeomagnetic studies show no rotation of the eastern Rhodope–Thrace region from the Oligocene onwards (Kissel *et al.* 1986; Kissel & Laj 1988), implying that the kinematic direction (*c.* 22°) recorded in the basement rocks in the footwall of detachments reflects the primary attitude of the NNE–SSW-oriented extension direction. In the case of the Kazdağ Massif, the NNE–SSW-oriented kinematic direction (*c.* 15°) in the basement rocks parallels the NNE–SSW trend of palaeomagnetic directions in the overlying Oligocene–Miocene volcanic rocks, showing a very minor component of clockwise rotation (Kissel & Laj 1988). Hence, the kinematic direction recorded at the latitude of the Kazdağ Massif is very close to the original attitude of the extension direction. With regard to the kinematic direction of extension in the northern Biga Peninsula, local variations appear compared with the Kazdağ Massif. More specifically, this refers to the more easterly oriented kinematic direction observed in the Kemer micaschists (*c.* 45°). In these metamorphic rocks, the trend of the kinematic direction appears to be drag against late faults, and/or to record differential block rotations. Both processes are linked to the neotectonic and active deformation along the NAFS, whose anastomosing splays propagated into the Biga Peninsula. Indeed, the kinematic direction strikes more easterly at the vicinity of the fault contact bounding the micaschists strip in the south and along the coast, where it is close to the fault splays of the NAFS and the Marmara Sea pull-apart basin (Armijo *et al.* 1999).

Overall, we consider that the bulk NNE–SSW kinematic direction of ductile extension as recorded in the basement rocks both in the eastern Rhodope–Thrace and the Biga Peninsula generally fits its original attitude, and thus reflects the extension direction for the whole study region. Interestingly, the kinematic direction in the northernmost Aegean region, as outlined in this study, parallels the trend of the restored latest Oligocene extension direction (*c.* 23°) in the central Aegean (Walcott & White 1998). This suggests that the extension direction probably remained close to the NNE–SSW orientation during Eocene–Oligocene to Pliocene times in the central–northern Aegean region. It also coincides with the NNE–SSW-trending corridor of no significant rotation in the north Aegean that divides the rotated western and eastern Aegean domains since the Eocene–Oligocene (Kissel *et al.* 2003).

From syn- to post-orogenic extension

Temporal constraints on the kinematic framework of the Tertiary extension in the northernmost Aegean region can be best understood in terms of combined stratigraphic information in supradetachment basins, intrusion ages of late- to post-tectonic granitoids, and radiometric data on the metamorphic–cooling history of the basement rocks in the footwalls of detachments and/or in the vicinity of extensional shear zones. A summary of the distinct events in the investigated areas, outlined in the previous sections, is given in Figure 16.

In the eastern Rhodope–Thrace of Bulgaria, the Maastrichtian to Paleocene–Early Eocene (Goranov & Atanasov 1992) supradetachment syntectonic deposits in the hanging wall of the Tokachka detachment north of the Kesebir–Kardamos dome indicate an Early Eocene minimum age for the extension (Bonev *et al.* 2006*a*). In northeastern Greece, between the Kesebir–Kardamos and the Byala reka–Kechros domes, Middle Eocene (Lutetian–Bartonian) clastic deposits and limestones are transgressive onto the metamorphic basement, filling the base of a fault-bounded basin NE of Komotini (Karfakis & Doutsos 1995; see fig. 3). In the Biga Peninsula, the oldest sediments unconformably overlying the Kemer micaschists are Early Eocene clastic deposits and volcanic rocks (Siyako *et al.* 1989). Eclogite-facies metamorphism in both the upper high-grade unit of the eastern Rhodope at 73–71 Ma and the Çamlıca micaschists at 69–65 Ma in the Biga Peninsula indicates a very close, time-equivalent high-pressure metamorphic history, which is largely coeval with the upper crustal extension-related sedimentation accompanying this earlier exhumation stage of metamorphic rocks from deeper crustal levels. The intrusion age of the

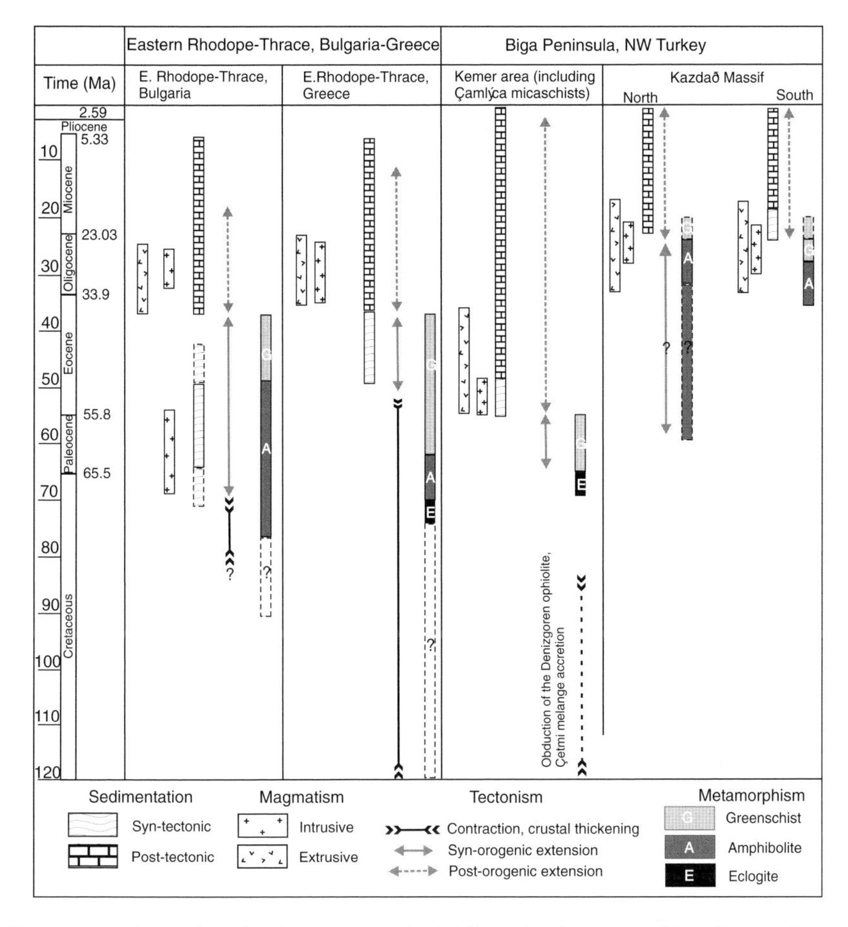

Fig. 16. Summary correlative chart showing age-constrained deformational, metamorphic and magmatic events, and sedimentation in the eastern Rhodope–Thrace and the Biga Peninsula.

earliest late- to post-tectonic granitoids at *c.* 69–53 Ma in eastern Rhodope, marking the cessation of contraction and the initiation of extension in this area, is indistinguishable from the 53 Ma age of the post-tectonic Karabiga granite, which constrains the age of ductile extension in the Kemer micaschists as pre-Eocene. The latter are lithologically and structurally comparable with the Çamlıca micaschists, and the two occurrences form a continuous metamorphic belt in the Biga Peninsula, separated by Tertiary volcanic rocks and sedimentary successions. As the peak metamorphism of the Çamlıca micaschists occurred in latest Maastrichtian time, the extensional deformation of the Kemer micaschists in medium-grade metamorphic conditions post-dated the high-pressure event, and thus must be Paleocene in age. Furthermore, the ductile extension in the Kemer micaschists as constrained above coincides in time with the pressure decrease to greenschist-facies metamorphic conditions at 61 Ma in the eastern Rhodope–Thrace area.

Therefore, there is a continuous Paleocene–early Eocene record of overlapping stratigraphic constraints and radiometric ages of high- to medium-pressure metamorphism and intrusive activity, for both the eastern Rhodope–Thrace and the Biga Peninsula. All these data collectively bracket an earlier phase of extension and exhumation processes in both areas.

We relate the Paleocene kinematic record of ductile extension in the Kemer micaschists, Paleocene–early Eocene syntectonic hanging-wall sedimentation and subsequent Middle Eocene cooling of the footwall of the Kesebir–Kardamos dome to an early synorogenic, NE-directed extension (Fig. 17). The synrogenic extension was NNE–SSW-directed in the Byala reka–Kechros dome, which shares with the previous units the same cooling history of the footwall, and contemporaneous syntectonic hanging-wall sedimentation (at least pre-Late Eocene, e.g. Boyanov & Goranov 2001) (see Fig. 3). The synorogenic nature of this

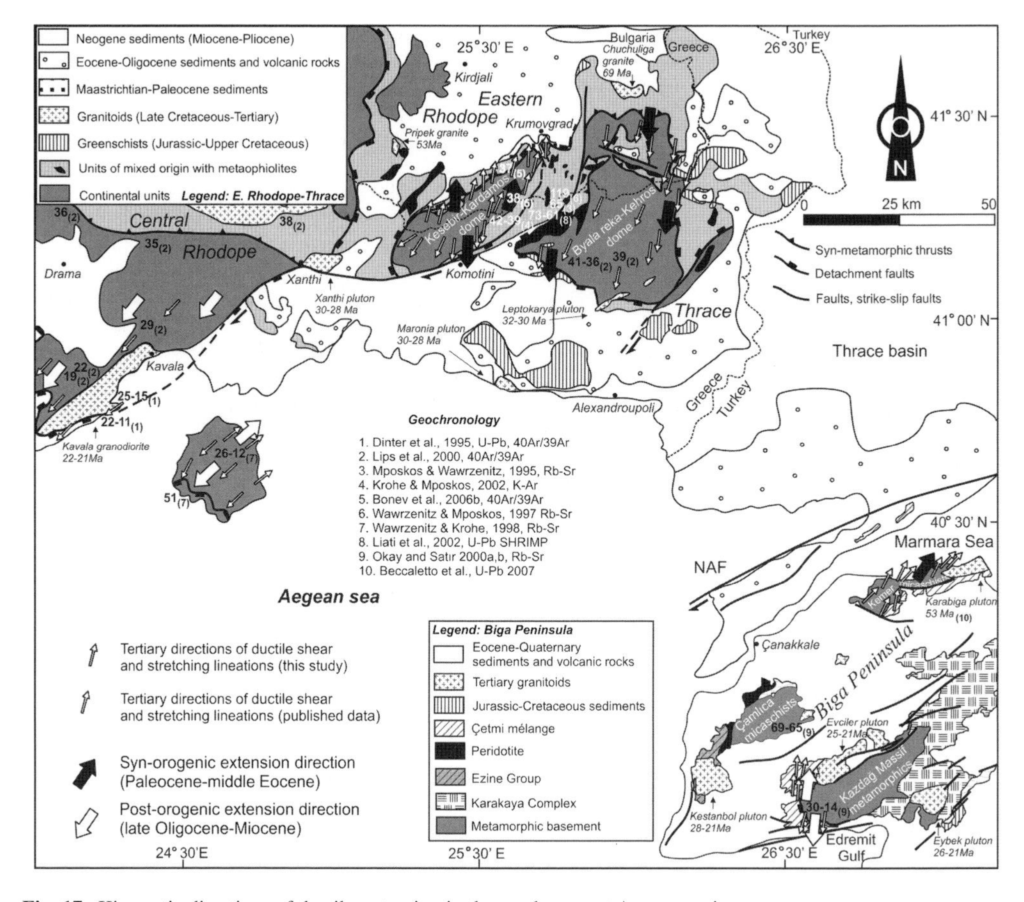

Fig. 17. Kinematic directions of ductile extension in the northernmost Aegean region.

extension is expressed by at least three critical points: (1) isothermal decompression of metamorphic rocks and preservation of the HP–LT paragenesis; (2) preservation of syn-stacking structures in the upper unit (e.g. hanging wall) of detachments where they are 'frozen', whereas these structures interfinger with coeval extensional fabrics in the lower unit (e.g. footwall) of detachments; (3) extensional deformation is time correspondent to the closure of the Vardar oceanic basin and the overall collisional context (see below).

In the eastern Rhodope–Thrace, post-detachment Late Eocene (Priabonian)–Oligocene sediments represent post-tectonic sequences (Bonev *et al.* 2006*a*). The sediments are interstratified with widespread volcanic rocks having a wide compositional range, and volcano-sedimentary successions with similar ages (37–25.5 Ma) (Lilov *et al.* 1987). The volcanic activity was accompanied between 32 and 28 Ma by the intrusion of post-tectonic granitoid bodies (e.g. Xanthi, Leptokarya–Kirki, Maronia; see fig. 3), that systematically cut through the metamorphic basement or the volcano-sedimentary successions (Del Moro *et al.* 1988; Koukouvelas & Pe-Piper 1991). The post-tectonic sedimentation persisted in Miocene to Pliocene–Quaternary times, extending southward into the southern Rhodope and Thrace basins.

Equivalents of the late Eocene–Oligocene sedimentation and magmatic activity described in the eastern Rhodope–Thrace are also found in NW Turkey. The earliest dated volcanic rocks at 37–28 Ma (Ercan *et al.* 1995) occur in the northern Biga Peninsula, accompanied by contemporaneous clastic and volcaniclastic successions. The latest Oligocene period also corresponds to the intrusion of several granitic plutons into shallow crustal levels, dated at 20.5–28 Ma for the Kestanbol pluton (Fytikas *et al.* 1976; Birkle & Satır 1995; Delaloye & Bingöl 2000), at 31–21 Ma for the Evciler pluton (Birkle & Satır 1995; Delaloye & Bingöl 2000; Okay & Satır 2000*b*, at 21–26 Ma for the Eybek pluton (Krushensky 1976; Delaloye & Bingöl 2000), and at 30 Ma for small granitoid bodies in the southern Kazdağ Massif (Beccaletto

& Steiner 2005) (Figs 10 and 17). These granitoids intruded the extensional shear zones or were cut by detachments. The widespread Oligocene–Miocene calc-alkaline volcanic rocks (Aldanmaz *et al.* 2000; Yılmaz *et al.* 2001), cogenetic with the plutonic bodies, are accompanied by time-equivalent sedimentation, which persisted into the Pliocene–Quaternary period (Siyako *et al.* 1989; Yılmaz & Karacık 2001). The early Miocene syntectonic sedimentation in the hanging wall of the Şelale detachment in the southern flank of the Kazdağ Massif, coeval with extrusive and intrusive magmatism in both flanks of the massif, is an expression of the subsequent latest Oligocene–early Miocene extensional stage in the southern Biga Peninsula. Therefore, we relate the kinematic pattern of the basement rocks of the Kazdağ Massif to post-orogenic extension following the earlier phase of Paleocene–early Eocene extension described above. In NW Turkey, the history of the sedimentary record in the Neogene basins, as well as the timing of the intrusive magmatism, is similar to the sedimentary basin fill, igneous activity and metamorphic basement cooling history of the late, second phase of extension in southern Greek Rhodope (Figs 16 and 17). There, the Thasos metamorphic core complex (Wawrzenitz & Krohe 1998) and the so-called Rhodope core complex (Dinter & Royden 1993; Dinter 1998, and references therein), have been related to latest Oligocene–Miocene post-orogenic Aegean back-are extension.

Close spatial and temporal relationships between the magmatic bodies and the shear deformation of the studied metamorphic terranes reveal interplay between magmatism and extensional tectonics. We consider that the granitic magmatism has induced thermal weakening of the thickened crust, thus facilitating ductile flow at depth. This is particularly true for the latest Cretaceous–earliest Eocene granitoids in the eastern Rhodope. There, coeval or subsequent partial melting of the crustal pile is expressed by diatexites–metatexites and migmatitic vein networks occurring respectively in the footwall and hanging wall of detachments. This role appears to be prominent also for the latest Oligocene–early Miocene granitoids associated with the late extensional phase. They are more abundant at the regional scale, implying that the advanced partial melting of the subcontinental lithosphere was broadly coeval with this latter extensional deformation.

Geodynamic context

Following and/or coeval with the Late Cretaceous crustal thickening in the region, the NE–SW-oriented extension in the north Aegean commenced earlier to the north, in Paleocene–early Eocene times (i.e. eastern Rhodope, northern Biga Peninsula). This early, synorogenic extension is contemporaneous with the closure of the Vardar Ocean, whose suture lies south of the study area, and the subsequent collision between the Pelagonia terrane and the Rhodope margin (e.g Stampfli & Borel 2004). The Vardar Ocean finally sutured in the Early Tertiary, before the intrusion of the Early Eocene Sithonia granite in the Vardar Zone (51 Ma Rb–Sr isochron, Christofides *et al.* 1990) (Fig. 18a). In western Turkey, the counterpart Izmir–Ankara suture zone encompasses time-equivalent Early Eocene intrusive events, following the Paleocene closure of the oceanic space north of the Sakarya continent, and the subsequent collision of the latter with the Anatolide–Tauride block and related southward thrusting (Harris *et al.* 1994; Okay & Tüysüz 1999). Then, the synorogenic extension was followed in the study area by a latest Oligocene–Miocene post-orogenic north Aegean extension, as exemplified by the Kazdağ massif in the southern Biga Peninsula (see Okay & Satir 2000*b*), and expressed in a similar way in the southern Rhodope (Fig. 18b). This later extension, which evolved further south compared with the synorogenic extension, is easier to identify because of its association with the Miocene detachments (e.g. Thasos detachment system, Wawrzenitz & Krohe 1998; Strymon valley detachment, Dinter & Royden 1993). This post-orogenic extension displays timing of tectono-metamorphic and magmatic patterns corresponding to the well-known extension in the Aegean domain that occurred from the latest Oligocene onwards (e.g. Jolivet & Patriat 1999; Gautier *et al.* 1999, and references therein).

The origin of the latest Oligocene–present Aegean extension has been variously attributed to distinct models, including back-arc spreading, tectonic escape, gravitational orogenic collapse, or a combination of two or more of these mechanisms (e.g. Le Pichon & Angelier 1979; Şengör *et al.* 1985; Jolivet *et al.* 1994; Seyitoğlu & Scott 1996; Koçyiğit *et al.* 1999).

The synorogenic extension in the eastern Rhodope, at the scale of the Kesebir–Kardamos dome, has been related to the gravitational adjustment of the orogenic wedge (e.g. Platt 1986; Dewey 1988), created in the subduction–collisional setting of the Vardar Zone. This gravitational collapse thus accommodated the northward extension and unroofing of this part of the Rhodope nappe stack (Bonev *et al.* 2006*a*). When restored to the situation before the post-Pliocene displacement along the NAFS, along which the Anatolia moved westward to southwestward (*c.* 85 km, Armijo *et al.* 1999), the Biga Peninsula would approximately be aligned in latitude with the eastern Rhodope–Thrace area, as indicated by

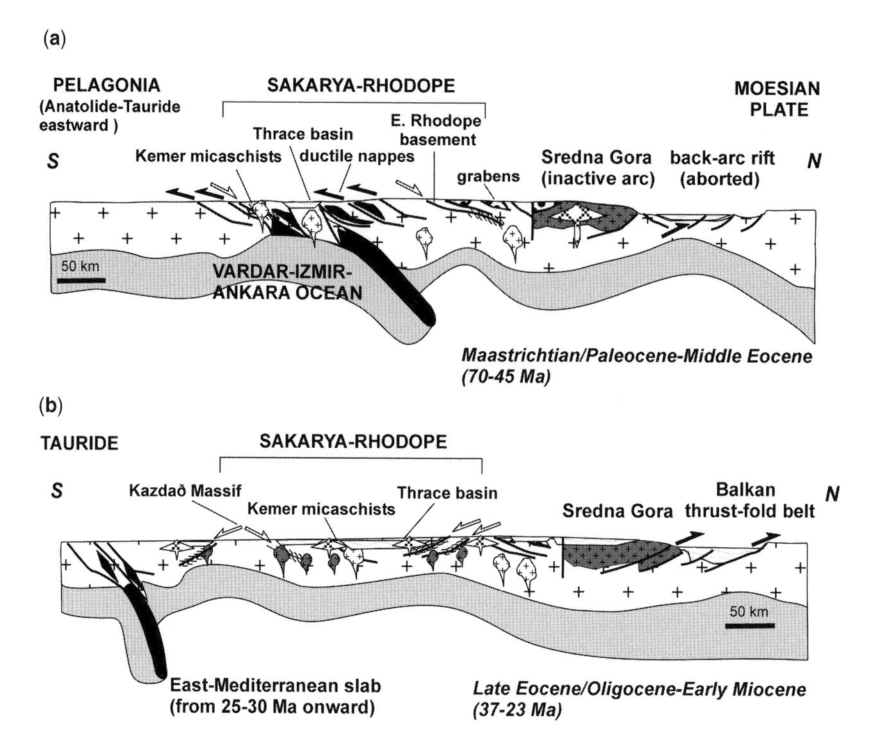

Fig. 18. Tectonic scenario for the Tertiary syn- to post-orogenic extension in the northernmost Aegean region.

the Miocene palaeogeography around the Marmara Sea (Sakınç & Yaltırak 2005, fig. 4). This implies that the ductile extension recorded in the Kemer micaschists and eastern Rhodope account for a common extensional process, along the same strike as that of the Alpine orogen. We therefore consider that both the eastern Rhodope–Thrace and northern Biga Peninsula represent a comparable crustal domain, which has accommodated hinterland-directed extensional unroofing and exhumation of the orogenic stack at the beginning of the Tertiary. Gravitationally induced orogenic collapse and coeval erosion may be important mechanisms during this early stage of extension. Paleocene–Eocene sedimentation in fault-bounded grabens (up to 1 km sedimentary fill) of the eastern Rhodope persisted into thick Eocene successions (>2 km) in the Thrace basin, which separate this area from the Biga Peninsula (Fig. 17). The sedimentation in the Thrace basin continued during the Miocene and to the present under conditions of continuing extension and active strike-slip faulting. In NW Turkey, there are petrological indications for significant crustal thickening prior to the Oligocene–Miocene post-collisional magmatism (Aldanmaz *et al.* 2000; Yılmaz *et al.* 2001). Collectively, the above data suggest that significant erosion, consistent with the removal of relatively high topography, and substantial reduction of crustal thickness took place prior to the phase of post-orogenic extension.

It is widely recognized that the post-orogenic Aegean extension resulted from back-arc spreading above the roll-back of the subducting East Mediterranean slab. From the latest Oligocene–Early Miocene to the present, the southward trench retreat has been accompanied by the migration of the extension, magmatism and sedimentation in the same direction (e.g. Innocenti *et al.* 1984; Jolivet *et al.* 1999). Other evidence accumulated in recent years indicates that the slab retreat at the Eurasian plate margin may have started earlier in the Cretaceous. The volcanic arc activity of the Sredna Gora Zone shows southward younging of the magmatism from 92 to 78 Ma (von Quadt *et al.* 2005) across the corresponding arc system. Following the latest Cretaceous cessation of volcanic activity, the intrusive magmatic activity subsequently transferred southward to the Rhodope. From Campanian (80 Ma granitoids in the central Rhodope, Peytcheva *et al.* 1998) to Maastrichtian–Early Eocene (69–53 Ma granitoids in the eastern Rhodope, Ovtcharova *et al.* 2003; Marchev *et al.* 2004*a*; and in the Biga Peninsula, Beccaletto *et al.*

2007), and finally Oligocene–Early Miocene time (32–21 Ma), granitoids in the Rhodope–Thrace area and NW Turkey systematically intruded the gradually cooling and exhuming metamorphic pile (Fig. 17). This implies the retreat of the southward-migrating subduction boundary, and also possible slab break-off of the subducting slab, as shown by seismic studies indicating the remnant of subducted fragments under the Rhodope (Shanov *et al.* 1992). Mantle seismic tomography has also documented the primary role of slab roll-back in the tectonic evolution of the Aegean since the Oligocene (Wortel & Spatman 2000). Spatial and temporal southward migration of extension accompanied by magmatism and sedimentation, starting from the northern slope of the Rhodope and extending to the central–northern Aegean region, calls for a progressive shift of the subduction boundary along the Eurasian plate margin at least from the latest Cretaceous onwards. In addition, geophysical data indicate comparable crustal thickness (*c.* 30 km) for the eastern Rhodope–Thrace and the Biga Peninsula, as well as uniform distribution of the vertical extension in their lithosphere (Papazachos & Skordilis 1998; Saunders *et al.* 1998). The latter is stretched by a factor of 1.2–2 (Saunders *et al.* 1998), recording the Tertiary extension in that part of the northern Aegean region.

Overall, the Paleocene–Miocene extensional kinematic framework defines the study region as an important extended domain in the north Aegean, which underwent syn- and post-orogenic extension. This pattern is similar to that of the central–southern Aegean and other parts of the Mediterranean region (e.g. Tyrrhenian Sea) (Jolivet & Goffé 2000). The geodynamic setting, which controlled the long-lasting (latest Cretaceous to present) record of extension in the north Aegean, therefore combines both gravitational collapse of overthickened crust and back-arc spreading above the retreating subduction boundary. The two processes are not exclusive, and complement each other as modes of crustal response during the syn- to post-orogenic evolution of the present-day Aegean region.

Conclusions

The main results of the investigations on the kinematic pattern of the metamorphic basement in the eastern Rhodope–Thrace and the Biga Peninsula regions of the northern Aegean region may be summarized as follows:

(1) The orientation of the stretching lineations and associated asymmetric shear fabrics in the exhumed ductile crust of the studied areas delineates a regionally consistent NNE–SSW to NE–SW kinematic direction. Generally, as indicated by palaeomagnetic data, this kinematic direction reflects the original attitude of the extension direction, relative to the present-day reference frame. Extensional shear fabrics records ductile–semiductile to brittle deformation conditions below the low-angle extensional detachments. The extensional deformation follows the peak metamorphism that is consistent with late orogenic stage extension and exhumation.

(2) This kinematic framework, combined with data from the regional geochronology (metamorphic–cooling ages in the basement rocks, crystallization ages of magmatic rocks) and stratigraphic evidence in the hanging wall of the detachments (syn- and post-tectonic sedimentary fill), provides constraints for the timing of extensional deformation in the region. The deformation includes both syn- and post-orogenic extensional phases that took place from the Paleocene to the Miocene.

(3) The kinematic analysis indicates a NE–SW-directed synorogenic extension in Paleocene–Early Eocene times recorded in the northern domain of the Biga Peninsula and metamorphic domes in eastern Rhodope–Thrace. In the latter region, the synorogenic extension was also NNE–SSW-directed in continuity with the kinematic direction of the nappe stacking, but in lower-temperature deformation conditions, involving retrogression during extension. Synorogenic extension relates to the latest Cretaceous–Early Tertiary subduction–collisional history of the Vardar Ocean, being coeval with its Paleocene closure. Hence, this episode of extension accommodated an early hinterland-directed exhumation of the Alpine orogenic stack.

(4) Post-orogenic extension is recorded in the south in the Kazdağ Massif. It occurred during latest Oligocene–Miocene times, following suturing and collision within the region. Kinematically, it involves SSW- and NNE-directed ductile–brittle displacements in the opposite flanks of the massif, assisted respectively by a low-angle detachment and a mylonitic shear zone. The history of this extensional phase is consistent with that of the Aegean post-orogenic back-arc extension, widely recognized in the central Aegean region. It is also time equivalent and kinematically similar to the metamorphic core complexes in the southern Greek Rhodope north of the Aegean region.

(5) Regional-scale correlations of the studied metamorphic terranes in the northern Aegean suggest that they were kinematically coupled during the Tertiary extension. They also show many other common features associated with the extensional tectonics, such as coeval or subsequent magmatism and syn- to post-tectonic sedimentation. The spatial and temporal consistency of structures, kinematics and deformational events indicates that the eastern Rhodope–Thrace and

Biga Peninsula collectively represent an extensional domain, which underwent syn- and post-orogenic extension in the northernmost part of the Late Alpine Aegean extensional province.

Work in Turkey was supported by the University of Lausanne 450th Anniversary Fund and the Academic Society of Vaud Canton, Switzerland. The University of Lausanne supported the work in Greece of N.B. The manuscript was improved following reviews by B. C. Burchfiel, E. Bozkurt, and I. Koukouvelas, and by support on editorial matters at every stage of this paper T. Taymaz, and we thank them for helpful suggestions and comments.

References

ALDANMAZ, E., PEARCE, J. A., THIRLWALL, M. F. & MITCHELL, J. G. 2000. Petrogenetic evolution of late Cenozoic, post-collisional volcanism in western Anatolia, Turkey. *Journal of Volcanology and Geothermal Research*, **102**, 67–95.

ALTINER, D., KOÇYIĞIT, A., FARINACCI, A., NICOSIA, U. & CONTI, M. A. 1991. Jurassic, Lower Cretaceous stratigraphy and paleogeographic evolution of the southern part of the northwestern Anatolia. *Geologica Romana*, **18**, 13–80.

ANIL, M., SAUPE, F., ZIMMERMANN, J. L. & ÖNGEN, S. 1989. K/Ar age determination of the Oligo-Miocene Nevruz–Çakıroba (Yenice–Çanakkale) quartz-monzonite stocks. *In*: *43rd Geological Congress of Turkey, Abstracts*, 25–26.

ARMIJO, R., MEYER, B., HUBERT, A. & BARKA, A. 1999. Westward propagation of the north Anatolian fault into the northern Aegean: timing and kinematics. *Geology*, **27**, 267–270.

BECCALETTO, L. & JENNY, C. 2004. Geology and correlation of the Ezine Zone: a Rhodope fragment in NW Turkey? *Turkish Journal of Earth Sciences*, **13**, 145–176.

BECCALETTO, L. & STEINER, C. 2005. Evidence of two-stage extensional tectonics from the northern edge of the Edremit Graben, NW Turkey. *Geodinamica Acta*, **18**, 283–297.

BECCALETTO, L., BARTOLINI, A.-C., MARTINI, R., HOCHULI, P. A. & KOZUR, H. 2005. Biostratigraphic data from the Çetmi melange, northwest Turkey: palaeogeographic and tectonic implications. *Palaeogeography, Paleoclimatology, Palaeoecology*, **221**, 215–244.

BECCALETTO, L., BONEV, N., BOSCH, D. & BRUGUIER, O. 2007. Record of a Paleogene syn-collisional extension in the north Aegean region: evidence from the Kemer micaschists (NW Turkey). *Geological Magazine*, **144**, 393–400.

BINGÖL, E., AKYÜREK, B. & KORMAZER, B. 1975. Geology of the Biga peninsula and some characteristics of the Karakaya blocky series. *In*: ENSTITÜSÜ, M. T. (ed.) *Proceedings of the Congress of Earth Sciences on the Occasion of the 50th Anniversary of the Turkish Republic*. General Directorate of Mineral Research and Exploration of Turkey (MTA), Ankara, 71–77.

BIRKLE, P. & SATIR, M. 1995. Dating, geochemistry and geodynamic significance of the Tertiary magmatism of the Biga Peninsula, NW Turkey. *In*: ERLER, A., ERCAN, T., BINGÖL, E. & ÖRÇEN, S. (eds) *Geology of the Black Sea Region*. Mineral Research Exploration Institute, Ankara, 171–180.

BONEV, N. 1996. Tokachka shear zone southwest of Krumovgrad in the Eastern Rhodopes, Bulgaria: an extensional detachment. *Annuaire de l'Universite de Sofia, Faculté de Géologie et Géographie, Livre 1 Géologie*, **89**, 97–106.

BONEV, N. G. 2001. Extension of syn-metamorphic thrust system in a part of Eastern Rhodope in the area north of Veykata summit, South Bulgaria. *Comptes Rendus de l'Académie Bulgare des Sciences*, **54**, 61–66.

BONEV, N. 2006. Cenozoic tectonic evolution of the eastern Rhodope Massif (Bulgaria): basement structure and kinematics of syn- to postcollisional extensional deformation. *In*: DILEK, Y. & PAVLIDES, S. (eds) *Post-collisional Tectonics and Magmatism in the Mediterranean Region and Asia*. Geological Society of America, Special Papers **409**, 211–235.

BONEV, N. & BECCALETTO, L. 2005. Northeastward ductile shear in the Kemer micaschists, Biga Peninsula (NW Turkey). *In*: TAYMAZ, T. (ed.) *International Symposium on the Geodynamics of Eastern Mediterranean: Active Tectonics of the Aegean Region*. Tübitak, Istanbul, Turkey, Abstracts 65.

BONEV, N. G. & STAMPFLI, G. M. 2003. New structural and petrologic data on Mesozoic schists in the Rhodope (Bulgaria): geodynamic implications. *Comptes Rendus, Géosciences*, **335**, 691–699.

BONEV, N., BURG, J.-P. & IVANOV, Z. 2006*a*. Mesozoic–Tertiary structural evolution of an extensional gneiss dome—the Kesebir–Kardamos dome, eastern Rhodope (Bulgaria–Greece). *International Journal of Earth Sciences*, **95**, 318–340.

BONEV, N., MARCHEV, P. & SINGER, B. 2006*b*. $^{40}Ar/^{39}Ar$ geochronology constraints on the Middle Tertiary basement extensional exhumation, and its relation to ore-forming and magmatic processes in the eastern Rhodope (Bulgaria). *Geodinamica Acta*, **19**, 265–280.

BOZKURT, E. 2001. Neotectonics of Turkey—a synthesis. *Geodinamica Acta*, **14**, 3–13.

BOZKURT, E. & MITTWEDE, S. K. 2001. Introduction to the geology of Turkey—a synthesis. *International Geology Review*, **43**, 578–594.

BOZKURT, E. & PARK, R. G. 1994. Southern Menderes Massif: an incipient metamorphic core complex in western Anatolia, Turkey. *Journal of the Geological Society, London*, **151**, 213–216.

BOZKURT, E. & PARK, R. G. 1997. Evolution of a mid-Tertiary extensional shear zone in the southern Menderes Massif, western Turkey. *Bulletin de la Société Géologique de France*, **168**, 3–14.

BOYANOV, I. & GORANOV, A. 2001. Late Alpine (Palaeogene) superimposed depressions in parts of Southeast Bulgaria. *Geologica Balcanica*, **31**, 3–36.

BOYANOV, I. & RUSSEVA, M. 1989. Lithostratigraphy and tectonic position of the Mesozoic rocks in the East Rhodopes. *Geologica Rhodopica*, **1**, 22–34.

BURCHFIEL, B. C., NAKOV, R. & TZANKOV, T. 2003. Evidence from the Mesta half-graben, SW Bulgaria, for the Late Eocene beginning of Aegean extension in the central Balkan Peninsula. *Tectonophysics*, **375**, 61–76.

BURG, J.-P., RICOU, L.-E., IVANOV, Z., GODFRIAUX, I., DIMOV, D. & KLAIN, L. 1996. Syn-metamorphic nappe complex in the Rhodope Massif. Structure and kinematics. *Terra Nova*, **8**, 6–15.

CHRISTOFIDES, G., D'AMICO, C., DEL MORO, A., ELEFTERIADIS, G. & KYRIAKOPULOS, C. 1990. Rb/Sr geochronology and geochemical characteristics of the Sithonia plutonic complex (Greece). *European Journal of Mineralogy*, **2**, 79–87.

DELALOYE, M. & BINGÖL, E. 2000. Granitoids from western and northwestern Anatolia: geochemistry and modeling of geodynamic evolution. *International Geology Review*, **42**, 241–268.

DEL MORO, A., INNOCENTI, F., KYRIAKOPOULOS, K., MANETTI, P. & PAPADOPOULOS, P. 1988. Tertiary granitoids from Thrace (northern Greece): Sr isotopic and petrochemical data. *Neues Jahrbuch für Mineralogie Abhandlungen*, **159**, 113–135.

DERCOURT, J., ZONENSHAIN, L. P., RICOU, L. -E. *ET AL.* 1986. Geological evolution of the Tethys belt from the Atlantic to the Pamirs since the Lias. *Tectonophysics*, **123**, 241–315.

DEWEY, J. 1988. Extensional collapse of orogens. *Tectonics*, **7**, 1123–1139.

DINTER, D. A. 1998. Late Cenozoic extension of the Alpine collisional orogen, northeastern Greece: origin of the north Aegean basin. *Geological Society of America Bulletin*, **110**, 1208–1230.

DINTER, D. A. & ROYDEN, L. 1993. Late Cenozoic extension in northeastern Greece: Strymon valley detachment system and Rhodope metamorphic core complex. *Geology*, **21**, 45–48.

DINTER, D. A., MACFARLANE, A. M., HAMES, W., ISACHSEN, C., BOWRING, S. & ROYDEN, L. 1995. U–Pb and $^{40}Ar/^{39}Ar$ geochronology of the Symvolon granodiorite: implications for the thermal and structural evolution of the Rhodope metamorphic core complex, northeastern Greece. *Tectonics*, **14**, 886–908.

DURU, M., PEHLIVAN, Ş., ŞENTÜRK, Y., YAVAŞ, F. & KAR, H. 2004. New results on the lithostratigraphy of the Kazdag Massif in northwest Turkey. *Turkish Journal of Earth Sciences*, **13**, 177–186.

ERCAN, T., SATIR, M., STEINITZ, G. *ET AL.* 1995. [Characteristics of Tertiary Volcanism in Biga Peninsula and Gökçeada, Bozccada and Tavçan Islands]. *Bulletin of Mineral Research and Exploration Institute of Turkey*, **117**, 55–86 [in Turkish with English abstract].

FYTIKAS, M., GUILIANO, O., INNOCENTI, F., MARINELLI, G. & MAZZUOLI, R. 1976. Geochronological data on recent magmatism of the Aegean Sea. *Tectonophysics*, **31**, 29–34.

GAUTIER, P. & BRUN, J.-P. 1994. Ductile crust exhumation and extensional detachments in the central Aegean (Cyclades and Evvia Islands). *Geodinamica Acta*, **7**, 57–85.

GAUTIER, P., BRUN, J.-P., MORICEAU, R., SOKOUTIS, D., MARTINOD, J. & JOLIVET, L. 1999. Timing, kinematics and cause of Aegean extension: a scenario based on a comparison with simple analogue experiments. *Tectonophysics*, **315**, 31–72.

GORANOV, A. & ATANASOV, G. 1992. Lithostratigraphy and formation conditions of Maastrichtian–Paleocene deposits in Krumovgrad District. *Geologica Balcanica*, **22**, 71–82.

GÜÇTEKIN, A., KÖPRÜBASI, N. & ALDANMAZ, E. 2004. Geochemistry of the Karabiga (Çanakkale) granitoid. *Yerbilimleri (Bulletin of Earth Sciences Application and Research Centre of Hacettepe University)*, **29**, 29–38.

HARKOVSKA, A., YANEV, Y. & MARCHEV, P. 1989. General features of the Palaeogene orogenic magmatism in Bulgaria. *Geologica Balcanica*, **19**, 37–72.

HARRIS, N. B. W., KELLEY, S. & OKAY, A. I. 1994. Post-collision magmatism and tectonics in northwest Anatolia. *Contributions to Mineralogy and Palaeontology*, **117**, 241–252.

HETZEL, R., PASSCHIER, C., RING, U. & DORA, O. 1995. Bivergent extension in orogenic belts: the Menderes massif (southwestern Turkey). *Geology*, **23**, 455–458.

INNOCENTI, F., KOLIOS, N., MANETTI, P., MAZZUOLI, R., PECCERILLO, A., RITA, F. & VILLARI, L. 1984. Evolution and geodynamic significance of Tertiary orogenic volcanism in northeastern Greece. *Bulletin of Volcanology*, **47**, 25–37.

IVANOV, R. & KOPP, K. O. 1969. Das Alttertiär Thrakiens und der Ostrhodope. *Geologica et Palaeontologica*, **3**, 123–153.

JACKSON, J. 1994. Active tectonics of the Aegean region. *Annual Review of Earth and Planetary Sciences*, **22**, 239–271.

JOLIVET, L. 2001. A comparison of geodetic and finite strain pattern in the Aegean, geodynamic implications. *Earth and Planetary Science Letters*, **187**, 95–104.

JOLIVET, L. & FACCENNA, C. 2000. Mediterranean extension and the Africa–Eurasia collision. *Tectonics*, **19**, 1095–1106.

JOLIVET, L. & GOFFÉ, B. 2000. Les dômes métamorphiques extensifs dans les chaînes de montagne. Extension syn-orogénique et post-orogénique. *Comptes Rendus de l'Académie des Sciences*, **330**, 739–751.

JOLIVET, L. & PATRIAT, M. 1999. Ductile extension and the formation of the Aegean Sea. *In*: DURAND, B., JOLIVET, L., HORVÁTH, F. & SÉRRANE, M. (eds) *The Mediterranean Basins: Tertiary Extension within the Alpine Orogen*. Geological Society, London, Special Publications, **156**, 427–456.

JOLIVET, L., BRUN, J.-P., GAUTIER, P., LALLEMANT, S. & PATRIAT, M. 1994. 3-D kinematics of extension in the Aegean from the Early Miocene to the present, insights from the ductile crust. *Bulletin de la Société Géologique de France*, **165**, 195–209.

KARFAKIS, I. & DOUTSOS, T. 1995. Late orogenic evolution of the Circum-Rhodope belt, Greece. *Neues Jahrbuch für Geologie und Paläontologie, Monatshefte*, **5**, 305–319.

KAUFFMANN, G., KOCKEL, F. & MOLLAT, H. 1976. Notes on the stratigraphic and paleogeographic position of the Svoula Formation in the innermost zones of the Hellenides (Northern Greece). *Bulletin de la Société Géologique de France*, **18**, 225–230.

KILIAS, A., FALALAKIS, G. & MOUNTRAKIS, D. 1999. Cretaceous–Tertiary structures and kinematics of the Serbomacedonian metamorphic rocks and their relation to the exhumation of the Hellenic hinterland

(Macedonia, Greece). *International Journal of Earth Sciences*, **88**, 513–531.

KISSEL, C. & LAJ, C. 1988. The Tertiary geodynamical evolution of the Aegean arc: a paleomagnetic reconstruction. *Tectonophysics*, **146**, 183–201.

KISSEL, C., KONDOPOULOU, D., LAJ, C. & PAPADOPOULOS, P. 1986. New paleomagnetic data from Oligocene formations of Northern Aegean. *Geophysical Research Letters*, **13**, 1039–1042.

KISSEL, C., LAJ, C., POISSON, A. & GÖRÜR, N. 2003. Paleomagnetic reconstruction of the Cenozoic evolution of the Eastern Mediterranean. *Tectonophysics*, **362**, 199–217.

KOCKEL, F., MOLLAT, H. & WALTHER, H. W. 1977. *Erlauterungen zur geologicschen Karte der Chalkidiki und angrenzender Gebiete 1/100.000 (Nord Griechenland)*. Bundesanstalt für Geowissenschaften und Rohstoffe, Hanover.

KOÇYIĞIT, A., YUSUFOĞLU, H. & BOZKURT, E. 1999. Evidence from the Gediz graben for episodic two-stages extension in western Turkey. *Journal of the Geological Society, London*, **156**, 605–616.

KOUKOUVELAS, I. K. & AYDIN, A. 2002. Fault structure and related basins of the North Aegean Sea and its surroundings. *Tectonics*, **21**, 10-1–10-17.

KOUKOUVELAS, I. & DOUTSOS, T. 1990. Tectonic stages along a traverse crosscutting the Rhodopian zone (Greece). *Geologische Rundschau*, **79**, 753–776.

KOUKOUVELAS, I. & PE-PIPER, G. 1991. The Oligocene Xanthi pluton, northern Greece: a granodiorite emplaced during regional extension. *Journal of the Geological Society, London*, **148**, 749–758.

KOZHOUKHAROVA, E. 1998. Eclogitization of serpentinites into narrow shear zones from the Avren syncline, Eastern Rhodopes. *Geochemistry, Mineralogy and Petrology*, **35**, 29–46.

KROHE, A. & MPOSKOS, E. 2002. Multiple generations of extensional detachments in the Rhodope Mountains (northern Greece): evidence of episodic exhumation of high-pressure rocks. *In*: BLUNDELL, D. J., NEUBAUER, F. & VON QUADT, A. (eds) *The Timing and Location of Major Ore Deposits in an Evolving Orogen*. Geological Society, London, Special Publications, **204**, 151–178.

KRUSHENSKY, R. D. 1976. Neogene calc-alkaline extrusive and intrusive rocks of the Karalar–Yeşiler area, northwest Anatolia. *Bulletin of Volcanology*, **40**, 336–360.

LE PICHON, X. & ANGELIER, J. 1979. The Hellenic Arc and trench system: a key to the neotectonic evolution of the eastern Mediterranean area. *Tectonophysics*, **60**, 1–42.

LE PICHON, X., CHAMOT-ROOKE, N., LALLEMANT, S. L., NOOMEN, R. & VEIS, G. 1995. Geodetic determination of kinematics of Central Greece with respect to Europe: implications for eastern Mediterranean tectonics. *Journal of Geophysical Research*, **100**, 12675–12690.

LIATI, A. 2005. Identification of repeated Alpine (ultra) high-pressure metamorphic events by U–Pb SHRIMP geochronology and REE geochemistry of zircon: the Rhodope zone of Northern Greece. *Contributions to Mineralogy and Petrology*, **150**, 608–630.

LIATI, A., GEBAUER, D. & WYSOCZANSKI, R. 2002. U–Pb SHRIMP-dating of zircon domains from UHP garnet-rich mafic rocks and late pegmatoids in the Rhodope zone (N Greece): evidence for Early Cretaceous crystallization and Late Cretaceous metamorphism. *Chemical Geology*, **184**, 281–299.

LILOV, P., YANEV, Y. & MARCHEV, P. 1987. K/Ar dating of the Eastern Rhodope Paleogene magmatism. *Geologica Balcanica*, **17**, 49–58.

LIPS, A. L. W. 1998. *Temporal constraints on the kinematics of the destabilization of an orogen—syn- to post-orogenic extensional collapse of the northern Aegean region*. PhD thesis, Vrije Universiteit, Amsterdam.

LIPS, A. L. W., WHITE, S. H. & WIJBRANS, J. R. 2000. Middle–Late Alpine thermotectonic evolution of the southern Rhodope Massif, Greece. *Geodinamica Acta*, **13**, 281–292.

LISTER, G. S., BANGA, G. & FEENSTRA, A. 1984. Metamorphic core complexes of cordilleran-type in the Cyclades, Aegean Sea, Greece. *Geology*, **12**, 221–225.

LYBERIS, N. 1984. Tectonic evolution of the North Aegean Trough. *In*: DIXON, J. E. & ROBERTSON, A. H. F. (eds), *The Geological Evolution of the Eastern Mediterranean*. Geological Society, London, Special Publications, **17**, 709–725.

MACHEVA, L. A. 1998. 3T-phengites in the rocks of Biala reka metamorphic group: an indicator for high-pressure metamorphism. *Geochemistry, Mineralogy and Petrology* **35**, 17–28.

MARCHEV, P., SINGER, B., ANDREW, C., HASSON, S., MORITZ, R. & BONEV, N. 2003. Characteristics and preliminary $^{40}Ar/^{39}Ar$ and $^{87}Sr/^{86}Sr$ data of the Upper Eocene sedimentary-hosted low-sulfidation gold deposits, Ada Tepe and Rosino, SE Bulgaria: possible relation with core complex formation. *In*: ELIOPOULOS, D. G., BAKER, T., BAUCHOT, V. ET AL. (eds) *Mineral Exploration and Sustainable Development, 2*. Millpress, Rotterdam, 1193–1196.

MARCHEV, P., RAICHEVA, R., DOWNES, H., VASELLI, O., CHIARADIA, M. & MORITZ, R. 2004*a*. Compositional diversity of Eocene–Oligocene basaltic magmatism in the Eastern Rhodopes, SE Bulgaria: implications for genesis and tectonic setting. *Tectonophysics*, **393**, 301–328.

MARCHEV, P., SINGER, B., JELEV, D., HASSON, S., MORITZ, R. & BONEV, N. 2004*b*. The Ada Tepe deposit: a sediment-hosted, detachment fault-controlled, low-sulfidation gold deposit in the Eastern Rhodopes, SE Bulgaria. *Schweizeriche Mineralogishe und Petrographishe Mitteilungen*, **84**, 59–78.

MCCLUSKY, S., BALASSANIAN, S., BARKA, A. ET AL. 2000. Global Positioning System constraints on plate kinematics and dynamics in the eastern Mediterranean and Caucasus. *Journal of Geophysical Research*, **105**, 5695–5720.

MCKENZIE, D. 1978. Active tectonics of the Alpine–Himalayan Belt, the Aegean Sea and surrounding regions. *Geophysical Journal of the Royal Astronomical Society*, **55**, 217–254.

MEULENKAMP, J. E., WORTEL, M. J. R., WAN WAMEL, W. A., SPAKMAN, W. & HOOGERDUYN STRATING, E. 1988. On the Hellenic subduction zone and the geodynamic evolution of Crete since the late Middle Miocene. *Tectonophysics*, **146**, 203–215.

MEYER, W. 1968. Alterstellung des Plutonismus im Südteil der Rila–Rhodope-Masse. *Geologie und Paläontologie*, **2**, 177–192.

MPOSKOS, E. D. & KOSTOPOULOS, D. K. 2001. Diamond, former coesite and supersilicic garnet in metasedimentary rocks from the Greek Rhodope: a new ultrahigh-pressure metamorphic province established. *Earth and Planetary Science Letters*, **192**, 497–506.

MPOSKOS, E & LIATI, A. 1993. Metamorphic evolution of metapelites in the high-pressure terrane of the Rhodope zone, northern Greece. *Canadian Mineralogist*, **31**, 401–424.

MPOSKOS, E. & WAWRZENITZ, N. 1995. Metapegmatites and pegmatites bracketing the time of HP-metamorphism in polymetamorphic rocks of the E. Rhodope, northern Greece: petrological and geochronological constraints. Geological Society of Greece Special Publication, **4**, 602–608.

MUKASA, S., HAYDOUTOV, I., CARRIGAN, C. & KOLCHEVA, K. 2003. Thermobarometry and $^{40}Ar/^{39}Ar$ ages of eclogitic and gneissic rocks in the Sredna Gora and Rhodope terranes of Bulgaria. *Journal of the Czech Geological Society*, **48**, 94–95.

OKAY, A. I. & GÖNCÜOGLU, M. C. 2004. The Karakaya Complex, a review of data and concepts. *Turkish Journal of Earth Sciences*, **13**, 77–95.

OKAY, A. I. & SATIR, M. 2000*a*. Upper Cretaceous eclogite-facies metamorphic rocks from the Biga Peninsula, northwest Turkey. *Turkish Journal of Earth Sciences*, **9**, 47–56.

OKAY, A. I. & SATIR, M. 2000*b*. Coeval plutonism and metamorphism in a latest Oligocene metamorphic core complex in northwest Turkey. *Geological Magazine*, **137**, 495–516.

OKAY, A. I. & TÜYSÜZ, O. 1999. Tethyan sutures of northern Turkey. *In*: DURAND, B., JOLIVET, L., HORVÁTH, F. & SÉRRANE, M. (eds) *The Mediterranean Basins: Tertiary Extension within the Alpine Orogen*. Geological Society, London, Special Publications, **156**, 475–515.

OKAY, A. I., SIYAKO, M. & BÜRKAN, K. A. 1991. Geology and tectonic evolution of the Biga Peninsula, northwest Turkey. *Bulletin of Technical University Istanbul*, **44**, 191–256.

OKAY, A. I., SATIR, M., MALUSKI, H., SIYAKO, M., MONIE, P., METZGER, R. & AKYÜZ, S. 1996. Paleo- and Neo-Tethyan events in northwestern Turkey: geologic and geochronologic constraints, *In*: YIN, A. & HARRISON, T. M. (eds) *The Tectonic Evolution of Asia*. Cambridge University Press, Cambridge, 420–441.

OKAY, A. I., SATIR, M., TÜYSÜZ, O., AKYÜZ, S. & CHEN, F. 2001. The tectonics of Strandja Massif: late-Variscan and mid-Mesozoic deformation and metamorphism in the northern Aegean. *International Journal of Earth Sciences*, **90**, 217–233.

OVTCHAROVA, M., QUADT, A. V., HEINRICH, C. A., FRANK, M., KAISER-ROHMEIER, M., PEYCHEVA, I. & CHERNEVA, Z. 2003. Triggering of hydrothermal ore mineralization in the Central Rhodopean Core Complex (Bulgaria)—insight from isotope and geochronological studies on tertiary magmatism and migmatisation, *In*: ELIOPOULOS, D. G., BAKER, T., BAUCHOT, V. ET AL. (eds) *Mineral Exploration and Sustainable Development*, 1, Millpress, Rotterdam, 367–370.

PAPANIKOLAOU, D. 1997. The tectonostratigraphic terranes of the Hellenides. *Annales Géologique des Pays Hellénique*, **37**, 495–514.

PAPAZACHOS, C. B. & SKORDILIS, E. M. 1998. Crustal structure of the Rhodope and surrounding area obtained by non-linear inversion of P and S travel times and its tectonic implications. *Acta Vulcanologica*, **10**, 339–345.

PE-PIPER, G. & PIPER, D. J. W. 2002. *The Igneous Rocks of Greece: the Anatomy of an Orogen*. Beiträge zur regionalen Geologie der Erde, **30**.

PEYTCHEVA, I. & QUADT, A. V. 1995. U–Pb zircon dating of metagranites from Byala Reka region in the east Rhodopes, Bulgaria. Geological Society of Greece Special Publication, **4**, 637–642.

PEYTCHEVA, I., KOSTITSIN, Y., SALNIKOVA, E., KAMENOV, B. & KLAIN, L. 1998. Rb–Sr and U–Pb isotope data for the Rila–West-Rhodopes batholith. *Geochemistry, Mineralogy and Petrology*, **35**, 93–105.

PICKETT, E. A. & ROBERTSON, A. H. F. 1996. Formation of the Late Paleozoic–Early Mesozoic Karakaya Complex and related ophiolites in NW Turkey by Paleotethyan subduction–accretion. *Journal of the Geological Society, London*, **153**, 995–1009.

PLATT, J. P. 1986. Dynamics of orogenic wedges and the uplift of high-pressure metamorphic rocks. *Geological Society of America Bulletin*, **97**, 1037–1053.

RICOU, L.-E., BURG, J.-P., GODFRIAUX, I. & IVANOV, Z. 1998. Rhodope and Vardar: the metamorphic and the olistostromic paired belts related to the Cretaceous subduction under Europe. *Geodinamica Acta*, **11**, 285–309.

RING, U., LAWS, S. & BERNET, M. 1999. Structural analysis of a complex nappe sequence and late-orogenic basisns from the Aegean Island of Samos, Greece. *Journal of Structural Geology*, **21**, 1575–1601.

ROBERTSON, A. H. F. & DIXON, J. E. 1984. Introduction: aspects of the geological evolution of the Eastern Mediterranean. *In*: DIXON, J. E. & ROBERTSON, A. H. F. (eds) *The Geological Evolution of the Eastern Mediterranean*. Geological Society, London, Special Publications, **17**, 1–74.

ROBERTSON, A. H. F., DIXON, J. E., BROWN, S. ET AL. 1996. Alternative tectonic models for the Late Palaeozoic–Early Tertiary development of Tethys in the Eastern Mediterranean region. *In*: MORRIS, A. & TARLING, D. H. (eds) *Palaeomagnetism and Tectonics of the Mediterranean Region*. Geological Society, London, Special Publications, **105**, 239–263.

ROYAJ, F. B. & ALTINER, D. 1998. Middle Jurassic–Lower Cretaceous biostratigraphy in the Central Pontides (Turkey): remarks on paleogeography and tectonic evolution. *Rivista Italiana di Paleontologia e Stratigrafia*, **104**, 167–180.

SAKINÇ, M. & YALTIRAK, C. 2005. Messinian crisis: what happened around the northeastern Aegean? *Marine Geology*, **221**, 423–436.

SAUNDERS, P., PRIESTLEY, K. & TAYMAZ, T. 1998. Variatons in the crustal structure beneath western Turkey. *Geophysical Journal International*, **134**, 373–389.

SCHERMER, E. R. 1993. Geometry and kinematics of continental basement deformation during the Alpine orogeny, Mount Olympos region, Greece. *Journal of Structural Geology*, **15**, 571–591.

ŞENGÖR, A. M. C. & YILMAZ, Y. 1981. Tethyan evolution of Turkey: a plate tectonic approach. *Tectonophysics*, **75**, 181–241.

ŞENGÖR, A. M. C., GÖRÜR, N. & ŞAROĞLU, F. 1985. Strike-slip deformation, basin formation and sedimentation: strike-slip faulting and related basin formation in zones of tectonic escape. *In*: BIDDLE, K. T. & CHRISTIE-BLICK, N. (eds), *Strike-slip Faulting and Basin Formation*, Society of Economic Paleontologists and Mineralogists, Special Publications, **37**, 227–264.

SEYITOĞLU, G. & SCOTT, B. C. 1996. The cause of N–S extensional tectonics in western Turkey: tectonic escape vs back-arc spreading vs orogenic collapse. *Journal of Geodynamics*, **22,** 145–153.

SHANOV, S., SPASSOV, E. & GEORGIEV, T. 1992. Evidence for the existence of a paleosubduction zone beneath the Rhodopean massif (Central Balkans). *Tectonophysics*, **206**, 307–314.

SIYAKO, M., BÜRKAN, K. A. & OKAY, A. I. 1989. Tertiary geology and hydrocarbon potential of the Biga and Gelibolu Peninsula. *Turkish Association of Petroleum Geologist, Bulletin*, **1/3**, 183–199.

SOKOUTIS, D., BRUN, J.-P., VAN DEN DRIESSCHE, J. & PAVLIDES, S. 1993. A major Oligo-Miocene detachment in southern Rhodope controlling north Aegean extension. *Journal of the Geological Society, London*, **150**, 243–246.

SOLDATOS, T. & CHRISTOFIDES, G. 1986. Rb–Sr geochronology and origin of the Elatia Pluton, Central Rhodope, North Greece. *Geologica Balcanica*, **16**, 15–23.

STAMPFLI, G. M. 2000. Tethyan oceans. *In*: BOZKURT, E., WINCHESTER, J. A. & PIPER, J. D. A. (eds) *Tectonics and Magmatism in Turkey and the Surrounding Area*. Geological Society, London, Special Publications, **173**, 1–23.

STAMPFLI, G. M. & BOREL, J. D. 2004. The TRANSMED transects in space and time: constraints on the paleotectonic evolution of the Mediterranean domain. *In*: CAVAZZA, W., ROURE, F. M., SPAKMAN, W., STAMPFLI, G. M. & ZIEGLER, P. A. (eds) *The TRANSMED Atlas—The Mediterranean Region from Crust to Mantle*. Springer, Berlin, 53–80.

TAYMAZ, T., JACKSON, J. & MCKENZIE, D. 1991. Active tectonics of the north and central Aegean Sea. *Geophysical Journal International*, **106**, 433–490.

VANDENBERG, L. C. & LISTER, G. S. 1996. Structural analysis of basement tectonites from the Aegean metamorphic core complex of Ios, Cyclades, Greece. *Journal of Structural Geology*, **18**, 1437–1454.

VON QUADT, A., MORITZ, R., PEYTCHEVA, I. & HEINRICH, C. A. 2005. Geochronology and geodynamics of Late Cretaceous magmatism and Cu–Au mineralization in the Panagyurishte region of the Apuseni–Banat–Timok–Srednogorie belt, Bulgaria. *Ore Geology Reviews*, **27**, 95–126.

WALCOTT, C. R. & WHITE, S. H. 1998. Constraints on the kinematics of post-orogenic extension imposed by stretching lineations in the Aegean region. *Tectonophysics*, **298**, 155–175.

WAWRZENITZ, N. & KROHE, A. 1998. Exhumation and doming of the Thasos metamorphic core complex (S Rhodope, Greece): structural and geochronological constraints. *Tectonophysics*, **285**, 301–332.

WAWRZENITZ, N. & MPOSKOS, E. 1997. First evidence for Lower Cretaceous HP/HT-metamorphism in the Eastern Rhodope, North Aegean Region, North-East Greece. *European Journal of Mineralogy*, **9**, 659–664.

WHEELER, J. & BUTLER, R. 1994. Criteria for identifying structures related to true crustal extension in orogens. *Journal of Structural Geology*, **16**, 1023–1027.

WORTEL, J. R. & SPAKMAN, W. 2000. Subduction and slab detachment in the Mediterranean–Carpathian region. *Science*, **290**, 1910–1917.

YANEV, Y. & BARDINTZEFF, J.-M. 1997. Petrology, volcanology and metallogeny of Palaeogene collision-related volcanism of the Eastern Rhodopes (Bulgaria). *Terra Nova*, **9**, 1–8.

YILMAZ, Y. & KARACIK, Z. 2001. Geology of the northern side of the Gulf of Edremit and its tectonic significance for the development of the Aegean grabens. *Geodinamica Acta*, **14**, 1–14.

YILMAZ, Y., GENÇ, Ş. C., GÜRER, F. *ET AL*. 2000. When did the western Anatolian grabens begin to develop? *In*: BOZKURT, E., WINCHESTER, J. A. & PIPER, J. D. A. (eds) *Tectonics and Magmatism in Turkey and the Surrounding Area*. Geological Society, London, Special Publications, **173**, 353–384.

YILMAZ, Y., GENÇ, Ş. C., KARACIK, Z. & ALTINKAYNAK, S. 2001. Two contrasting magmatic associations of NW Anatolia and their tectonic significance. *Journal of Geodynamics*, **31**, 243–271.

Geodetic constraints on kinematics of southwestern Bulgaria from GPS and levelling data

I. GEORGIEV[1], D. DIMITROV[1], T. BELIJASHKI[1], L. PASHOVA[1], S. SHANOV[2] & G. NIKOLOV[2]

[1]*Central Laboratory of Geodesy, Bulgarian Academy of Sciences, Acad. G. Bonchev Str. Bl. 1, Sofia 1113, Bulgaria (e-mail: ivan@bas.bg)*

[2]*Geological Institute, Bulgarian Academy of Sciences, Acad. G. Bonchev Str. Bl. 24, Sofia 1113, Bulgaria*

Abstract: Southwestern Bulgaria belongs to the southern marginal parts of the central Balkan neotectonic region and borders the northern side of the highly seismic north Aegean region. The recent horizontal and vertical motion of the tectonic structures is controlled primarily by the collision stage, caused by continuing ENE palaeosubduction in the Ionian and Adriatic seas and extensional processes northwards of the North Aegean Trough. The present-day generation of small to moderate seismicity and crustal faulting suggests that the complex tectonic processes in the SW Bulgaria region are active. To monitor and study the tectonic deformation in SW Bulgaria, a global positioning system (GPS) network was established in early 2001. Analysis of GPS data from 1996 to 2004 resulted in a horizontal velocity field representing active surface deformations. Horizontal velocities at 38 GPS sites with respect to stable Eurasia are obtained. A new map of the recent vertical velocities is compiled, based on recomputed data from the repeated precise levelling for the period 1929–1991. We obtained evidence of recent active faulting. Based on the geological and geodetic data the SW Bulgaria is separated into five blocks with homogeneous kinematic behaviour. The average motion of each block varies from 1.3 to 3.4 mm a^{-1}, and the whole region velocity is *c.* 1.8 $\pm$ 0.7 mm a^{-1} in a direction N154° with respect to stable Eurasia. Geodetic data correlate well with the geological data on neotectonic motions in SW Bulgaria.

Southwestern Bulgaria is the most active tectonic area of Bulgaria, with high seismicity. The area belongs to the southern marginal parts of the central Balkan neotectonic region, a zone of recent extension of the Earth's crust with complex interference of horizontal and vertical motion of the geological structures (Zagorchev 1992, 2001). Geological and geophysical data confirm the recent activity of the fault structures, which were formed in Late Neogene and Quaternary times. The area is in the north of the north Aegean region, and is greatly influenced by its tectonics and high seismicity. The strongest earthquake for the last two centuries in Europe occurred in SW Bulgaria in the Krupnik–Kresna region, with magnitude M. *c.* 7.8. The recent seismicity, as monitored by the National Operational Telemetric System for Seismic Information (NOTSSI) network, indicates concentration of events in the Krupnik–Kresna and Mesta river areas (Botev 2000). Although SW Bulgaria has been the focus of intense geological and geophysical studies the available geodetic data for the crustal movement have been insufficient to constrain the kinematics and dynamics of the tectonic structures.

Geodetic monitoring constrains geophysical and geological models that resolve the kinematics and dynamics of the deformations. The global positioning system (GPS) provides a tool to quantify deformations of the crust to a precision and on a scale unprecedented in the Earth Sciences (McClusky *et al.* 2000). Together with the vertical motion, obtained from repeated precise levelling, these new constraints on the kinematics of deformation in turn provide constraints on rheological models of the lithosphere and the forces responsible for active deformation.

Recent crustal motion in SW Bulgaria results from extensional motion in the inner part of the Aegean region, although the kinematic model of the north Aegean and central Balkan neotectonic regions is still under debate. SW Bulgaria is under the influence of the current collision stage, caused by ENE palaeosubduction in the Ionian and Adriatic seas (Papazachos 1999). McClusky *et al.* (2000) estimated the complex influence of the SW-directed motion of the Anatolian and North Aegean plates along the North Anatolian Fault Zone (NAFZ), in the Aegean Sea, which causes the formation of an extensional province northwards of the North

From: TAYMAZ, T., YILMAZ, Y. & DILEK, Y. (eds) *The Geodynamics of the Aegean and Anatolia*. Geological Society, London, Special Publications, **291**, 143–157.
DOI: 10.1144/SP291.7 0305-8719/07/$15.00

Aegean Trough (NAT) (Taymaz *et al.* 2004). Those researchers suggested that northern Greece and SW Bulgaria are undergoing slight motion relative to Eurasia.

In this paper we present the results from GPS campaigns performed over the last few years in a specially established geodynamic network in SW Bulgaria. We have also recomputed data from the repeated precise levelling for the period 1929–1991 and compiled a new map of recent vertical motion in the region. Joint analysis of the geodetic and geological data is performed to investigate the recent tectonic processes in SW Bulgaria. Based on this analysis we try to determine the recent deformation pattern in the region and its connection with the north Aegean.

Tectonic framework of the Eastern Mediterranean

The tectonic setting of the Eastern Mediterranean is dominated by the collision of the Arabian and African plates with Eurasia (McKenzie 1970; Jackson & McKenzie 1984, 1988). Plate movement models (DeMets *et al.* 1994; Jestin *et al.* 1994) based on analysis of global sea-floor spreading, fault systems and earthquake slip vectors indicate that the Arabian plate is moving in NNW relative to Eurasia at a velocity of about 18–25 mm a^{-1}, averaged over about 3 Ma (Fig. 1). These models also indicate that the African plate is moving in a northerly direction relative to Eurasia at a velocity of about 10 mm a^{-1}. The leading edge of the African plate is being subducted along the Hellenic arc at a higher rate than the relative northward movement of the African plate itself, as a result of the slab rollback effect (Royden 1993).

The complexity of the plate interactions and associated crustal deformation in the Eastern Mediterranean region is reflected in the many destructive earthquakes that have occurred throughout recorded history. McKenzie (1970), Jackson & McKenzie (1988) and Jackson (1992) developed a plate tectonic framework for understanding deformation and examined the principles that control continental tectonics in the region. They proposed the existence of an Aegean plate that moves at a distinctly different rate to the Anatolian plate, and the separation of these two plates by a zone of north–south extension in western Turkey. Other work as have argued for the existence of a number of

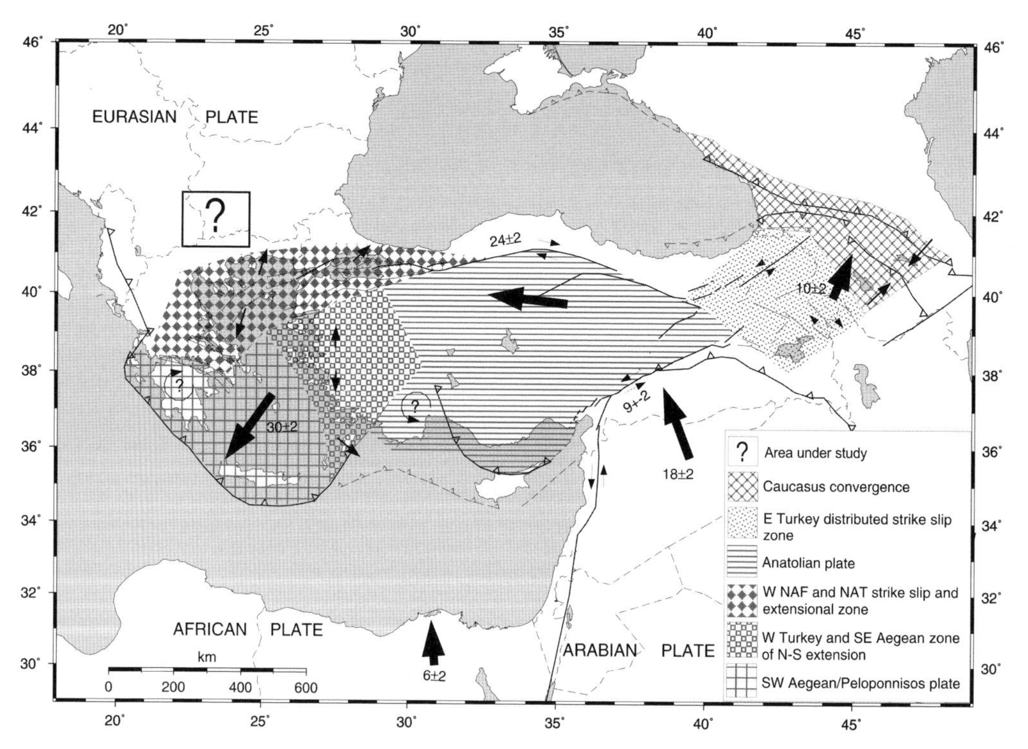

Fig. 1. Schematic illustration of the tectonic setting of the Eastern Mediterranean. Shading shows areas of coherent motion. Bold arrows indicate generalized regional motion (in mm a^{-1} and errors after McClusky *et al.* 2000). NAF, North Anatolian Fault; NAT, North Aegean Trough.

plates (McKenzie 1970, 1972; Dewey *et al.* 1973; Dewey & Şengör 1979; Bonchev 1980). Some of the investigators have presented the region as the southern ductile deformed edge of Eurasia (Tapponnier 1977; Mercier *et al.* 1979) or as a zone dominated by obduction (Zeilinga de Boer 1989). The researchers differ only slightly in their conclusions, and demonstrate different features of the current process of subduction along the Hellenic arc. The available data favour a regional extension with a dominant north–south direction, and all researchers agree on the existence of an extensional regime in the Earth's crust northwestward from the North Anatolian Fault (NAF).

As a result of GPS measurements for the period 1988–1997 a quantitative map of present-day plate motion and deformations for the Eastern Mediterranean plate collision zone has been produced (Fig. 1; McClusky *et al.* 2000). The SW Aegean–Peloponnese moves towards the SSW relative to Eurasia at 35 ± 2 mm a^{-1} in a coherent fashion with low internal deformation (<2 mm a^{-1}). The SE Aegean region deviates significantly from this coherent movement, rotating anticlockwise and moving towards the Hellenic trench (i.e. towards the SE) at 9 ± 1 mm a^{-1} relative to the SW Aegean. Right-lateral strike-slip deformation associated with the NAF extends into the north Aegean and terminates near the Gulf of Corinth. The NAT and Gulf of Corinth form the principal northern boundary of the SW Aegean plate. The common kinematic deformation pattern provided by GPS for Anatolia and the Aegean is qualitatively similar to that proposed by McKenzie (1970).

This relative movement towards the west and its transformation to well-expressed southward displacement with velocities higher than 30 mm a^{-1} in the Aegean denotes the complicated crustal and mantle processes of this part of Eurasia. One of the reasons for this deviation could be the palaeosubduction zone beneath the Rhodopes (Shanov *et al.* 1992). The remaining slab detachment in the mantle hinders the westward movement of the Anatolian plate, and transforms it to southward movement.

The tectonic processes in south Bulgaria during Neogene and Quaternary times are the result of the collapse of the Late Alpine orogen, the extensional environment in the Aegean back-arc, and the complex interference of intense vertical and horizontal motion (Zagorchev 2001).

Geological background of SW Bulgaria

SW Bulgaria comprises the territory south of the Vitosha and Plana Mountains and west of the Central Rhodope Mountains. The deeply cut relief of this region includes the highest Bulgarian mountains of alpine type (Rila and Pirin), many moderately high to low mountain massifs, and hilly and depressed regions. SW Bulgaria is an area with varied relief structures, which at present are subject to horizontal and vertical motion of different intensities. After the Early Miocene transgression episode, the area underwent prolonged tectonic quiescence and peneplanation in Early and early Middle Miocene times. It became again an area of intense tectonic activity in late Middle Miocene and Late Miocene times, when a complex fault pattern and graben-and-horst block structure was formed, and the present northern peri-Aegean fluvial systems developed (Zagorchev 2001). The analysis of the geological information for the period from the Late Miocene to the present shows that the region is subject to relative extension. In the period of the Young Alpine tectonic stage the main blocks were divided into parts. Horst and graben structures formed, and are characterized by intensive horizontal and vertical motion in conditions of dominant extension throughout the neotectonic stage.

The main block structures, along with the active fault structures according to the available geological data (Shanov *et al* 2001; Zagorchev 2001), are shown in Figure 2. The vertical motions, which are the most intensive, form the pattern of the area. The motion of the Pirin block is especially intensive: during the neotectonic stage it has been uplifted with respect to the Sandanski graben by about 2000–3000 m. The displacements between the Rila horst and the Blagoevgrad graben have also been intensive. In recent times, after the Pliocene, the Belasitsa horst has been undergoing intensive uplift relative to the Strumeshnitsa graben.

The region has a key position with regard to the regional neotectonics and recent tectonics, as it is situated at the crossing points of several important fault belts (lineaments): the NNW–SSE-striking Strouma (Kraishtid) lineament, the WNW–ESE-striking Maritsa Lineament, and the northernmost western feathering faults of the North Anatolian Fault Zone (NAFZ) (Zagorchev 1992).

GPS geodynamic network in SW Bulgaria

A GPS network for monitoring the present-day kinematics in SW Bulgaria was established in early 2001. The network consists of 38 points chosen to cover the main tectonic structures and to provide spatial coverage throughout this region (see Fig. 6). The choice of each point location was made after field geological investigations. The points are fixed into solid bedrock with

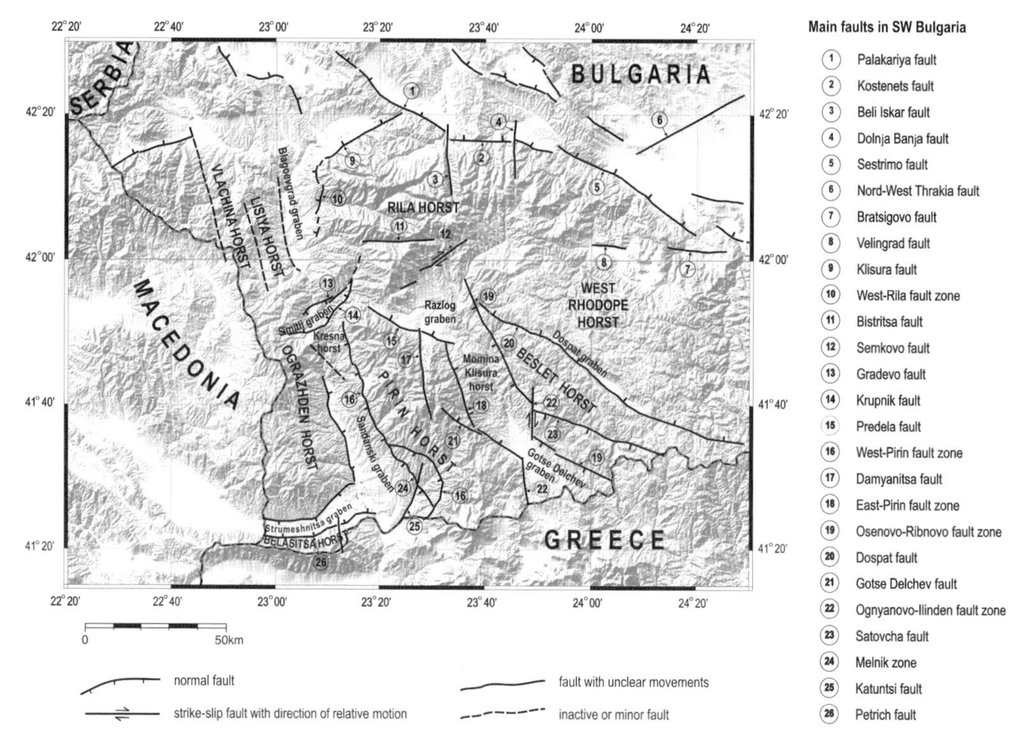

Fig. 2. Scheme of main tectonic structures in SW Bulgaria. Topography is from USGS–NOAA (http://seamless.usgs.gov).

bronze and stainless steel bolts. The network includes three EUREF points and two triangulation first-order points: the highest peaks in the Balkans (Musala, 2925 m, and Vihren, 2918 m). One point of the regional network is a point from a local geodynamic network in Krupnik, and is a concrete pillar with an enforced centring device.

GPS campaigns

All GPS sites have been measured in campaigns every year from 2001 to 2004. Taking into account that the permanent IGS site coordinates incorporated into the solution have millimetre accuracy (Altamimi & Boucher 2001) and that the expected velocities in the region are up to 3–4 mm a^{-1} (McClusky *et al.* 2000; Kotzev *et al.* 2001), we decided to adopt simultaneous observations, for a minimum of 36 h (usually 48 h) every year, using dual-frequency GPS receivers. The reason for measuring the geodynamic networks at a 1 year interval is that SW Bulgaria is the most seismically active area of the country and it is possible to try to determine pre- and post-seismic motion associated with comparatively strong earthquakes. This is especially true for the Krupnik–Kresna seismogenic zone, where the seismic activity is regular (Botev 2000).

Simultaneous GPS observations were performed with Trimble 4000SSE, Trimble 4400 and Trimble 5700 receivers equipped with L1/L2 ground plane antennas. All observations are made with a sampling rate of 30 s and 10° elevation masks. Data from previous GPS campaigns performed in 1996, 1997 and 1998 were used for some points: three EUREF points (SATO, PETR and SAPA), and FROL, PADA, MALA and BELM (Milev *et al.* 1999). The typical occupation time for each site in 1996, 1997 and 1998 was 48 h. The only exception was the occupation time for the first-order triangulation point Vihren (VIHR), which was *c.* 16 h in 2001 and 2003 because of difficult meteorological conditions and logistics problems.

GPS data processing

Station positions are conventionally given in an Earth fixed reference frame, which is the International Terrestrial Reference Frame (ITRF) in satellite geodesy. As a result of the improvement in observation quality, model refinement and density, the reference frame is regularly updated. The most recent International Terrestrial Reference System (ITRS) realization, namely ITRF2000, appears to be the most accurate and extensive ITRF version ever developed (Altamimi & Boucher 2001).

All SW Bulgaria GPS campaigns have been processed in ITRF2000. The processing of the GPS data was achieved using the Bernese Software 4.2 (Hugentobler *et al.* 2001). The software was developed 'as a tool for highest accuracy requirements' in the field of GPS applications. Data from 10 IGS stations have been integrated in the analysis to constrain the network into a global reference frame: Ankara (ANKR), Bucharest (BUCU), Zimmerwald (ZIMM), Graz (GRAZ), Wettzell (WTZR), Zwenigorod (ZWEN), Ohrid (ORID), Matera (MATE), Jozegoslav (JOZE) and Golosiiv (GLSV). Ionosphere models for Europe are used to help the ambiguity resolution and troposphere parameters are estimated for each station (Mervart 1995).

GPS observations are organized in campaigns of some days, which are repeated on a regular timespan. The data are processed along with a selected set of the IGS permanent sites for one campaign and the campaigns are then combined to estimate velocities. We compute daily solutions, which are combined in campaign and multi-campaign solutions. The adopted strategy for the processing is summarized in Table 1.

The ionosphere free linear combination L3 is used as the basic observable. Ambiguities that are fixed in the previous runs are introduced as known parameters, and unresolved ambiguities are eliminated before the inversion of the normal system. Only one IGS site is constrained on its coordinates for numerical stability (Hugentobler *et al.* 2001) and all other stations are loosely constrained. The correlation between the double differences is handled correctly and ± 2 mm *a priori* sigma is assumed to one zero difference observation. A standard Saastamoinen troposphere model is used as the *a priori* model and 12 troposphere parameters for each station per day are estimated. The troposphere parameters are eliminated before inversion of the normal system and the equations with site coordinates are saved for later combination.

The normal equations of each campaign are combined to obtain the final solution. The IGS sites GLSV, GRAZ, JOZE, MATE, WTZR, ZIMM and ZWEN are constrained with 0.0001 m sigma to their ITRF values for coordinates and 0.05 mm a^{-1} for velocities, whereas ANKR, BUCU, ORID and Sofia (SOFI) coordinates and velocities are estimated to check the solution.

The r.m.s. of the individual epoch solutions are of the order of 1.5 mm (Table 2). To check the accuracy of the solution, the combined solution is transformed to the ITRF2000 coordinate frame at each epoch by Helmert transformation. The results are satisfactory considering the solutions are derived from episodic campaigns (Rothacher *et al.* 1997).

The error estimates of the GPS coordinates are usually overestimated and a covariance rescaling factor of about 10 is used (Brockman 1996; Becker *et al.* 2002) to convert the internal precision of solution to an estimate of the accuracy. In this study the final error estimates from the adjustments are multiplied by a factor to match the errors obtained from the r.m.s. repeatability of the single campaign solution. Thus, the accuracy of a single campaign is described by comparing repeatability of site coordinates. Based on the r.m.s. repeatability a covariance rescaling factor of 7.5 was applied to coordinate and velocity error estimates in our study.

Table 1. *GPS data processing strategy*

Bernese GPS Software, Version 4.2
IGS orbits and IGS pole coordinates (transformed into ITRF2000 if before 2000)
ITRF2000 Reference Frame
IGS_01 phase eccentricity file (elevation-dependent phase centre corrections)
Sampling rate 30 s
15° elevation cutoff
Elevation-dependent weighting of phase observations
Ionosphere models for Europe for each day from CODE
Float and Fixed Solution (QIF-Ambiguity resolution)
Estimation of one troposphere parameter every 2 h with Saastamoinen *a priori* model (no meteorological data were used)
Daily solutions and normal equations are computed for analysis of the repeatability, loosely constrained solution for each observation session with an *a priori* sigma of 1 m
Combined normal equation files for each epoch generated from the daily normal equations
IGS/EUREF permanent tracking sites as reference with coordinates and velocities constrained in ITRF2000 (absolute velocities) and relative to stable part of Eurasia

Table 2. *Characteristics of processing for GPS campaigns*

Epoch of campaign	Number of new sites	r.m.s. (mm)	Reference sites	r.m.s. of component (mm)		
				North	East	Up
1996	4	±1.5	GRAZ, JOZE, MATE, WTZR, ZIMM, ZWEN, ANKR	2.7	3.1	4.3
1997	7	±1.4	GRAZ, JOZE, MATE, WTZR, ZIMM, ZWEN, ANKR, SOFI	3.1	3.9	5.0
1998	4	±1.4	GRAZ, JOZE, MATE, WTZR, ZIMM, ZWEN, SOFI	0.5	1.2	4.3
SWBG 2001	14	±1.6	GRAZ, JOZE, MATE, WTZR, ZIMM, ZWEN, ANKR, BUCU, ORID, SOFI	3.5	4.2	7.2
SWBG 2002	26	±1.6	GRAZ, JOZE, MATE, WTZR, ZIMM, ANKR, BUCU, ORID, SOFI	4.8	5.9	11.4
SWBG 2003	28	±1.6	GLSV, GRAZ, JOZE, MATE, WTZR, ZIMM, ANKR, BUCU, ORID, SOFI	6.5	7.3	14.1

Velocity solution

Plate motion is described either in an absolute sense (absolute movement of an individual plate with respect to a fixed hotspot reference frame) or relative to a plate. For the East Mediterranean region the Eurasia plate is assumed to be stable and velocities are given relative to a rigid Eurasia. These velocities are suitable for studying crustal movement between the African, Aegean and Eurasian plates.

Application of space-geodetic techniques to the studies of regional intra-plate deformations requires elimination of the rigid plate movement from the obtained results. This can be achieved by fixing the reference frame to a tectonic plate, either by using a global plate movement model or by determining the site displacements relative to several sites assumed to be moving rigidly with the plates.

Until 2001, the rigid movement of Eurasia was accounted for by the movement predicted by NUVEL-1A-NNR (DeMets *et al.* 1994). With the last realization of the ITRS (ITRF2000; Altamimi & Boucher 2001) a new rotation vector for the rigid Eurasia was computed, which is significantly different from the one in NUVEL-1A-NNR.

All results presented below are relative to the Eurasian plate after the rigid rotation of the plate is removed according to the Altamimi & Boucher (2001) rotation pole. Velocities of 38 points relative to Eurasia in SW Bulgaria are shown in Figure 6. Velocities are not plotted for the points Medeni poljani (MEDI) and Padesh (PADE) because we have only two epochs of observations with a 1 year timespan. As a result of local effects (landslide), the Dospat (DOSP) point is also excluded from the velocity estimation.

Vertical motion from repeated levelling measurements

The recent vertical motion of the Earth's crust is studied using the existing first- and second-order levelling lines in SW Bulgaria (Fig. 3). These lines form loop XXII of the State Geodetic Levelling Network from 1980 (Burilkov & Belyashky 1980). Within the framework of this loop, 16 second-order levelling lines are measured. The first- and second-order levelling lines are measured in three cycles with average epochs as follows: 1929.5, 1960.5 and 1981 for the first order; 1931.5, 1958.5 and 1986 for the second order.

The measurements are corrected for average rod scale and temperature. Before and after each field season the rods were calibrated. The measurements are not corrected for refraction because the requirements for levelling (Anonymous 1980) are strictly followed. The levelling data used in the analysis are presented in Table 3. The last measurements (third cycle) of the first- and second-order levelling lines (respectively in 1978–1984 and 1983–1991) are used as the second epoch t_2. The average measurement epochs for the 10 polygons (see Fig. 3) are given in Table 4. For the whole network the average epochs are $t_1 = 1946$ and $t_2 = 1981$.

The choice of the two epochs in the case of more than two cycles of measurements depends on several factors, including the time intervals between measurements, the accuracy, the methods and instruments used in the measurements, the preserved nodal benchmarks, and the number of preserved identical intermediate benchmarks. During the processing of the first-order levelling the first

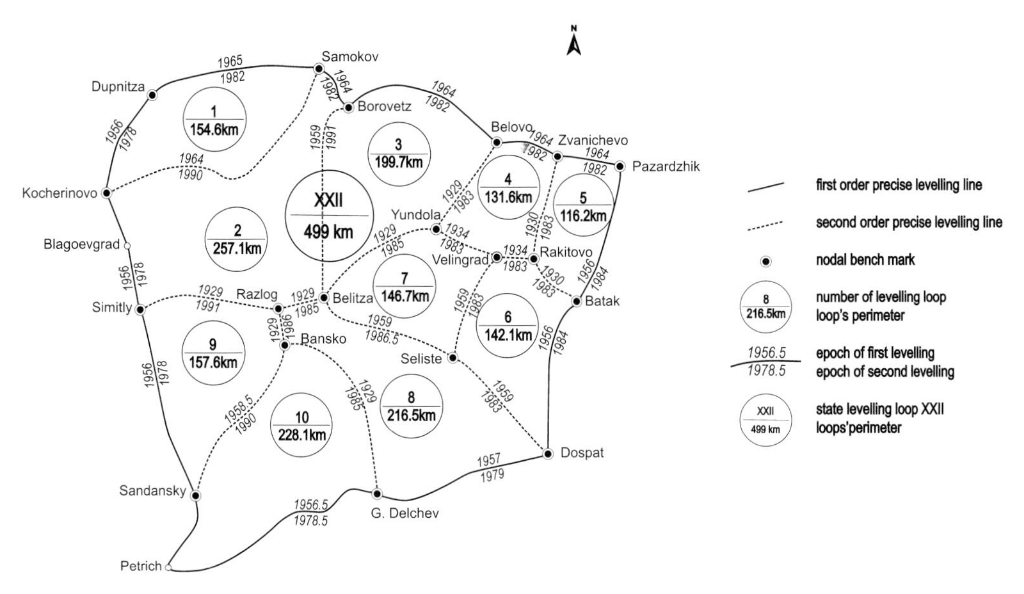

Fig. 3. Scheme of the precise levelling lines in SW Bulgaria.

cycle (1929.5) is not used because of the great number of benchmarks destroyed during road reconstruction.

The number of nodal benchmarks is 19, of which 12 are situated on the first-order lines and seven in the internal part of the network. The sections of the first-order lines between nodal benchmarks are considered as single lines. The vertical velocities are calculated assuming linear variation of vertical movement in time.

The nodal benchmark in Pazardzhik is chosen as the initial site for the levelling network adjustment for the whole SW Bulgaria region. Its absolute velocity is $v_{abs} = -0.5 \pm 0.2$ mm a^{-1} with respect to the basic benchmark of the State Levelling Network, Fundamental Subsurface benchmark (FSB) 28 near the Varna tide gauge, for which zero vertical velocity is accepted. When referring to absolute velocities we mean the velocities of all benchmarks assigned to FSB 28. It is the initial vertical point for heights in Bulgaria for the period 1930–1958, the Black Sea vertical datum, after which the Baltic datum was adopted. The reasons to assume $v = 0$ mm a^{-1} for FSB 28 are as

Table 3. *Content of database for precise levelling network in SW Bulgaria*

Characteristics	Precise levelling		
	First order	Second order	Generally
Number of levelling lines	12	16	28
Average length of the levelling line (km)	42.5	39.5	40.7
Number of benchmarks:			
basic	12	7	19
intermediate	105	119	224
total	117	126	243
Loops			10
Average epoch of levelling:			
t_1	1959.5	1941.3	1949.1
t_2	1980.8	1985.7	1983.6
Average timespan (years)	21.3	44.4	34.5
Mean error m (mm km$^{-1/2}$):			
for epoch t_1	±0.78	±0.76	±0.77
for epoch t_2	±0.41	±0.65	±0.54

Table 4. *Characteristics of levelling measurement*

No. of loop	Average epoch of the measurement		Time interval (years)	Misclosures (mm)		Loops measuring interval (years)	
	t_2	t_1	$t_2 - t_1$	W_{t_2}	W_{t_1}	Epoch t_2	Epoch t_1
1	1983	1962	21	20.3	6.6	12	9
2	1986	1950	36	−108.1	−63.6	13	36
3	1985	1944	41	35.7	61.3	9	36
4	1983	1938	45	−23.1	−26.2	1	35
5	1983	1945	38	−21.1	28.8	2	34
6	1983	1948	35	−50.7	10.1	1	29
7	1984	1945	39	33.2	−21.1	3	30
8	1984	1944	40	27.7	−32.4	8	30
9	1986	1943	43	32.7	19.4	13	30
10	1988	1944	44	−24.8	23.0	12	30
Average	1981	1946	38	−7.8	0.6	7	30

follows: (1) this benchmark was built in 1932 after studies of the stability of the region; (2) numerous control measurements have shown that its altitude does not change; (3) the values obtained for its height from the adjustments in 1958 and 1983 of the Unified Precise Levelling Network in Eastern Europe differ only by 2 mm.

The relative velocities along the levelling lines are obtained as difference between heights of common benchmarks dividing by the timespan relative to one of the nodal benchmarks of the line. These relative velocities are input to the adjustment procedure. As a result, the relative velocities of the nodal benchmarks are obtained as well as their absolute velocities using the velocity of the initial benchmark in Pazardzhik. The r.m.s. of the adjusted relative velocities is ± 0.14 mm a^{-1} per km and the rms for the whole network calculated from the loop's misclosures is ± 0.13 mm a^{-1}. The absolute vertical velocities are within the range of -3.0 to $+1.2$ mm a^{-1} and -1.6 to $+2.4$ mm a^{-1} for the first- and second-order levelling lines, respectively.

The presence of the second-order levelling lines allows a surface interpolation of the vertical velocities on the basis of the whole levelling network. It should be noted that the interpolation between levelling lines cannot guarantee revealing the velocity anomalies between the lines but gives a common idea of the vertical velocities of the tectonic structure. The absolute values of vertical velocities for all benchmarks are plotted on a map and the isolines are drawn by interpolation (Fig. 4).

An attempt has been made to find a correlation between the sharp change in the relative vertical velocity sign at the boundary between two adjacent sections along the first-order levelling line and the presence of faults in these places. An example with four profiles is presented in Figure 5. The fitted straight lines, which express the trends in the changes of the velocity for the given section or line, are drawn. Comparing the obtained peaks of the velocity changes and geological data on fault activity, remarkable coincidences are revealed and these are listed in Table 4.

Geodetic constraints in SW Bulgaria

The horizontal velocities of 38 GPS points in SW Bulgaria indicate movement in a SSE direction, with respect to stable Eurasia. The average velocity is of about 1.8 ± 0.7 mm a^{-1} (Fig. 6). This movement is in good agreement with the dextral movement along the NAFZ. However, considering that the Anatolian plate movement is several times faster than the movement of SW Bulgaria, it is evident that the anticlockwise rotation of the Anatolian plate is accompanied by its southward detachment from stable Europe. This is expressed exactly by the north–south tectonic extension described in the structures of the north Aegean region.

The results confirm that the recent crustal motion in SW Bulgaria maintains the same tendencies as revealed for the whole period of neotectonic motion (since late Middle Miocene times) when the present tectonic pattern of the area formed.

As can be seen on the map of vertical motion, the zero velocity isoline divides SW Bulgaria into to an eastern part with predominantly negative values and a western part with predominantly positive values. There is clearly expressed uplift of the western part of the Rila horst, which is separated from the horst itself by the Beli Iskar fault.

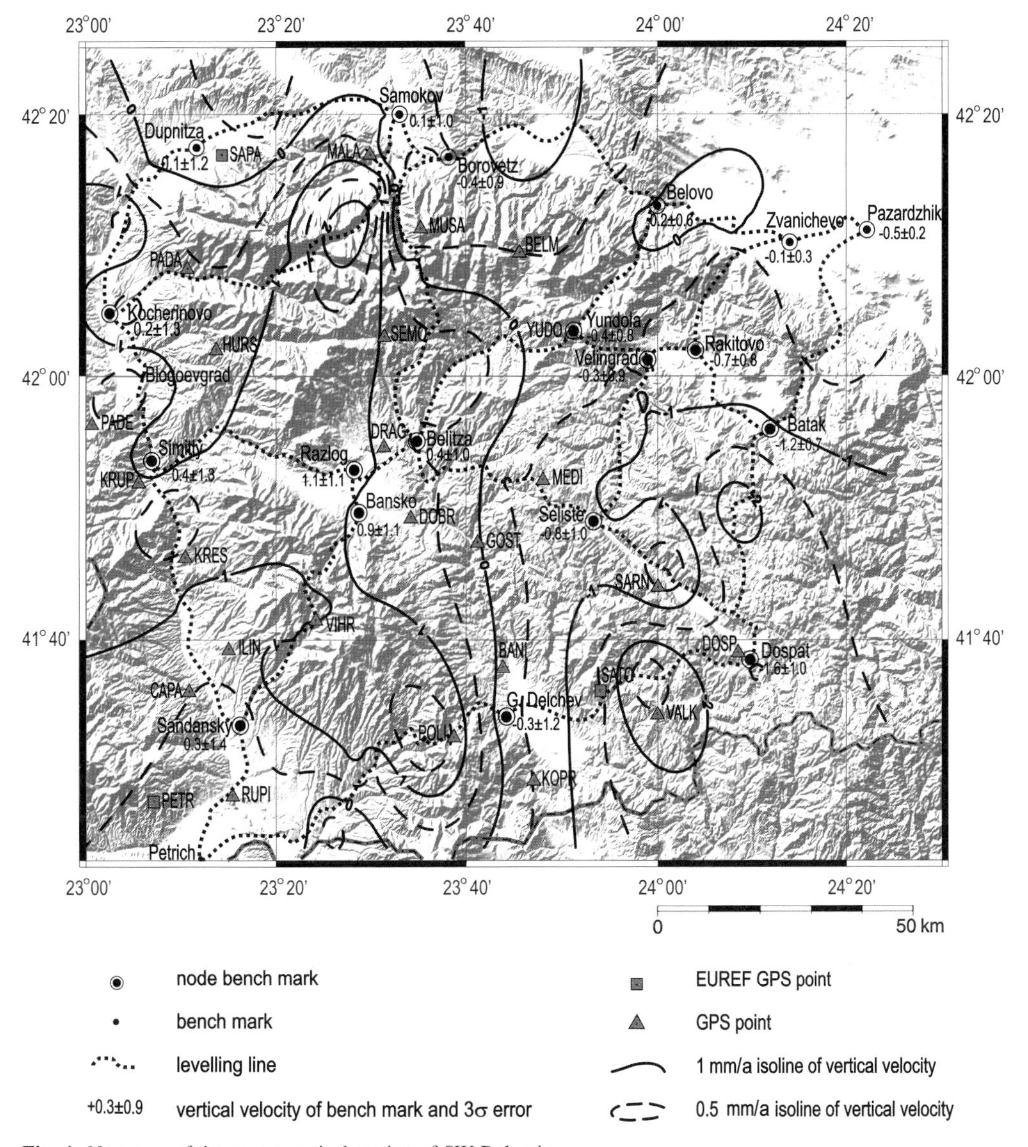

Fig. 4. New map of the recent vertical motion of SW Bulgaria.

This fault is known as a strike-slip fault and the levelling data confirm that the eastern part of the Rila horst is subject to subsidence. Intensive uplift is also recognized in part of the Pirin horst. Local uplifts are found to the south of Blagoevgrad and Krupnik. According to the geological data these coincide with the Momina–Klisura horst and Kresna horst, respectively, which have undergone uplight during the Neogene. Negative velocities are predominant for the whole west Rhodope region, with a minimum value of -3 mm a^{-1} in the region of Satovcha.

For the western part of the study area, the Krupnik fault is the tectonic line of regional significance. The velocity of the site KRUP (geodetic point from local geodynamic network in Krupnik) is in a northerly direction at 3 mm a^{-1}. This displacement is compatible with the processes of NNW-directed normal faulting (dipping N60–70°) along this fault (strike N50–60°), as well as with a reconstruction of the contemporary tectonic stress field (Shanov & Dobrev 2000). The velocity of 3.4 mm a^{-1} of total deformation along the fault calculated by an extensometer installed on a

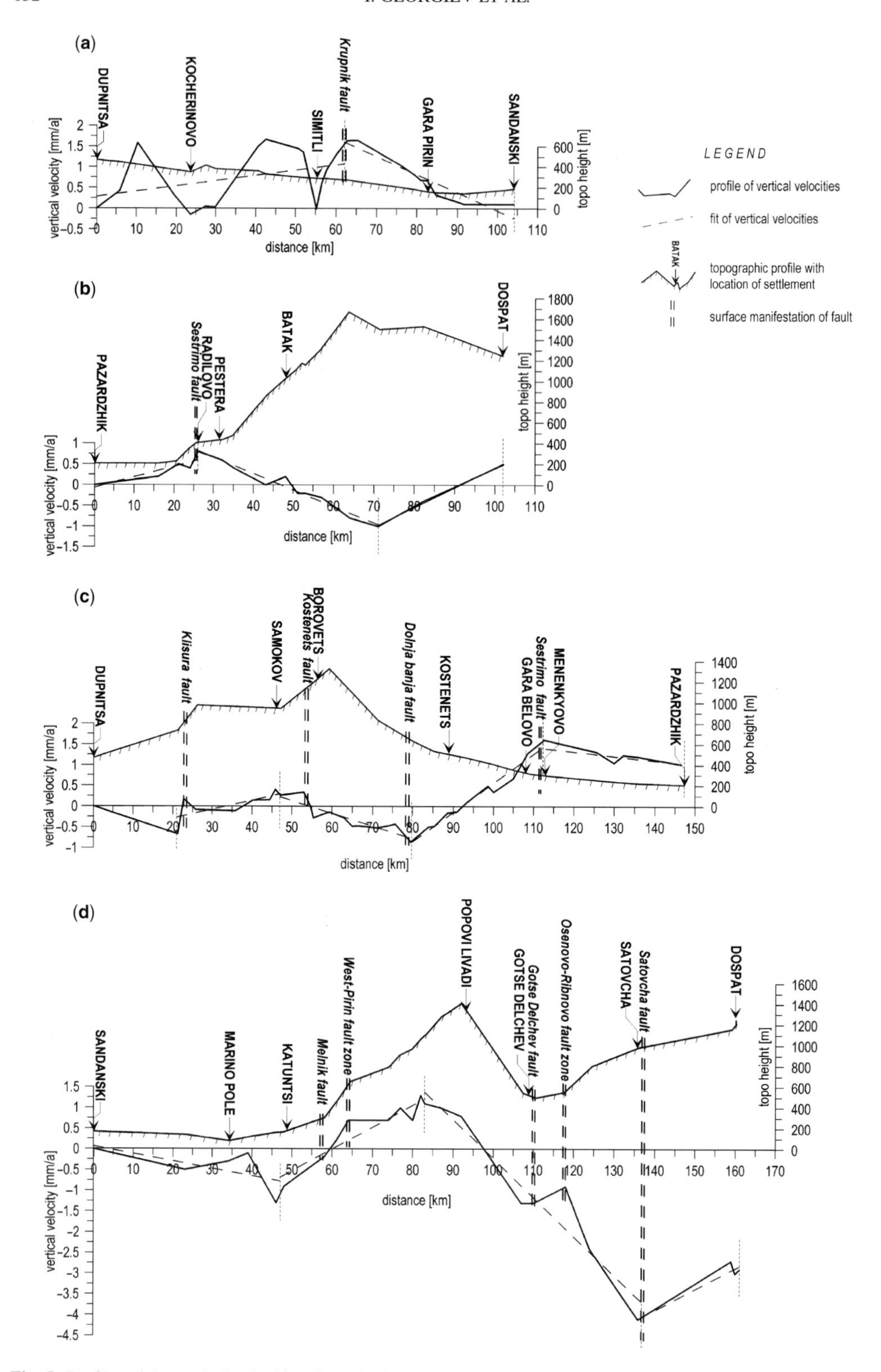

Fig. 5. Profiles of the vertical velocities along the levelling lines and active faulting.

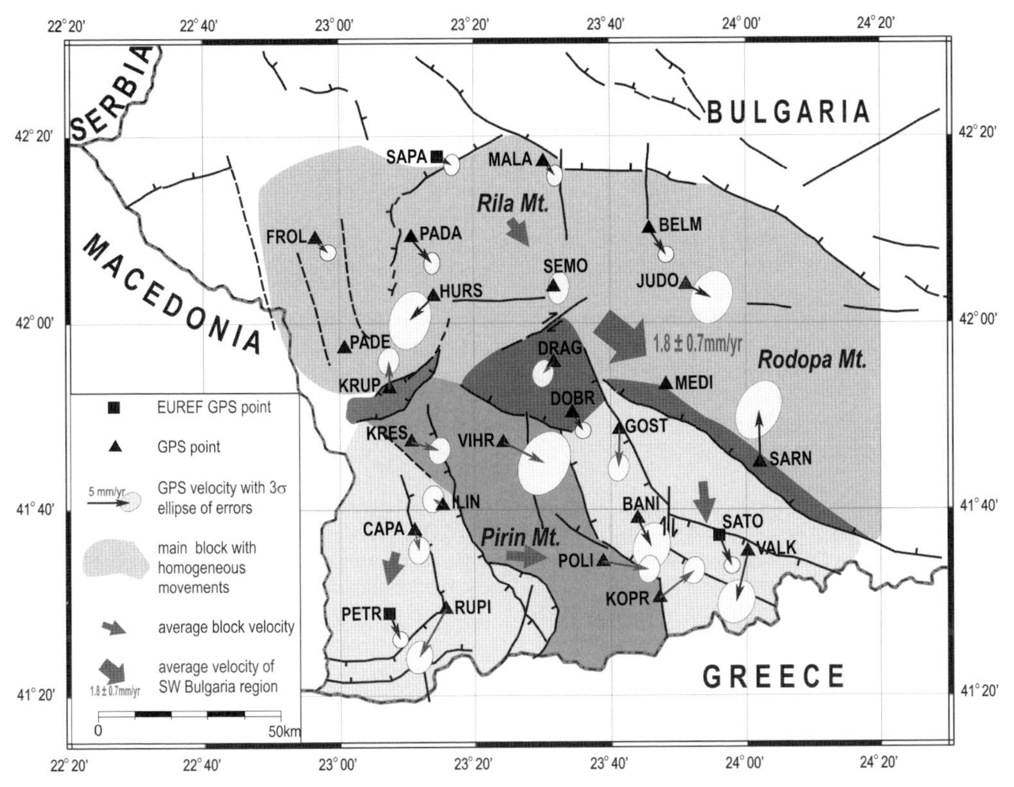

Fig. 6. GPS-derived velocities with 95% confidence ellipses relative to Eurasia for sites in SW Bulgaria and main blocks with homogeneous movements.

segment of the fault is in good agreement with the GPS result. Extensometric observation indicates a sinistral component of movement along the fault. This indicates a higher velocity of displacement of the southern block; that is, the northern block is retarded in its movement toward the SE or is undergoing dextral rotation.

The situation is different for the Predela fault: the southwesterly movement of the site DRAG (situated in the Razlog graben) is related to the NNE-directed subsidence of the predominant hanging wall of the Predela fault and the consecutive 'delay' of the sedimentary complex of the graben relative to the general tendency of the area. The lower velocity of displacement of this site, as well as that of the site DOBR compared with the site VIHR, gives a reason to suppose normal faulting along the Predela fault. All points north of the line of the Krupnik fault–Predela fault show lower velocities compared with points south of it. Northward from this line the processes along the fault structures are less expressive. The difference in velocity between the site FROL (Lisia horst) and the site PADA (Rila horst) is about 1–2 mm a^{-1}, which indicates normal faulting along the West Rila fault zone.

All GPS sites located in the Rila Mountain (PADA, BELM, MALA and SAPA) have small and uniform velocities. It could be assumed that the whole Rila horst is moving homogeneously and no conclusions can be reached about fault activity within this block. The activity of the bordering structures of the Razlog graben is expressed by right-lateral movement along the Semkovo fault according to geological data. It is difficult to investigate the Semkovo fault using the existing GPS data because the velocities of sites SEMO and DRAG are insignificant.

The available geodetic data support the proposal that the Rila and West Rhodope horsts are a jointly moving tectonic unit. The area west of the Rila horst can also be integrated to this unit because of the uniform velocities of sites FROL, SAPA, MALA and PADA. Only the vertical displacement along the Beli Iskar fault separates the western part of the Rila Horst from this quasi-homogeneous unit. The velocity of site HURS shows different directions of movement relative to the main tendency but the reason for this is probably local effects that have not yet been sufficiently well studied.

South from the line of the Krupnik and Predela faults, an extensional regime can be seen in the Sandanski graben. The site VIHR in the Pirin Mountain is diverging from the site ILIN, and this signifies that the normal faulting along the West Pirin fault zone is directed westward. The velocities of the site ILIN, within the 3σ error ellipse, can be explained as resulting from two contradictory tendencies: (1) The movement of the whole area to the SE; (2) subsidence of the Sandanski graben along the hanging wall of the West Pirin fault zone (strike N160°, WSW-dipping at 80° for the northern part and 35° for the southern segment); with the second movement being dominant. Similar displacements are observed along the Melnik fault.

The tectonic units south of the Krupnik fault and east of the Ograzhden block can be characterized by higher velocities of displacement toward the SE, especially the northern parts (sites KRES, GOST and VIHR) compared with the southern parts (sites POLI and KOPR). Indications for an anticlockwise rotation can be also seen here, confirming that the Krupnik and Predela faults are contemporary active tectonic structures.

A similar movement to the SE, following the general trend of the area, is revealed in the Ograzhden horst. No significant motion can be derived along the Ograzhden fault. According to palaeomagnetic studies in the Republic of Macedonia (FYROM) (Pavlides & Kondopoulou 1987), this part of the Serbo-Macedonian massif has been rotated clockwise, from Oligocene time until the present, at about 10°. A similar result was found by the analysis of the neotectonic and the contemporary tectonic stress fields (Shanov 1997). The sites CAPA, PETR and RUPI generally support the existence of this rotation.

The Pirin horst shows faster SE movement and vertical uplift of 2 mm a^{-1}, obtained from precise levelling. The azimuths of the horizontal velocity vectors of sites VIHR, POLI and KOPR give an indication of anticlockwise rotation. According to the palaeomagnetic data of P. Nojarov (pers. comm.), southern Pirin has been rotated anticlockwise 120° from the Neogene to the present. A similar rotation has been derived on the basis of tectonic reconstructions and interpretations (Zagorchev 1971).

The Dospat fault activity can be described as normal faulting along the fault surface striking N120° and dipping northeastwards N70–80°. The NE movement of the hanging block is supported by the velocities of sites SARN and MEDI. The Satovcha fault is characterized by normal faulting along a N110°-striking plane, and a SSW dip direction. The hanging block is displaced southwards according to the velocities of sites SATO and VALK.

The Gotse Delchev graben located between the Pirin and Beslet horsts is 'compressed' by these two structures and this can be seen especially for the most southern part of the graben. The sites POLI and KOPR, located on the Pirin horst, show an ENE movement towards the graben, and sites SATO and VALK indicate SSW-directed displacement, caused by 'push' by the Beslet horst. The fault activity is mostly expressed along the Satovcha fault as normal faulting.

Based on the tectonic structures, and the obtained horizontal and vertical crustal motion, we can separate SW Bulgaria into five homogeneous blocks (Fig. 6). The average motion of each block ranges from 1.3 to 3.4 mm a^{-1}. The geological and geodetic data for the recent active faults in SW Bulgaria are summarized in Table 5. These new results constrain the geological data on the recent tectonics of SW Bulgaria.

Implications

A GPS geodynamic network of 38 sites has been established in SW Bulgaria, and has been measured every year since 2001. Present-day motions obtained by GPS and levelling data are in agreement with geological information on neotectonic deformation. The study shows that SW Bulgaria can be separated into five blocks with homogeneous kinematic behaviour. The whole region moves with an average velocity of about 1.8 ± 0.7 mm a^{-1} in a SE direction (N154°) relative to stable Eurasia. The GPS-derived velocities gradually increase from north to south (Fig. 6). Analysis of levelling data for the timespan 1929–1991 shows that the western part of the Rila horst is still undergoing uplift whereas the eastern part is subsiding. Intensive uplift is also manifested in part of the Pirin horst. Negative vertical velocities are predominant for the whole west Rhodope region.

Geological, GPS and levelling data help to constrain the locations and behaviour of the recent active faults. The results obtained reveal new characteristics of current crust movements in SW Bulgaria. Based on the horizontal and vertical velocities, we can speculate that there is tectonic activity along the West Pirin fault (which marks the eastern margin of the Sandanski graben), the Melnik fault and probably also the Katountsi fault. The Krupnik fault is also recently active. The Predela and Dospat faults are marked as active normal faults. Activity is indicated along the NE border of the Gotse Delchev graben, at the Satovcha normal fault.

Table 5. *Recent fault activity according to geological and geodetic data for SW Bulgaria*

Fault	Geological data				Activity according to	
	Structural characteristics					
	Type	Strike (deg)	Dip	Tectonics	GPS	Levelling
Klisura fault	Normal	60	NW	NW block moving down	No	Yes
Beli Iskar fault	Strike-slip	0	Vertical	Strike-slip and normal motion	No	Yes
Bistritsa fault	Strike-slip	90	Vertical	Strike-slip and normal motion	No	No
West Rila fault zone	Normal	0–10	W	West block moving down	No	No
Gradevo fault	Normal	60–70	SSE	SE block moving down	No	No
Krupnik fault	Normal	50–60	NNW	NW block moving down	Yes, normal	Yes
Semkovo fault	Normal	60	SSE	SE block moving down	Yes, strike-slip	No
Predela fault	Normal	120	NE	NE block moving down	No	No
Damyanitsa fault	Strike-slip	0	Vertical	Strike-slip and normal motion	Yes, strike-slip	Yes
East Pirin fault zone	Normal	160	ENE	East block moving down	No	No
Osenovo–Ribnovo fault zone	Normal	160–120	WSW	SW block moving down	No	No
Dospat fault	Normal	120	NE	NE block moving down	Yes, normal	Yes
Satovcha fault	Normal	110	SSW	SW block moving down	Yes, normal	Yes
Ognyanovo–Ilinden fault zone	Strike-slip	0	Vertical	Strike-slip and normal motion	No	No
Gotse Delchev fault	Normal	120	NE	NE block moving down	No	No
West Pirin fault zone	Normal	160	WSW	SW block moving down	Yes, normal	No
Melnik fault	Normal	150	WSW	SW block moving down	Yes, normal	No
Kostenets fault	Normal	100	N	North block moving down	No	Yes
Katuntsi fault	Normal	30	NW	NW block moving down	Yes, normal	No

The results obtained constrain the complicated tectonic model of SW Bulgaria and contribute to understanding of the dominant regional north–south extensional regime in the north Aegean. They also confirm that the main reason for the recent crustal motion is the same endogenous forces under whose influence the main tectonic structures in SW Bulgaria were formed and the neotectonic motion of the crust took place.

The current movements in SW Bulgaria are in agreement with those in northern Greece, as well with the dextral movement along the NAF (McClusky *et al.* 2000). Considering that the movement of the Aegean plate is several times faster than that in SW Bulgaria, it is evident that the anticlockwise rotation of the Anatolian and Aegean plates is accompanied by their southward detachment from stable Europe. This is expressed exactly by the north–south tectonic extension in the structures of the northern Aegean Sea area: the west NAF and North Anatolian trough strike-slip and extensional zone. The orientations of the present crustal motion in northern Greece and SW Bulgaria can explain the existence of the predominant extensional regime in the southern part of the central Balkans, as confirmed by geological data (Shanov *et al.* 2001; Zagorchev 2001).

This research was financially supported by the Centre for Scientific Investigations and Design at the University of Architecture, Civil Engineering and Geodesy (Grant BN-2/2001) and the National Council 'Scientific Investigations' at the Ministry of Education and Science (Grant NZ-1105/2001). We are grateful to all colleagues and

PhD students who participated in the GPS campaigns. We thank the anonymous reviewers for constructive suggestions and comments, and T. Taymaz for his editorial assistance.

References

ALTAMIMI, Z. & BOUCHER, C. 2001. The ITRS and ETRS89 relationship: new results from ITRF2000. *In*: *Proceedings of the Symposium of the IAG Subcommission for Europe (EUREF), Dubrovnik, Croatia, 16–18 May 2001*. *In*: Mitteilungen des Bundesamtes für Kartographie and Geodasie, Band 23, 2003.

ANONYMOUS 1980. *Instruction for levelling first and second order*. Committee of Architecture and Public Works at the Council of Ministers, Head Office of Geodesy, Cartography and Cadastre.

BECKER, M., ZERBINI, S., BAKER, T., *ET AL.* 2002. Assessment of height variations by GPS at Mediterranean and Black Sea coast tide gauges from SELF projects. *Global and Planetary Change*, **34**, 5–35.

BONCHEV, E. 1980. Basic geodynamic problems of our lands. *In*: NACHEV, I. & IVANOV, R. (eds) *Geodynamics of the Balkans*. Technique, Sofia, 83–93 [in Bulgarian].

BOTEV, E. 2000. On the seismicity of South-West Bulgaria during 1980–1999. *Reports on Geodesy*, **4**, 81–89.

BROCKMAN, E. 1997. *Combination of solutions for geodetic and geodynamic applications of the global positioning system (GPS)*. Geodatisch-geophysikalische Arbeiten in der Schweiz, Schweizerischen Geodatischen Kommission, vol. **55**, PhD thesis, Zürich.

BURILKOV, T. & BELYASHKY, T. 1980. *A project of the State levelling network first and second order of the People's Republic of Bulgaria*. Ministry of Regional Development and Public Works, Sofia [in Bulgarian].

DEMETS, C., GORDON, R. G., ARGUS, D. F. & STEIN, S. 1994. Effect of recent revisions to the geomagnetic reversal time scale on estimates of current plate motion. *Geophysical Research Letters*, **21**, 2191–2194.

DEWEY, J. & ŞENGÖR, A. 1979. Aegean and surrounding regions: complex multiplate and continuum tectonics in a convergent zone. *Geological Society of America Bulletin*, **90**, 84–92.

DEWEY, J., PITMAN, W., RYAN, W. & BONNIN, J. 1973. Plate tectonics and the evolution of the Alpide System. *Geological Society of America Bulletin*, **84**, 3137–3180.

HUGENTOBLER, U., SCHAER, S. & FRIDEZ, P. (eds) 2001. *Bernese GPS Software Version 4.2*. Astronomical Institute, University of Bern.

JACKSON, J. 1992. Partitioning of strike-slip and convergent movement between Eurasia and Arabia in eastern Turkey. *Journal of Geophysical Research*, **97**, 12471–12479.

JACKSON, J. & MCKENZIE, D. P. 1984. Active tectonics of the Alpine–Himalayan Belt between western Turkey and Pakistan. *Geophysical Journal of the Royal Astronomical Society*, **77**, 185–264.

JACKSON, J. & MCKENZIE, D. P. 1988. The relationship between plate motion and seismic moment tensors, and the rates of active deformation in the Mediterranean and Middle East. *Geophysical Journal*, **93**, 45–73.

JESTIN, F., HUCHON, P. & GAULIER, J. M. 1994. The Somalia Plate and the East African rift system; present-day kinematics. *Geophysical Journal International*, **116**, 637–654.

KOTZEV, V., NAKOV, R., BURCHFIEL, B. C., KING, R & REILINGER, R. 2001. GPS study of active tectonics in Bulgaria: results from 1996 to 1998. *Journal of Geodynamics*, **31**, 189–200.

MCCLUSKY, S., BALASSANIAN, S., BARKA, A. *ET AL.* 2000. Global positioning system constraints on plate kinematics and dynamics in the Eastern Mediterranean and Caucasus. *Journal of Geophysical Research*, **105**, 5695–5719.

MCKENZIE, D. P. 1970. Plate tectonics of the Mediterranean region. *Nature*, **226**, 239–243.

MCKENZIE, D. P. 1972. Active tectonics of the Mediterranean region. *Geophysical Journal of the Royal Astronomical Society*, **30**, 109–185.

MERCIER, J., DELIBASSIS, N., GAUTHIER, A. *ET AL.* 1979. La néotectonique da l'Arc Egéen. *Revue de Géologie Dynamique et de Géographie Physique*, **21**, 67–92.

MERVART, I. 1995. *Ambiguity resolution techniques in geodetic and geodynamic applications of the global positioning system*. Geodatisch-geophysikalische Arbeiten in der Schweiz, Schweizerischen Geodatischen Kommission, vol. **53**, PhD thesis, Zürich.

MILEV, G., MINCHEV, M. & KOTZEV, V. 1999. [International projects connected with main application of GPS on the territory of Bulgaria.] *In*: *Proceedings of the Symposium 'Modern Information and GPS Technologies'*. Union of Surveyors and Land Managers in Bulgaria, Sofia, 11–12 November, 1999, 93–104 [in Bulgarian].

PAPAZACHOS, C. 1999. Seismological and GPS evidence for the Aegean–Anatolia interaction. *Geophysical Research Letters*, **17**, 2653–2656.

PAVLIDIS, S. B. & KONODOUPOULOU, D. P. 1987. Neotectonic and paleomagnetic results from Neogene basins of Macedonia (N Greece) and their geodynamic implications. *Annals of the Hungarian Geological Institute*, **70**, 253–258.

ROTHACHER, M., SPRINGER, T. A., SCHAER, S. & BEUTLER, G. 1997. Processing strategies for regional GPS networks. *In*: BRUNNER, F. K. (ed.) *Advances in Positioning and Reference Frames*. IAG Symposium, **118**, 93–100.

ROYDEN, L. 1993. The tectonic expression of slab pull at continental convergent boundaries. *Tectonics*, **12**, 303–325.

SHANOV, S. 1997. [*Contemporary tectonic stress field in the eastern part of the Balkan peninsula.*] DSc thesis, Geological Institute of Bulgarian Academy of Sciences [in Bulgarian].

SHANOV, S. & DOBREV, D. 2000. Tectonic stress field in the epicentral area of 04.04.1904 Krupnik earthquake from streak on slickensides. Geodynamic investigations on the territory of Bulgaria. Investigations of the Krupnik–Kresna region related to the 1904 earthquake. *Reports of Geodesy*, **4**, 117–122.

SHANOV, S., SPASSOV, E. & GEORGIEV, T. 1992. Evidence for the existence of the paleosubduction zone beneath the Rhodopean massif (Central Balkans). *Tectonophysics*, **206**, 307–314.

SHANOV, S., KURTEV, K., NIKOLOV, G., BOYKOVA, A & RANGELOV, B. 2001. Seismotectonic characteristics of the western periphery of the Rhodope Mountain region. *Geologica Balcanica*, **31**, 53–66.

TAPPONNIER, P. 1977. Évolution tectonique du système alpin en Mediterranée: poinçonnement et écrasement rigide–plastique. *Bulletin de la Société Géologique de France*, **19**, 437–460.

TAYMAZ, T., WESTAWAY, R. & REILINGER, R. 2004. Active faulting and crustal deformation in the Eastern Mediterranean region. *Tectonophysics*, **391**, 1–9.

ZAGORCHEV, I. 1971. Some features of the Young Alpine block structure of a part of Southwest Bulgaria. *Bulletin of the Geological Institute, Series Geotectonics*, **20**, 17–27 [in Bulgarian].

ZAGORCHEV, I. 1992. Neotectonic development of the Struma (Kraistid) Lineament, Southwest Bulgaria and Northern Greece. *Geological Magazine*, **129**, 197–222.

ZAGORCHEV, I. 2001. Geology of SW Bulgaria: an overview. *Geologica Balcanica*, **21**, 3–52.

ZEILINGA DE BOER, J. 1989. The Greek enigma: is development of the Aegean Orogen dominated by forces related to subduction or obduction? *Marine Geology*, **87**, 31–54.

Shear velocity structure in the Aegean region obtained by joint inversion of Rayleigh and Love waves

E. E. KARAGIANNI & C. B. PAPAZACHOS

Aristotle University of Thessaloniki, Geophysical Laboratory, PO Box 352-1, GR 54124 Thessaloniki, Greece (e-mail: elkarag@geo.auth.gr)

Abstract: We present a shear velocity model of the crust and uppermost mantle under the Aegean region by simultaneous inversion of Rayleigh and Love waves. The database consists of regional earthquakes recorded by portable broadband three-component digital stations that were installed for a period of 6 months in the broader Aegean region. For each epicentre–station ray path group velocity dispersion curves are measured using appropriate frequency time analysis (FTAN). The dispersion measurements for more than 600 Love wave paths have been used. We have also incorporated previous results for *c.* 700 Rayleigh wave paths for the study area. The single-path dispersion curves of both waves were inverted to regional group velocity maps for different values of period (6–32 s) via a tomographic method. The local dispersion curves of discrete grid points for both surface waves were inverted nonlinearly to construct 1D models of shear-wave velocity v. depth. In most cases the joint inversion of Rayleigh and Love waves resulted in a single model (from the multiple models compatible with the data) that could interpret both Rayleigh and Love wave data. Around 60 local dispersion curves for both Rayleigh and Love waves were finally jointly inverted. As expected, because of the complex tectonic environment of the Aegean region the results show strong lateral variations of the S-wave velocities for the crust and uppermost mantle. Our results confirm the presence of a thin crust typically less than 28–30 km in the whole Aegean Sea, which in some parts of the southern and central Aegean Sea becomes significantly thinner (20–22 km). In contrast, a large crustal thickness of about 40–45 km exists in western Greece, and the remaining part of continental Greece is characterized by a mean crustal thickness of about 35 km. A significant sub-Moho upper mantle low-velocity zone (LVLmantle) with velocities as low as 3.7 km s^{-1}, is clearly identified in the southern and central Aegean Sea, correlated with the high heat flow in the mantle wedge above the subducted slab and the related active volcanism in the region. The results obtained results are compared with independent body-wave tomographic information on the velocity structure of the study area and exhibit a generally good agreement, although significant small-scale differences are also identified.

In the present study we present a new high-resolution shear velocity model for the crust and uppermost mantle of the Aegean area obtained by the simultaneous inversion of Rayleigh and Love wave group velocities. The Aegean region (Fig. 1) lies at the convergence zone of the Eurasian and African lithospheric plates and is characterized by a complex tectonic setting. The Eastern Mediterranean plate is subducting under the Aegean, and this results in the formation of a well-defined Benioff zone (Papazachos & Comninakis 1969, 1971; Caputo *et al.* 1970; McKenzie 1970, 1978; Le Pichon & Angelier 1979). The Aegean microplate is moving at an average velocity of *c.* 35–40 mm a^{-1} towards the SW with respect to Eurasia (McKenzie 1972; Jackson 1994; Papazachos *et al.* 1998; McClusky *et al.* 2000) and is characterized by high tectonic activity, with volcanic activity (e.g. Georgalas 1962), magnetic anomalies and positive isostatic anomalies (e.g. Fleischer 1964; Vogt & Higgs 1969; Makris 1976), high heat flow (e.g. Fytikas *et al.* 1984) and high attenuation of seismic energy (e.g. Papazachos & Comninakis 1971; Hashida *et al.* 1988).

The main topographic features of tectonic origin and the local stress field of the study area are shown in Figure 1 (modified from Papazachos & Papazachou 1997). Two of the main sedimentary basins in Greece are depicted, namely the Axios basin with a maximum thickness of sediments of about 10 km (Roussos 1994), and the North. Aegean trough, with a sedimentary thickness of about 6 km (Kiriakidis 1988). In the same figure the main characteristics of the Hellenic arc, such as the volcanic arc, the sedimentary arc (Hellenides mountain range), the southern Aegean basin and the Hellenic trench, are also shown. The main zone of compression, with thrust faults that are observed along the Hellenic arc and along the western coast of northern Greece and Albania, is associated with the subduction of the Eastern Mediterranean beneath the Aegean (Papazachos & Delibasis

From: TAYMAZ, T., YILMAZ, Y. & DILEK, Y. (eds) *The Geodynamics of the Aegean and Anatolia.* Geological Society, London, Special Publications, **291**, 159–181.
DOI: 10.1144/SP291.8 0305-8719/07/$15.00

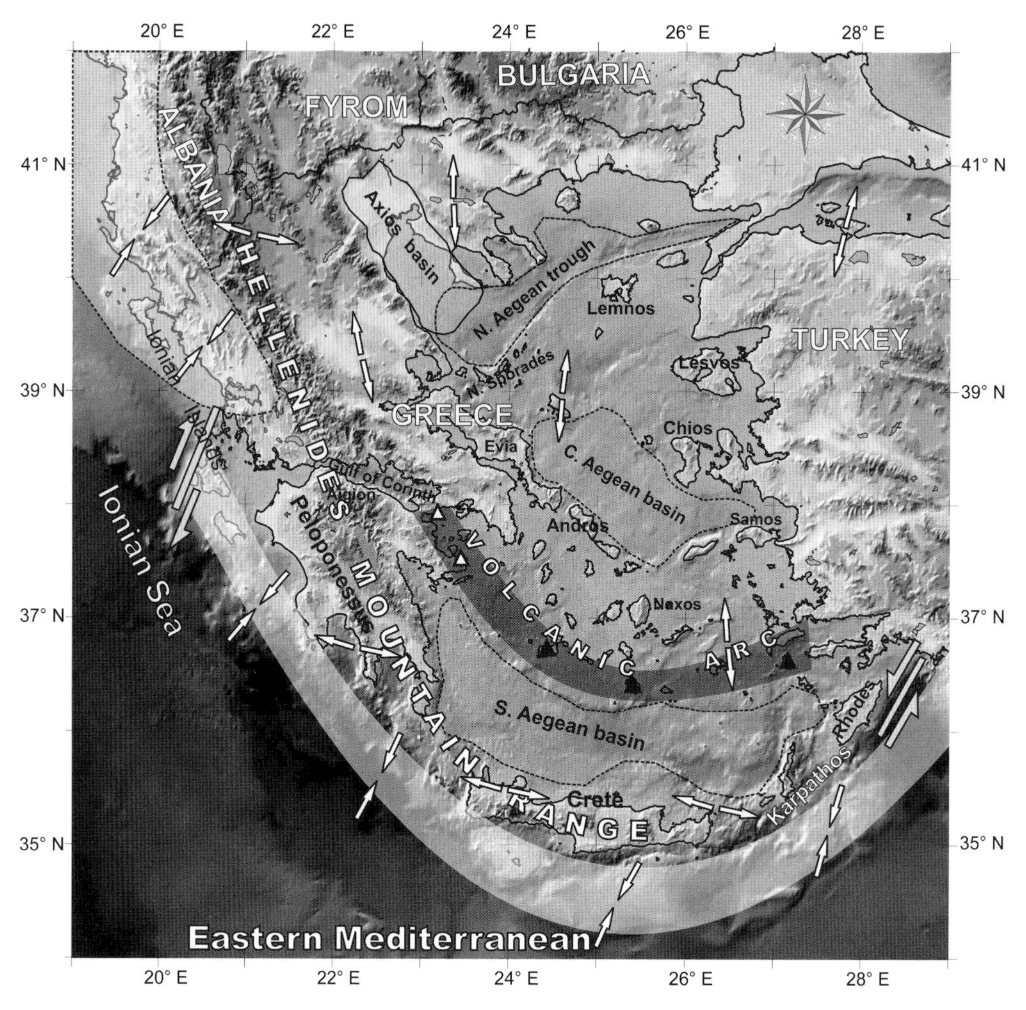

Fig. 1. Main seismotectonic features of the broader Aegean region (modified from Papazachos *et al.* 1998).

1969) and the continental–continental-type collision between the Adriatic (Apulia) microplate and the western Greek–Albanian coasts (Anderson & Jackson 1987). A dextral strike-slip zone (North Anatolian Fault; NAF) is observed in the North Anatolia–North Aegean trough (McKenzie 1970, 1972; Taymaz *et al.* 1991) and the Kefallonia area (Scordilis *et al.* 1985). The North Anatolia Fault was studied by Canitez & Toksoz (1971), who combined body- and surface-wave data and estimated the focal depth and the source parameters for earthquakes located at the fault. Later Saatcilar *et al.* (1999) mapped the active faults in the north Aegean by using reprocessed seismic reflection data, and suggested that the area is dominated by normal faults in an extensional regime, whereas for the east–central Aegean area they assumed active normal faults that instantaneously act as strike-slip faults and generate earthquakes of $M > 5.0$. The largest part of the back-arc Aegean area is dominated by normal faults with an east–west trend, suggesting north–south extension (e.g. Kurt *et al.* 1999). Finally, a narrow zone of east–west extension lies between the thrust faults of the outer Hellenic arc and the normal faults in the back-arc area, as has been identified using fault-plane solutions (e.g. Papazachos *et al.* 1984, 1998; Kiratzi *et al.* 1987), as well as by recent global positioning system (GPS) measurements (McClusky *et al.* 2000). Taymaz *et al.* (1991) used improved focal mechanisms of earthquakes constrained by P- and SH-body-wave modelling and first motions and showed that the western Aegean area is dominated by normal faults whereas the central and eastern Aegean is dominated by right-lateral strike-slip faults.

The velocity structure of the crust and upper mantle in the Aegean area has been extensively studied mainly using either body waves (usually P-waves) and partly surface-wave data. Travel times of body waves generated either by earthquakes (Panagiotopoulos 1984; Panagiotopoulos & Papazachos 1985; Plomerova *et al.* 1989; Taymaz 1996) or by explosions (Makris 1973, 1978; Delibasis *et al.* 1988; Voulgaris 1991; Bohnhoff *et al.* 2001) have been used for the study of the velocity structure under the Aegean region. An example is the work of Clement *et al.* (2004), who studied the seismicity and crustal deformation in the Gulf of Corinth by the use of seismic reflection and wide-angle reflection–refraction profiles. Their results depict the Moho discontinuity at a depth of about 40 km under the western end of the Gulf north of Aigion, which rises to a depth of about 32 km under the northern coast in the eastern part of the Gulf. Moreover, their results show a sedimentary layer beneath the Gulf of no more than 2.7 km.

An overall description of the 3D lithosphere and upper mantle P-wave structure has been presented in various tomographic studies (e.g. Spakman 1986; Christodoulou & Hatzfeld 1988; Spakman *et al.* 1988, 1993; Drakatos 1989; Drakatos *et al.* 1989; Ligdas *et al.* 1990; Drakatos & Drakopoulos 1991; Ligdas & Main 1991; Ligdas & Lees 1993; Papazachos *et al.* 1995; Tiberi *et al.* 2000). Papazachos & Nolet (1997) used travel-time data from local earthquakes in Greece and the surrounding areas and presented detailed results for the 3D P- and S-velocity structure of the Aegean lithosphere.

The published results on the lithospheric structure of the broader Aegean region based on measurements of the dispersion of surface waves are more limited. The early works of Papazachos *et al.* (1967) and Papazachos (1969) used group and phase velocities of Love and Rayleigh waves and studied the structure of the SE and Eastern Mediterranean region, respectively. Payo (1967, 1969) studied the structure of the Mediterranean Sea region using dispersion measurements of surface waves and identified a significant difference in the crust between the Eastern and Western Mediterranean, whereas Calcagnile & Panza (1980) and Calcagnile *et al.* (1982) focused on the regional study of the lithosphere–asthenosphere structure in the Mediterranean region. More recently, Kalogeras (1993) and Kalogeras & Burton (1996) used group velocity measurements of Rayleigh waves and produced 1D shear-wave velocity models down to 60–70 km along some seismic ray paths in the broader Aegean region. Marquering & Snieder (1996) used a waveform inversion method, in which the synthetic seismograms are calculated using surface-wave coupling, and studied the S-wave velocity structure beneath Europe, western Asia and the NE Atlantic to a depth of 670 km. They showed that in Europe, where the ray density is highest, small-scale structures are recovered, such as the presence of high velocities associated with the Hellenic subduction zone. Saunders *et al.* (1998) used teleseismic receiver functions to investigate the crustal structure at two locations in western Turkey using seismic data recorded on small arrays of temporary broadband seismological stations, and their results showed that western Turkey is characterized by a crust of *c.* 30 km thickness whereas a thicker crust of about 34 km is observed in eastern Turkey.

Yanovskaya *et al.* (1998) studied the shear-wave velocity structure under the Black Sea and surrounding regions using surface-wave group velocities. Recently, Martinez *et al.* (2000) studied the shear velocity structure of the lithosphere–asthenosphere system under the broader Mediterranean region using group velocities derived from the Rayleigh wave fundamental mode, and Pasyanos *et al.* (2001) investigated the broader southern Eurasia and Mediterranean Sea region using group velocities of both Rayleigh and Love waves. Karagianni *et al.* (2002) applied a surface-wave tomography method to group velocities of Rayleigh waves to connect group velocity variations with tectonic features of the crust and the upper mantle in the Aegean region. Raykova & Nikolova (2000, 2003) used the dispersion properties of Rayleigh and Love waves to study the anisotropy and constructed 1D models for the crust and uppermost mantle in southeastern Europe by the simultaneous inversion of both dispersion curves. Meier *et al.* (2004) used the two-station method to calculate the phase velocity curves of the fundamental Rayleigh mode and inverted them for 1D models of S-wave velocity. They estimated an average depth of the Moho beneath Crete at *c.* 50 km depth and above this discontinuity they found very low S-wave velocities of about 3.5 km s^{-1}. More recently, Karagianni *et al.* (2005), using the group velocities of the fundamental mode of Rayleigh waves, derived a 3D tomographic image of the shear-wave velocity structure of the crust–uppermost mantle in the Aegean region and estimated a new 3D image of the crustal thickness at this area. Surface-wave tomography on teleseismic events has been used by Bourova *et al.* (2005) to provide a 3D image of the lithosphere beneath the Aegean Sea. Their resolution analysis, which varied from 200 to 800 km, shows that only large-scale lateral anomalies greater than 200 km can be revealed from their dataset.

In general, the previous studies in the broader Aegean region show the existence of strong variations of the crustal structure. A thin crust of about 20–30 km has been proposed for the back-arc area, whereas a significant crustal

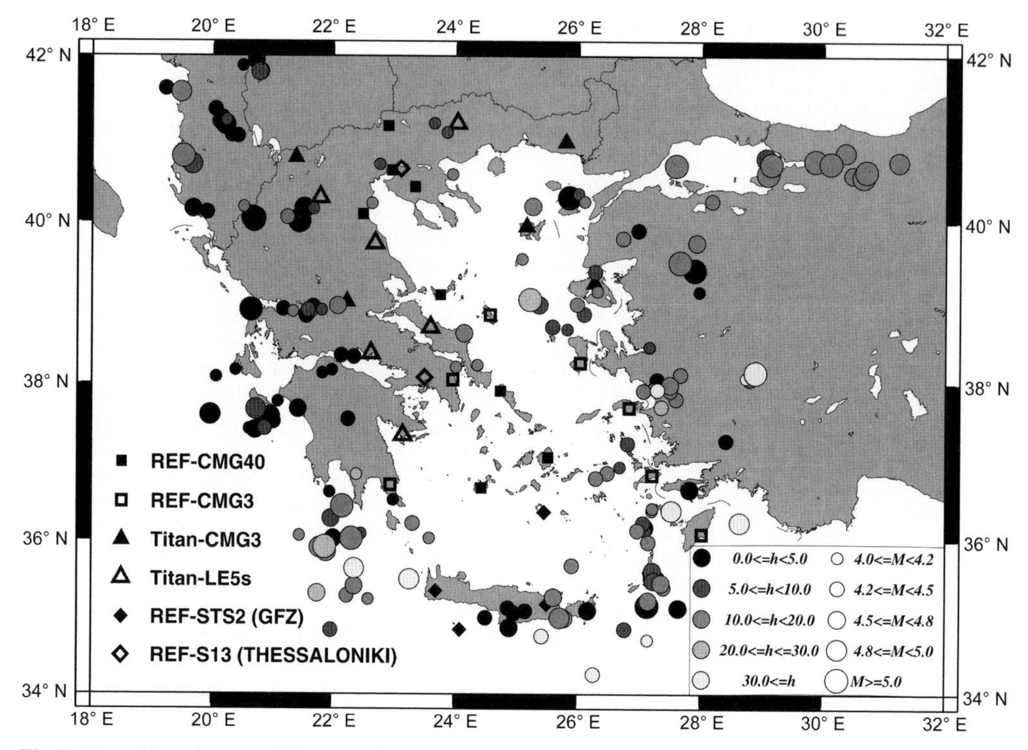

Fig. 2. Map of the epicentres of the events that were used in the present study and the locations of the portable stations. The earthquake depth and magnitude are shown by the grey shading and variable radius of the circles, respectively.

thickness (40–47 km) has been identified along the Hellenides mountain range. The crust has a normal thickness (28–37 km) in the eastern part of the Greek mainland, in the northern and central Aegean, in western Turkey and in Crete.

Unfortunately, 3D S-wave velocities for the Aegean region are not well known because of the limited number of body-wave tomographic studies on the S-velocity structure. The main purpose of this paper is to obtain a detailed 3D tomographic image of S-wave velocity for the uppermost *c.* 50 km in the Aegean region by combining Rayleigh and Love wave group velocities along different ray paths. These velocities were used to construct group velocity tomographic maps for different periods, ranging from 6 to 32 s, and jointly invert them to determine the 3D S-wave velocity structure of the study area.

Data

In the present study group velocity maps have been used as the main dataset for the S-velocity model determination. These maps represent the variation of the local group velocities of the fundamental mode of Rayleigh and Love waves for each type of wave at different periods and have been obtained by a tomographic inversion of single-path dispersion curves. These curves were computed for different single ray paths crossing the Aegean region, using the method of frequency time analysis (FTAN) of Levshin *et al.* (1972, 1989, 1992). We have considered 185 regional earthquakes within the area defined by 34–42° N and 19–31° E, which were recorded by 35 broadband stations of a temporary network that has been installed in the broader Aegean region. The locations of the events, as well as of the portable stations are shown in Figure 2, and a detailed description of these data has been given elsewhere (e.g. Hatzfeld *et al.* 2001).

Around 600 observed Love wave group velocity curves have been determined in this work, and we have also incorporated previous results for around 700 Rayleigh wave curves of the study area (Karagianni *et al.* 2002). The observed dispersion curves of Rayleigh waves have been calculated using the vertical component, Z, of the waveforms. The two horizontal components (east–west and north–south) of each record were rotated to a radial R (along the great-circle path) and transverse T component (perpendicular to the great-circle path) and the transverse component was used to estimate the dispersion curves of Love waves. An example of the vertical and transverse components of a seismic record for an earthquake located in the Ionian Islands and used in the present study is shown in Figure 3. The ability of the analyst to identify the direct fundamental mode arrival and to distinguish it from higher modes, other reflections or coda waves is critical for the frequency bandwidth for which a dispersion curve can be estimated. A detailed description of the FTAN method has been presented in an earlier study (Karagianni *et al.* 2002). As the mean path length is of the order of 400 km, the Rayleigh and Love waves have been well recorded in the period range from 5 s to 30–35 s. This is also shown in Figure 4, where two FTAN maps for the components of Figure 3 for both Rayleigh and Love waves are shown. The FTAN method was applied for different (crossing) ray paths, and the coverage of the study area for a period of 10 s for both Rayleigh and Love waves is shown in Figure 5. As can be seen, the azimuthal distribution of the paths for both Rayleigh and Love waves is uniform and the coverage is satisfactory, especially in the central Aegean, where a large number of portable stations were located. Poorer density of seismic ray paths is observed in western Greece, NNE Greece and SW Crete because of the lack of earthquakes and recording stations.

Methods of analysis of surface waves

Tomography method

The group velocity curves of Rayleigh and Love waves for different ray paths estimated from the FTAN program have been used to construct the group velocity tomographic maps for different periods. For this purpose we have used a generalized 2D linear inversion method of Ditmar & Yanovskaya (1987) and Yanovskaya & Ditmar (1990), which is a generalization to two dimensions of the classical 1D method of Backus & Gilbert (1968) and is appropriate for regional studies as a solution is constructed on a plane surface obtained after transformation from spherical coordinates (Yanovskaya 1982; Jobert & Jobert 1983; Yanovskaya *et al.* 2000). The result of the surface-wave tomography is the estimation of local values of group velocity for each wave type at different grid points over the study area, which is used to obtain group velocity maps for different periods.

In tomography, the knowledge of the resolution is important to estimate the minimum resolvable features for a given sample and to determine those features that may be a numerical artefact. Yanovskaya (1997) and Yanovskaya *et al.* (1998) proposed the use of two parameters as resolution measures.

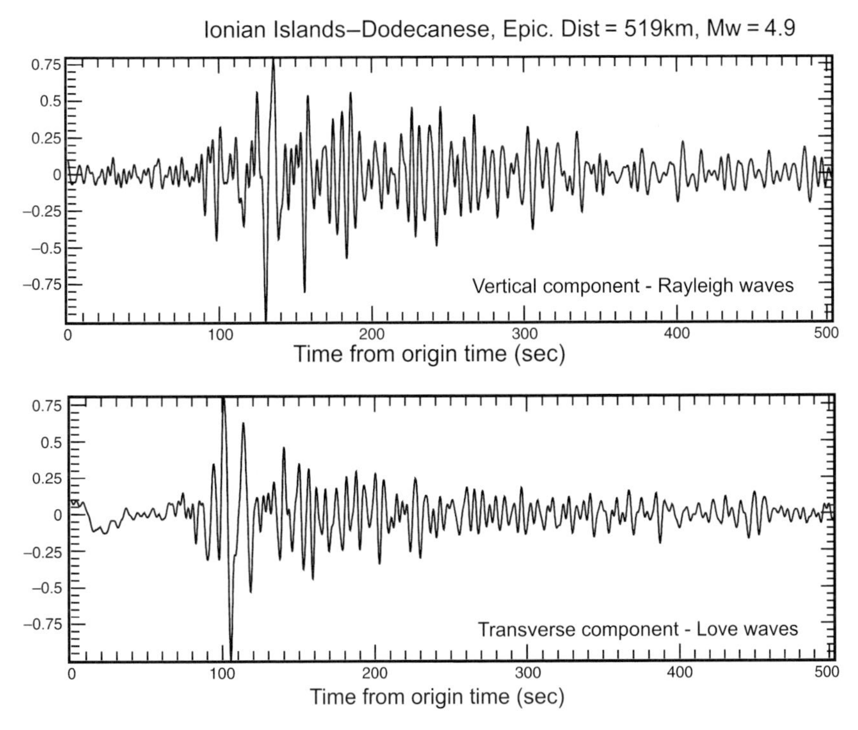

Fig. 3. Vertical and transverse components of a seismic record for a selected earthquake located in the Ionian Islands, used in this study to calculate the dispersion curves of Rayleigh and Love waves, respectively.

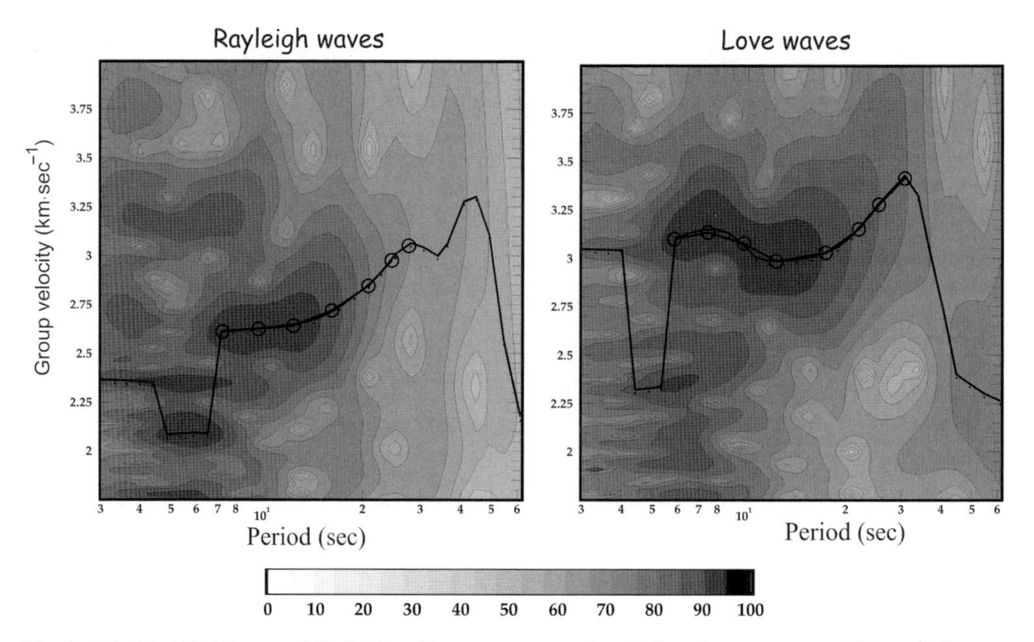

Fig. 4. Calculated FTAN maps of Rayleigh and Love waves, as estimated from the components of Figure 3. The continuous line indicates the dispersion curve that was identified by the analyst.

The first resolution parameter is the mean size of the averaging area, L, given by

$$L = [s_{\min}(x, y) + s_{\max}(x, y)]/2 \qquad (1)$$

where $s_{\min}(x, y)$ and $s_{\max}(x, y)$ are respectively the smallest and largest axes of an ellipse, which the averaging area can be approximated to, centred at each examined point (x, y). As the resolution is closely correlated to the density of the crossing ray paths in each cell, it is clear that small values of the mean size of the averaging area (corresponding to high resolution) should appear in the areas that are crossed by a large number of ray paths and vice versa.

The second parameter is the stretching of the averaging area, which provides information on the azimuthal distribution of the ray paths and is given by the ratio

$$2[s_{\max}(x, y) - s_{\min}(x, y)]/[s_{\max}(x, y) + s_{\min}(x, y)]. \qquad (2)$$

Small values of the stretching parameter imply that the paths are more or less uniformly distributed along all directions, hence the resolution at each point can be represented by the mean size of the averaging area. In contrast, large values of this parameter (usually >1) mean that the paths have a preferred orientation and that the resolution along this direction is likely to be small (Yanovskaya 1997).

Inversion method

In this study we employed the Hedgehog nonlinear inversion method for the determination of the S-wave velocity structure using the dispersion curves of surface waves (Keilis-Borok & Yanovskaya 1967; Valyus 1968; Knopoff 1972; Biswas & Knopoff 1974; Calcagnile & Panza 1980; Panza 1981). The Earth model is represented by a set of parameters (P- and S-wave velocities, and densities in a layered Earth), which may be varied or held fixed in the inversion and can be either independent or dependent, on the basis of *a priori* knowledge. For the independent parameters acceptable models are sought, whereas the dependent parameters maintain a fixed relationship with the independent ones. The ratio between S- and P-wave velocities, as well as the P-wave velocity and the density, are kept fixed in the inversion process because the phase velocity of the fundamental mode of Love waves is mainly sensitive to the shear-wave velocity (e.g. Urban *et al.* 1993). For that reason, only the S-wave velocities and the layer thicknesses were chosen as independent parameters.

For the inversion each parameter was specified to vary within a particular range, with upper and lower limits. The bounds within which we have allowed the independent parameters to vary are large enough that a large number of models are tested but at the same time these limits have been chosen so that the resulting velocities do

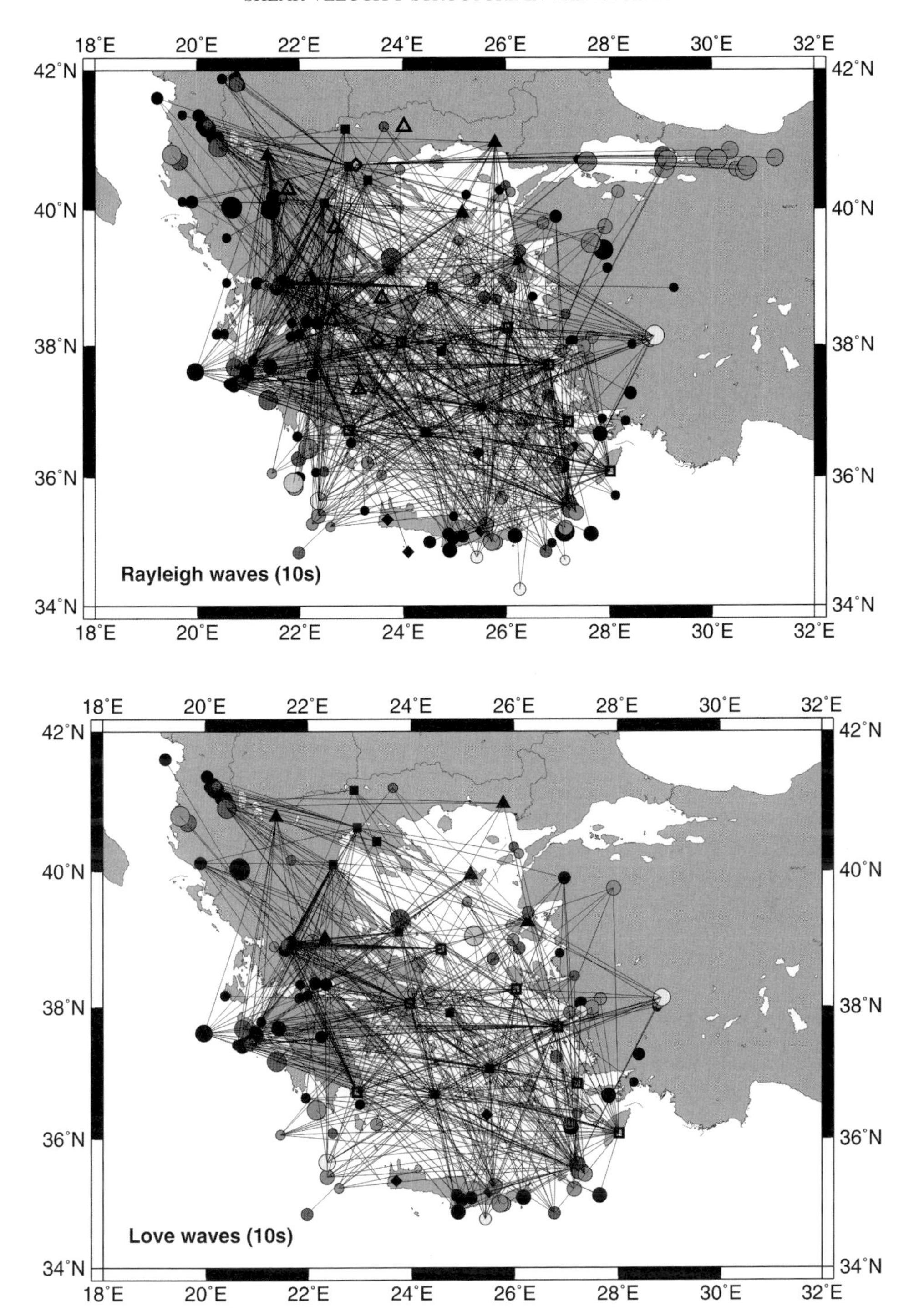

Fig. 5. Coverage of the ray paths used in this study for a period of 10 s for both Rayleigh and Love waves. Symbols as in Figure 2.

not reach unrealistic values. In addition, the steps used for the parameter variation should be properly chosen because large steps may exclude examining some models, whereas steps that are too small could result in an unnecessarily large number of tested models. For that reason the steps for the variation of the velocity parameters have been estimated according to the resolving power of the information contained in the available data (Panza 1981).

After perturbing the chosen set of model parameters within the pre-assigned parameters space, a set of theoretical values of group velocities (of Rayleigh wave first in the present study) are computed using the Knopoff method (Knopoff 1964; Schwab & Knopoff 1972; Schwab *et al.* 1984). Starting from the largest period, the theoretical group velocity is computed and compared with the observed value for the specific period. If the difference lies within the observational errors depending upon the quality of the measurements (single point error) the inversion proceeds to test the next shorter period, and so on. If the test is successful at all periods of the dispersion curve, the r.m.s. difference (root mean square deviation) between theoretical and observed values of group velocity of Rayleigh waves is computed and compared with a value defined *a priori* on the basis of the quality of the data. After tests, we set this value to 60% of the mean value of the single point experimental error of group velocity values, to avoid artificial large jumps of the theoretical dispersion curves or solutions with a systematic bias with respect to the experimental curve. On the other hand, if the test fails at any period, the model is rejected and a new model in the neighbourhood of the previous one is tested.

Models that pass both above-mentioned criteria were accepted for the case of Rayleigh waves and were then tested with the observed dispersion curves of Love waves using the same procedure and quality criteria. In general, Love wave data exhibit larger errors than Rayleigh waves, as Love waves are recorded in the horizontal component, which is more noisy than the vertical one, and hence it is more difficult for the analyst to accurately identify the Love wave dispersion curve. For grid points with a good path coverage the typical single point error for the case of Rayleigh waves is usually less than $0.06\ \mathrm{km\ s^{-1}}$, and is larger for the case of Love waves. These errors become generally larger for both wave types at the borders of the study area, where the number of crossing rays was limited. The same process was repeated until the neighbourhood around each satisfactory combination of the search parameters was explored.

Tomographic images of group velocity variations

As the group velocity variations of Rayleigh waves have been discussed in detail in previous studies (Karagianni *et al.* 2002, 2005), we present here only the group velocity maps of Love waves, shown in Figure 6 for six period values. In this figure we present results only for areas where the mean size of the averaging area is less than 200 km. Moreover, we have also plotted on each map the 100 km contour for the mean size of the averaging length for comparison. The variations are shown as percentages, relative to the corresponding mean velocities, which are also listed for each period. In general, the group velocity maps show significant lateral velocity heterogeneity, with variations up to 30% in the study area. At shorter periods (7–14 s) Love wave maps show significant low-velocity anomalies in western Greece under the Hellenides mountain range and in northern Greece in the Axios basin, as a result of the thick sedimentary formations in these areas, also identified in the tomographic maps of Rayleigh waves (Karagianni *et al.* 2002). Whereas for this period range a clear low-velocity layer in the southern Aegean Sea was dominant for the case of Rayleigh waves, the corresponding maps of Love waves for this period range do not exhibit a similar pattern. A possible explanation for this difference is that the structure at a certain depth affects Love wave dispersion curves at larger periods than Rayleigh waves. As the period increases an increase of Love group velocities is observed all over the Aegean Sea, as a result of the thin crust of the region. In contrast, low-velocity anomalies are clearly observed in western Greece, suggesting that the crust there is thick. At a period of 28 s the group velocity maps of Love waves show high velocities in the south and central Aegean Sea, whereas for larger periods (32 s) a low-velocity anomaly is found between Crete and the volcanic arc. A similar velocity anomaly in this area was depicted in the group velocity maps of Rayleigh waves for a period of 28 s.

The resolution parameters associated with the local group velocities of Love waves are shown for two periods (7 and 32 s) in Figure 7, and it can be seen that their values are more or less the same as those for Rayleigh waves (Karagianni *et al.* 2002). The mean size of the averaging area (quantifying the resolution of the local group velocity distribution) is of the order of 40–90 km in the central part of the Aegean and becomes larger (*c.* 150–200 km) only at the borders of the maps, where the path coverage is limited. The values of the stretching parameter of the averaging area generally indicate that the azimuthal distribution of the

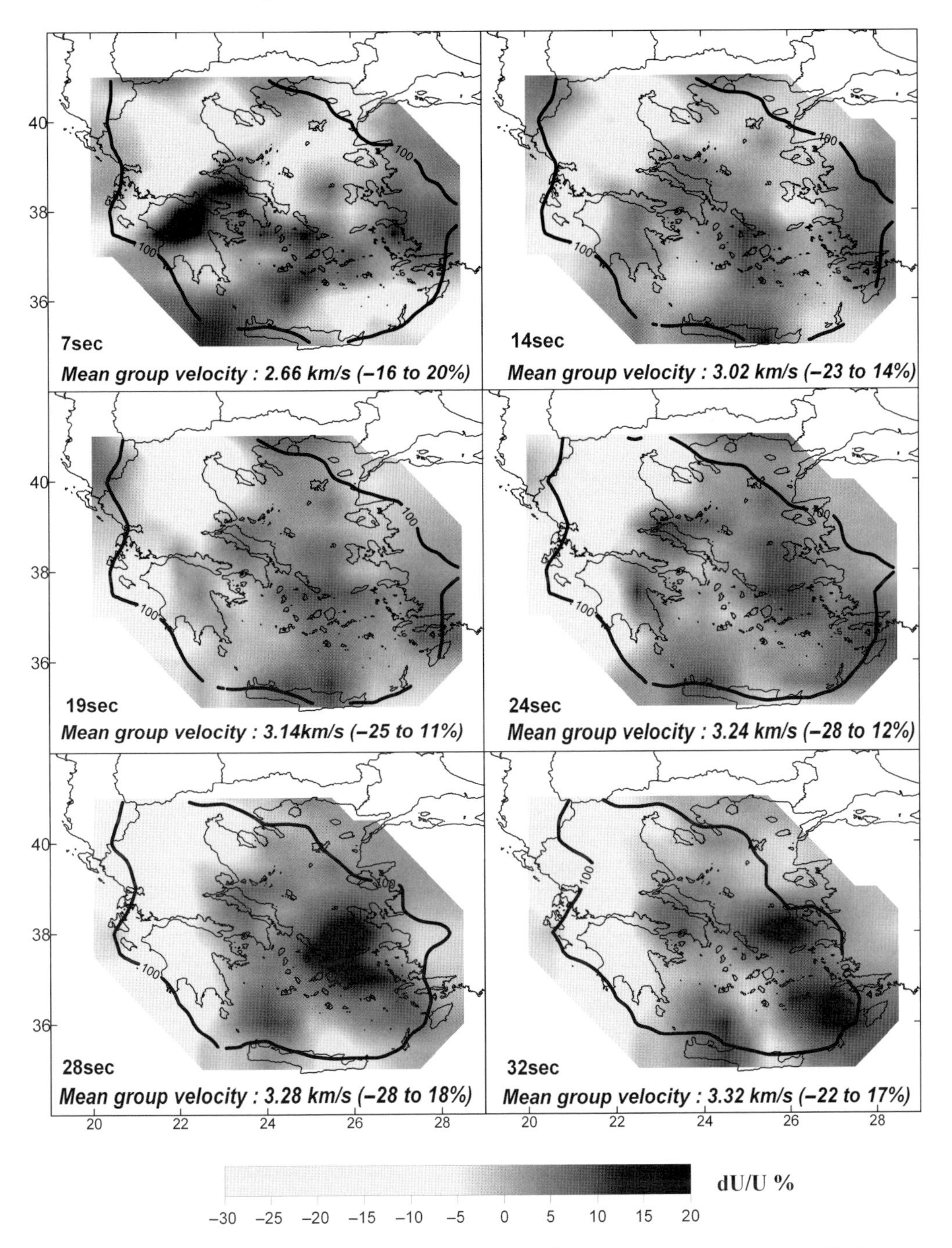

Fig. 6. Estimated Love wave group velocity maps at selected periods. Maps represent lateral variations (in %) of group velocity, relative to the average group velocity across each map.

paths is uniform, as this parameter has values smaller than one for most parts of the study area, and only at the borders of the area where the path coverage is poorer do we obtain values of one or larger. The standard errors of the local group velocities of Love waves are larger than those of Rayleigh waves and range between 0.07 and 0.12 km s^{-1} for all the examined periods.

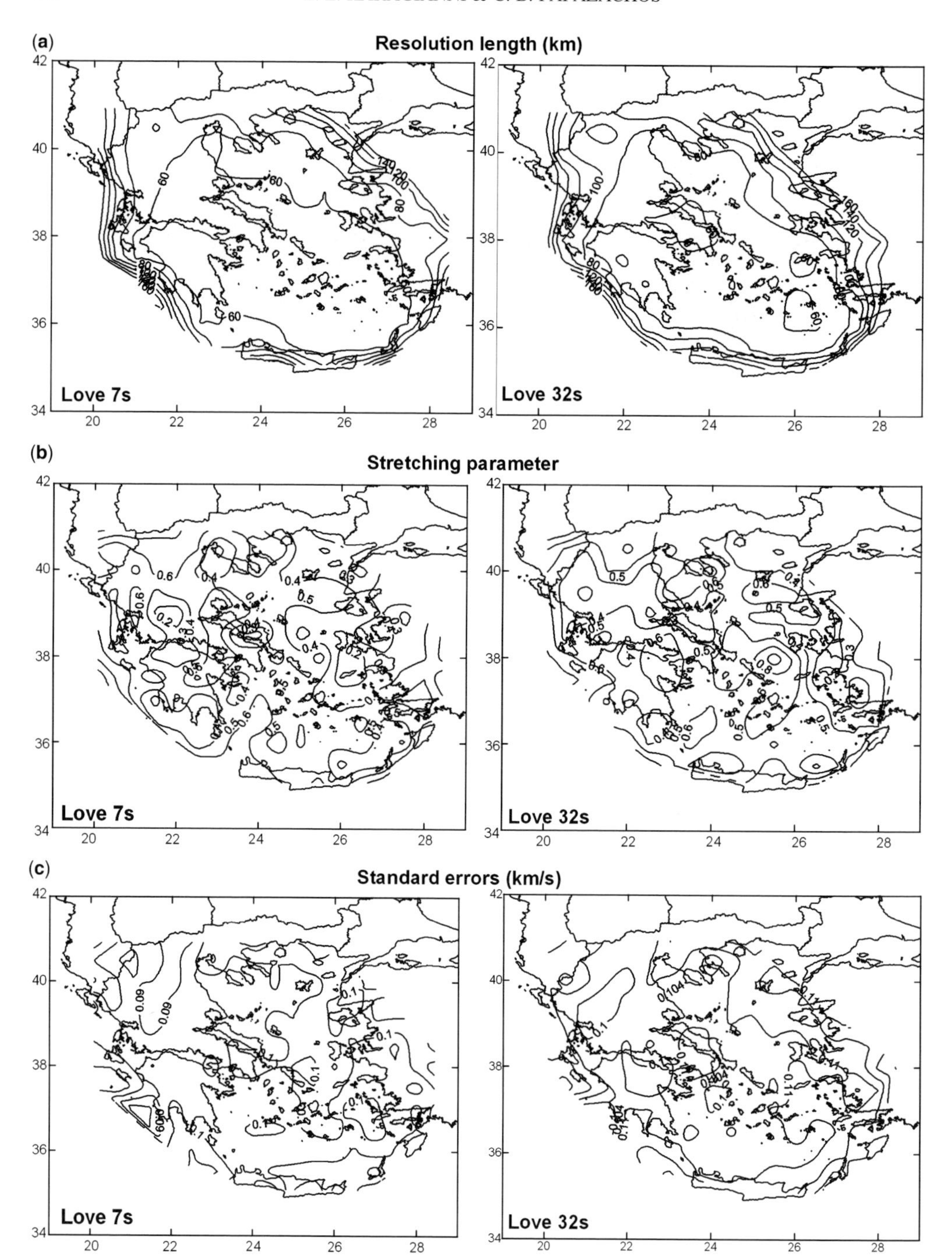

Fig. 7. Resolution information for the smaller (7 s) and the larger period (32 s) used in the present study: (**a**) resolution length (in km); (**b**) distribution of the elongation of the averaging area; (**c**) standard error (in km s^{-1}) associated with the estimated group velocity maps.

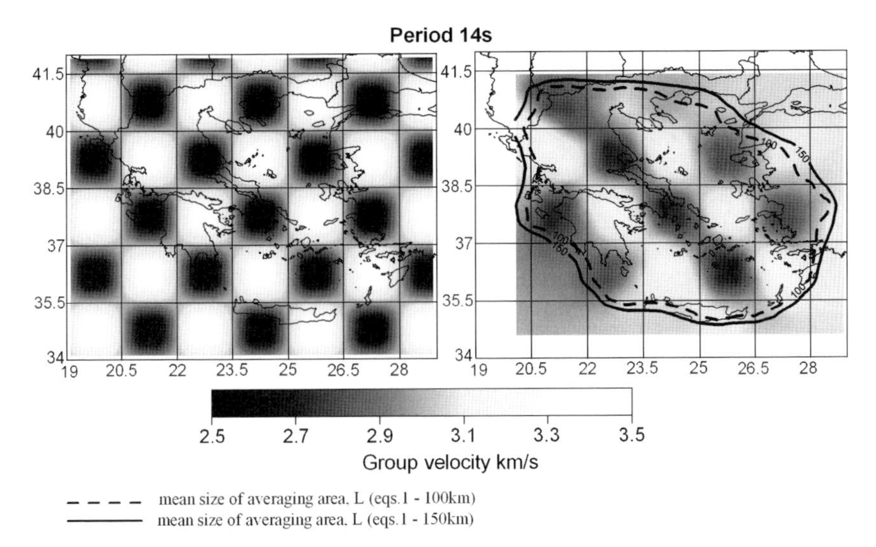

Fig. 8. Chequerboard test for 14 s Love wave using a cell of $1.5^\circ \times 1.5^\circ$ (typical anomaly size *c.* 150km), using a sinusoidal velocity perturbation with a peak amplitude of $\pm 20\%$.

To further explore the resolving information of the tomographic results of the joint inversion of Rayleigh and Love waves we have performed additional resolution tests. For this reason, we divided the study area into cells and applied a sinusoidal velocity perturbation with respect to the average group velocity for each examined period. Using this model, travel times along paths linking source and recording stations were calculated, adopting the same ray coverage configuration as in Figure 5b. A solution for these velocity perturbations was obtained using the same regularization parameters as for the real data in the tomographic inversion. The chequerboard test was performed for different cell sizes, as well as for different values of the velocity perturbation. In Figure 8 we show the results of the chequerboard test for a period of 14 s using perturbation anomalies with a peak amplitude of 20% and a length of *c.* 150 km. Moreover, the contour lines of 100 and 150 km for the mean size of the averaging area for the real data are also superimposed. We observe that the well-resolved area (according to the chequerboard tests with anomalies *c.* 150 km) is in very good agreement with the corresponding contour of 150 km for the mean size of the averaging area, which verifies the reliability of the resolution estimates shown in Figure 7.

Inversion results

From the estimated group velocity maps a local group velocity of Rayleigh (Karagianni *et al.* 2002) and Love waves has been constructed for different grid points over the Aegean region. Finally, a mean local dispersion curve with its standard error for each type of wave was assigned to each $0.5^\circ \times 0.5^\circ$ grid cell using the dispersion curves that corresponded to the four corners of each cell. Because the patterns of paths for Rayleigh and Love waves are slightly different for different periods, we may expect a difference between the velocity sections obtained from Rayleigh and Love waves separately. Moreover, it is well known that the velocities determined from Rayleigh and Love waves can differ because of the possible anisotropic properties of the crust and upper mantle. To reduce the errors arising from differences of the pattern paths, we have jointly inverted the Rayleigh and Love wave dispersion curves. Sixty mean local dispersion curves of Rayleigh and Love waves were inverted, producing vertical 1D shear-wave velocity models for different examined points. After the inversion, a simple bilinear interpolation scheme was adopted and a 3D image of the S-wave structure was derived for depths ranging from 5 to 45 km.

Despite the fact that with the adopted nonlinear inversion scheme the starting model does not significantly affect the final results, the starting model of S-wave velocities for the inversion has been taken from the work of Papazachos & Nolet (1997) and Martinez *et al.* (2000), who gave detailed information on the 3D S-wave velocity structure of the Aegean lithosphere. Furthermore, information on the shear velocity structure of the study area from Karagianni *et al.* (2005), based on the inversion only of Rayleigh waves, has been also considered. For the cases where the inversion was performed at grid points located in the sea, the starting 1D model was overlain by a water layer of variable thickness (0.2–2.5 km), according to the bathymetric map for the broader Aegean region. This was important only for Rayleigh waves, as the water layer does not affect the propagation of Love waves. Our data are not adequate to resolve the elastic properties of the very shallow layers, or those of the deep mantle (greater than 45–50 km). For this reason these properties were kept fixed, using existing results from the previously mentioned literature. In particular, the elastic properties (the S-wave velocities and the layer thickness for the upper 3–5 km) have been fixed on the basis of information coming mainly from seismic soundings by the Greek Public Petroleum Company (N. Roussos, pers. comm.) and from other existing geophysical investigations (e.g. Makris 1976, 1977; Martin 1987; Roussos 1994). The P-wave velocities have been defined on the basis of S-wave velocity values, using the V_P/V_S values given by Papazachos & Nolet (1997). The density values that were used in the present study have been calculated from the values of P-wave velocities using the relationship of Barton (1986) for crustal and upper mantle material.

For each examined grid point, 8–10 parameters have been allowed to vary in the inversion scheme; namely, the S-wave velocities in 4–5 layers reaching a depth of about 40–45 km, as well as their related thicknesses. In the inversion process for each examined point we considered the depth of 45–50 km as a maximum depth at which the independent parameters were allowed to vary. This is based upon the variation of the group velocity sensitivity kernels on the shear-wave velocity v. depth (Panza 1981; Urban *et al.* 1993), as has already shown in detail in the previous work of Karagianni *et al.* (2002).

In Figure 9 we present the inversion results for a common depth range (26–30 km) after the independent inversion of Rayleigh and Love waves. In general, these images show high S-wave velocities (4.0–4.3 km s^{-1}) in the Aegean Sea (typical upper mantle velocities), whereas crustal S-wave velocities (*c.* 3.5 km s^{-1}) are observed in western Greece. Love wave inversion exhibits slightly higher S-wave velocities compared with the Rayleigh wave data, especially in the Aegean Sea region (lower crust, uppermost mantle). However, the high S-wave velocities resulting from Rayleigh wave inversion extend over a larger area in the Aegean Sea compared with the S-wave velocities of Love wave inversion. The critical point is whether this bias reflects specific model properties (e.g. anisotropy) or whether a subset of the models derived from the Hedgehog inversion could explain both Rayleigh and Love wave group velocity data. To examine this issue, we performed a joint inversion of Rayleigh and Love wave data to search for models that satisfy both experimental dispersion curves of Raleigh and Love waves. As an example, we present the results for an examined grid point in the central Aegean Sea in Figure 10, where the experimental and the theoretical dispersion curves of both waves are depicted. It can be observed that Love wave group velocities are larger than the velocities of Rayleigh waves for all periods, and also exhibit larger errors. However, the experimental–theoretical dispersion curve fit for both types of waves is fairly good and does not appear to follow a systematic pattern (e.g. systematic underestimation of Love group velocities). These results are representative of almost all the examined grid points, with larger discrepancies observed only for a few grid points at the margins of the examined area.

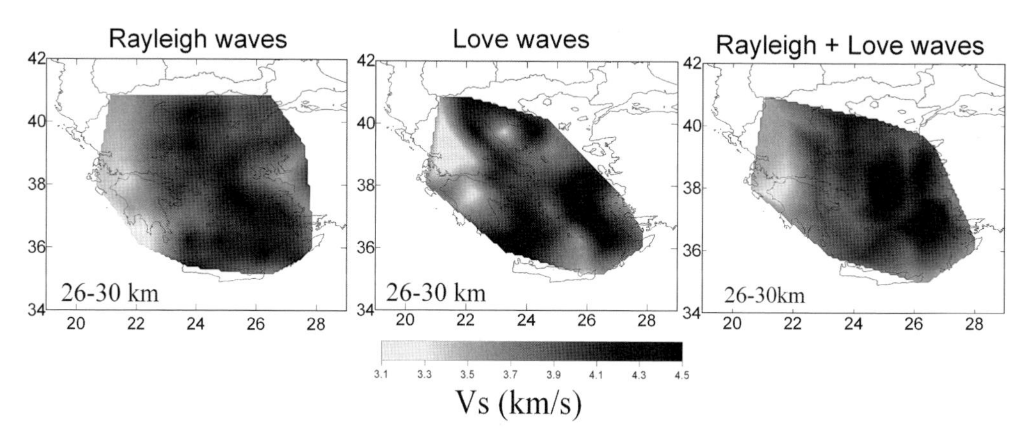

Fig. 9. Horizontal cross-section of the shear-wave velocity model for the depth range 26–30 km in the Aegean region, as derived from the independent inversion of Rayleigh and Love waves, as well as the joint inversion of the both waves.

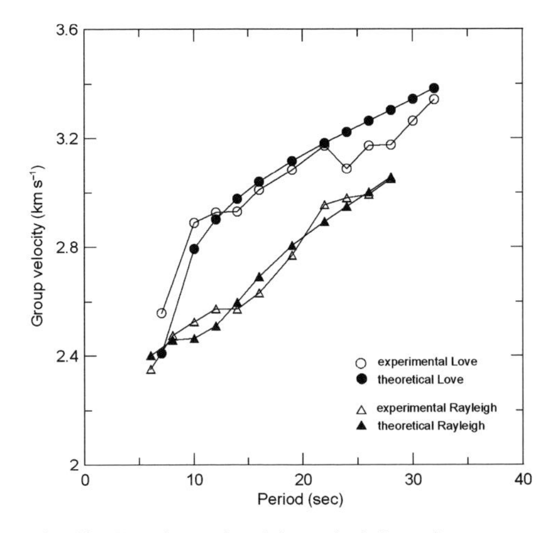

Fig. 10. Experimental and theoretical dispersion curves for a selected grid point.

On the basis of these results we suggest that the joint inversion of both Rayleigh and Love waves can give more reliable results by selecting a small number of velocity models that are compatible with both datasets. This is confirmed in Figure 11, where for a random grid point in the study area we show the inversion results using only Rayleigh wave data, Love wave data and jointly Rayleigh and Love wave data. The solutions that are estimated from the joint inversion of Rayleigh and Love waves do not reflect the average of the individual solutions but correspond to models that satisfy both Rayleigh and Love wave group dispersion data. This is why the number of accepted solutions in the simultaneous inversion of Rayleigh and Love wave data has been significantly decreased compared with the solutions of the independent inversion of Rayleigh and Love wave data. Nevertheless, most of the solutions of the above three inversion schemes show similar structural

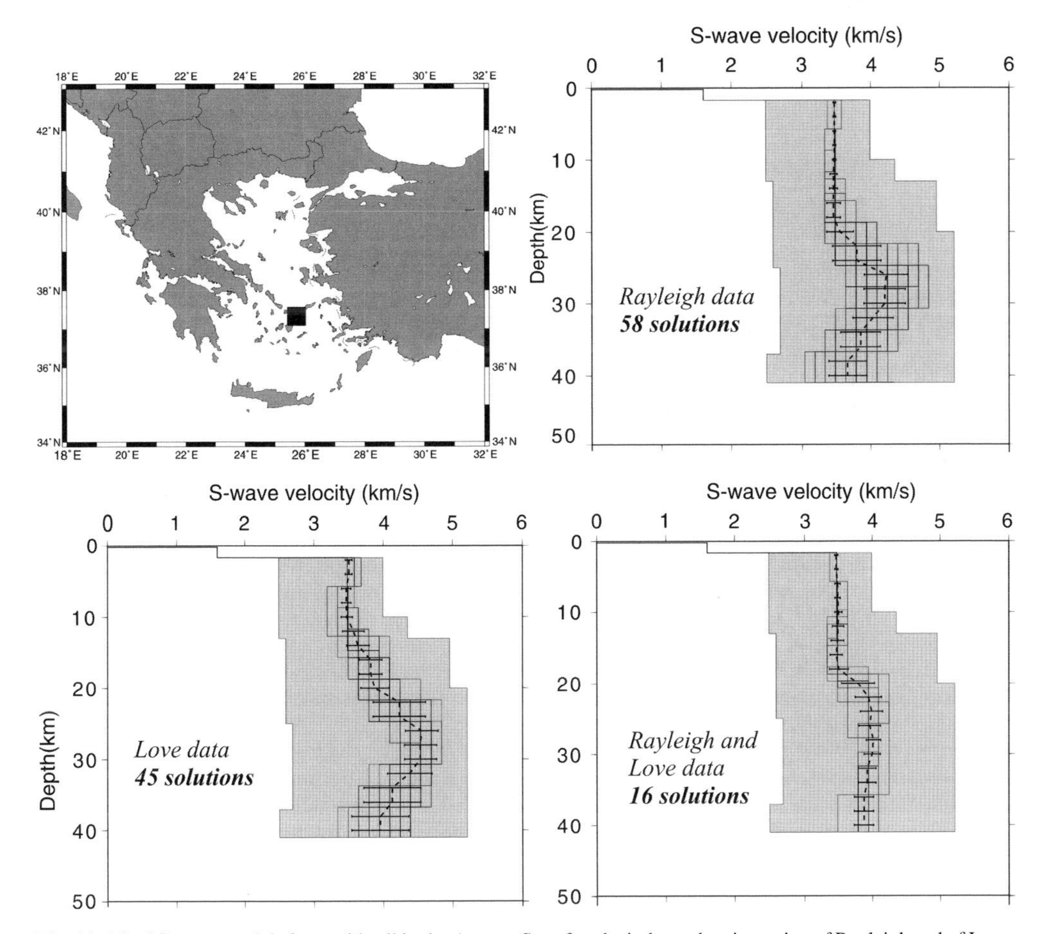

Fig. 11. Final S-wave models for a grid cell in the Aegean Sea after the independent inversion of Rayleigh and of Love waves, and the joint inversion of both types of surface waves.

features. These are the Moho discontinuity, which is shown at a depth of about 20–22 km, as well as a decrease in velocity just below the Moho, which is interestingly more evident when Rayleigh and Love wave are inverting separately.

S-wave velocity variations

A 3D image of the Moho discontinuity for the study area was constructed using the local 1D vertical models of shear-wave velocity, as derived from the inversion process for each geographical point examined. Using all the solutions of the inversion the average S-wave velocity was estimated using a depth interval of 2 km to define the average S-velocity depth profile at each point. We have assumed that a shear-wave velocity of about 3.9–4.0 km s^{-1} is a reasonable approximation for the S-wave velocity at the bottom of the lower crust. The depth at which we observed a significant jump in the S-wave velocity that 'crossed' this limit of 3.9–4.0 km s^{-1} (e.g. from 3.8 to 4.3 km s^{-1}) was assumed to be the Moho discontinuity. However, for some solutions a smooth transition rather than a steep jump was observed in the S-wave velocity from the lower crust to the upper mantle, in which cases a Moho depth was not assigned. On the other hand, because for each grid point several solutions existed, this resulted in not assigning a Moho depth only for a small number of solutions for each grid point (typically *c*. 20%).

Significant lateral change in the crustal thickness can be observed in Figure 12a, as is expected from the tectonic complexity of the Aegean region. Western Greece, with the presence of the Hellenides mountains, is dominated by a thick crust of the order of 40–45 km. As we move to the east the crust becomes thinner, and most of continental Greece is characterized by a normal thickness of crust of about 34–38 km. Some parts of eastern continental Greece (e.g. the island of Evia) have a thinner crust of about 28 km. The most interesting feature is that the whole Aegean Sea is characterized by a thin crust, typically less than 28–30 km. In some parts in the southern and central Aegean Sea a significantly thinner crust of about 20–22 km is observed, in agreement with the north–south extensional field (Fig. 1), as well as with the rise of hot mantle material owing to the subduction of the eastern Mediterranean plate under the Aegean microplate. The north Aegean Sea is also characterized by a thin crust, and a crustal thickness of about 25 km is observed near the North Sporades islands. Only in some parts of the north Aegean Sea (between the islands of Lemnos and Lesvos) does the crust show a thickness of about 32–34 km. Results are not shown for the borders of the Aegean region (NE Greece and Crete) because the inversion process did not give models that satisfied both experimental Rayleigh and Love wave data, resulting either from the larger errors of the experimental curves at the borders of the study area or from the presence of anisotropy. However, the above-mentioned results are in agreement with those in previous investigations (e.g. Makris 1976; Brooks & Kiriakidis 1986; Chailas *et al.* 1993; Papazachos 1994, 1998; Karagianni *et al.* 2005).

Using all the estimates for the Moho depth at each grid point where the joint inversion of Rayleigh and Love waves has been performed (as described above), we have also estimated (Fig. 12b) the corresponding standard deviation of the Moho depths and used it as an error estimate of the Moho depth map of Figure 12a. It is observed that for the largest part of the study area the estimated error is less than 4 km, which is acceptable considering the wide range within which the Moho depth discontinuity varies (more than 25 km). The relatively small error of the Moho depth discontinuity confirms that most of the acceptable solutions of the inversion give similar values for the Moho depth, despite the large parameter space that the inversion was searching for solutions. Only at the borders of the study area does the error become larger than 4 km, as the data coverage is poor (group velocity along different ray paths), resulting in low resolution. Of course, the presented errors reflect only the data variability and cannot account for any additional errors in our estimation as a result of the limitations or the assumptions of the inversion approach.

Using the 1D S-wave models derived in this study we have constructed several horizontal velocity sections for different depths, shown in Figure 13. The shallow depth sections (6–16 km) are characterized by relatively low S-wave velocities for most of the study area. The direction of these low-velocity anomalies more or less follows the known Dinaric trend (NNW–SSE) in NW Greece and then changes to ENE–WSW in the central and southern Aegean Sea, becoming almost east–west in western Turkey. These depicted anomalies roughly coincide with the active fault zones in the Aegean region, which are the result of the NNW–SSE extension process in the broader Aegean region, as well as with the westward escape of the Anatolian plate (e.g. McKenzie 1972; Taymaz *et al.* 1991).

The S-wave velocity image changes as we move to greater depths (16–20 km), as velocities larger than 3.5 km s^{-1} (which are typical values of the lower crust) are found in the northern and southern Aegean Sea, whereas smaller velocities are observed for the remaining part of the study area.

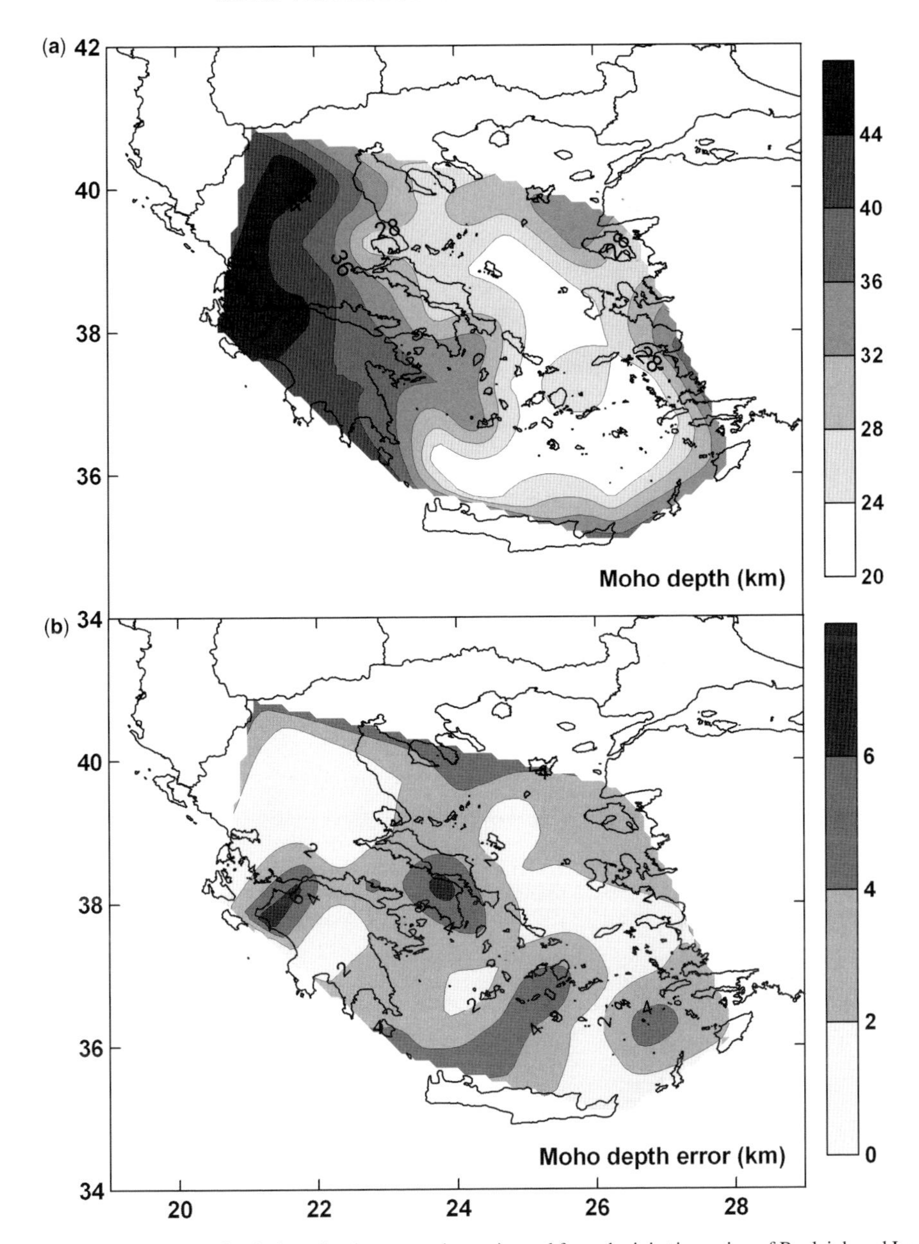

Fig. 12. (**a**) Moho depths (km) for the broader Aegean region estimated from the joint inversion of Rayleigh and Love waves. (**b**) Estimated error (in km) of the Moho discontinuity.

For the depth range of 20–26 km typical velocities of the lower crust (3.6–3.8 km s^{-1}) are observed in eastern continental Greece, whereas larger velocities of about 4.1–4.2 km s^{-1} are observed in SE Aegean Sea, indicating the presence of uppermost mantle. Similar S-wave upper mantle velocities have been found by Kalogeras (1993) along three paths that cross the south Aegean Sea (Karpathos–Athens, Rhodes–Athens and SW Turkey–Athens). Lower S-wave velocities are still present in western Greece, which probably correspond to middle–lower crustal levels.

To a depth of about 30 km S-wave upper mantle velocities (4.0–4.4 km s^{-1}) dominate throughout the Aegean Sea, whereas crustal S-wave velocities (*c.* 3.5 km s^{-1}) are still observed in western Greece, confirming the difference in crustal thickness between continental Greece and the Aegean

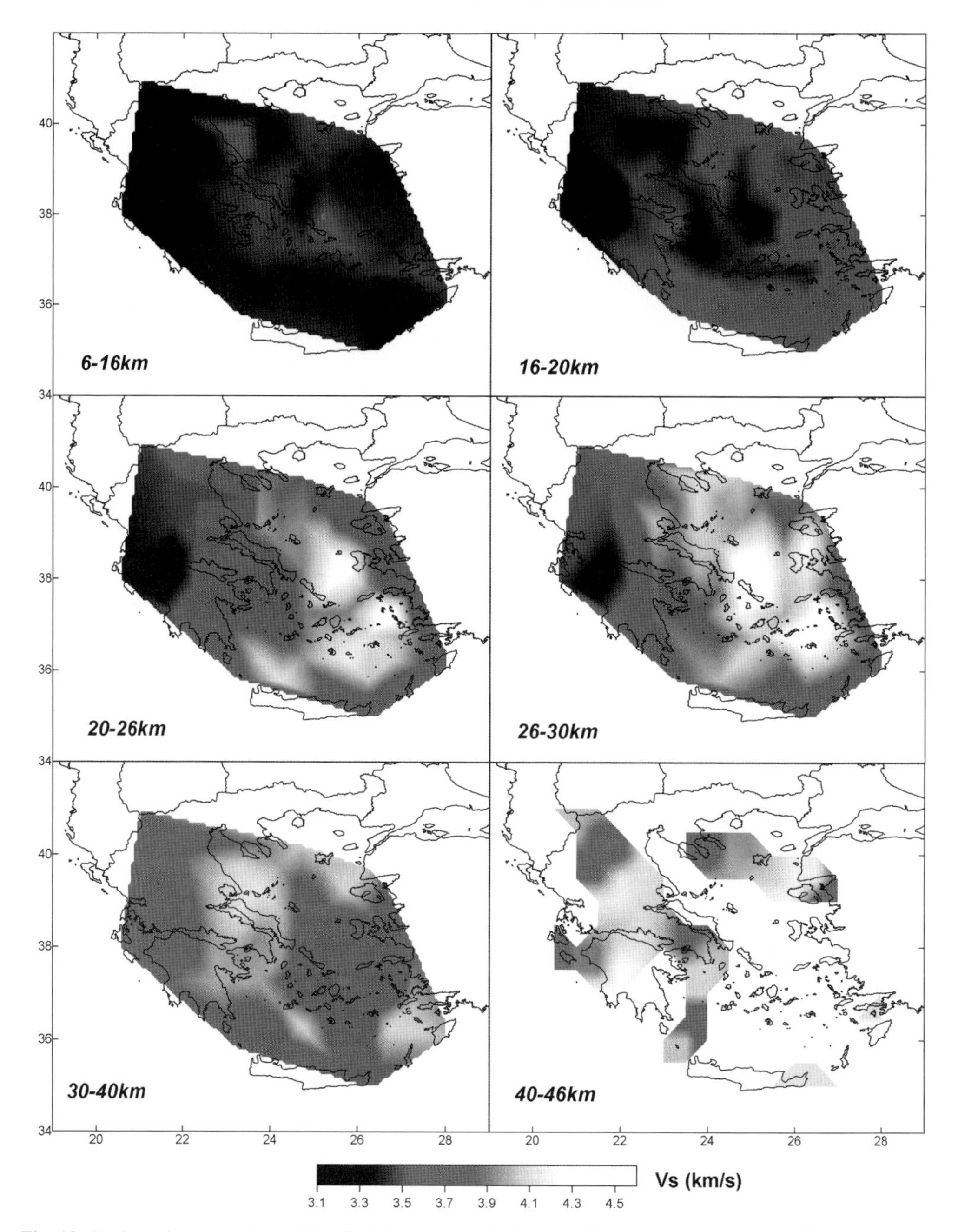

Fig. 13. Horizontal cross-sections of the final shear-wave velocity model for the Aegean region at different depth ranges.

Sea. The S-wave upper mantle values that have been observed in some parts of the study area are lower than typical upper mantle values throughout most of Europe. Similarly low S-wave upper mantle velocities have been found also for Turkey by Mindevalli and Mitchell (1989).

For depths ranging from 30 to 40 km the high upper mantle S-wave velocities in the southern and central Aegean Sea change to relatively low S-wave velocities (3.6–3.8 km s^{-1}). These low S-wave velocities are observed just below the Moho discontinuity and can be correlated with the uplift of hot mantle materials and high surface heat flow (Hurtig *et al.* 1991). Moreover, low upper mantle velocities have been also found in body-wave tomographic studies (Spakman 1986; Spakman *et al.* 1993; Papazachos & Nolet 1997). A similar situation has been observed along the active (seismically and volcanically) side of the Tyrrhenian Sea and bordering land (Pontevivo & Panza 2002; Panza *et al.* 2004). Kalogeras & Burton (1996) detected similar low S-wave velocities at a depth of about 30 km along three paths (Karpathos–Athens, Rhodes–Athens and SW Turkey–Athens) that cross the southern Aegean Sea. Taymaz (1996) suggested that both P- and S-wave velocities in the crust and upper mantle of the Aegean are notably lower than those beneath the Hellenic Trench near Crete, which he partly attributed to high heat flow values in the Aegean back-arc in contrast to low heat flow values in the Sea of Crete. This is also in agreement with the work of Gok *et al.* (2000), who provided an attenuation study of regional shear waves (Sn and Lg) in the northern Aegean and Greek mainland area and showed that an inefficient Sn propagation along the volcanic arc is due to the low upper mantle velocities and high attenuation observed in this area. The very low S-wave velocity anomalies in the mantle wedge (*c.* 10%, locally up to *c.* 15%) obtained in the present study (V_S *c.* 3.7 km s^{-1}) in combination with the P-wave velocity anomalies from body-wave tomography (*c.* 5–6%, locally 7% with V_P *c.* 7.3–7.4-km s^{-1}) verify the presence (Karagianni *et al.* 2005) of an unusually high V_P/V_S ratio of 1.9–2.0 in the mantle wedge of the southern Aegean Arc. The combination of such low velocities with high V_P/V_S values corresponds to the presence of a high fraction (*c.* 15–20%) of partial melt in the mantle wedge for a typical mantle composition (e.g. Birch 1969), in agreement with existing results from petrogenetic analyses of volcanic rocks (Zelimer 1998), which also suggest a high degree of mantle melting (*c.* 15–20%) beneath the volcanic arc in the Aegean region.

Discussion and conclusions

The results of the joint inversion of Rayleigh and Love waves verify the existence of strong S-wave velocity lateral variations in the Aegean. Our inversion results suggest that there is no significant Rayleigh–Love discrepancy in the Aegean region and there are several isotropic models that can satisfy both Rayleigh and Love wave data. It should be noted that the amount of anisotropy in the upper mantle and lower crust is a matter of debate, especially in continental regions and in subduction zones (e.g. Montagner & Tanimoto 1991) Hatzfeld *et al.* (2001) studied the shear-wave anisotropy in the upper mantle beneath the Aegean region and found that little azimuthal anisotropy is observed along the Hellenic arc and in continental Greece, whereas significant anisotropy is observed in the north Aegean Sea. On the other hand, any observed Rayleigh–Love anisotropy should be mainly caused by radial anisotropy. Preliminary results show that radial anisotropy seems to be small or of very local character in the Aegean region at crustal–upper mantle levels, with typical values of *c.* 0–3% (e.g. Boschi *et al.* 2004; Endrun *et al.* 2006). Such small anomalies will have a weak influence on the joint Rayleigh–Love interpretation and in any case are much smaller than the corresponding velocity anomalies observed in the final 3D model (Fig. 13), which locally exceed 20%.

The most prominent feature confirmed in this work is the large crustal thickness contrast between western Greece (along the Hellenides mountain range) and the Aegean Sea (e.g. Fig. 13). In the southern Aegean Sea, as well as in parts of the central Aegean Sea, the crust has a thickness of about 20–22 km. This crustal thinning is associated with the extensional tectonics of the Aegean back-arc area and the rise of mantle material, as a result of the convergence between the Eurasian and African plates. Generally, the inner Aegean Sea shows a crustal thickness less than 28–30 km, whereas in western Greece a significant crustal thickness of about 40–46 km is observed along the Hellenides mountain range. A normal crust of thickness *c.* 30–34 km is observed in eastern continental Greece. The thin crust (20–22 km) between the volcanic arc and Crete has also been found in previous studies (e.g. Makris 1976; Bohnhoff *et al.* 2001) but the present study suggests for the whole Aegean Sea a thinner crust compared with previous works, in agreement with gravity data (Papazachos 1994; Tsokas & Hansen 1997; Tirel *et al.* 2004), which have indicated the presence of a thin crust (*c.* 25 km) throughout the Aegean Sea.

A low-velocity crustal layer at depths from 10 to 20 km along the Hellenic arc was observed by Papazachos (1994) and Papazachos *et al.* (1995), and was later confirmed by Papazachos & Nolet (1997) for the P-wave structure. This layer was correlated to the Hellenides mountain range and the Alpine orogenesis, in accordance with the ideas of weakening mid-crustal intrusions and the associated LVL, which was proposed by Mueller (1977). More recently, Karagianni *et al.* (2005) showed that this crustal low-velocity layer is also found in the S-wave structure, as obtained by the inversion of Rayleigh waves. The results of the joint inversion of Rayleigh and Love waves confirm the existence of this low-velocity crustal layer along the outer Hellenic arc, as is also shown by the vertical cross-sections in Figure 14.

The results of the present study also confirm the presence of a strong low-velocity mantle wedge layer (MW LVZ) with S-wave velocities about of 3.7–3.8 km s^{-1} above the subducted slab in the southern Aegean Sea, as well as in a small part of the central Aegean Sea, whereas the remaining Aegean Sea and continental Greece exhibit more typical uppermost mantle velocities (*c*. 4.2–4.4 km s^{-1}). However, these values are still slightly lower than previously considered values mainly based on joint P- and S-wave travel-time inversions (e.g. Papazachos & Nolet 1997). The very low S-wave velocity anomaly that is found in the mantle wedge (associated with partial melting in the southern Aegean Sea) has also been observed in similar geotectonic environments, such as the Tonga, Alaska and Japan volcanic arcs (e.g. Zhao *et al.* 1995, 1997), with S-wave low-velocity anomalies locally exceeding 15%.

Our results seem to be in a very good agreement with independent information on the velocity structure (e.g. Papazachos 1993) as determined from several published results (deep seismic profiles, tomography, gravity inversion, etc.). In general, the surface-wave results tend to give a stronger crustal thinning for the central and southern Aegean Sea compared with previously published results. Moreover, there is relatively good agreement between our results and independent information, such as the work of Li *et al.* (2003), who used data from broadband seismographs on five islands in the southern Aegean Sea to construct receiver function images of the crust and upper mantle for the areas south of Crete and into the Aegean. They estimated a thin crust of about 25 km in the central Aegean Sea near Naxos, and a thicker crust of about 28 km for the SW coast of Asia Minor. For stations located in Crete, Li *et al.* showed that the depth of the Moho discontinuity varies from 32 to 39 km, which is in good agreement with our results. The only significant difference from our results is the Moho depth of 32 km estimated for Santorini, which is much thicker than our estimate of 22 km in the present study. However, more recent results (Endrun *et al.* 2005) show that the receiver functions in Santorini have been misinterpreted and a typical thickness of 23 km is determined from the receiver function data, in very good agreement with the thickness found in our study.

The results of this study were also compared against the independently determined 3D P-wave velocity model for the crust and uppermost mantle of the Aegean area (Papazachos & Nolet 1997). In Figure 14, the general features of the crustal thickness variation from P-waves (travel-time inversion of body waves) and S-waves (from surface-wave group velocity inversion) are also in good agreement for both examined cross-sections but again surface-wave results tend to show smaller crustal thickness values for the central part of the southern Aegean Sea. This is probably due to the poorer accuracy of catalogue phase-picks used in travel-time inversion, which often suffer from significant picking errors (up *c*. 0.5–1 s at regional distances, perhaps larger for S-waves), as well as from the lack of recording stations (and partly events) in the southern Aegean Sea region. Furthermore, the high upper-mantle velocities immediately below the Moho in the same area are clearly identified in both models. On the other hand, the P-wave tomography shows the low velocities associated with the mantle wedge above the subducted slab at depths of about 55–70 km (and locally at larger depths), whereas surface-wave models start to identify this low-velocity layer in the mantle at much greater depths (*c*. 35–40 km). Although the V_S model determined in the present study has no resolving power at greater depths because of the limited frequency range of the observed group velocity curves, it is expected that the effect of partial melt on S-wave velocities is much more pronounced compared with that on P-wave results. Therefore, this discrepancy should be attributed to the stronger melt signature in S-models, probably in combination with the crustal thickness overestimation for the thin crust areas from tomographic results owing to the damping and smoothing constraints usually applied when inverting travel-time data for large-scale models.

It should be pointed out that the amplitude of S-wave velocity anomalies from surface-wave inversion are significantly larger (up to more than 200% higher) when considering the S-wave structure from travel times (Papazachos & Nolet 1997). This effect is especially prominent for the southern Aegean mantle low-velocity layer, corresponding to the mantle wedge above the southern Aegean subduction zone, where very low upper-mantle S-velocities are detected from surface-wave inversion, in agreement with independent information (Gok *et al.* 2000). These results suggest a possible bias not only of the tomographic approach in travel-time inversions but also of the original travel-time data, as large S-residuals are often rejected during the preliminary relocation procedures followed by most seismological networks. Moreover, additional factors such as anisotropy may need to be taken into account for such discrepancies, as well as for local Rayleigh–Love differences. However, the results obtained in the present study suggest that the V_S model obtained here is generally in good agreement with existing 3D

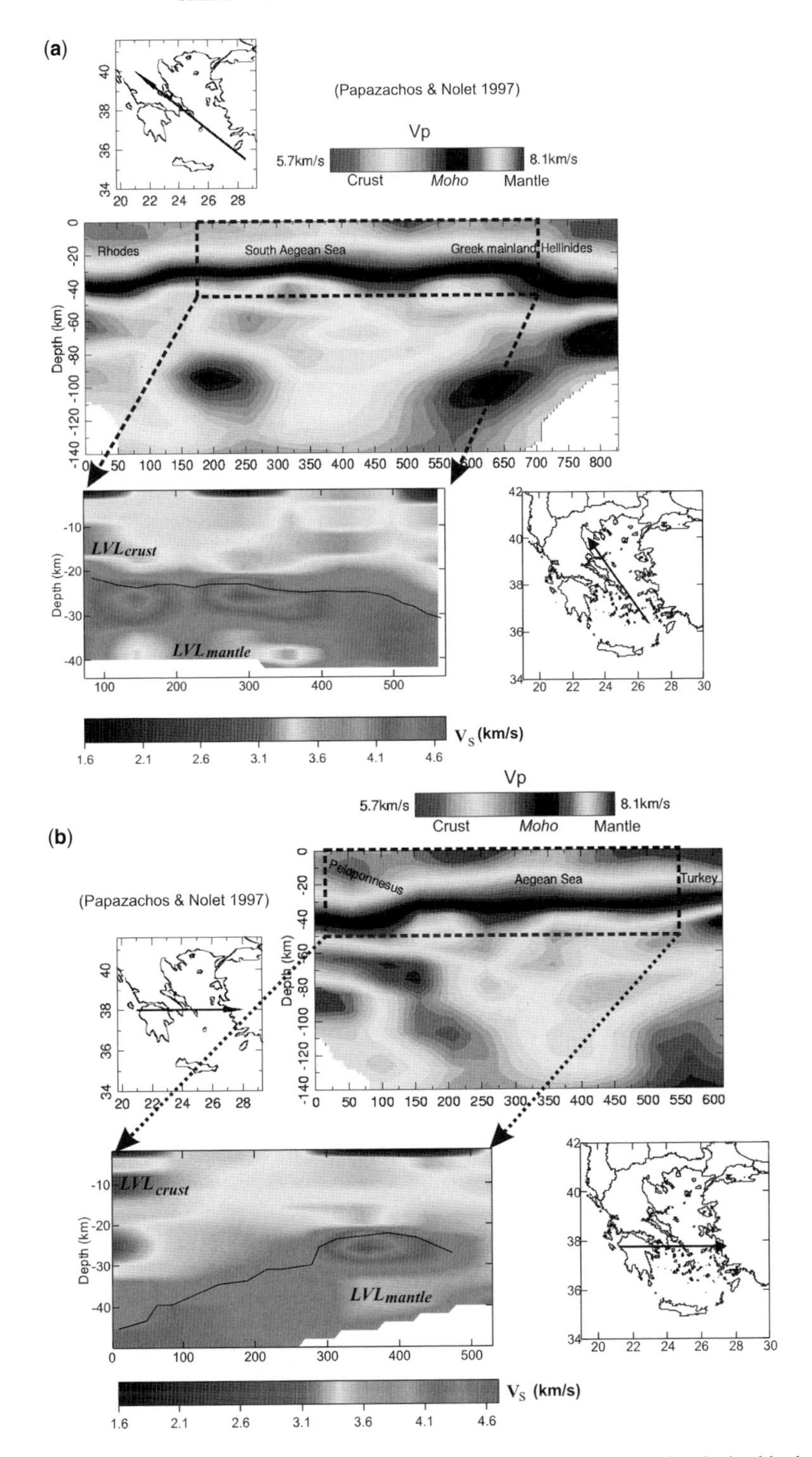

Fig. 14. (**a**) Comparison of the V_P model of Papazachos & Nolet (1997) with the V_S results obtained in the present study for a SE–NW cross-section in the study area. The similar crustal thickness variation in the two models and the high sub-Moho velocities in the southern Aegean Sea should be noted. However, the mantle LVL is detected at shallower depths in the V_S models than in the V_P models. (**b**) Same for a typical east–west cross-section.

P- and S-models (at least regarding its principal features and characteristics) and provides additional information on the true amplitude of S-wave anomalies that can be identified. Furthermore, this observation suggests that a joint inversion of travel-time and surface-wave data may be able to provide a single V_P–V_S model that can simultaneously interpret both datasets by reducing the non-uniqueness of their inversion.

The data collection for this work was financed by the EC Environment and Climate project (contract ENV4-CT96-0277). We thank D. Hatzfeld and a large number of people for their effort to make the field experiments successful. We are grateful to A. Levshin for providing the FTAN code, and to T. Yanovskaya and P. Ditmar from the University of St. Petersburg, Russia, for providing the tomographic inversion program. We would like to thank the GeoForschungsZentrums, Potsdam, Germany, for providing the records from their network in the southern Aegean Sea. We would also like to thank Ş. Barış, an anonymous reviewer and T. Taymaz for their constructive comments and suggestions, which helped to improve this work. This work was partly financed by the project Pythagoras-Environment founded by EPEAEK (21945) and PEP of Crete (81106).

References

ANDERSON, H. & JACKSON, J. 1987. Active tectonics of the Adriatic Region. *Geophysical Journal of the Royal Astronomical Society*, **91**, 937–983.

BACKUS, G. & GILBERT, F. 1968. The resolving power of gross Earth data. *Geophysical Journal of the Royal Astronomical Society*, **16**, 169–205.

BARTON, P. J. 1986. The relationship between seismic velocity and density in the continental crust—a useful constraint? *Geophysical Journal of the Royal Astronomical Society*, **87**, 195–208.

BIRCH, F. 1969. Density and composition of the upper mantle: first approximation as an olivine layer: *In*: HART, P. J. (ed.) *The Earth's Crust and Upper Mantle Structure, Dynamic Processes, and their relation to Deep-Seated Geological Phenomena.* Geophysical Monograph Series, **13**, 18–36. AGU, Washington D.C.

BISWAS, N. N. & KNOPOFF, L. 1974. The structure of the upper mantle under the United States from the dispersion of Rayleigh waves. *Geophysical Journal of the Royal Astronomical Society*, **36**, 515–539.

BOHNHOFF, M., MAKRIS, J., PAPANIKOLAOU, P. & STAVRAKAKIS, G. 2001. Crustal investigation of the Hellenic subduction zone using wide aperture seismic data. *Tectonophysics*, **343**, 239–262.

BOSCHI, L., EKSTRÖM, G. & KUSTOWSKI, B. 2004. Multiple resolution surface wave tomography: the Mediterranean basin. *Geophysical Journal International*, **157**, 293–304.

BOUROVA, E., KASSARAS, I., PEDERSEN, H. A., YANOVSKAYA, T., HATZFELD, D. & KIRATZI, A. 2005. Constraints on absolute S velocities beneath the Aegean Sea from surface wave analysis. *Geophysical Journal International*, **160**, 1006–1019.

BROOKS, M. & KIRIAKIDIS, L. 1986. Subsidence of the North Aegean Trough: an alternative view. *Journal of the Geological Society, London*, **143**, 23–27.

CALCAGNILE, G. & PANZA, G. F. 1980. The main characteristics of the lithosphere–asthenosphere system in Italy and surrounding regions. *Pure and Applied Geophysics*, **119**, 865–879.

CALCAGNILE, G., D'INGEO, F., FARRUGIA, P. & PANZA, G. F. 1982. The lithosphere in the central–eastern Mediterranean area. *Pure and Applied Geophysics*, **120**, 389–406.

CANITEZ, N. & TOKSOZ, M. N. 1971. Focal mechanism and source depth of earthquakes from body and surface wave data. *Bulletin of the Seismological Society of America*, **61**, 1369–1379.

CAPUTO, M., PANZA, G. F. & POSTPISCHL, D. 1970. Deep structure of the Mediterranean basin. *Journal of Geophysical Research*, **75**, 4919–4923.

CHAILAS, S., HIPKIN, R. G. & LAGIOS, E. 1993. Isostatic studies in the Hellenides. *In*: 2nd *Congress of the Hellenic Geophysical Union, 5–7 May, Florina, Greece*, **2**, 492–504.

CHRISTODOULOU, A. & HATZFELD, D. 1988. Three-dimensional crustal and upper mantle structure beneath Chalkidiki (northern Greece). *Earth and Planetary Science Letters*, **88**, 153–168.

CLEMENT, C., SACHPAZI, M., CHARVIS, P., GRAINDERGE, D., LAIGLE, M., HIRN, A. & ZAFIROPOULOS, G. 2004. Reflection–refraction seismics in the Gulf of Corinth: hints at deep structure and control of the deep marine basin. *Tectonophysics*, **391**, 97–108.

DELIBASIS, N., MAKRIS, J. & DRAKOPOULOS, J. 1988. Seismic investigation of the crust and the upper mantle in western Greece. *Annales Géologiques de Pays Helléniques*, **33**, 69–83.

DITMAR, P. G. & YANOVSKAYA, T. B. 1987. A generalization of the Backus–Gilbert method for estimation of lateral variations of surface wave velocity. *Physics of the Solid Earth, Izvestia Academy of Sciences, USSR*, **23**, 470–477.

DRAKATOS, G. 1989. *Seismic tomography—determination of high and low velocity zones beneath Greece and surrounding regions.* PhD thesis, University of Athens. XXIII, **3**, 157–172.

DRAKATOS, G. & DRAKOPOULOS, J. 1991. 3-D velocity structure beneath the crust and upper mantle of the Aegean sea region. *Pure and Applied Geophysics*, **135**, 401–420.

DRAKATOS, G., LATOUSSAKIS, J., STAVRAKAKIS, G., PAPANASTASIOU, D. & DRAKOPOULOS, J. 1989. 3-Dimensional velocity structure of the north–central Greece from inversion of travel times. *In*: *3rd Congress of the Geological Society of Greece.* Geological Society of Greece, Athens XXIII, **3**, 157–172.

ENDRUN, B., CERANNA, L., MEIER, T., RISCHE, M., BOHNHOFF, M. & HARJES, H.-P. 2005. Structural properties of the Hellenic subduction zone derived from receiver functions and surface wave dispersion. *Geophysical Research Abstracts*, **7**, 7123.

ENDRUN, B., MEIER, T., LEBEDEV, S., BOHNHOFF, M. & HARJES, H.-P. 2006. Constraints on S-velocity, radial and azimuthal anisotropy in the Aegean region from surface wave dispersion, *Geophysical Research Abstracts*, **8**, 7830.

FLEISCHER, U. 1964. Schwerestorungen im ostlichen Mittelmeer: nach Messungen mit einem Askania-Seegravimeter. *Deutsche Hydrographische Zeitschrift* **17**, 153.

FYTIKAS, M., INNOCENTI, F., MANETTI, P., MAZZUOLI, R., PECCERILLO, A. & VILLARI, L. 1984. Tertiary to Quaternary evolution of the volcanism in the Aegean region. *In*: DIXON, J. E. & ROBERTSON, A. H. F. (eds) *The Geological Evolution of the Eastern Mediterranean*. Geological Society, London, Special Publications, **17**, 687–699.

GEORGALAS, G. 1962. *Catalogue of the Active Volcanoes and Solfatara Fields in Greece, Part 12*. International Association of Volcanology, Rome.

GOK, R., TURKELLI, N., SANDVOL, E., SEBER, D. & BARAZANGI, M. 2000. Regional wave propagation in Turkey and surrounding regions. *Geophysical Research Letters*, **27**, 429–432.

HASHIDA, T., STAVRAKAKIS, G. & SHIMAZAKI, K. 1988. Three-dimensional seismic attenuation beneath the Aegean region and its tectonic implication. *Tectonophysics*, **145**, 43–54.

HATZFELD, D., KARAGIANNI, E., KASSARAS, I., ET AL. 2001. Shear wave anisotropy in the upper mantle beneath the Aegean related to internal deformation. *Journal of Geophysical Research*, **106**, 30737.

HURTIG, E., CERMAK, V., HAENEL, R. & ZUI, V. 1991. *Geothermal Atlas of Europe*. Hermann Haack, Gotha.

JACKSON, J. 1994. Active tectonics of the Aegean region. *Annual Review of Earth and Planetary Sciences*, **22**, 239–271.

JOBERT, N. & JOBERT, J. 1983. An application of the ray theory to the propagation of waves along a laterally heterogeneous spherical surface. *Geophysical Research Letters*, **10**, 1148–1151.

KALOGERAS, J. S. 1993. *A contribution of surface seismic waves in the study of the crust and upper mantle in the area of Greece*. PhD thesis, University of Athens.

KALOGERAS, J. S. & BURTON, P. W. 1996. Shear-wave velocity models from Rayleigh-wave dispersion in the broader Aegean area. *Geophysical Journal International*, **125**, 679–695.

KARAGIANNI, E. E., PANAGIOTOPOULOS, D. G., PANZA, G. F., ET AL. 2002. Rayleigh wave group velocity tomography in the Aegean area. *Tectonophysics*, **358**, 187–209.

KARAGIANNI, E. E., PAPAZACHOS, C. B., PANAGIOTOPOULOS, D. G., SUHADOLC, P., VUAN, A. & PANZA, G. F. 2005. Shear velocity structure in the Aegean area obtained by inversion of Rayleigh waves. *Geophysical Journal International*, **160**, 127–143.

KEILIS-BOROK, V. I. & YANOVSKAYA, T. B. 1967. Inverse problems of seismology. *Geophysical Journal of the Royal Astronomical Society*, **13**, 223–234.

KIRATZI, A. A., PAPADIMITRIOU, E. E. & PAPAZACHOS, B. C. 1987. A microearthquake survey in the Steno dam site in northwestern Greece. *Annales Geophysicae*, **5**, 161–166.

KIRIAKIDIS, L. G. 1988. The Vardar ophiolite: a continuous belt under the Axios basin sediments. *Geophysical Journal International*, **98**, 203–212.

KNOPOFF, L. 1964. A matrix method for elastic wave problems. *Bulletin of the Seismological Society of America*, **54**, 431–438.

KNOPOFF, L. 1972. Observation and inversion of surface wave dispersion. *Tectonophysics*, **13**, 497–519.

KURT, H., DEMIRBAG, E. & KUSCU, I. 1999. Investigation of the submarine active tectonism in the gulf of Gokova, southwest Anatolia–southeast Aegean Sea, by multi-channel seismic reflection data. *Tectonophysics*, **305**, 477–496.

LE PICHON, X. & ANGELIER, J. 1979. The Hellenic arc and trench system: a key to the neotectonic evolution of the eastern Mediterranean area. *Tectonophysics*, **60**, 1–42.

LEVSHIN, A. L., RATNIKOVA, L. I. & BERTEUSSEN, K. A. 1972. On a frequency–time analysis of oscillations. *Annales Geophysicae*, **28**, 211–218.

LEVSHIN, A. L., YANOVSKAYA, T. B., LANDER, A. V., BUKCHIN, B. G., BARMIN, M. P., RATNIKOVA, L. I. & ITS, E. N. 1989. Recording, identification and measurement of surface wave parameters. *In*: KEILIS-BOROK, V. I. (ed.) *Seismic Surface Waves in a Laterally Inhomogeneous Earth*. Kluwer, Dordrecht, 131–182.

LEVSHIN, A. L., RATNIKOVA, L. I. & BERGER, J. 1992. Peculiarities of surface-wave propagation across central Eurasia. *Bulletin of the Seismological Society of America*, **82**, 2464–2493.

LI, X., BOCK, G., VAFIDIS, A., ET AL. 2003. Receiver function study of the Hellenic subduction zone: imaging crustal thickness variations and the oceanic Moho of the descending African lithosphere. *Geophysical Journal International*, **155**, 733–748.

LIGDAS, C. N. & LEES, J. M. 1993. Seismic velocity constrains in the Thessaloniki and Chalkidiki areas (northern Greece) from a 3D tomographic study. *Tectonophysics*, **228**, 97–121.

LIGDAS, C. N. & MAIN, I. G. 1991. On the resolving power of tomographic images in the Aegean area. *Geophysical Journal International*, **107**, 197–203.

LIGDAS, C. N., MAIN, I. G. & ADAMS, R. D. 1990. 3D structure of the lithosphere in the Aegean Sea region. *Geophysical Journal International*, **102**, 219–229.

MAKRIS, J. 1973. Some geophysical aspects of the evolution of the Hellenides. *Bulletin of the Geological Society of Greece*, **10**, 206–213.

MAKRIS, J. A. 1976. A dynamic model of the Hellenic arc deduced from geophysical data. *Tectonophysics*, **36**, 339–346.

MAKRIS, J. A. 1977. *Geophysical Investigation of the Hellenides. Hamburger Geophysikalische Einzelschriften*, **34**, 124.

MAKRIS, J. A. 1978. The crust and upper mantle of the Aegean region from deep seismic soundings. *Tectonophysics*, **46**, 269–284.

MARQUERING, H. & SNIEDER, R. 1996. Shear-wave velocity structure beneath Europe, the northeastern Atlantic and western Asia from waveform inversions including surface-wave mode coupling. *Geophysical Journal International*, **127**, 283–304.

MARTIN, L. 1987. *Structure et évolution récente de la Mer Egée*. Thèse de Doctorat, Université Paris-Sud.

MARTINEZ, M. D., LANA, X., CANAS, J. A., BADAL, J. & PUJADES, L. 2000. Shear-wave velocity tomography of the lithosphere–asthenosphere system beneath the Mediterranean area. *Physics of the Earth and Planetary Interiors*, **122**, 33–54.

MCKENZIE, D. P. 1970. The plate tectonics of the Mediterranean region. *Nature*, **226**, 239–243.

MCKENZIE, D. P. 1972. Active tectonics of the Mediterranean region. *Geophysical Journal of the Royal Astronomical Society*, **30**, 109–185.

MCKENZIE, D. P. 1978. Active–tectonics of the Alpine–Himalayan belt: the Aegean Sea and surrounding regions. *Geophysical Journal of the Royal Astronomical Society*, **55**, 217–254.

MCCLUSKY, S., BALASSANIAN, S., BARKA, A., ET AL. 2000. Global Positioning System constraints on plate kinematics and dynamics in the eastern Mediterranean and Caucasus. *Journal of Geophysical Research*, **105**, 5695–5719.

MEIER, T., DIETRICH, K., STOCKHERT, B. & HARJES, H.-P. 2004. One-dimensional models of shear wave velocity for the eastern Mediterranean obtained from the inversion of Rayleigh wave phase velocities and tectonic implications. *Geophysical Journal International*, **156**, 45–58.

MINDEVALLI, O. Y. & MITCHELL, B. J. 1989. Crustal structure and possible anisotropy in Turkey from seismic surface wave dispersion. *Geophysical Journal International*, **98**, 93–106.

MONTAGNER, J. P. & TANIMOTO, T. 1991. Global upper mantle tomography of seismic velocities and anisotropies. *Journal of Geophysical Research*, **96**, 20337–20351.

MUELLER, S. 1977. A new model of the continental crust. *In*: HEACOOK, J. G. (ed.) *The Earth's Crust: its Nature and Physical Properties*. Geophysical Monograph, American Geophysical Union, **20**, 289–317.

PANAGIOTOPOULOS, D. G. 1984. *Travel time curves and crustal structure in the southern Balkan region*. PhD thesis, University of Tessaloniki.

PANAGIOTOPOULOS, D. G. & PAPAZACHOS, B. C. 1985. Travel times of Pn waves in the Aegean and surrounding area. *Geophysical Journal of the Royal Astronomical Society*, **80**, 165–176.

PANZA, G. F. 1981. The resolving power of seismic surface waves with respect to crust and upper mantle structural models. *In*: CASSINIS, R. (ed.) *The Solution of the Inverse Problem in Geophysical Interpretation*. Plenum, New York, 39–77.

PANZA, G. F., PONTEVIVO, A., SARAÓ, A., AOUDIA, A. & PECCERILLO, A. 2004. Structure of the lithosphere–asthenosphere and volcanism in the Tyrrhenian Sea and surroundings. *Memorie del Servizio Geologico*, **XLIV**, 29–56.

PAPAZACHOS, B. C. 1969. Phase velocities of Rayleigh waves in southeastern Europe and eastern Mediterranean Sea. *Pure and Applied Geophysics*, **75**, 47–55.

PAPAZACHOS, B. C. & COMNINAKIS, P. E. 1969. Geophysical features of the Greek islands Arc and Eastern Mediterranean Ridge. *Comptes Rendus Séance Conference Réunie Madrid*, **16**, 74–75.

PAPAZACHOS, B. C. & COMNINAKIS, P. E. 1971. Geophysical and tectonic features of the Aegean arc. *Journal of Geophysical Research*, **76**, 8517–8533.

PAPAZACHOS, B. C. & DELIBASIS, N. D., 1969. Tectonic stress field and seismic faulting in the area of Greece. *Tectonophysics*, **7**, 231–255.

PAPAZACHOS, B. C. & PAPAZACHOU, C. B. 1997. *The Earthquakes of Greece*. Ziti, Thessaloniki.

PAPAZACHOS, B. C., POLATOU, M. & MANDALOS, N. 1967. Dispersion of surface waves recorded in Athens. *Pure and Applied Geophysics*, **67**, 95–106.

PAPAZACHOS, B. C., KIRATZI, A., HATZIDIMITRIOU, P. & ROCCA, A. 1984. Seismic faults in the Aegean area. *Tectonophysics*, **106**, 71–85.

PAPAZACHOS, B. C., PAPADIMITRIOU, E. E., KIRATZI, A. A., PAPAZACHOS, C. B. & LOUVARI, E. K. 1998. Fault plane solutions in the Aegean Sea and the surrounding area and their tectonic implications. *Bolletino di Geofisica Teorica ed Applicata*, **39**, 199–218.

PAPAZACHOS, C. B. 1993. Determination of crustal thickness by inversion of travel times with an application in the Aegean area. *In*: *2nd Congress of the Hellenic Geophysical Union, 5–7 May, Florina, Greece*, **3**, 483–491.

PAPAZACHOS, C. B. 1994. *Structure of the crust and upper mantle in SE Europe by inversion of seismic and gravimetric data*. PhD thesis, University of Thessaloniki. [in Greek]

PAPAZACHOS, C. B. 1998. Crustal and upper mantle P and S velocity structure of the Serbomacedonian massif (Northern Greece). *Geophysical Research Letters*, **134**, 25–39.

PAPAZACHOS, C. B. & NOLET, G. 1997. P and S deep structure of the Hellenic area obtained by robust nonlinear inversion of travel times. *Journal of Geophysical Research*, **102**, 8349–8367.

PAPAZACHOS, C. B., HATZIDIMITRIOU, P. M., PANAGIOTOPOULOS, D. G. & TSOKAS, G. N. 1995. Tomography of the crust and upper mantle in southeast Europe. *Journal of Geophysical Research*, **100**, 405–422.

PASYANOS, M. E., WALTER, W. R. & HAZLERT, S. E. 2001. A surface wave dispersion study of the Middle East and North Africa for monitoring the comprehensive nuclear-test-ban treaty. *Pure and Applied Geophysics*, **158**, 1445–1474.

PAYO, G. 1967. Crustal structure of the Mediterranean Sea by surface waves. Part I, Group velocity. *Bulletin of the Seismological Society of America*, **57**, 151–172.

PAYO, G. 1969. Crustal structure of the Mediterranean sea by surface waves. Part II, Phase velocity and travel times. *Bulletin of the Seismological Society of America*, **59**, 23–42.

PLOMEROVA, J., BABUSKA, V., PUJDUSAK, P., HATZIDIMITRIOU, P., PANAGIOTOPOULOS, D., KALOGERAS, J. & TASSOS, S. 1989. Seismicity of the Aegean and surrounding areas in relation to topography of the lithosphere–asthenosphere transition. *In*: *Proceedings of the 4th International Symposium on Analysis Seismicity and Seismic Risk, Bechyne, Czechoslovakia, 4–9 September*, 209–215.

PONTEVIVO, A. & PANZA, G. F. 2002. Group velocity tomography and regionalization in Italy and bordering areas. *Physics of the Earth and Planetary Interiors*, **134**, 1–15.

RAYKOVA, R. B. & NIKOLOVA, S. B. 2000. Shear wave velocity models of the Earth's crust and uppermost mantle from the Rayleigh waves in Balkan Peninsula and adjacent areas. *Bulgarian Geophysical Journal*, **26**, 11–27.

RAYKOVA, R. B. & NIKOLOVA, S. B. 2003. Anisotropy in the Earth's crust and uppermost mantle in southeastern Europe obtained from the Rayleigh and Love surface waves. *Journal of Applied Geophysics*, **54**, 247–256.

ROUSSOS, N. 1994. Stratigraphy and paleogeographic evolution of the Paleogene Molassic basins of the North Aegean area. *Bulletin of the Geological Society of Greece*, **XXX**, 275–294.

SAATCILAR, R., ERGINTAV, S., DEMIRBAG, E. & INAN, S. 1999. Character of active faulting in the North Aegean Sea. *Marine Geology*, **160**, 339–353.

SAUNDERS, P., PRIESTLEY, K. & TAYMAZ, T. 1998. Variations in the crustal structure beneath western Turkey. *Geophysical Journal International*, **134**, 373–389.

SCHWAB, F. A. & KNOPOFF, L. 1972. Fast surface wave and free mode computations. *In*: BOLT, B. A (ed.) *Methods in Computational Physics*. Academic Press, New York, 86–180.

SCHWAB, F. A., NAKANISHI, K., CUSCITO, M., PANZA, G. F. & LIANG, G. 1984. Surface-wave computations and the synthesis of theoretical seismograms at high frequencies, *Bulletin of the Seismological Society of America*, **74**, 1555–1578.

SCORDILIS, E. M., KARAKAISIS, G. F., KARACOSTAS, B. G., PANAGIOTOPOULOS, D. G., COMNINAKIS, P. E. & PAPAZACHOS, B. C. 1985. Evidence for transforming faulting in the Ionian Sea: the Cefalonia island earthquake sequence of 1983. *Pure and Applied Geophysics*, **123**, 388–397.

SPAKMAN, W. 1986. Subduction beneath Eurasia in connection with the Mesozoic Tethys. *Geologie en Mijnbouw*, **65**, 145–153.

SPAKMAN, W., WORTEL, M. J. R. & VLAAR, N. J. 1988. The Hellenic subduction zone: atomographic image and its dynamic implications. *Geophysical Research Letters*, **15**, 60–63.

SPAKMAN, W., VAN DER LEE, S. & VAN DER HILST, R. D. 1993. Travel-time tomography of the European–Mediterranean mantle down to 1400 km. *Physics of the Earth and Planetary Interiors*, **79**, 3–74.

TAYMAZ, T. 1996. S–P-wave traveltime residuals from earthquakes and lateral inhomogeneity in the upper mantle beneath the Aegean and the Hellenic Trench near Crete. *Geophysical Journal International*, **127**, 545–558.

TAYMAZ, T., JACKSON, J. & MCKENZIE, D. 1991. Active tectonics of the north central Aegean Sea. *Geophysical Journal International*, **106**, 433–490.

TIBERI, C., LYON-CAEN, H, HATZFELD, D., ET AL. 2000. Crustal and upper mantle structure beneath the Corinth rift (Greece) from a teleseismic tomography study. *Journal of Geophysical Research*, **105**, 28159–28171.

TIREL, C., GUEYDAN, F., TIBERI, C. & BRUN, J.-P. 2004. Aegean crustal thickness inferred from gravity inversion. Geodynamical implications. *Earth and Planetary Science Letters*, **228**, 267–280.

TSOKAS, G. N. & HANSEN, R. O. 1997. Study of the crustal thickness and the subducting lithosphere in Greece from gravity data. *Journal of Geophysical Research*, **102**, 20585–20597.

URBAN, L., CICHOWICZ, A. & VACCARI, F. 1993. Computation of analytical partial derivatives of phase and group velocities for Rayleigh waves with respect to structural parameters. *Studia Geophysica et Geodaetica*, **37**, 14–36.

VALYUS, V. P. 1968. Determining seismic profiles from a set of observations. *Vychislitelnaya Seismologiya*, **4**, 3–14. [in Russian]. (English translation in: KELLIS-BOROK, V. I. (ed.) *Computational Seismology*. Consultants Bureau, New York, 1972, 114–118.)

VOGT, P. & HIGGS, P. 1969. An aeromagnetic survey of the eastern Mediterranean Sea and its interpretation. *Earth and Planetary Science Letters*, **5**, 439–448.

VOULGARIS, N. 1991. *Investigation of the crustal structure in western Greece (Zakinthos–NW Peloponessus area)*. PhD thesis, University of Athens. [in Greek]

YANOVSKAYA, T. B. 1982. Distribution of surface wave group velocities in the North Atlantic. *Izvestiya Akademi Nauk SSSR, Fizika Zemli*, **2**, 3–11.

YANOVSKAYA, T. B. 1997. Resolution estimation in the problems of seismic ray tomography. *Izvestia, Physics of the Solid Earth*, **33**, 762–765.

YANOVSKAYA, T. B. & DITMAR, P. G. 1990. Smoothness criteria in surface wave tomography. *Geophysical Journal International*, **102**, 63–72.

YANOVSKAYA, T. B., KIZIMA, E. S. & ANTOMOVA, L. M. 1998. Structure of the crust in the Black Sea and adjoining regions from surface wave data. *Journal of Seismology*, **2**, 303–316.

YANOVSKAYA, T. B., ANTOMOVA, L. M. & KOZHEVNIKOV, V. M. 2000. Lateral variations of the upper mantle structure in Eurasia from group velocities of surface waves. *Physics of the Earth and Planetary Interiors*, **122**, 19–32.

ZELIMER, G. F. 1998. *Petrogenetic processes and their timescales beneath Santorini, Aegean Volcanic Arc, Greece*. PhD thesis, The Open University, Milton Keynes.

ZHAO, D., CHRISTENSEN, D. & PULPAN, H. 1995. Tomographic imaging of the Alaska subduction zone. *Journal of Geophysical Research*, **100**, 6487–6504.

ZHAO, D., XU, Y., WIENS, D. A., DORMAN, L., HILDEBRAND, J. & WEBB, S. 1997. Depth extent of the Lau back-arc spreading center and its relation to subduction processes. *Science*, **278**, 254–257.

A model for the Hellenic subduction zone in the area of Crete based on seismological investigations

T. MEIER[1], D. BECKER[1], B. ENDRUN[1], M. RISCHE[1], M. BOHNHOFF[2], B. STÖCKHERT[1] & H.-P. HARJES[1]

[1]*Institute of Geology, Mineralogy and Geophysics, Ruhr-University Bochum, NA 3/173, Universitätsstr. 150, D-44780 Bochum, Germany (e-mail: meier@geophysik.rub.de)*

[2]*GeoForschungsZentrum, Telegrafenberg, D-14473 Potsdam, Germany*

Abstract: The island of Crete represents a horst structure located in the central forearc of the retreating Hellenic subduction zone. The structure and dynamics of the plate boundary in the area of Crete are investigated by receiver function, surface wave and microseismicity using temporary seismic networks. Here the results are summarized and implications for geodynamic models are discussed. The oceanic Moho of the subducted African plate is situated at a depth of about 50–60 km beneath Crete. The continental crust of the overriding Aegean lithosphere is about 35 km thick in eastern and central Crete, and typical crustal velocities are observed down to the upper surface of the downgoing slab beneath western Crete. A negative phase at about 4 s in receiver functions occurring in stripes parallel to the trend of the island points to low-velocity slices within the Aegean lithosphere. Interplate seismicity is spread out about 100 km updip from the southern coastline of Crete. To the south of western Crete, this seismically active zone corresponds to the inferred rupture plane of the magnitude 8 earthquake of AD 365. In contrast, interplate motion appears to be largely aseismic beneath the island. The coastline of Crete mimics the shape of a microseismically quiet realm in the Aegean lithosphere at 20–40 km depth, suggesting a relation between active processes at this depth range and uplift. The peculiar properties of the lithosphere and the plate interface beneath Crete are tentatively attributed to extrusion of material from a subduction channel, driving differential uplift of the island by several kilometres since about 4 Ma.

Since the Late Cretaceous, the tectonics of the Eastern Mediterranean region has been controlled by (1) convergence between Africa and Eurasia and (2) subduction of oceanic lithosphere of narrow small oceanic basins separating Gondwana-derived terranes (e.g. Dercourt *et al.* 1986; Gealey 1988; Stampfli & Borel 2004). Oceanic lithosphere is now almost completely subducted, with remaining remnants beneath the Ionian basin and in the Eastern Mediterranean south of western Turkey. The closure of the earlier subducted oceanic basins and the collision of the intervening terranes with the Eurasian active margin resulted in a number of distinct orogenic belts and accretion of continental crust. In the Aegean, rollback of the active continental margin has been important since at least 30 Ma (e.g. Angelier *et al.* 1982; Thomson *et al.* 1998; ten Veen & Postma 1999), collision between the African passive margin and the Aegean continental lithosphere is incipient (Mascle & Chaumillon 1997; Mascle *et al.* 1999; ten Veen & Kleinspehn 2003; Meier *et al.* 2004*a*).

The position of subducted oceanic lithosphere in the Hellenic subduction zone was imaged by seismic tomography in the upper mantle (Spakman *et al.* 1988; Ligdas *et al.* 1990; Papazachos *et al.* 1995; Taymaz 1996; Alessandrini *et al.* 1997; Papazachos & Nolet 1997; Piromallo & Morelli 1997, 2003; Marone *et al.* 2004) and well into the lower mantle (Spakman *et al.* 1993; Bijward *et al.* 1998; Bijward & Spakman 2000; Karason & van der Hilst 2000; Faccenna *et al.* 2003), down to depths of about 1700 km. Only the upper part of the slab represents African lithosphere that was subducted at the present active continental margin south of Crete. Subduction south of Crete started at about 20–15 Ma, when the plate boundary shifted to the southern border of an accreted microcontinent building most of the continental crust of present Crete (Thomson *et al.* 1998). The active margin has since retreated by about 350–500 km in a southwestward direction (ten Veen & Meijer 1998; ten Veen & Kleinspehn 2003). The coeval convergence between the African and the Eurasian plates amounts to about 150 km (Dercourt *et al.* 1986; Gealey 1988; Jolivet & Faccenna 2000; Faccenna *et al.* 2003; Stampfli & Borel 2004). This means that a slab of about 500–650 km length entered the subduction zone since the plate boundary shift to the south of Crete (Meier *et al.* 2004*a*). In accordance with tectonic reconstructions (Dercourt *et al.* 1986; Gealey 1988; Stampfli &

From: TAYMAZ, T., YILMAZ, Y. & DILEK, Y. (eds) *The Geodynamics of the Aegean and Anatolia*. Geological Society, London, Special Publications, **291**, 183–199.
DOI: 10.1144/SP291.9 0305-8719/07/$15.00

Borel 2004), the slab of more than 2000 km length imaged by seismic tomography must be composed of lithosphere once underlying different oceanic basins that were successively subducted at different active margins. After closure of each oceanic basin and collision of the trailing continental terrane with the Eurasian margin, a new active margin developed to the south of the accreted terrane, and subduction of the next stripe of oceanic lithosphere commenced. To explain the continuous slab imaged by seismic tomography, Meier *et al.* (2004*a*) proposed that continental crust of the terranes was delaminated from the underlying continental mantle lithosphere, as proposed by Thomson *et al.* (1998, 1999) in the 'buoyant escape' scenario based on the geological record of Crete. If this is true, the slab imaged by seismic tomography would be composed of alternating sections of oceanic and continental mantle lithosphere (Meier *et al.* 2004*a*; van Hinsbergen *et al.* 2005) of variable width, corresponding to the oceanic basins and continental terranes depicted in the palaeogeographical reconstructions (Dercourt *et al.* 1986; Gealey 1988; Stampfli & Borel 2004). It is possible that the presence of sections of delaminated subcontinental mantle devoid of hydrated material is the reason for the missing deep seismicity in the present-day Hellenic subduction zone.

The recent kinematics of the Eastern Mediterranean are characterized by westward drift and counter-clockwise rotation of the Anatolian microplate and southward escape of the Aegean lithosphere (McKenzie 1970, 1972, 1978; Le Pichon *et al.* 1995; McClusky *et al.* 2000). Global positioning system (GPS) surveys demonstrate that the rate of SSW motion of the Aegean lithosphere with respect to stable Eurasia increases towards the south (McClusky *et al.* 2000). This implies extension in the overriding plate, which is attributed to rollback and retreat of the subduction zone (Le Pichon & Angelier 1979; Angelier *et al.* 1982; ten Veen & Meijer 1998; Armijo *et al.* 2003). Within the given kinematic framework, the high curvature of the active continental margin implies that subduction becomes increasingly oblique from west to east. In the western forearc, the vectors of motion are oriented normal to the continental margin. Towards the east, in the area of Rhodos, the angle between the motion vector and the active margin decreases to about 20°. The relative velocity between the subducting African plate and the overriding Aegean lithosphere is about 4.5 cm a^{-1} (McClusky *et al.* 2000) at the active margin.

Seismic activity in the forearc of the Hellenic subduction zone is considerable. In the last 2.5 ka several events reached a magnitude of eight (Papazachos *et al.* 2003). In AD 365, an event with an estimated magnitude of 8.3 occurred in the western forearc, to the SW of Crete. The late Holocene uplift, of up to 9 m at the southwestern coast of Crete, is attributed to co-seismic deformation during this event (Pirazzoli *et al.* 1982; Stiros 2001). According to the global catalogue of Engdahl *et al.* (1998), eight events with a magnitude of six or above occurred in the Hellenic subduction zone in the second half of the 20th century.

One peculiarity of the Hellenic forearc is the geologically recent rapid uplift in the area of Crete (Meulenkamp *et al.* 1994; Lambeck 1995), of up to several kilometres since about 4 Ma with an average rate close to 1 mm a^{-1}. This Plio-Pleistocene uplift has produced a marked relief, with Crete rising to nearly 2.5 km above sea level, close to narrow furrows with water depths of more than 3 km to the south. These deep topographic furrows to the south of Crete do not mark the position of the plate boundary (Le Pichon & Angelier 1979; Angelier *et al.* 1982). The widely used term 'Hellenic trench', commonly used to refer to the distinct topographic feature marking the site where the subducted plate disappears beneath the forearc at a convergent plate boundary, may be misleading here. In the Hellenic subduction zone, the accretionary complex (e.g. Kastens 1991; Mascle & Chaumillon 1997) is located to the south of these 'trenches' beneath the Mediterranean ridge, and the trenches mark tectonic features within the forearc, and hence within the overriding Aegean lithosphere. The Cretan horst structure shows the following characteristic features: (1) it is narrow in the north–south direction; (2) it is bounded by normal faults on all sides; (3) uplift commenced abruptly in the early Pliocene; (4) uplift is contemporaneous with continuing forearc-parallel extension of the Aegean lithosphere. The question is: what drives Crete up?

In this paper, recent results of receiver function, surface-wave and microseismicity studies using temporary local networks are reviewed. The implications for the structure and dynamics of the Hellenic subduction zone are discussed and a model for the rapid localized uplift of Crete is proposed.

Slab segmentation in the Hellenic subduction zone

The seismicity of the Hellenic subduction zone between 1964 and 1998 according to the global relocated ISC catalogue (Engdahl *et al.* 1998) is shown in Figure 1. In this region the catalogue is complete for events with magnitudes greater than *c.* 4.7. Clusters of events are observed within the part of the forearc characterized by intense shallow seismicity. The trend of these clusters is oriented NE–SW (Fig. 1). The depth of hypocentres increases with distance from the plate boundary and reaches

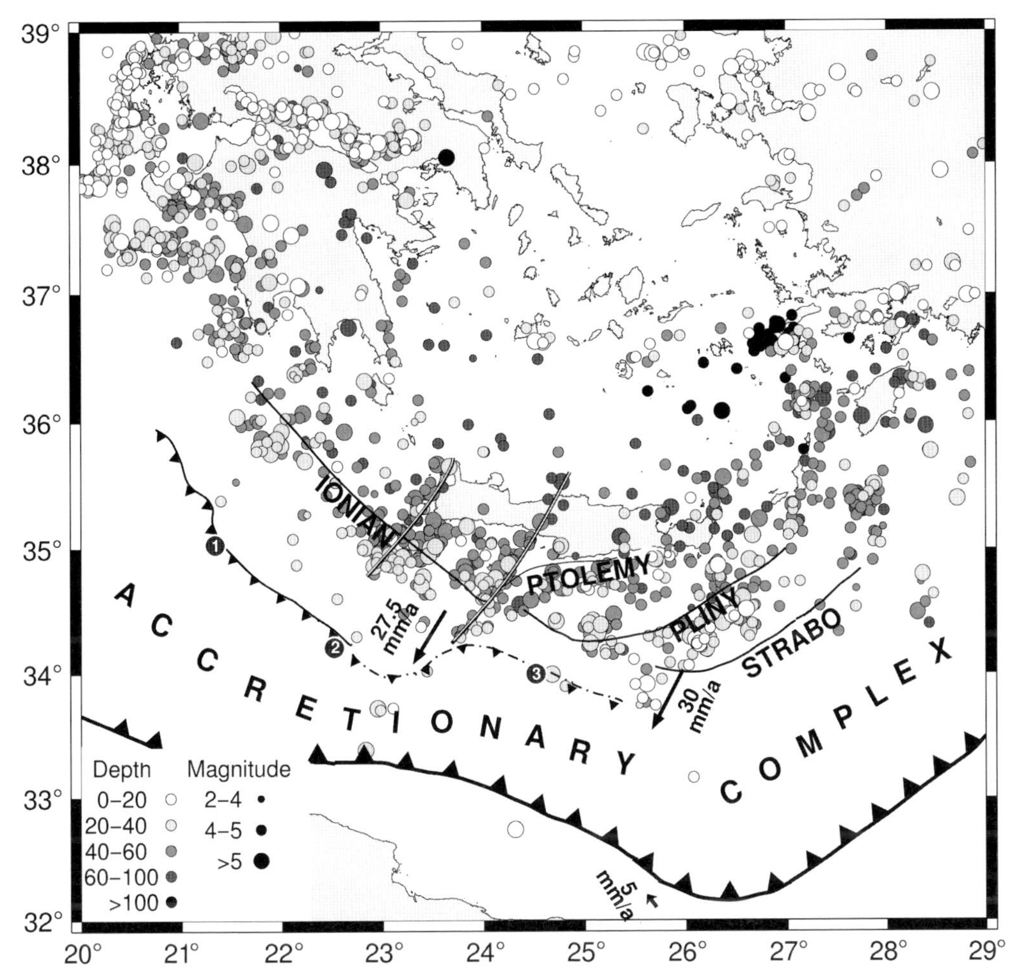

Fig. 1. Hypocentres of events in the Aegean region listed in the relocated ISC catalogue (Engdahl *et al.* 1998) for the years 1964–1998. The position of the deformation front of the Hellenic subduction zone (Le Pichon *et al.* 1995) and the border between the central and the inner units of the Mediterranean Ridge (continuous line with triangles, Lallemant *et al.* 1994; dashed–dotted line with triangles, Mascle *et al.* 1999) are indicated. Points labelled with numbers mark the southern border of the Aegean crust as determined by seismic studies (1, Lallemant *et al.* 1994; 2, Brönner 2003; 3, Bohnhoff *et al.* 2001), corresponding to the transition between the inner and the central units of the Mediterranean Ridge. Relative horizontal velocities with respect to Eurasia (McClusky *et al.* 2000) are indicated by arrows. In addition, the deep furrows ('trenches') to the south of Crete and the NE–SW-striking linear patterns of seismicity in western and central Crete are emphasized.

about 100 km in the west and about 160 km in the east. These intermediate depth events define a strongly curved Benioff zone, with its lower termination approximately beneath the volcanic arc (Papazachos & Comninakis 1971; Comninakis & Papazachos 1980; Makropoulos & Burton 1984; Taymaz *et al.* 1990; Hatzfeld & Martin 1992; Knapmeyer 1999; Papazachos *et al.* 2000).

The hypocentres of the relocated ISC catalogue (Engdahl *et al.* 1998) are projected onto vertical cross-sections in Figure 2. The figure shows that there is not much intermediate-depth Benioff zone seismicity compared with, for example, the western Pacific subduction zones. As is evident from Figure 2d and e, the Benioff zone can be traced to a depth of about 160 km, with a faint seismic gap between about 100 and 125 km (Papazachos *et al.* 2004). In the east, the Benioff zone dips more steeply than in the west. As is evident from Figure 2, data from local networks are required to image the seismogenic zones beneath the forearc in greater detail.

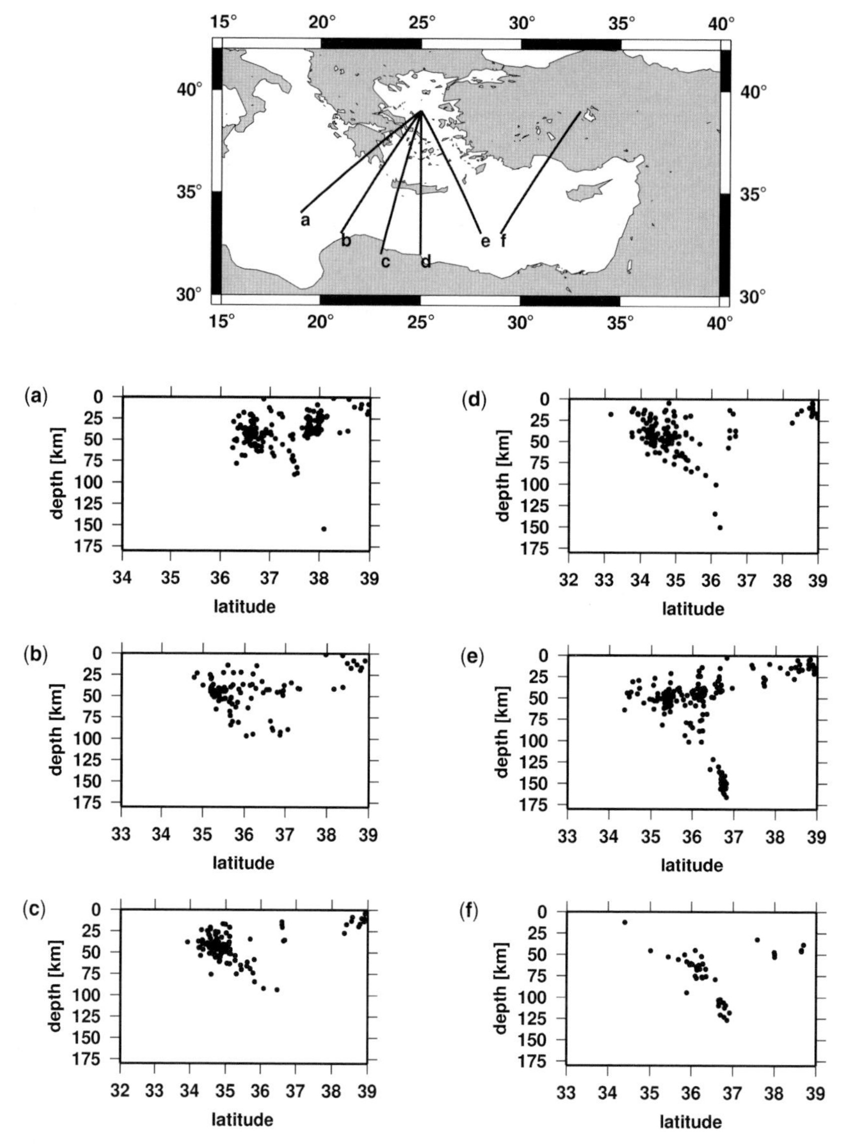

Fig. 2. Hypocentres from the global catalogue by Engdahl *et al.* (1998) projected onto vertical planes along the profile lines indicated in the map.

In the western Cyprus arc (section f in Fig. 2), the hypocentres define a Benioff zone with subduction towards the NE beneath Anatolia. As the slab in the eastern part of the Hellenic subduction zone is inclined towards the NW, a slab window is expected to exist beneath western Turkey, which probably affects deformation and magmatism of this region. The schematic sketch in Figure 3 illustrates the situation, as suggested by intermediate-depth seismicity (e.g. Engdahl *et al.* 1998).

In the western segment of the Hellenic subduction zone, the dip of the slab increases at a depth of about 75 km (Hatzfeld 1994; Papazachos *et al.* 1995, 2000; Taymaz *et al.* 1990; Papazachos & Nolet 1997). Based on a regional tomographic study, Papazachos & Nolet (1997) concluded that at about 200 km depth the western and eastern slab segments meet along a rather sharp edge with a right angle. Wortel & Spakman (2000) assumed a tear within the slab, propagating horizontally from the Ionian Sea along the forearc, which causes incipient slab detachment. Apart from the strong curvature enhanced by rollback (e.g. ten Veen & Kleinspehn 2003), the segmentation of

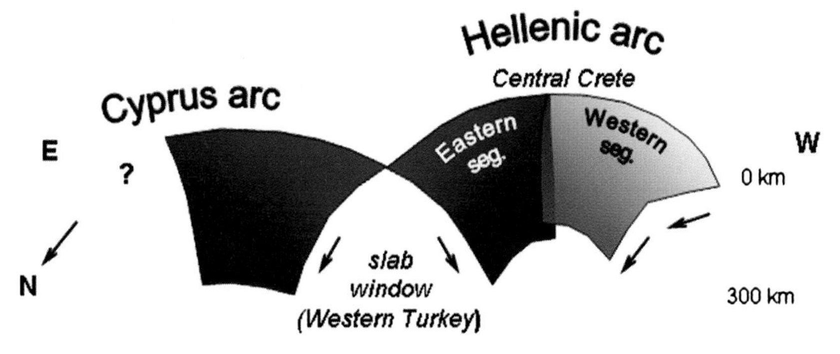

Fig. 3. Schematic illustration of the first-order segments of the African slab in the Eastern Mediterranean down to about 300 km (view from the north).

the subducting slab may result from lateral heterogeneity of the African lithosphere, which is indicated by body-wave and surface-wave studies for the Libyan Sea south of Crete (Piromallo & Morelli 1997; Bijward & Spakman 2000; Marone *et al.* 2004). To the south of western Crete, seismic velocities in the mantle lithosphere are lower than those typically observed for oceanic lithosphere (Marone *et al.* 2004; Meier *et al.* 2004*a*). This indicates that the passive margin of northern Africa is entering the Hellenic subduction zone to the south of western Crete (Meier *et al.* 2004*a*). In contrast, to the south of eastern Crete, according to tomographic studies (Bijward & Spakman 2000; Marone *et al.* 2004), remnants of oceanic lithosphere seem still to be present.

Topography and tectonics in the area of Crete

Crete is situated in the central forearc of the Hellenic subduction zone. Figure 4 depicts the topography of the Hellenic convergent margin along a north–south section at 25.12° E, from the Cyclades over Crete to northern Africa. The volcanic arc is underlain by continental crust of the Cyclades, which has undergone considerable stretching since mid-Tertiary times and is characterized by metamorphic core complexes. The average water depth in the volcanic arc region is a few hundred metres. It increases to more than 1 km in the Cretan Sea north of Crete, where crustal thickness is reduced to well below 20 km (Bohnhoff *et al.* 2001). As is evident from Figure 4, Crete forms a narrow spike in topography, abruptly rising on both sides.

The Ionian trench is a NW–SE-trending furrow barely 30 km from the southwestern corner of Crete (Fig. 1). According to wide-angle seismic data, the southern border of the Aegean lithosphere is located about 150 km to the SSW of this deep furrow (Truffert *et al.* 1993; Lallemant *et al.* 1994; Bohnhoff *et al.* 2001). Consequently, the Hellenic trench system does not mark the plate boundary but constitutes tectonic features within the forearc. Further to the east, to the south of central Crete, the Ionian trench system bifurcates into three branches trending NE–SW, referred as the Ptolemy, Pliny and Strabo trenches (Fig. 1). These furrows are at an angle of about 30° to the direction of relative motion between the Aegean and the subducted African plate. The Ptolemy and Pliny trenches are therefore interpreted to represent transtensional structures within the Aegean lithosphere (Huchon *et al.* 1982; Huguen *et al.* 2001; Brönner 2003; ten Veen & Kleinspehn 2003). South of eastern Crete, the Strabo trench marks the position of the backstop, where sediments of the Mediterranean Ridge accretionary complex are backthrust onto Aegean continental crust (Huguen *et al.* 2001). According to Mascle & Chaumillon (1997), the southern border of the Aegean lithosphere appears to be offset by about 50 km to the south of central Crete (Fig. 1).

The southern deformation front of the Hellenic convergent margin is located to the south of the Mediterranean Ridge (e.g. Le Pichon *et al.* 1995; Mascle & Chaumillon 1997; Mascle *et al.* 1999). To the south of western Crete, the deformation front bends northward (Fig. 1). This is attributed to incipient collision of the accretionary complex with the passive continental margin of Africa (LePichon *et al.* 1995; Mascle & Chaumillon 1997; Mascle *et al.* 1999). The arc-normal contraction in the western forearc, indicated by focal mechanism solutions for events down to about 20 km depth (Lyon-Caen *et al.* 1988; Taymaz *et al.* 1990; Hatzfeld *et al.* 1993*a*; Baker *et al.* 1997; Benetatos *et al.* 2004; Bohnhoff *et al.* 2005), is consistent with incipient collision.

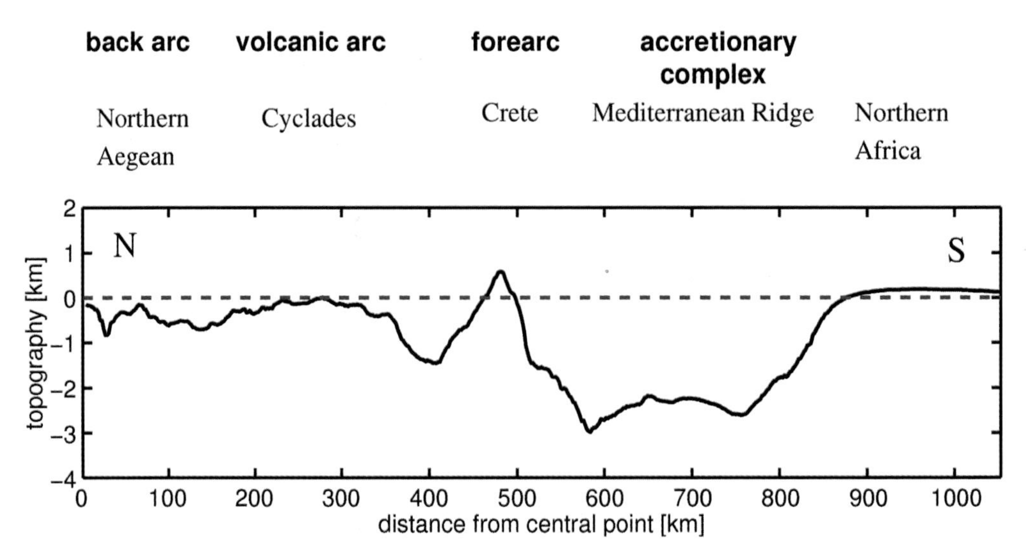

Fig. 4. Average topography along a profile from the Cyclades to northern Africa along 25.12° E from 39.5° N to 29° N.

Temporary seismic networks

To gain more detailed information on the structure and seismicity of the convergent margin, several temporary local seismic networks were installed in the Aegean in recent decades. In 1988, microseismic activity was investigated using a temporary seismic network covering large parts of the southern Aegean (Hatzfeld *et al.* 1993*a*). Regionally focused studies of microseismic activity were dedicated to the Peloponnese (Hatzfeld *et al.* 1989; Sachpazi *et al.* 2000; Makris *et al.* 2004) and to the eastern part of the volcanic arc (Makris *et al.* 2000). In 2002–2004, a temporary seismic network provided insight into microseismic activity in the Cyclades (Bohnhoff *et al.* 2004). The island of Crete was the base for a temporary local network of seismic stations, first installed in eastern Crete and then displaced to the western part of the island for a second month of observation (de Chabalier *et al.* 1992; Hatzfeld *et al.* 1993*b*). This study was complemented by a further campaign in central Crete in 1995 (Delibasis *et al.* 1999). Since 1996, a more extensive database was obtained by operating five temporary local onshore networks on Crete with up to 47 stations and recording intervals of up to 18 months (Becker 2000; Meier *et al.* 2004*b*). Data from these five networks are considered in this paper. Recently, these onshore networks were complemented by an offshore ocean-bottom seismograph (OBS) network in the Libyan Sea south of central Crete, with the aim of gaining information on microseismicity near the Ptolemy and Pliny trenches and localizing interplate microseismicity with higher accuracy. The location of the temporary stations deployed between 1996 and 2004 is shown in Figure 5. Moreover, nine permanent broadband stations are currently operated in the region of Crete by the GEOFON network (Hanka & Kind 1994), the National Observatory of Athens and MEDNET (Boschi & Morelli 1994). This dense station coverage yields an extensive database for receiver function, surface-wave and microseismicity studies.

Structure of the subduction zone in the area of Crete

The most important of the structural features of the subduction zone is the position of the subducted African lithosphere. To obtain this first-order information, Knapmeyer & Harjes (2000) used receiver functions (Fig. 6) for short-period stations in western Crete, and Li *et al.* (2003) studied receiver functions of broadband stations on the island. The results indicate that the crust–mantle discontinuity within the subducting slab is at about 55 km depth beneath Crete. This was confirmed by Meier *et al.* (2004*a*), who studied surface-wave propagation between two broadband stations on Crete and found the African Moho at a similar depth of 50 km. A WNW–ESE-oriented cross-section through migrated receiver functions of short-period stations in western and central Crete (Endrun *et al.* 2004) is shown in Figure 6a, and clearly depicts the discontinuity interpreted to represent the African Moho within the slab at about 50–55 km depth.

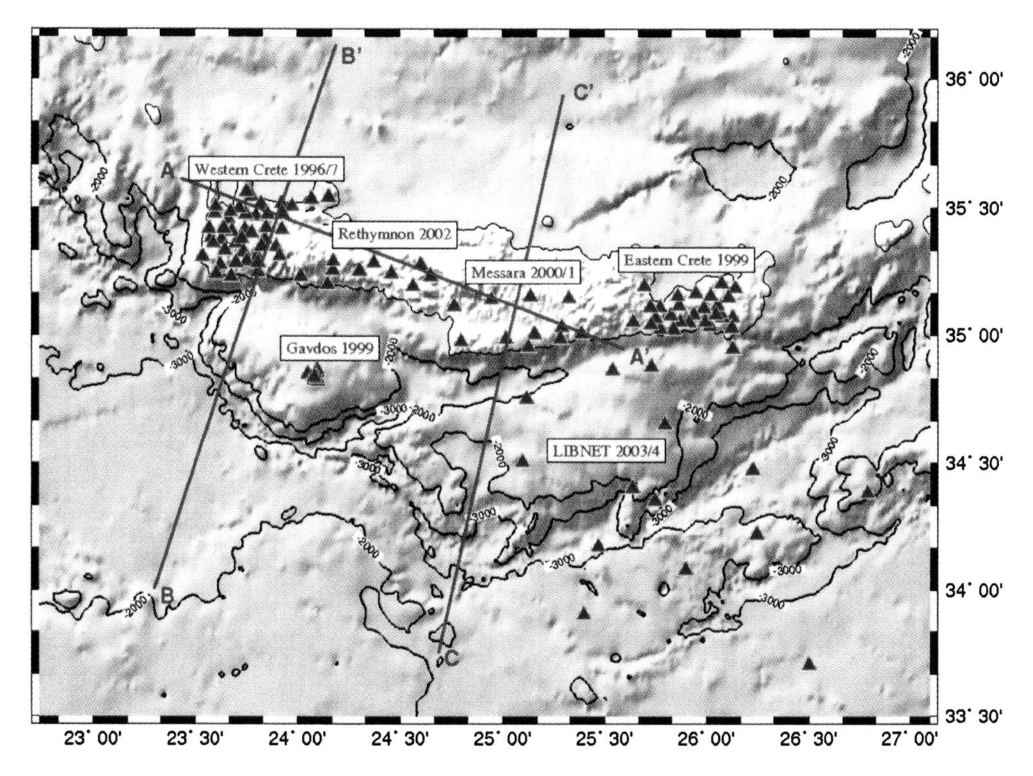

Fig. 5. Temporary short-period networks deployed in the area of Crete between 1996 and 2004. Western Crete: six stations in May–August 1996 and 47 stations in September–December 1997; Gavdos: 12 stations in May–August 1999; Messara: eight stations in April 2000–December 2001; Rethymnon: seven stations in May–November 2002; LIBNET: up to eight OBS deployed in four 2 month campaigns in 2003–2004; eastern Crete: 27 stations in January–March 1999. Location of receiver function profiles shown in Figure 6 are indicated by red lines.

The notable differences in the structure between the eastern and western parts of section A (Fig. 6a) suggest a significant lateral heterogeneity within the overriding Aegean lithosphere. A discontinuity observed at about 35 km beneath central Crete is interpreted as the crust–mantle discontinuity of the Aegean lithosphere. This interpretation is in accordance with wide-angle seismic data (Bohnhoff *et al.* 2000) and surface-wave studies (Endrun *et al.* 2004; Meier *et al.* 2004*a*). According to wide-angle seismic data, the depth of the Aegean Moho strongly decreases northwards and is located at about 17 km depth beneath the Sea of Crete (Bohnhoff *et al.* 2000), probably as a result of extensional thinning of the crust in the recent geological past. A negative phase in the receiver function of a broadband station in northern central Crete (Li *et al.* 2003) and in the receiver function image along section C (Fig. 6c) may be caused by the strong Moho topography (Endrun *et al.* 2005). In contrast, in western Crete the receiver function image shown in Figure 6a does not reveal a positive discontinuity at about 35 km depth that could be attributed to a transition from crustal to mantle velocities. Wide-angle seismic data reveal high reflectivity and a complex structure of the Aegean lithosphere at this depth (Bohnhoff *et al.* 2000). Li *et al.* (2003) found a negative phase in the receiver function of a broadband station in western Crete at about 4 s, corresponding to about 35 km depth. The detailed analysis of the spatial distribution of this negative phase (Endrun *et al.* 2004) revealed a marked lateral heterogeneity at 35 km depth. Arc-parallel stripes with a negative receiver function phase at 4 s delay time in western Crete point to slices with varying velocity including distinct low-velocity bodies. Finally, Figure 6b shows a NNE–SSW-oriented section of migrated receiver functions in western Crete. Negative phases are present at 20–40 km depth beneath northwestern Crete. Analysis of surface-wave propagation within the short-period network in western Crete indicates typical crustal velocities for the Aegean lithosphere down to the plate contact at about 45 km depth (Endrun *et al.* 2004). If so, the crustal thickness in western Crete would be

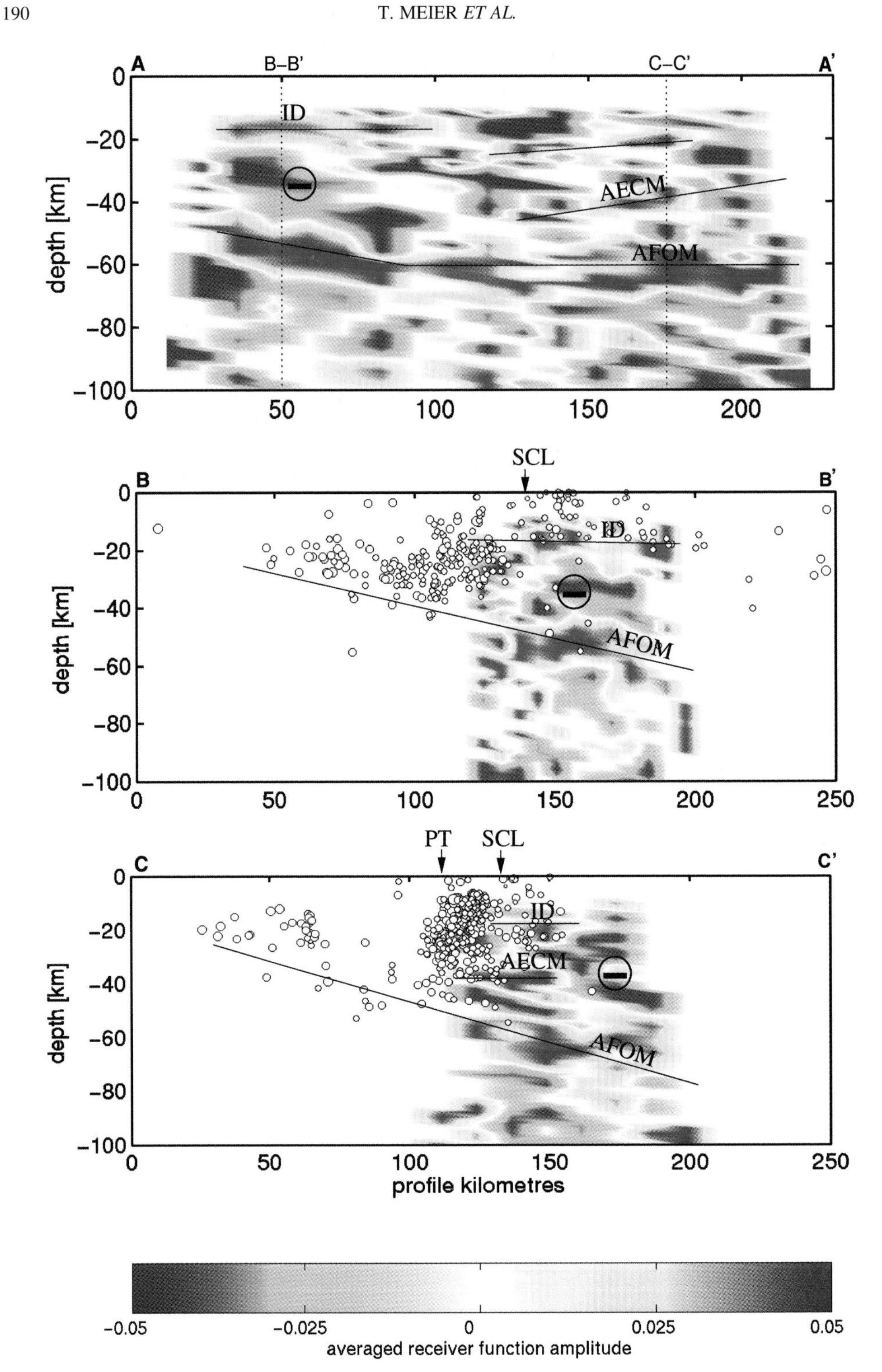

Fig. 6. Cross-sections through migrated receiver functions of short-period stations (location of the profiles shown by red lines in Fig. 5). The hypocentres of microseismicity are projected onto arc-normal cross-sections BB′ and CC′. AECM, Aegean continental Moho; AFOM, African oceanic Moho; ID, intracrustal discontinuity; PT, Ptolemy trench; SCL, southern coastline of Crete.

comparable with that beneath the Peleponnese. The persistent low-velocity layer at mid-crustal depths identified by Papazachos *et al.* (1995) beneath the Peleponnese seems to be absent beneath western Crete, however (Endrun *et al.* 2004). Instead, low-velocity zones indicated by negative receiver function phases seem to represent individual slices of limited lateral extent within the Aegean lithosphere.

McClusky *et al.* (2000) demonstrated that the southeastern part of the overriding Aegean lithosphere is moving relative to the southwestern part. Segmentation of the subducted slab and lateral heterogeneity of the overriding Aegean lithosphere in the area of Crete are indicated by several features: (1) the change in the dip and the ultimate depth of the Benioff zone; (2) the NE–SW-oriented cluster of seismic events in central Crete (Fig. 1); (3) the lateral heterogeneity of the Aegean lithosphere beneath central Crete found by receiver function, surface-wave and seismicity studies; (4) changes in the structure of the Hellenic trench system along the forearc; (5) changes in the position of the southern border of the Aegean lithosphere south of central Crete (Fig. 1).

Microseismicity

The catalogue of global seismicity published by Engdahl *et al.* (1998) covers a timespan of about 40 years and an earthquake magnitude range of about 4–6.5. Localization uncertainties can be as high as 20 km in the Aegean. Temporary onshore local networks reduce these uncertainties to about 5–10 km (Meier *et al.* 2004*b*). Localization accuracy of microseismicity south of Crete will be further improved when data from offshore networks become available. The observation periods of temporary networks are much smaller, however, than that of the global catalogue by Engdahl *et al.* (1998). For the networks shown in Figure 5 observation times vary between 3 and 18 months. Nevertheless, about 5000 events with magnitudes between −1 and 4.5 were detected and localized. The persistent high level of microseismic activity in the forearc allowed investigation of the spatial distribution of seismically active zones despite the relatively short observation times inherent in the temporary networks. Clearly, an assessment of temporal variations in the seismic activity would require longer observation periods. The occurrence of swarm-like clusters of microseismicity in the forearc was detected by Becker *et al.* (2006).

Within the observation period, microseismic activity is observed between about 20 and 40 km depth (Fig. 6b and c) along the inferred plate contact. The current width of the interplate seismogenic zone is about 100 km to the south of western Crete, close to 50 km to the south of central Crete, and again up to about 100 km in the east. At the southern border of the overriding Aegean lithosphere, the plate contact is characterized by a microseismically quiet zone about 50 km wide (Meier *et al.* 2004*b*). This zone may represent a décollement horizon with aseismic deformation localized in unconsolidated sediments.

Particularly intense microseismicity at the plate contact is observed to the south of central Crete (Fig. 6c). The northern border of this seismically active zone along the plate interface coincides with the southern coastline of Crete (Meier *et al.* 2004*b*). The microseismically active zone along the plate interface to the south of western Crete is currently about 100 km wide (Meier *et al.* 2004*b*) and coincides with the fault plane proposed by Papazachos *et al.* (1999) for the historical magnitude 8.3 event that occurred on 25 July AD 365. The length of the fault plane activated in this event is estimated to be about 200 km, based on palaeoseismological studies (Pirazzoli *et al.* 1982; Stiros 2001). If the microseismically active zone corresponds to the part of the fault activated in the AD 365 event, the ruptured area would be about 200 km × 100 km, roughly compatible with the expected rupture area of a magnitude 8 event. Laigle *et al.* (2004) proposed that the rupture area exceeded the width of the currently microseismically active zone in a down-dip direction.

Beneath Crete, the plate contact is almost devoid of microseismicity. In this area, microseismicity is restricted to the upper *c.* 20 km of the Aegean lithosphere, with a low level of microseismic activity down to the inferred upper interface of the subducted African plate (Fig. 6b and c). In addition, a vertical zone of localized seismic activity follows the Ptolemy trench and cuts through the entire Aegean lithosphere, from the plate interface through the uppermost crust (Meier *et al.* 2004*a*).

Microseismic activity in the area of Crete is concentrated in a certain depth range, which is not uniform over the area. Owing to complete coverage of Crete by the successively operated local networks, the median of the vertical depth distributions of the events could be derived for the entire island and its surroundings; the pattern is shown in Figure 7. To the south of Crete the median depth of the microseismicity corresponds approximately to the inferred depth of the plate contact. The shift towards a shallower median depth of microseismicity correlates with the southern coastline of Crete (Fig. 7). Seismic activity at the inferred level of the plate interface is almost absent beneath Crete. The coastline of Crete delineates a seismically quiet zone at a depth between about 20 and 40 km beneath Crete. This coincidence suggests that the

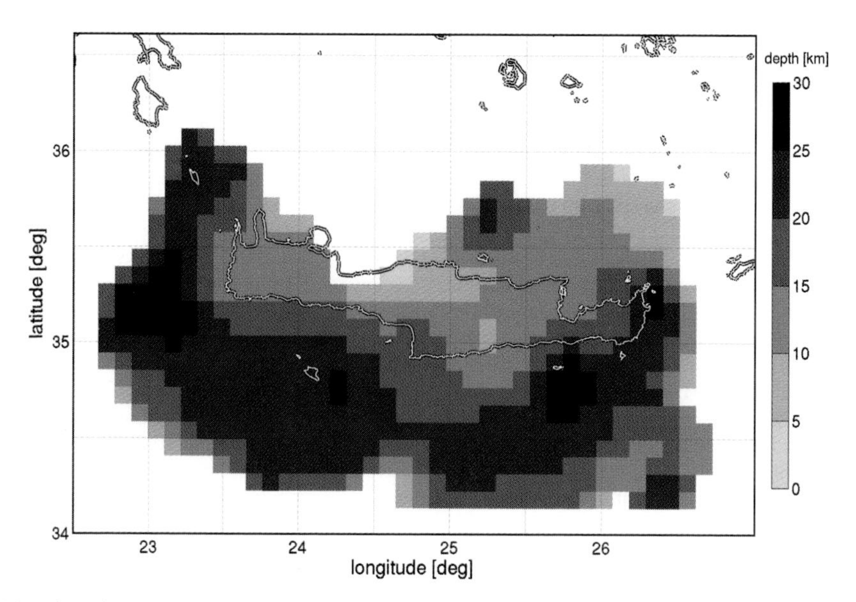

Fig. 7. Map view of average hypocentre depths of microseismicity. Hypocentres of *c.* 5000 microseismic events were detected and localized using data from temporary stations shown in Figure 5.

recent uplift of Crete may be related to processes currently active at these depths.

Implications for geodynamic models of the Hellenic subduction zone around Crete

The inferred structure of the Hellenic subduction zone in the area of Crete, and the distribution of seismicity, is shown by the schemes in Figure 8. Figure 8a displays the situation in western Crete, viewing to the NW, and Figure 8b that in eastern Crete, viewing to the NE. As shown in Figure 8, there are some differences in the structure and distribution of seismicity in the two areas. In western Crete, the leading edge of the passive African continental margin has entered the subduction zone and collision has commenced, whereas oceanic lithosphere is still subducted south of eastern Crete with rollback being effective (ten Veen & Meijer 1998; ten Veen & Postma 1999; ten Veen & Kleinspehn 2003). Beneath eastern Crete, the slab dips at an angle of about 19° compared with about 15° beneath western Crete.

As emphasized in Figure 8, interplate microseismicity is intense to the south of Crete. In contrast, the plate interface appears to be microseismically inactive further to the north, where the interface is at depths greater than *c.* 40 km, with a sharp cutoff beneath the southern coastline of Crete. This finding is similar to observations in other subduction zones (e.g. Tichelaar & Ruff 1993). Fault-plane solutions (Taymaz *et al.* 1990; Bohnhoff *et al.* 2005) show that the seismically active upper part of the Aegean lithosphere beneath western Crete undergoes contraction normal to and extension parallel to the plate boundary. A similar situation holds for the forearc beneath the Peloponnese (Lyon-Caen *et al.* 1988; Hatzfeld *et al.* 1993*a*; Baker *et al.* 1997; Benetatos *et al.* 2004), indicating coupling between the plates in the frontal part of the forearc. The current overall extensional tectonics observed at the surface (e.g. Angelier *et al.* 1982) is in apparent contrast to the forearc-normal compressional regime at upper crustal depths indicated by focal mechanisms. Focal mechanisms of intermediate-depth events point to forearc-parallel compression and down-dip extension (Papazachos *et al.* 2000). These intermediate-depth events are supposed to occur within the subducting plate and suggest decoupling between the plates below about 50 km depth.

To the south of Crete, the NW–SE-trending Hellenic trench system bifurcates eastwards into WSW–ENE-trending transtensional graben structures (Fig. 1). For the Ptolemy trench (Fig. 6), to the south of eastern Crete, microseismic activity indicates a nearly vertical fault system that penetrates the entire Aegean lithosphere. Therefore, the Ptolemy trench is interpreted as a boundary with an independently moving forearc sliver (Fig. 8b) driven by oblique subduction (ten Veen & Kleinspehn 2003). The sense of motion along the Ptolemy trench is sinistral. The same probably holds for the Pliny trench, a transtensional structure

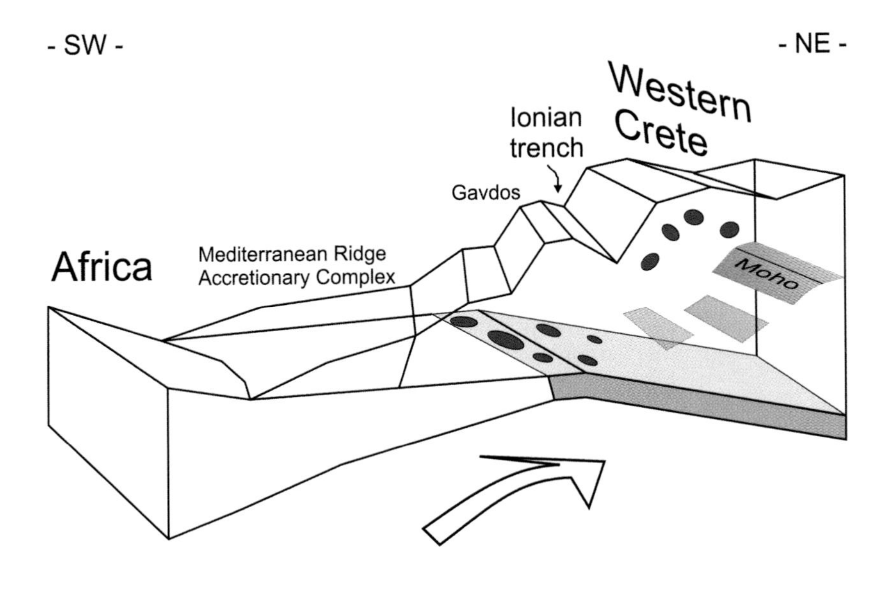

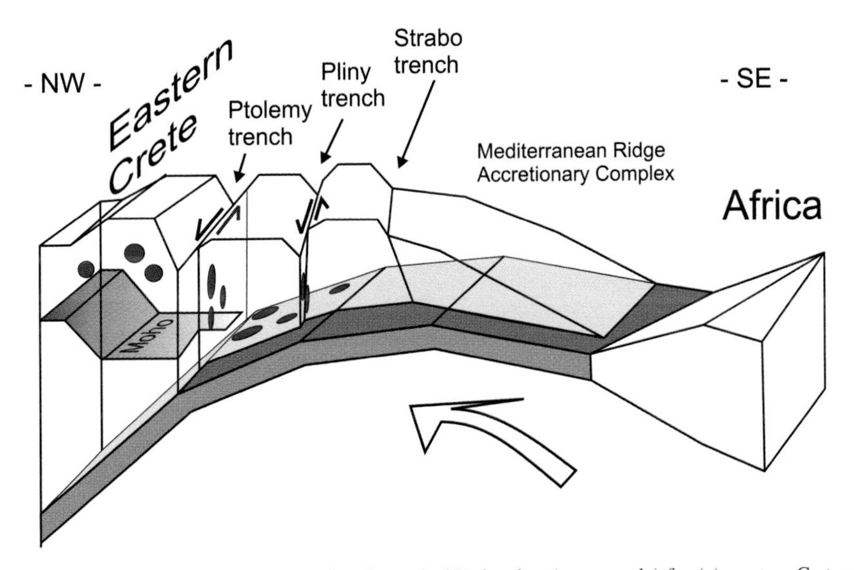

Fig. 8. Interpretation of the seismological results, shown in 3D sketches (not to scale) for (**a**) western Crete and (**b**) eastern Crete. The location of seismic events in the observation period is schematically indicated by red spots. The inferred position of the Aegean Moho based on receiver functions is marked by a red surface. Discontinuous blue surfaces schematically depict the negative phases observed in receiver functions beneath western Crete, suggesting low-velocity bodies in the overriding lithosphere (see Fig. 6).

paralleling the Ptolemy trench to the SE. Based on the observed microseismicity, motion between the forearc slivers is active. Sinistral displacement between these slivers created a basin south of central Crete, filled by a thick sedimentary sequence (Casten & Snopek 2005). The position of this basin is discernible from the offset in the front of the overriding Aegean lithosphere (Mascle & Chaumillon 1997) and the distribution of interplate seismicity (Fig. 1).

Integration of the tectonic structure of the forearc and the information on present-day kinematics is hampered by the fact that a marked change in the mechanical state of the system must have occurred a few million years ago in the Pliocene, when the rapid localized uplift of the narrow Cretan horst structure commenced. The onset of collision in the area of western Crete is expected to have caused further modifications. Notably, there is ample evidence for significant crustal thinning by arc-normal extension in the area of the Sea of Crete, but GPS surveys demonstrate no active extension in this part of the forearc (e.g. Le Pichon *et al.* 1995; McClusky *et al.* 2000). Recent extensional structures observed at the surface on Crete could therefore be the result of gravitational spreading of a mountain belt undergoing rapid localized uplift.

Several possible causes for this localized uplift have been proposed. Angelier *et al.* (1982) assumed that the uplift is caused by underplating of subducted sediments. Alternatively, buckling of the forearc as a result of coupling of the plates has been proposed (e.g. Savage 1983). Giunchi *et al.* (1996) introduced an additional arc-normal force in their numerical models to achieve the observed uplift rates. The buckling models seem to be in conflict with the narrow and discontinuous character of the horst structure, being bounded by normal faults. Forearc buckling by arc-normal compression would be expected to result in a continuous bulge of longer wavelength. As an alternative, based on the results summarized in the present paper, the driving mechanism for the localized uplift of Crete is suspected to be extrusion of material from a subduction channel (Cloos & Shreve 1988*a*, *b*). This hypothesis is schematically depicted in Figure 9.

Rapid return flow of material along the hanging wall of the subducted slab has been proposed to explain the rapid exhumation (e.g. Rubatto & Hermann 2001) of high- and ultrahigh-pressure metamorphic rocks. Numerical simulations (Gerya *et al.* 2002; Gerya & Stöckhert 2007) support the feasibility of this concept, predicted particle trajectories and pressure–temperature–time paths being in accordance with the natural record in collisional belts. In these simulations, material extruded from the subduction channel is shown to spread out beneath the forearc. Such material, including hydrated mantle material, may build up part of the lower Aegean lithosphere (Fig. 9), which is characterized by typical crustal velocities, a complex internal structure, and the absence of seismicity. If so, the sudden onset of uplift in the Pliocene (e.g. Le Pichon *et al.* 1995) is tentatively interpreted to be the surface expression of the onset of extrusion from the subduction channel, after a period without return flow. This period, during which return flow and extrusion were not effective, could be envisaged to be a consequence of the mid-Miocene accretion of the microcontinent (Thomson *et al.* 1999) now building up most of the continental

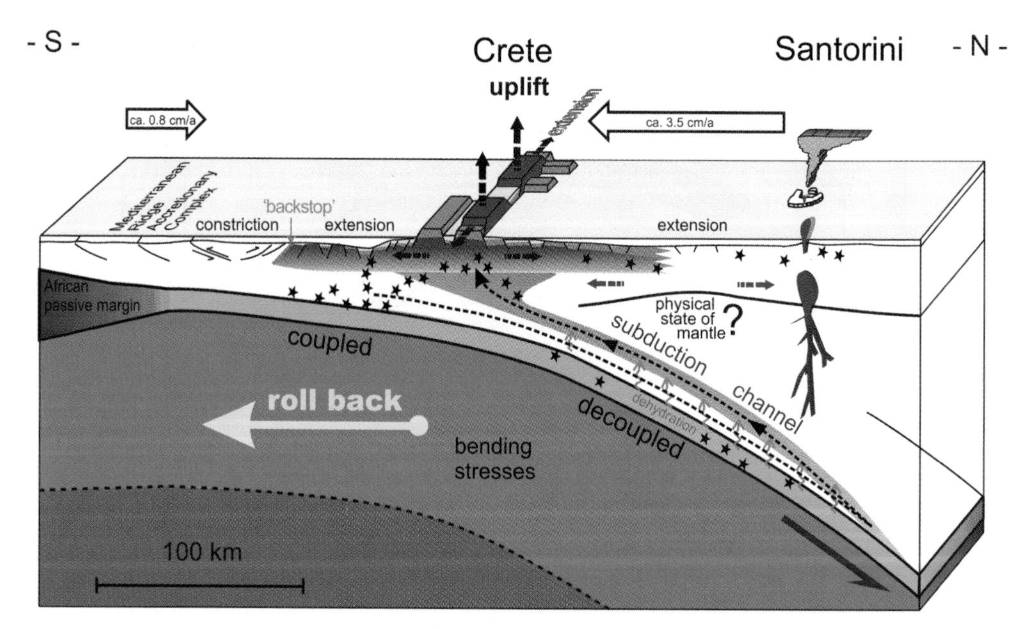

Fig. 9. Three-dimensional schematic illustration of the principal features of the Hellenic subduction zone in the area of Crete, and the proposed model for uplift of the island. The location of seismic activity within the observation period is shown schematically by stars. The bold black dashed arrows and labels 'extension' and 'constriction' refer to the geological record since about 4 Ma, when the uplift of Crete commenced. The proposed realm of backflow in the upper portion of a subduction channel is shown in purple.

crust beneath Crete. The simulations by Gerya *et al.* (2002) have demonstrated the importance of hydration of the mantle wedge atop the subducting slab for return flow. Subduction of dry subcontinental lithosphere, from which the buoyant continental crust of the accreted microcontinent had been detached (Thomson *et al.* 1999), is probably not favourable for return flow in a subduction channel. It is possible that a sufficient length of partially hydrated oceanic lithosphere of the African plate, trailing the continental lithosphere of the microcontinent, had to be subducted before return flow could be established. Assuming convergence and rollback with a constant rate close to 3 cm a^{-1}, about 400 km of oceanic lithosphere have been subducted between the time of accretion of the microcontinent at around 20 Ma and the onset of rapid uplift at about 4 Ma. A timespan of some 10 Ma would be sufficient to circulate material to a depth of about 100 km and back into the lower crust by return flow with plate velocity. It is therefore suspected that subduction of barren subcontinental mantle may have caused a temporary absence and subsequent re-establishment of return flow in a subduction channel, with the sudden onset of Cretan uplift as the surface expression of extrusion into the deeper crust becoming effective.

The recent extensional structures at the surface on Crete would then be a response of the upper crust to uplift, whereas contraction normal to the forearc, as indicated by the focal mechanisms, could persist in the middle crust because of coupling between the plates (Fig. 9). Notably, the addition of material extruded from the subduction channel would lead to growth of the forearc. This growth could partially compensate rollback and may have rendered lithospheric extension beneath the Sea of Crete ineffective since about 4 Ma, simultaneous with the sudden onset of uplift of Crete.

Summary and conclusions

Using data from short-period temporary networks, the lithospheric structure in the area of Crete was imaged by migrated receiver functions. A pronounced discontinuity at 50–60 km depth beneath Crete is interpreted to represent the African Moho within the subducting slab. Assuming a thickness of the African oceanic crust of about 5–10 km, the interplate contact is estimated to be at a depth of about 40–55 km beneath Crete. Within the overriding Aegean lithosphere, receiver function images display a marked forearc-parallel heterogeneity. In central and eastern Crete the Aegean continental crust is about 35 km thick and underlain by a *c.* 15–20 km thick Aegean mantle wedge. Beneath western Crete, a crust–mantle boundary in the overriding Aegean could not be identified. A study of Rayleigh wave propagation across western Crete revealed typical crustal velocities down to the plate interface. Although a negative phase in the receiver functions could be explained by Moho topography in central Crete, this feature probably indicates the presence of low-velocity bodies in the lithosphere beneath western Crete. The lateral distribution of the negative receiver function phase points to considerable heterogeneity within the thickened crust of western Crete, with low-velocity bodies that elongate parallel to the forearc. The low velocities possibly reflect serpentinization of mantle material (e.g. Oleskevich *et al.* 1999). In the area of western Crete, the passive continental margin of Africa has entered the subduction zone (Meier *et al.* 2004*a*), and the plate contact dips at a lower angle than in eastern Crete (Meier *et al.* 2004*b*).

Tomographic studies (e.g. Papazachos & Nolet 1997) and the distribution of intermediate-depth seismicity point to a segmentation of the subducting slab in the Hellenic subduction zone, and to a corresponding segmentation of the overriding Aegean lithosphere. This east–west segmentation is indicated by (1) changes in the structure of the Hellenic trench system, (2) an offset in the southern border of the Aegean lithosphere, and (3) marked lateral heterogeneity within the Aegean lithosphere. The tectonics in the area of eastern Crete is governed by oblique subduction, causing left-lateral movement between forearc slivers, separated by nearly vertical microseismically active faults systems, which cut through the entire Aegean lithosphere.

To the south of Crete intense seismicity along the plate interface is observed in a zone up to 100 km wide. In contrast, little microseismicity is observed at the level of the plate interface beneath Crete, where the interface is located at a depth greater than *c.* 40 km. A change in the orientation of the stress field at *c.* 40 km, indicated by focal-plane solutions of intermediate-depth seismicity (Papazachos *et al.* 2000), points to seismic coupling between the plates in the shallower part of the subduction zone south of Crete, and decoupling of the plates at deeper levels. This decoupling is consistent with the concept of a subduction channel as shown in Figure 9.

The level of microseismicity is very low at a depth of *c.* 20–40 km beneath the island of Crete. The coastline of Crete follows roughly the boundaries of this seismically quiet realm within the Aegean lithosphere, where typical crustal velocities are observed down to the plate interface beneath western Crete. These observations point to a relation between the uplift of Crete and the physical state of the lithosphere in this depth range. It is tentatively proposed that the uplift of Crete, which commenced at about 4 Ma, is driven by return flow in the upper portion of a subduction channel (Fig. 9) and by extrusion of material into the deeper crust.

We thank M. Laigle and an anonymous reviewer for their thoughtful comments and criticism that helped to improve the manuscript. We are grateful to T. Taymaz for his editorial help. This study was supported by the Collaborative Research Centre 526 of the Deutsche Forschungsgemeinschaft.

References

ALESSANDRINI, B., BERANZOLI, L., DRAKATOS, G., FALCONE, C., KARANTONIS, G., MELE, M. M. & STAVRAKAKIS, G. N. 1997. Tomographic image of the crust and the uppermost mantle of the Ionian and the Aegean regions. *Annali di Geofisica*, **XL**, 151–160.

ANGELIER, J., LYBERIS, N., LE PICHON, X., BARRIER, E. & HUCHON, P. 1982. The tectonic development of the Hellenic arc and the sea of Crete: a synthesis. *Tectonophysics*, **86**, 159–196.

ARMIJO, R., FLERIT, F., KING, G. & MEYER, B. 2003. Linear elastic fracture mechanics explains the past and present evolution of the Aegean. *Earth and Planetary Science Letters*, **217**, 85–95.

BAKER, C., HATZFELD, D., LYON-CAEN, H., PAPADIMITRIOU, E. & RIGO, A. 1997. Earthquake mechanisms of the Adriatic Sea and Western Greece: implications for the oceanic subduction–continental collision transition. *Geophysical Journal International*, **131**, 559–594.

BECKER, D. 2000. *Mikroseismizitaet und Deformation der Kruste Ostkretas*. Diploma thesis, University of Hamburg.

BECKER, D., MEIER, T., RISCHE, M., BOHNHOFF, M. & HARJES, H.-P. 2006. Spatio-temporal microseismicity clustering in the Cretan region. *Tectonophysics*, **423**, 3–16.

BENETATOS, C., KIRATZI, A., PAPAZACHOS, C. & KARAKAISIS, G. F. 2004. Focal mechanisms of shallow and intermediate depth earthquakes along the Hellenic Arc. *Journal of Geodynamics*, **37**, 253–296.

BIJWARD, H. & SPAKMAN, W. 2000. Nonlinear global P-wave tomography by iterated linearised inversion. *Geophysical Journal International*, **141**, 71–82.

BIJWAARD, H., SPAKMAN, W. & ENGDAHL, E. R. 1998. Closing the gap between regional and global travel time tomography. *Journal of Geophysical Research*, **103**, 30055–30078.

BOHNHOFF, M., MAKRIS, J., STAVRAKAKIS, G. & PAPANOKOLAOU, D. 2001. Crustal investigation of the Hellenic subduction zone using wide aperture seismic data. *Tectonophysics*, **343**, 239–262.

BOHNHOFF, M., RISCHE, M., MEIER, T., ENDRUN, B., BECKER, D., HARJES, H.-P. & STAVRAKAKIS, G. 2004. CYC-NET: a temporary seismic network on the Cyclades (Aegean Sea, Greece). *Seismological Research Letters*, **75**, 352–359.

BOHNHOFF, M., HARJES, H.-P. & MEIER, T. 2005. Deformation and stress regimes in the Hellenic subduction zone from focal mechanisms. *Journal of Seismology*, **9**, 341–366.

BOSCHI, E. & MORELLI, A. 1994. The MEDNET program. *Annali di Geofisica*, **37**, 1066–1070.

BRÖNNER, M. 2003. *Untersuchung des Krustenaufbaus entlang des Mediterranen Rückens abgeleitet aus geophysikalischen Messungen*. PhD thesis, University of Hamburg.

CASTEN, U. & SNOPEK, K. 2005. Gravity modelling of the Hellenic subduction zone—a regional study. *Tectonophysics*, **417**, 183–200.

CLOOS, M. & SHREVE, R. L. 1988*a*. Subduction-channel mode of prism accretion, melange formation, sediment subduction, and subduction erosion at convergent plate margins: 1. Background and description. *Pure and Applied Geophysics*, **128**, 455–500.

CLOOS, M. & SHREVE, R. L. 1988*b*. Subduction-channel mode of prism accretion, melange formation, sediment subduction, and subduction erosion at convergent plate margins: 2. Implications and discussion. *Pure and Applied Geophysics*, **128**, 501–545.

COMNINAKIS, P. E. & PAPAZACHOS, B. C. 1980. Space and time distribution of the intermediate focal depth earthquakes in the Hellenic arc. *Tectonophysics*, **70**, 35–47.

DE CHABALIER, J. B., LYON-CAEN, H., ZOLLO, A., DESCHAMPS, A., BERNARD, P. & HATZFELD, D. 1992. A detailed analysis of microearthquakes in western Crete from digital three-component seismograms. *Geophysical Journal International*, **110**, 347–360.

DELIBASIS, N., ZIAZIA, M., VOULGARIS, N., PAPADOPOULOS, T., STAVRAKAKIS, G., PAPANASTASSIOU, D. & DRAKATOS, G. 1999. Microseismic activity and seismotectonics of the Heraklion area (central Crete Island, Greece). *Tectonophysics*, **308**, 237–248.

DERCOURT, J., ZONENSHAIN, L. P., RICOU, L.-E. *ET AL.* 1986. Geological evolution of the Tethys belt from the Atlantic to the Pamirs since the Lias. *Tectonophysics*, **123**, 241–315.

ENDRUN, B., MEIER, T., BISCHOFF, M. & HARJES, H.-P. 2004. Lithospheric structure in the area of Crete constrained by receiver functions and dispersion analysis of Rayleigh phase velocities. *Geophysical Journal International*, **158**, 592–608.

ENDRUN, B., CERANNA, L., MEIER, T., BOHNHOFF, M. & HARJES, H.-P. 2005. Modeling the influence of Moho topography on receiver functions: a case study from the central Hellenic subduction zone. *Geophysical Research Letters*, **32**, doi:10.1029/2005GL023066.

ENGDAHL, E. R., VAN DER HILST, R. & BULAND, R. 1998. Global teleseismic earthquake relocation with improved travel times and procedures for depth determination. *Bulletion of the Seismological Society of America*, **88**, 722–743.

FACCENNA, C., JOLIVET, L., PIROMALLO, C. & MORELLI, A. 2003. Subduction and the depth of convection in the Mediterranean mantle. *Journal of Geophysical Research*, **108**, 2009, doi:10.1029/2001JB001690.

GEALEY, W. K. 1988. Plate tectonic evolution of the Mediterranean–Middle East region. *Tectonophysics*, **155**, 185–306.

GERYA, T. V. & STÖCKHERT, B. 2007. 2-D numerical modeling of tectonic and metamorphic histories at active continental margins. *International Journal of Earth Sciences*, **95**, 250–274.

GERYA, T. V., STÖCKHERT, B. & PERCHUK, A. L. 2002. Exhumation of high-pressure metamorphic rocks in a subduction channel—a numerical simulation. *Tectonics*, **21**, 6-21–6-19.

GIUNCHI, C., KIRATZI, A., SABADINI, R. & LOUVARI, E. 1996. A numerical model of the Hellenic subduction zone: active stress field and sea-level changes. *Geophysical Research Letters*, **23**, 2485–2488.

HANKA, W. & KIND, R. 1994. The GEOFON Program. *Annali di Geofisica*, **37**, 1060–1065.

HATZFELD, D. 1994. On the shape of the subducting slab beneath the Peloponnese, Greece. *Geophysical Research Letters*, **21**, 173–176.

HATZFELD, D. & MARTIN, C. 1992. Intermediate depth seismicity in the Aegean defined by teleseismic data. *Earth and Planetary Science Letters*, **113**, 267–275.

HATZFELD, D., PEDOTTI, G., HATZIDIMITRIOU, P., *ET AL*. 1989. The Hellenic subduction beneath the Peloponnesus: first results of a microearthquake study. *Earth and Planetary Science Letters*, **93**, 283–291.

HATZFELD, D., BESNARD, M., MAKROPOULOS, K., *ET AL*. 1993*a*. Subcrustal microearthquake seismicity and fault plane solutions beneath the hellenic arc. *Journal of Geophysical Research*, **98**, 9861–9870.

HATZFELD, D., BESNARD, M., MAKROPOULOS, K. & HATZIDIMITRIOU, P. 1993*b*. Microearthquake seismicity and fault-plane solutions in the southern Aegean and its geodynamic implications. *Geophysical Journal International*, **115**, 799–818.

HUCHON, P., LYBERIS, N., ANGELIER, J., LE PICHON, X. & RENARD, V. 1982. Tectonics of the Hellenic Trough: a synthesis of a Sea-Beam and submersible observations. *Tectonophysics*, **86**, 69–112.

HUGUEN, C., MASCLE, J., CHAUMILLON, E., WOODSIDE, J. M., BENKHELIL, J., KOPF, A. & VOLKONSKAYA, A. 2001. Deformation styles of the eastern Mediterranean Ridge and surroundings from combined swath mapping and seismic reflection profiling. *Tectonophysics*, **343**, 21–47.

JOLIVET, L. & FACCENNA, C. 2000. Mediterranean extension and the Africa–Eurasia collision. *Tectonics*, **19**, 1095–1106.

KARASON, H. & VAN DER HILST, R. 2000. Constraints on mantle convection from seismic tomography. *In*: RICHARDS, M. A., GORDON, R. G. & VAN DER HILST, R. (eds) *The History and Dynamics of Global Plate Motions*. American Geophysical Union, Washington D.C., 277–288.

KASTENS, K. A. 1991. Rate of outward growth of the Mediterranean Ridge accretionary complex. *Tectonophysics*, **199**, 25–50.

KNAPMEYER, M. 1999. Geometry of the Aegean Benioff zones. *Annali di Geofisica*, **42**, 27–37.

KNAPMEYER, M. & HARJES, H.-P. 2000. Imaging crustal discontinuities and the downgoing slab beneath western Crete. *Geophysical Journal International*, **143**, 1–22.

LAIGLE, M., SACHPAZI, M. & HIRN, A. 2004. Variation of seismic coupling with slab detachment and upper plate structure along the western Hellenic subduction zone. *Tectonophysics*, **391**, 85–95.

LALLEMANT, S., TRUFFERT, C., JOLIVET, L., HENRY, P., CHAMOT-ROOKE, N. & DE VOOGD, B. 1994. Spatial transition from compression to extension in the Western Mediterranean Ridge accretionary complex. *Tectonophysics*, **234**, 33–52.

LAMBECK, K. 1995. Late Pleistocene and Holocene sea-level change in Greece and south-western Turkey: a separation of eustatic, isostatic and tectonic contributions. *Geophysical Journal International*, **122**, 1022–1044.

LE PICHON, X. & ANGELIER, J. 1979. The Hellenic arc and trench system: a key to the neotectonic evolution of the Eastern Mediterranean area. *Tectonophysics*, **60**, 1–42.

LE PICHON, X., CHAMOT-ROOKE, N. & LALLEMANT, S. 1995. Geodetic determination of the kinematics of central Greece with respect to Europe: implications for eastern Mediterranean tectonics. *Journal of Geophysical Research*, **100**, 12675–12690.

LI, X., BOCK, G., VAFIDIS, A., *ET AL*. 2003. Receiver function study of the Hellenic Subduction Zone: imaging crustal thickness variations and the oceanic Moho of the descending African lithosphere. *Geophysical Journal International*, **155**, 733–748.

LIGDAS, C. N., MAIN, I. G. & ADAMS, R. D. 1990. 3-D structure of the lithosphere in the Aegean region. *Geophysical Journal International*, **102**, 219–229.

LYON-CAEN, H., ARMIJO, R., DRAKOPOULOS, J. *ET AL*. 1988. The 1986 Kalamata (south Peloponnesus) earthquake: detailed study of a normal fault and tectonic implications. *Journal of Geophysical Research*, **93**, 14967–15000.

MAKRIS, J., CHONIA, T., STAVRAKAKIS, G. & NIKOLOVA, S. 2000. Two- and three-D active and passive seismic studies of the Nisyros Volcano—East Aegean Sea. *Seismological Research Letters*, **71**, 216–220.

MAKRIS, J., PAPOULIA, J. & DRAKATOS, G. 2004. Tectonic deformation and microseismicity of the Saronikos Gulf, Greece. *Bulletin of the Seismological Society of America*, **94**, 920–929.

MAKROPOULOS, K. C. & BURTON, P. W. 1984. Greek tectonics and seismicity. *Tectonophysics*, **106**, 275–304.

MARONE, F., VAN DER LEE, S. & GIARDINDI, D. 2004. Three-dimensional upper-mantle S-velocity model for the Eurasia–Africa plate boundary region. *Geophysical Journal International*, **158**, 109–130.

MASCLE, J. & CHAUMILLON, E. 1997. Pre-collisional geodynamics of the Mediterranean Sea: the Mediterranean Ridge and the Tyrrhenian Sea. *Annali di Geofisica*, **XL**, 569–586.

MASCLE, J., HUGUEN, C., BENKHELIL, J. *ET AL*. 1999. Images may show start of European–African plate collision. *EOS Transactions, American Geophysical Union*, **80**, 421.

MCCLUSKY, S., BALASSANIAN, S., BARKA, A., *ET AL*. 2000. Global Positioning System constraints on plate kinematics and dynamics in the eastern Mediterranean and Caucasus. *Journal of Geophysical Research*, **105**, 5695–5719.

MCKENZIE, D. P. 1970. The plate tectonics of the Mediterranean region. *Nature*, **226**, 239–243.

McKenzie, D. P. 1972. Active tectonics of the Mediterranean region. *Geophysical Journal of the Royal Astronomical Society*, **30**, 109–185.

McKenzie, D. P. 1978. Active tectonics of the Alpine–Himalayan belt: the Aegean Sea and surrounding regions. *Geophysical Journal of Royal Astronomical Society*, **55**, 217–254.

Meier, T., Dietrich, K., Stöckhert, B. & Harjes, H.-P. 2004*a*. One-dimensional models of shear-wave velocity for the eastern Mediterranean obtained from the inversion of Rayleigh wave phase velocities and tectonic implications. *Geophysical Journal International*, **156**, 45–58.

Meier, T., Rische, M., Endrun, B., Vafidis, A. & Harjes, H.-P. 2004*b*. Seismicity of the Hellenic Subduction Zone in the area of western and central Crete observed by temporary local seismic networks. *Tectonophysics*, **383**, 149–169.

Meulenkamp, J. E., van der Zwaan, G. J. & van Wamel, W. A. 1994. On Late Miocene to recent vertical motions in the Cretan segment of the Hellenic arc. *Tectonophysics*, **234**, 53–72.

Oleskevich, D. A., Hyndman, R. D. & Wang, K. 1999. The updip and downdip limits to great subduction earthquakes; thermal and structural models of Cascadia, South Alaska, SW Japan, and Chile. *Journal of Geophysical Research*, **104**, 14965–14991.

Papazachos, B. C. & Comninakis, P. E. 1971. Geophysical and tectonic features of the Aegean arc. *Journal of Geophysical Research*, **76**, 8517–8533.

Papazachos, B. C., Papaioannou, C. A., Papazachos, C. B. & Savvaidis, A. S. 1999. Rupture zones in the Aegean region. *Tectonophysics*, **308**, 205–221.

Papazachos, B. C., Karakostas, V. G., Papazachos, C. B. & Scordilis, E. M. 2000. The geometry of the Wadati–Benioff zone and lithospheric kinematics in the Hellenic arc. *Tectonophysics*, **319**, 275–300.

Papazachos, B. C., Comninakis, P. E., Karakaisis, G. F., Karakostas, B. G., Papaioannou, C. A., Papazachos, C. B. & Scordilis, E. M. 2003. *A Catalogue of Earthquakes in Greece and the Surrounding Area for the Period 550BC–2002*. Geophysical Laboratory, University of Thessaloniki.

Papazachos, B. C., Dimitriadis, S. T., Panagiotopoulos, D. G., Papazachos, C. B. & Papadimitriou, E. E. 2004. Deep structure and active tectonics of the southern Aegean volcanic arc. *In*: Fytikas, M. (ed.) *The Southern Aegean Active Volcanic Arc: Present Knowledge and Future Perspectives*. Developments in Volcanology, **7**, 47–64.

Papazachos, C. B. & Nolet, G. 1997. P and S deep velocity structure of the Hellenic area obtained by robust nonlinear inversion of travel times. *Journal of Geophysical Research*, **102**, 8349–8367.

Papazachos, C. B., Hatzidimitriou, P. M., Panagiotopoulos, D. G. & Tsokas, G. N. 1995. Tomography of the crust and upper mantle in southeast Europe. *Journal of Geophysical Research*, **100**, 12405–12422.

Pirazzoli, P. A., Thommeret, J., Thommeret, Y., Laborel, J. & Montaggioni, L. F. 1982. Crustal block movements from Holocene shorelines: Crete and Antikythira (Greece). *Tectonophysics*, **86**, 27–43.

Piromallo, C. & Morelli, A. 1997. Imaging the Medditerranean upper mantle by P-wave travel time tomography. *Annali di Geofisica*, **XL**, 963–979.

Piromallo, C. & Morelli, A. 2003. P wave tomography of the mantle under the Alpine–Mediterranean area. *Journal of Geophysical Research*, **108**, 2065.

Rubatto, D. & Hermann, J. 2001. Exhumation as fast as subduction? *Geology*, **29**, 3–6.

Sachpazi, M., Hirn, A., Clement, C. *et al.* 2000. Western Hellenic subduction and Cephalonia transform: local earthquakes and plate transport and strain. *Tectonophysics*, **319**, 301–319.

Savage, J. C. 1983. A dislocation model of strain accumulation and release at a subduction zone. *Journal of Geophysical Research*, **88**, 4984–4996.

Spakman, W., Wortel, M. J. R. & Vlaar, N. J. 1988. The Hellenic subduction zone: A tomographic image and its geodynamic implications. *Geophysical Research Letters*, **15**, 60–63.

Spakman, W., van der Lee, S. & van der Hilst, R. 1993. Travel-time tomography of the European–Mediterranean mantle down to 1400 km. *Physics of the Earth and Planetary Interiors*, **79**, 3–74.

Stampfli, G. M. & Borel, G. 2004. The TRANSMED transects in space and time: constraints on the paleotectonic evolution of the Mediterranean Domain. *In*: Cavazza, W., Roure, F. M., Spakman, W., Stampfli, G. M. & Ziegler, P. A. (eds) *The TRANSMED Atlas—The Mediterranean Region from Crust to Mantle*. Springer, Berlin, 53–80.

Stiros, S. C. 2001. The AD 365 Crete earthquake and possible seismic clustering during the fourth to sixth centuries AD in the Eastern Mediterranean: a review of historical and archaeological data. *Journal of Structural Geology*, **23**, 545–562.

Taymaz, T. 1996. S–P-wave traveltime residuals from earthquakes and lateral heterogeneity in the upper mantle beneath the Aegean and the Hellenic Trench near Crete. *Geophysical Journal International*, **127**, 545–558.

Taymaz, T., Jackson, J. & Westaway, R. 1990. Earthquake mechanisms in the Hellenic Trench near Crete. *Geophysical Journal International*, **102**, 695–731.

ten Veen, J. H. & Kleinspehn, K. L. 2003. Incipient continental collision and plate-boundary curvature: Late Pliocene–Holocene transtensional Hellenic forearc, Crete, Greece. *Journal of the Geological Society, London*, **160**, 161–181.

ten Veen, J. H. & Meijer, P. T. 1998. Late Miocene to Recent tectonic evolution of Crete (Greece): geological observations and model analysis. *Tectonophysics*, **298**, 191–208.

ten Veen, J. H. & Postma, G. 1999. Roll-back controlled vertical movements of the outer-arc basins of the Hellenic subduction zone. *Basin Research*, **11**, 243–266.

THOMSON, S. N., STÖCKHERT, N. & BRIX, M. R. 1998. Thermochronology of the high-pressure metamorphic rocks of Crete, Greece: implications for the speed of tectonic processes. *Geology*, **26**, 259–262.

THOMSON, S. N., STÖCKHERT, N. & BRIX, M. R. 1999. Miocene high-pressure metamorphic rocks of Crete, Greece: rapid exhumation by bouyant escape. *In*: RING, U., BRANDON, M. J., LISTER, G. S. & WILLETT, S. D. (eds) *Exhumation Processes: Normal Faulting, Ductile Flow and Erosion*. Geological Society, London, Special Publications, **154**, 87–107.

TICHELAAR, B. & RUFF, L. 1993. Depth of seismic coupling along subduction zones. *Journal of Geophysical Research*, **98**, 2017–2037.

TRUFFERT, C., CHAMOT-ROOKE, N., LALLEMANT, S., DE VOOGD, B., HUCHON, P. & LE PICHON, X. 1993. The crust of the Western Mediterranean Ridge from deep seismic data and gravity modelling. *Geophysical Journal International*, **114**, 360–372.

VAN HINSBERGEN, D. J. J., HAFKENSCHEID, E., SPAKMAN, W., MEULENKAMP, J. E. & WORTEL, R. 2005. Nappe stacking resulting from subduction of oceanic and continental lithosphere below Greece. *Geology*, **33**, doi:10.1130/G20878.1, 325–328.

WORTEL, M. J. R. & SPAKMAN, W. 2000. Subduction and slab detachment in the Mediterranean–Carpathian region. *Science*, **290**, 1910–1917.

Understanding tsunamis, potential source regions and tsunami-prone mechanisms in the Eastern Mediterranean

S. YOLSAL[1], T. TAYMAZ[1] & A. C. YALÇINER[2]

[1]*Department of Geophysical Engineering, Seismology Section, the Faculty of Mines, Istanbul Technical University, Maslak TR-34469, Istanbul, Turkey (e-mail: yolsalse@itu.edu.tr)*

[2]*Department of Civil Engineering, Ocean Engineering Research Centre, Middle East Technical University, TR-06531, Ankara, Turkey*

Abstract: Historical tsunamis and tsunami propagation are synthesized in the Eastern Mediterranean Sea region, with particular attention to the Hellenic and the Cyprus arcs and the Levantine basin, to obtain a better picture of the tsunamigenic zones. Historical data of tsunami manifestation in the region are analysed, and compared with current seismic activity and plate interactions. Numerical simulations of potential and historical tsunamis reported in the Cyprus and Hellenic arcs are performed as case studies in the context of the nonlinear shallow-water theory. Tsunami wave heights as well as their distribution function are calculated for the Paphos earthquake of 11 May 1222 and the Crete earthquake of 8 August 1303 as illustrative examples depicting the characteristics of tsunami propagation, and the effects of coastal topography and near-shore amplification. The simulation studies also revealed that the long-normal distributions are compatible with reported damage. Furthermore, it is necessary to note that high-resolution bathymetry maps are a crucial component in tsunami wave simulations, and this aspect is rather poorly developed in the Eastern Mediterranean. The current study also demonstrates the role of bottom irregularities in determining the wave-height distribution near coastlines. Assuming the probability of occurrence of destructive tsunamigenic earthquakes, these studies will help us to evaluate the tsunami hazard for the coastal plains of the Eastern Mediterranean Sea region. We suggest that future oceanographic and marine geophysical research should aim to improve the resolution of bathymetric maps, particularly for the details of the continental shelf and seamounts.

The complexity of plate interactions and associated crustal deformation in the Eastern Mediterranean region is reflected in many destructive earthquakes that have occurred throughout its recorded history, many of which are well documented and studied. Catastrophic tsunamis have also been observed at most of the European coasts. The Eastern Mediterranean region, including the surrounding areas of western Turkey and Greece, is one of the most seismically active and rapidly deforming regions in the world. Thus, the region provides an excellent natural laboratory and offers a unique opportunity to improve our understanding of the complexities of continental tectonics in an active collisional orogen (Taymaz *et al.* 2004; Fig. 1). The major scientific observations from this natural laboratory have clearly helped us to better understand the tectonic processes in active collision zones, the mode and nature of continental growth, and the causes and distribution of seismic, volcanic and geomorphological events (e.g. tsunamis) and their impact on humans and civilization. A tsunami is a very large ocean- or sea-wave triggered by various large-scale disturbances of the ocean floor such as submarine earthquakes, volcanic activities or landslides. These waves have unusually long wavelength, in excess of 100 km, generated in the open ocean and transformed into a series of catastrophic oscillations on the sea surface close to coastal zones. At the vicinity of the earthquake source, multiple reflections owing to deep basins and partial wave trapping because of the complex sea-bottom morphology generate complicated wave-train patterns, and it is difficult to determine whether these are related to the source or path effects. On the other hand, there is a long record of tsunami occurrences and damaging tsunamis observed repeatedly in the oceans and seas. Future tsunamis could be even more catastrophic than past events, as a result of the increasing occupation of the coasts with the economic development of coastal countries in recent decades. Furthermore, protection from natural disasters and mitigation of their effects on the environment and societies are becoming more important issues throughout the world.

The seismicity of the Aegean and the Mediterranean regions in general has been recorded from the ancient world to the end of the Middle Ages by an

From: TAYMAZ, T., YILMAZ, Y. & DILEK, Y. (eds) *The Geodynamics of the Aegean and Anatolia.* Geological Society, London, Special Publications, **291**, 201–230.
DOI: 10.1144/SP291.10 0305-8719/07/$15.00

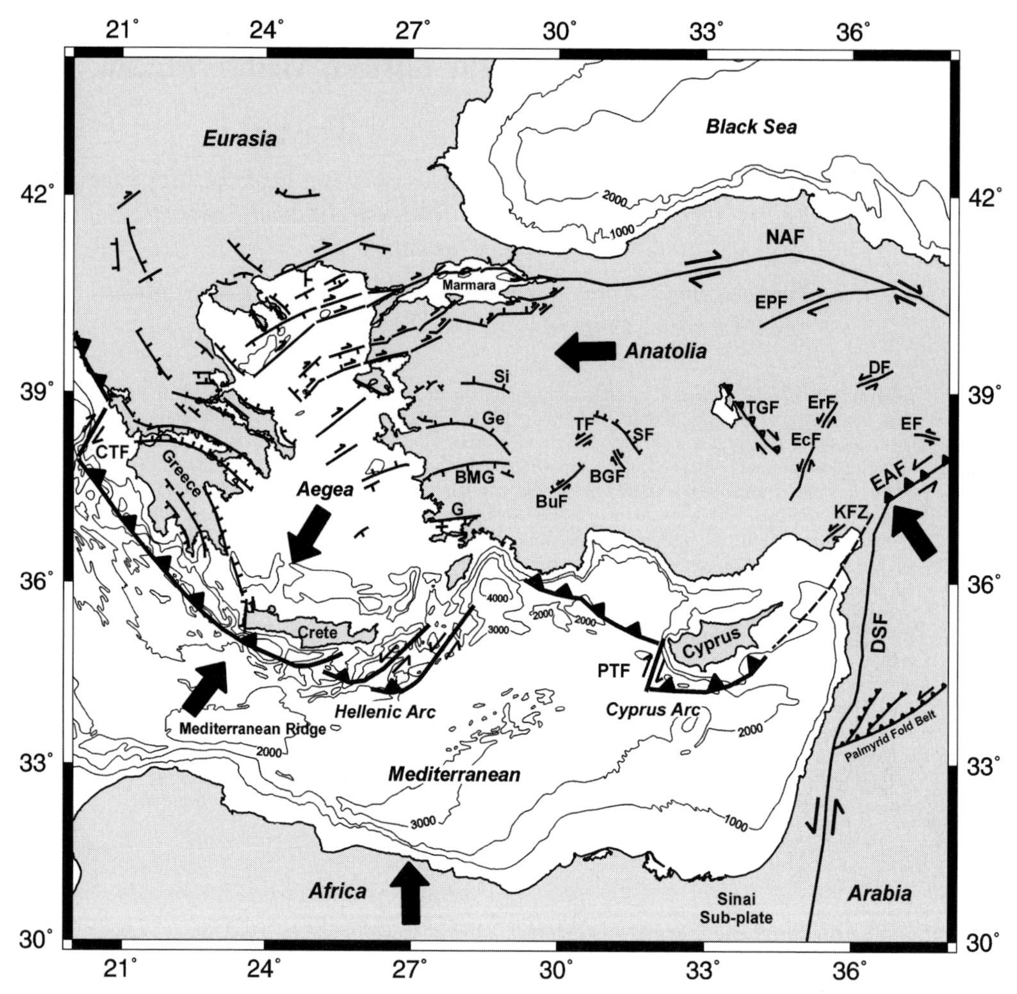

Fig. 1. Summary sketch map of the faulting and bathymetry in the Eastern Mediterranean Sea region, compiled from our observations and those of Le Pichon *et al.* (1984), Taymaz *et al.* (1990, 1991), Şaroğlu *et al.* (1992), Papazachos *et al.* (1998) and McClusky *et al.* (2000). NAF, North Anatolian Fault; EAF, East Anatolian Fault; DSF, Dead Sea Fault; EPF, Ezinepazarı Fault; PTF, Paphos Transform Fault; CTF, Cephalonia Transform Fault; G, Gökova; BMG, Büyük Menderes Graben; Ge, Gediz Graben; Si, Simav Graben; BuF, Burdur Fault; BGF, Beyşehir Gölü Fault; TF, Tatarlı Fault; SF, Sultandağ Fault; TGF, Tuz Gölü Fault; EcF, Ecemiş Fault; ErF, Erciyes Fault; DF, Deliler Fault; EF, Elbistan Fault; KFZ, Karataş–Osmaniye Fault Zone (see also Taymaz *et al.* 2007). Large black arrows show relative motions of plates with respect to Eurasia (McClusky *et al.* 2000, 2003). Bathymetric contours are shown at 1000 m interval, and are from GEBCO (1997) and Smith & Sandwell (1997*a*, *b*).

extraordinary wealth of written and epigraphic sources and this historical heritage is one of the most precious in the world. Thus, for the present study historical earthquakes and associated tsunamis are identified from verified catalogues (e.g. Guidoboni *et al.* 1994; Ambraseys & Melville 1995; Guidoboni & Comastri 2005*a*, *b*; Sbeinati *et al.* 2005). Understanding of the geometry and evolution of potential source (seismogenic) regions and the source rupture process along active zones has crucial implications for tsunami generation. Around the Mediterranean Sea, Marmara Sea and Black Sea there is a high potential risk for generation of tsunamis, and the most destructive events have occurred along the coasts of Portugal, Italy, Greece and Turkey. The impact of tsunamis on human societal life can be traced back in written history to late Minoan time (1600–1300 BC) in the Eastern Mediterranean, when the strongest tsunami caused by the volcanic eruption of Santorini resulted in degradation of the Minoan civilization, and it has been further concluded that this eruption and the following tsunami were widely observed on the coastal plains of western Turkey and Crete (Minoura *et al.* 2000). Hence, tsunamis have caused severe damage and flooded lowlands in many segments of the Mediterranean coasts. Historical documents, and geological, archaeological and many trench studies demonstrate that parts of the Turkish coastlines have suffered from disastrous sea-waves several times in the past (Yalçıner *et al.* 2002, 2004; Boschi *et al.* 2005; Guidoboni & Comastri 2005*a*, *b*; Scheffers & Kelletat 2005; Fokaefs & Papadopoulos 2006; Papadopoulos *et al.* 2007). The style of seismic deformation along the Cyprus and Hellenic arcs exhibits the characteristics and structural complexities associated with strike-slip, thrust and normal faulting as a result of convergence between the Aegean, Anatolia, Eurasia and Eastern Mediterranean lithosphere. Existing observations and inferred seismological results indicate that there are younger tectonic, slope failure features within the lower part of the continental slope that are tsunamis-prone locations along the neighbouring coastlines of the Mediterranean Sea

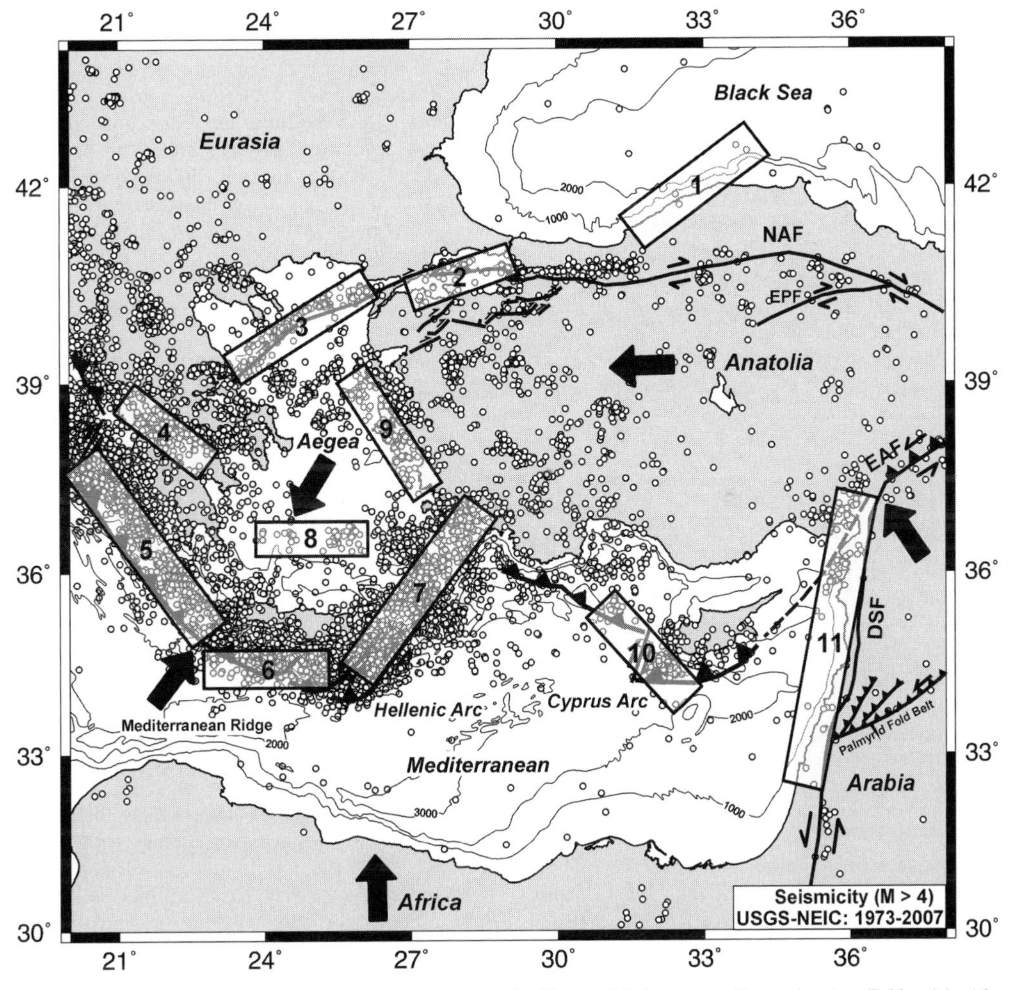

Fig. 2. A sketch map of the known tsunamigenic zones in the Eastern Mediterranean Sea region (see Tables A1–A3 for details). Rectangular boxes with numbers refer to the regions discussed in the text: 1, Bartın–Amasra shelf, SW Black Sea; 2, Sea of Marmara; 3, North Aegean trough; 4, Gulf of Corinth; 5, western Hellenic arc; 6, south of Crete; 7, eastern Hellenic arc; 8, the Cyclades; 9, Seferihisar-Kuşadası, W Turkey; 10, SW of Cyprus arc; 11, Dead Sea Fault Zone and Levantine Sea. Large black arrows show relative motions of plates with respect to Eurasia (McClusky *et al.* 2000, 2003). Bathymetric contours are shown at 1000 m interval, and are from GEBCO (1997). Seismicity of the region reported by USGS–NEIC during 1973–2007 for M > 4 is shown by small open circles. The display convention of major plate boundaries is the same as in Figure 1.

(Makris & Stobbe 1984; Taymaz *et al.* 1990, 1991; ten Veen *et al.* 2004; Yolsal & Taymaz 2004, 2005; Fig. 2). In addition to historical and geological information, and the distribution of active fault zones, volcanoes and other probable tsunami-prone sea-bottom morphological structures, there are numerous source regions that may be considered responsible for severe tsunamis. However, major tsunami recurrence in the Eastern Medierranean region is of the order of several decades and the memory of tsunamis is short lived. Thus, the compilation of reliable tsunami databases is of great importance for a wide range of tsunami research (e.g. statistics and hazard assessment, numerical modelling, risk assessment, early warning operations, public awareness).

Hence, we aim to concentrate on tsunami risk mapping for regions where no severe tsunami has occurred recently, but the geomorphological and topographic features, and the geodynamic and seismotectonic settings are similar to those of areas devastated by recent catastrophic tsunamis such as Sumatra (Barber *et al.* 2005) and where reliable historical records of tsunamis are available (Guidoboni & Comastri 2005*a*, *b*). In this paper, tsunami events known to have occurred in the Eastern Mediterranean Sea region are summarized, and synthetic tsunami simulations are presented as case studies to demonstrate preliminary tsunami risk estimates. Thus, this study deals only with earthquake-induced tsunamis.

Quantification of tsunamis

There have been considerable efforts early the 1930s to improve quantification of tsunamis observed globally (Sieberg 1932). However, this is still a puzzling aspect in tsunami research, as several intensity scales have been proposed to measure tsunami size (i.e. intensity and/or magnitude). On the other hand, earthquake magnitude is an objective physical parameter that defines the energy release radiated at the centroid, and does not directly reflect macroseismic effects although variable intensities at different geological locations can be observed. Nevertheless, Okal (1988) has already studied in detail the influence of the seismic source parameters (e.g. focal depth, source mechanism (geometry of faulting), seismic moment and directivity) on the generation of a tsunami, using the modal approach for laterally homogeneous, structural models. These boundary conditions certainly play an important role in defining the amplitude of tsunami waves, but so too do the other essential key parameters (i.e. the effects of the directivity as a result of rupture propagation along a fault, and enhanced tsunami excitation in a medium with weaker elastic properties, such as sedimentary layers). Thus, the quantification of tsunamis could easily be approached by analogy to general aspects of earthquake seismology (e.g: Abe 1979, 1981, 1985; 1989; Murty & Loomis 1980; Gasperini & Ferrari 2000; Papadopoulos & Fokaefs 2005; Papadopoulos & Satake 2005). In the Mediterranean region, tsunami intensity (*k*) is traditionally estimated using the Sieberg–Ambraseys scale (Ambraseys 1962), and tsunami magnitude (M_T or M_L) is generally calculated using the analytical formulae that have been developed (e.g. Murty & Loomis 1980). However, it should also be noted that the tsunami magnitude scales are usually based on direct measurements of tsunami wave heights at tide gauges located near coastlines. On the other hand, there are other effective parameters in tsunami generation such as coastal topography, variations in near-shore bathymetry, and the reflection, refraction, diffraction and resonance of sea-waves, to name a few. Thus, it is desirable to have a better calibration of analytical formulae based on both the quality and quantity of instrumental tide-gauge measurements.

Analysis of historical tsunamis in the Eastern Mediterranean

The descriptions of historical tsunamigenic earthquakes in the Eastern Mediterranean region are provided in valuable catalogues in various languages and recently compiled by Guidoboni & Comastri (2005*a*), who analysed sources in several languages (Greek, Latin, Arabic, Hebrew, Armenian, Italian, French, German, Ottoman and modern Turkish, etc.). Although the accounts of events gathered in catalogues cannot necessarily be described as definitive owing to the nature of research considered, they are very valuable documentation for researchers. Of course, the catalogues also contain much information on earthquakes, tsunamis, environmental effects, stories related to societal life and even religious belief, which should be carefully used by cross-correlating with other sources of information to gather self-consistent and complete datasets. Thus, in this study we have summarized tsunamigenic earthquakes in groups in the Appendix (Tables A1–A3) with relevant references.

In Figure 3, the locations of tsunamigenic earthquakes reported during the 11th–15th centuries in the Eastern Mediterranean region are plotted and are correlated with seismically active regions (see Table A1; Guidoboni & Comastri 2005*a*). Locations of tsunamigenic earthquakes

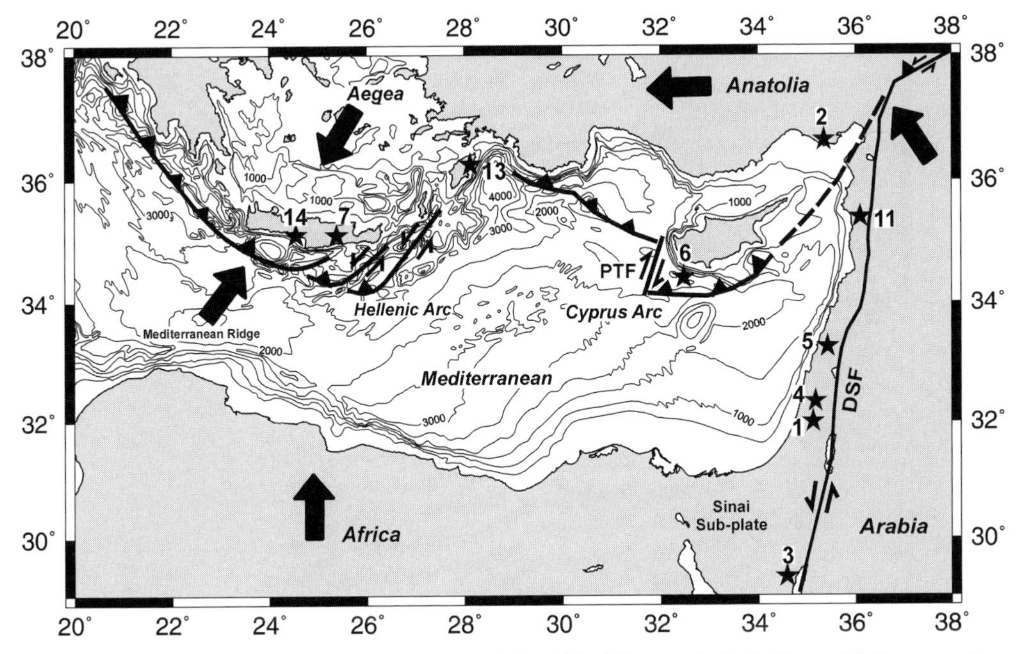

Fig. 3. Reported locations of tsunamigenic earthquakes during 11th–15th centuries in the Eastern Mediterranean Sea region (Guidoboni & Comastri 2005*a*). Large black arrows show relative motions of plates with respect to Eurasia (McClusky *et al.* 2000, 2003). Bathymetric contours are shown at 500 m interval, and are from GEBCO (1997). The display convention of major plate boundaries is the same as in Figure 1. Numbers refer to tsunamigenic events in Table A1.

reported in the Levantine basin and surrounding regions from 1365 BC to AD 1900 are shown in Figure 4, and a selected table of historical tsunamigenic earthquakes with estimated intensities at relevant locations and accompanying geomorphological effects are further summarized in Table A2 (see Sbeinati *et al.* 2005). It can be seen that there are about a dozen or so strong tsunami events in the Eastern Mediterranean, which reflects an apparent recurrence interval of about 150–200 years. It is also evident that tsunamigenic events are associated mainly either with seismogenic zones or with the active volcanic complex of Thera and seamounts of the Eastern Mediterranean Sea. However, some of the damaging historical tsunamis (e.g. 1303 and 1481) in the eastern Hellenic arc also threatened the coastal plains of the Cyprus, the Levantine and Alexandria–Nile Delta (Egypt) regions, and thus special care should be taken in evaluating of the tsunami risk of the region.

In Figure 5, locations of tsunamigenic earthquakes reported along the Hellenic arc and trench system are plotted, and they are correlated with relevant seismogenic zones (see Table A3; Papadopoulos *et al.* 2007).

Synthetic tsunami simulations

Geodynamic and seismotectonic setting

The Eastern Mediterranean Sea region is seismically active and its geodynamic and seismotectonic setting is mainly dominated by the Hellenic and the Cyprus arcs, the left-lateral strike-slip Dead Sea fault and the Levantine rift (Fig. 6). There are many historical documents available to correlate earthquakes and tsunamis along these seismogenic zones (e.g. Guidoboni *et al.* 1994; Ambraseys & Melville 1995; Papazachos *et al.* 1999; Guidoboni & Comastri 2005*a*, *b*; Sbeinati *et al.* 2005; Fokaefs & Papadopoulos 2006). In the Hellenic and Cyprus arcs crustal and intermediate-depth dip-slip faulting earthquakes often occur mostly in the submarine environment, and therefore damaging events are expected to generate strong tsunamis by co-seismic displacement. However, there are some cases where the generation mechanism of locally strong tsunamis associated with earthquakes reported clearly on land along the strike-slip Dead Sea fault and the Levantine rift remains unexplained. One possibility could be triggered slumping of unstable sediments on the shelf and/or

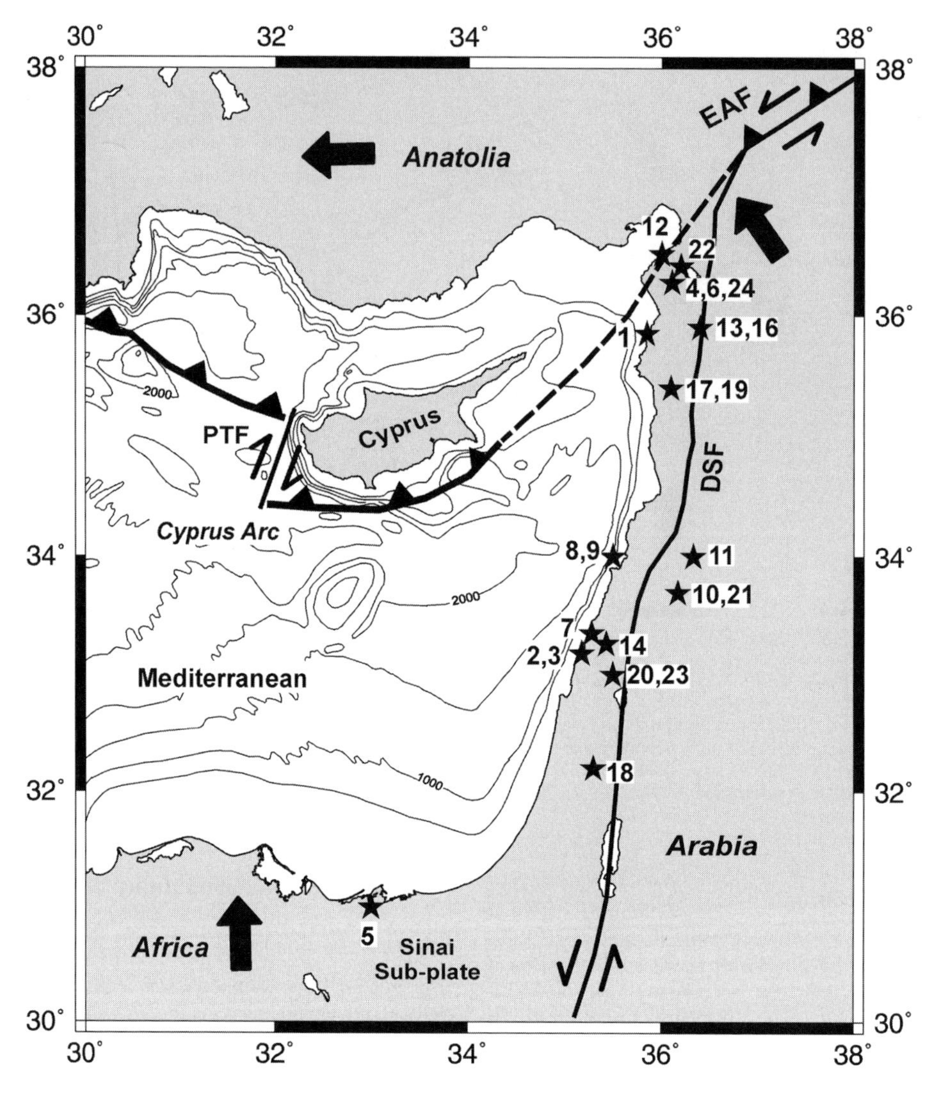

Fig. 4. Reported locations of tsunamigenic earthquakes in the Levantine basin and surrounding regions from 1365 BC to AD 1900 (Sbeinati *et al.* 2005). Large black arrows indicate relative motions of plates with respect to Eurasia (McClusky *et al.* 2000, 2003). Bathymetric contours are shown at 500 m interval, and are from GEBCO (1997). The display convention of major plate boundaries is the same as in Figure 1. Numbers refer to tsunamigenic events in Table A2.

propagating line-source and directivity effects of the rupture. It should also be noted that none of the reported tsunamis associated with seismic activity in the Dead Sea fault region propagated large distances. This may be due to very strong sea-wave attenuation, which is a known feature of landslide-generated tsunamis. In the Cyprus arc, earthquake activity is clearly recognized at offshore seismogenic zones west and SW of Cyprus, where tsunami generation by coseismic displacement in the submarine environment is possible (Figs 4 and 6). Therefore, tsunami potential and hazard in the Cyprus–Levantine region should not be neglected when considering the effects of the 1303 and 1481 tsunamis in the Hellenic arc (Tables A1–A3; Guidoboni & Comastri 2005*a*). In summary, the kinematics of the active sutures and plate boundaries and associated secondary structures in the Eastern Mediterranean Sea region is capable of generating damaging tsunamis.

In Figure 6, the seismicity of the Cyprus arc reported by the USGS–NEIC during 1973–2007

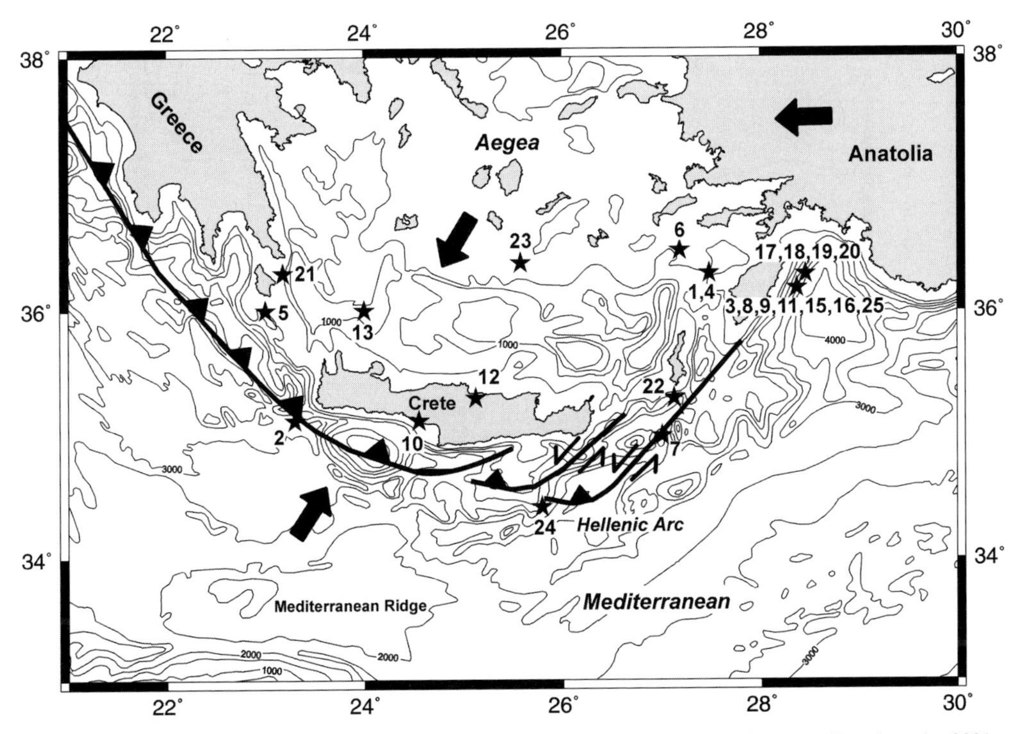

Fig. 5. Reported locations of tsunamigenic earthquakes along the Hellenic arc and trench system (Papadopoulos 2001; Papadopoulos *et al.* 2007). Large black arrows show relative motions of plates with respect to Eurasia (McClusky *et al.* 2000, 2003). Bathymetric contours are shown at 500 m interval, and are from GEBCO (1997). Major plate boundaries are as shown in Figure 1. Numbers refer to tsunamigenic events in Table A3.

for $M > 3$ is shown, including bathymetry data provided by GEBCO (1997). The current seismic activity is mainly concentrated on the southern flanks of the Troodos massif, and south and SW of Cyprus along the Paphos transform fault (PTF). The details of the clusters at crustal and intermediate depths and their source rupture properties have been described by Yolsal & Taymaz (2004, 2005). It is obvious from the distribution of histograms that not many large earthquakes ($M > 7$) have occurred in the region for about 30 years or so. This is not surprising when the recurrence periods in the historical catalogues of earthquakes and tsunamis are considered. Nevertheless, this region is capable of generating tsunamis by coseismic fault displacements provided that large earthquakes occur at relatively shallow depths.

The nonlinear shallow-water theory

Tsunamis are mainly generated by the sudden movement of the sea bottom as a result of submarine earthquakes, which causes long sea-waves for which the vertical acceleration of water particles is negligible compared with gravitational acceleration. The curvature of trajectories of water particles is relatively small except for oceanic propagation of a tsunami. Consequently, the vertical motion of water particles has no effect on the pressure distribution. Thus, it is a good approximation to assume that the pressure is hydrostatic. Furthermore, tsunami waves travel outwards in all directions from the source area, with the direction of the main energy propagation generally being orthogonal to the direction of the earthquake rupture zone, at various speeds depending on the depth of water propagated. Near shorelines, the tsunami-wave speed slows to just a few tens of kilometres per hour; however, the height of the waves increases to tens of metres. For the propagation of tsunami waves in shallow water, the horizontal eddy turbulence can be negligible compared with bottom friction except for the run-up on land. Therefore, it can be assumed that horizontal velocities of water particles are vertically uniform. In recent years, numerical models have been developed to simulate tsunami waves and their

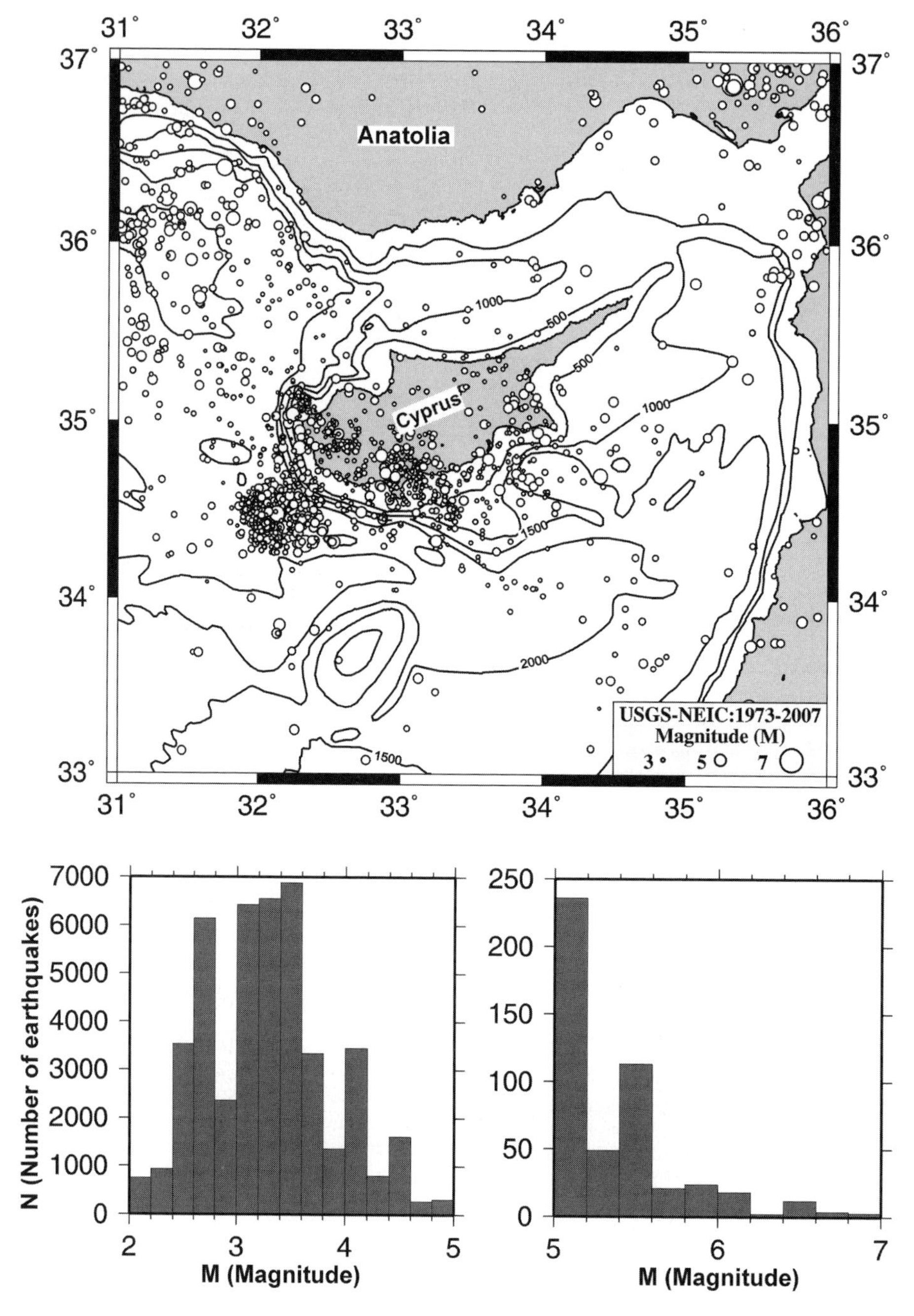

Fig. 6. Seismicity of the Cyprus arc and surroundings reported by USGS–NEIC during 1973–2007 scaled with respect to magnitudes for M > 3. Bathymetry data are derived from GEBCO/97-BODC, provided by GEBCO (1997) and Smith & Sandwell (1997*a*, *b*).

interaction with land masses based on long-wave equations with respect to related initial and boundary conditions. In the present study, the nonlinear shallow-water mathematical models TUNAMI-N2, AVI-NAMI and NAMI-DANCE are used to simulate the propagation of tsunami waves in the form

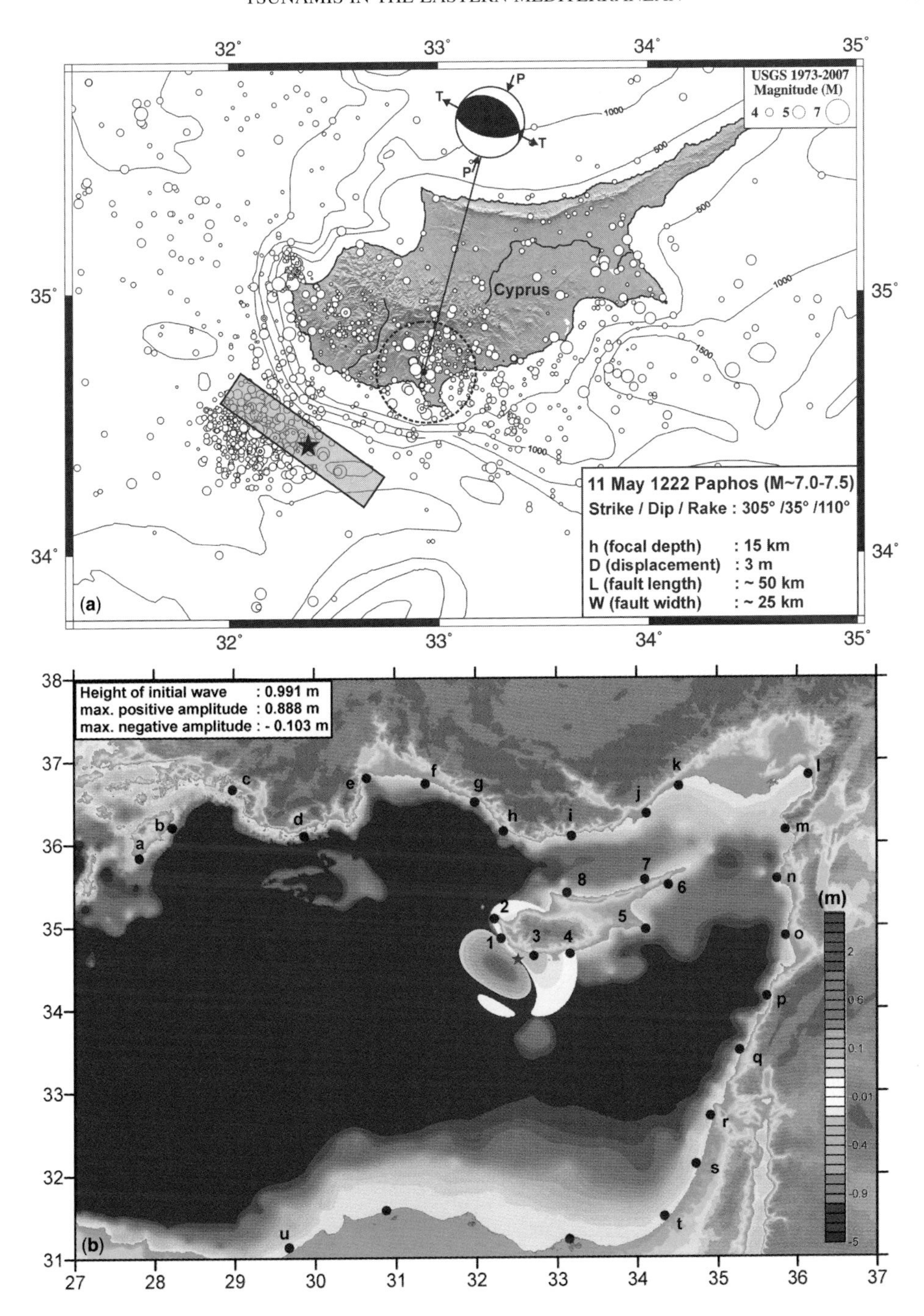

Fig. 7. (**a**) Seismicity of the Cyprus arc and surroundings as in Figure 6. Box with star refers to approximate location of the 11 May 1222 Paphos earthquake ($M \sim 7.5$) and tsunami, for which simulations are generated by applying an analogous representative earthquake derived from teleseismic P- and SH-wave modelling studies of current earthquakes. The parameters of tsunami simulations are shown in the box in the lower right corner, and a representative fault-plane solution obtained from current earthquake source mechanisms is shown above in a lower hemisphere projection. (**b**) Snapshot of the initial tsunami generated with the parameters given in (a). Letters and numbers refer to geographical locations where macroseismic observations have been partly reported (Guidoboni & Comastri 2005*a*), and synthetic tsunami mareograms generated at pseudo-tide-gauge locations, and details are shown in Figure 8a and b.

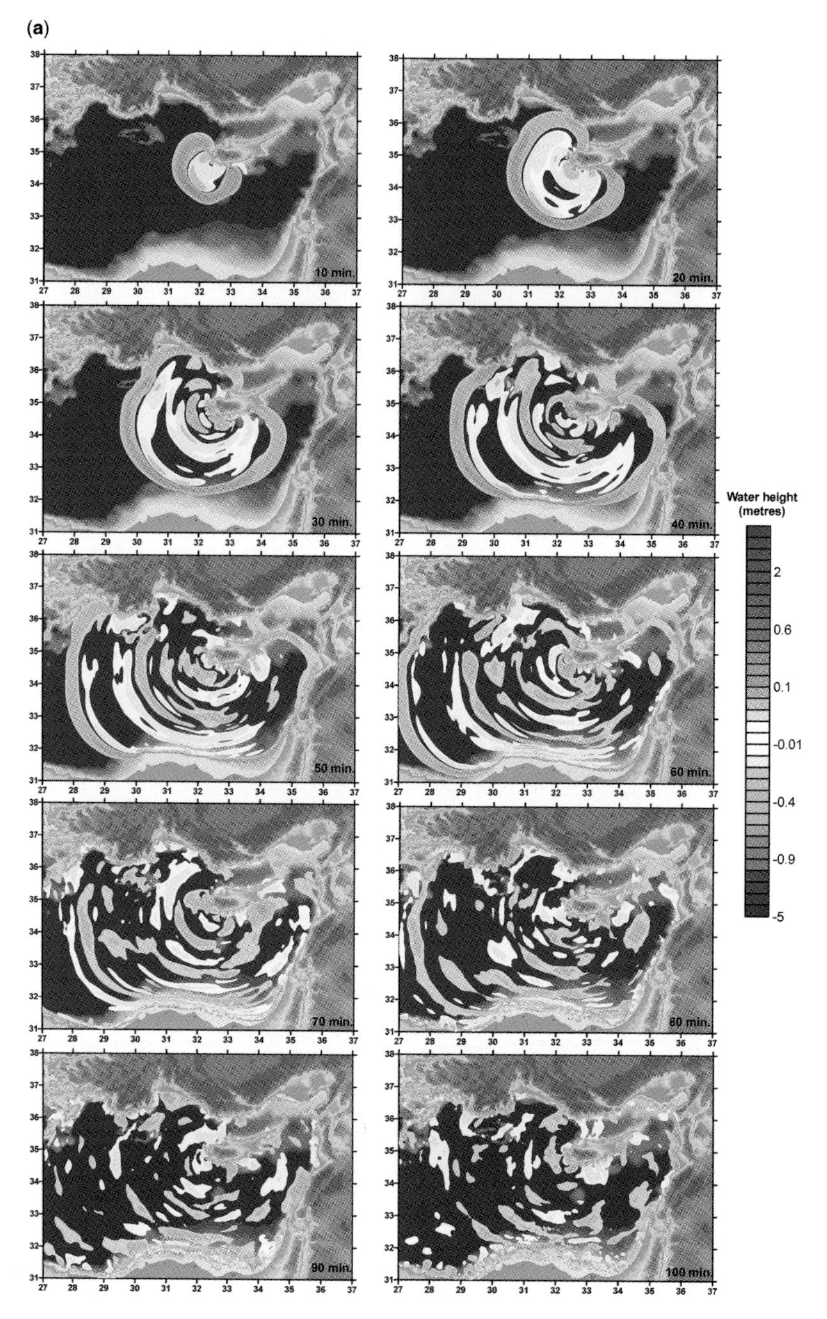

Fig. 8. (**a**) Snapshots of the tsunami wave propagation generated using nonlinear shallow-water theory for various times at 10 min intervals for the 11 May 1222 Paphos event. The water surface height is in metres and a colour scale is given on the right-hand side in metres. It should be noted that the bathymetric features of the Nile Delta (Fig. 7*b*) act as a natural barrier to slow down tsunami waves at shallower depths (<500 m) by refractions and diffractions of sea-waves. If we compare synthetic tsunami mareograms at 30, 40, 50 and 60 min, anomalous sea-wave heights are obvious at location *u* in (**b**) (Alexandria, Egypt), where there is *c.* 1 m mareogram amplitude. This could be due to a deep basin and sudden shelf break clearly reflected in the bathymetry. However, no significant sea-wave amplitudes were calculated at locations marked in (b) with filled circles and without letters. (b) Computed tsunami records at selected locations for the 11 May 1222 Paphos event. The vertical and horizontal scales show water surface elevation (wse) in centimetres and tsunami simulation time (t) in minutes, respectively. Above each tsunami mareogram, the maximum synthetic wave-heights (H) and theoretical arrival times (T) are given in centimetres and minutes, respectively.

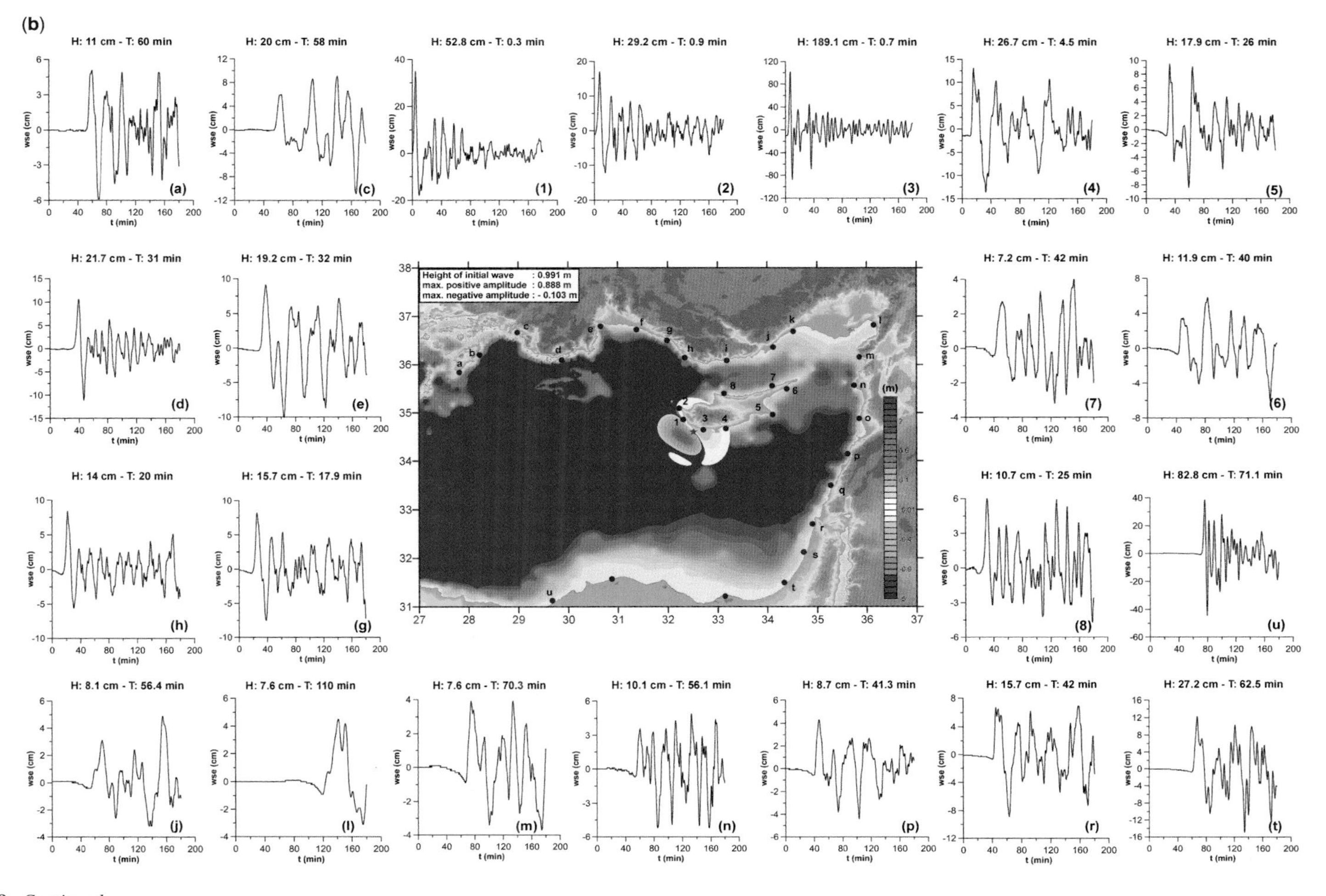

Fig. 8. *Continued.*

of Saint-Venant equations (Shuto *et al.* 1990; Pelinovsky *et al.* 2001; Zahibo *et al.* 2003; Yalçıner & Pelinovsky 2007).

Case studies

Paphos, Cyprus, 11 May 1222 (06:15 UT, latitude 34°42′N, longitude 32°48′E, Io = IX, Me ~ 7.0–7.5). In this section we present a case study of the 11 May 1222 Paphos, Cyprus earthquake and tsunami for which synthetic tsunami simulations are generated to analyse the importance of joint examination of earthquake source mechanism and tsunami simulation studies (Figs 6–8). The Paphos, Cyprus earthquake and related tsunami of 11 May 1222 is one of the most destructive events reported in many historical catalogues (Ergin *et al.* 1967; Ambraseys *et al.* 1994; Guidoboni & Comastri 2005*a*; Table A1). The towns of Limasol, Paphos and Nicosia were severely affected, especially Paphos, where the castle collapsed and there were many victims. The earthquake was also felt in regions as far as Alexandria (Egypt), and the harbour at Paphos was left completely without water.

The choice of the tsunami source is usually a complicated problem because it requires a good knowledge of the earthquake rupture mechanism. The related parameters for the 11 May 1222 Paphos event are adapted by analogy to current plate boundaries and earthquake source mechanisms obtained by inversion of teleseismic P- and SH-waveforms (Yolsal & Taymaz 2004, 2005). We have assumed that the initial wave elevation reflects instantaneously the bottom displacement, obtain rough estimates of the tsunami characteristics, although the earthquake magnitude, source depth and displacement are also critical parameters. However, the trade-off between the tsunami heights on different coastal plains should be more realistic, as it depends on the coastal topography and on very rough characteristics of the tsunami source (i.e. earthquake source orientation).

In the present study, we use the numerical models TUNAMI-N2 and AVI-NAMI based on the method of Okada (1985) for simulation and animation of tsunami generation and propagation, and of coastal amplification of nonlinear long waves in a given arbitrarily shaped bathymetry. TUNAMI-N2 and AVI-NAMI algorithms are the key tools for developing studies for the wave propagation and coastal amplification of tsunamis in relation to various initial conditions. The routines also compute the water surface fluctuations and velocities at all locations, even for shallow-water and land regions within the limitations of the bathymetric grid size used (Shuto *et al.* 1990; Goto *et al.* 1997; Yalçıner *et al.* 2003, 2004). The coseismic deformation is computed using an elastic dislocation model that yields the vertical deformation on the sea floor in the epicentral area as a function of the ground elastic parameters and the fault-plane geometry (Okada 1985). For the simulation, it is assumed that this deformation is instantaneous and fully transmitted to the sea surface. Hence, the earthquake source can then be modelled as a rupture of a single rectangular fault plane characterized by parameters describing location, orientation and rupture direction of the plane (i.e. rupture length, rupture width, focal depth, maximum displacement, strike, dip, rake angles). We have further gathered global bathymetric data provided by GEBCO (1987) and Smith & Sandwell (1997*a*, *b*) with a 1000 m grid size for the tsunami simulations and a time-step of $\Delta x/\Delta t = (2gh_{max})^{1/2}$, where h_{max} and g are the maximum still water depth and gravitational acceleration, respectively, in a concept of stability to provide stable and meaningful results.

Figure 8a shows snapshots of the tsunami wave propagation generated using nonlinear shallow-water theory for various times at 10 min intervals for the 11 May 1222 Paphos event. It should be noted that the bathymetric features of the Nile Delta (Fig. 8a and b) act as a natural barrier to slow down tsunami waves at shallower depths (<500 m), as a result of refractions and diffractions of sea-waves. The effects of tsunami waves are clearly visible when synthetic tsunami simulations at 30, 40, 50 and 60 min are compared. Anomalous sea-wave heights are obvious at location *u* (Alexandria, Egypt), where there is *c.* 1 m mareogram amplitude (Fig. 8b). This could be due to a deep basin and sudden shelf break reflected in the coastal bathymetry. However, there are no significant sea-wave amplitudes calculated at locations marked with filled circles without letters. The computed water surface elevations (wse) and theoretical arrival times are presented at selected locations for the 11 May 1222 event in Figure 8b.

Crete, 8 August 1303 (03:30 UT, latitude 35°11′N, longitude 25°38′N, Io = X, Me ~ 8.0). The earthquake of 8 August 1303 proves to be one of the largest and best-documented seismic events in the history of the Mediterranean area. The effects of this earthquake and associated tsunami waves were very destructive and in many ways comparable with other reported events of 29 May 1508 (Ambraseys *et al.* 1994) and 12 October 1856 (Sieberg 1932; Ambraseys *et al.* 1994). It has been suggested that the epicentre was probably near the island of Crete, and

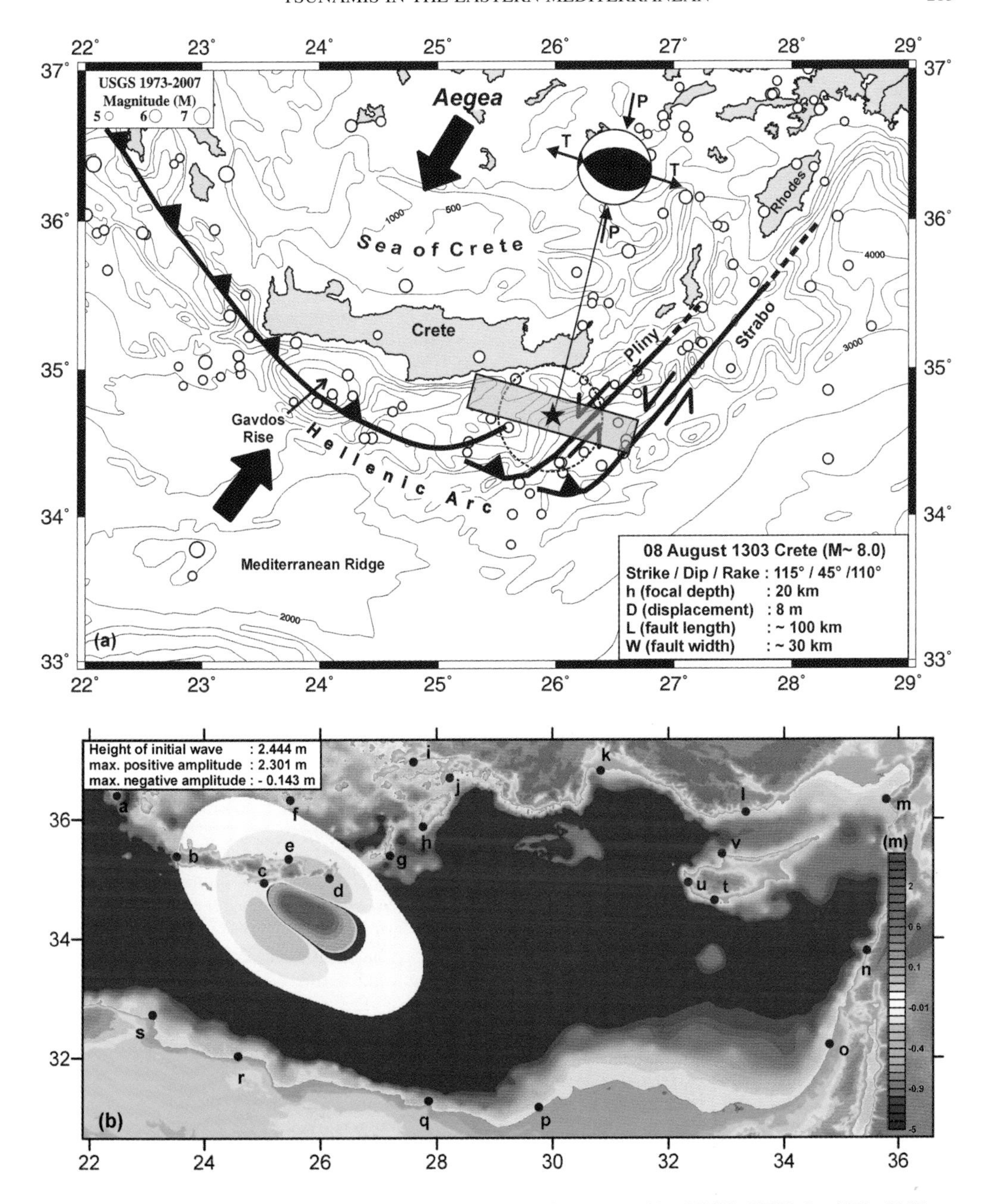

Fig. 9. (**a**) Seismicity of the Hellenic arc and surrounding region reported by USGS–NEIC for 1973–2007 scaled with respect to magnitudes for M > 5. The box with star indicates the approximate location of the 8 August 1303 Crete earthquake (M ~ 8.0) and tsunami, for which simulations are generated by applying an analogous representative earthquake derived from teleseismic P- and SH-wave modelling studies of current earthquakes. The parameters of tsunami simulations are shown in the box in the lower right corner, and a representative fault-plane solution obtained from current earthquake source mechanisms is shown above in a lower hemisphere projection. Large black arrows show relative motions of plates with respect to Eurasia (McClusky *et al.* 2000, 2003). Bathymetric contours are shown at 500 m interval, and are from GEBCO (1997). (**b**) Snapshot of the initial tsunami generated with the parameters given in (a). Letters and numbers refer to geographical locations where macroseismic observations were partly reported (Guidoboni & Comastri 2005*a*), and synthetic tsunami mareograms generated at pseudo-tide-gauge locations, and details are shown in Figure 10a and b.

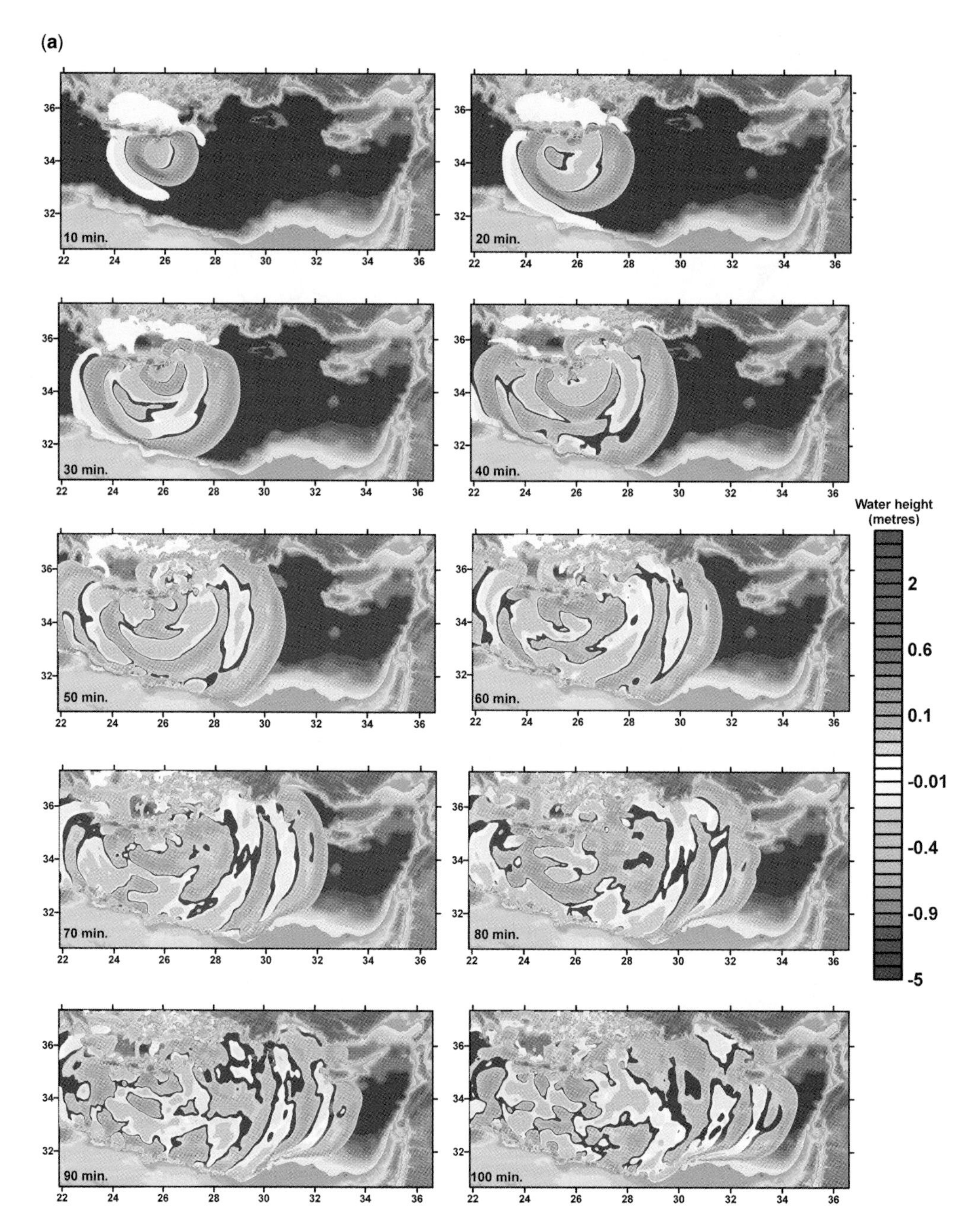

Fig. 10. (**a**) Snapshots of the tsunami wave propagation generated using nonlinear shallow-water theory for various times at 10 min intervals for the 8 August 1303 Crete event. The water surface height is in metres and a colour scale is given on the right-hand side in metres. (**b**) Computed tsunami records at selected locations for the 8 August 1303 Crete event. The vertical and horizontal scales show water surface elevation (wse) in metres and tsunami simulation time (t) in minutes, respectively. Above each tsunami mareogram are maximum synthetic wave-heights (H) and theoretical arrival times (T), given in centimetres and minutes, respectively.

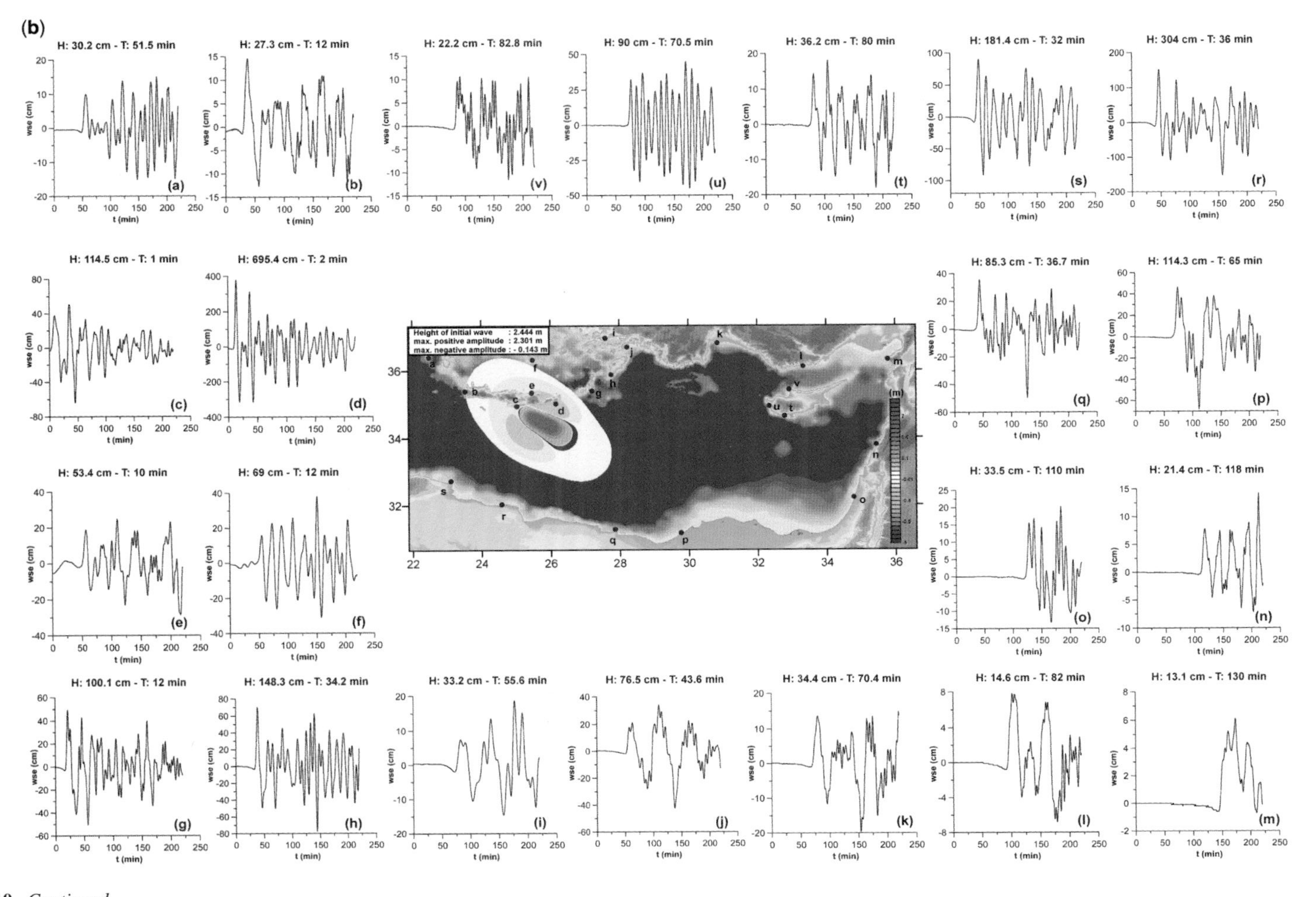

Fig. 10. *Continued.*

after this event tsunami waves were reported to be seen as far as the coastlines of Crete, the Pelepon-nese, Rhodes, Antalya (SW Turkey), Cyprus, Acre and Alexandria–Nile delta (Egypt). In addition, this earthquake and associated damage distributions are listed in most descriptive and parametric catalogues for the Mediterranean basin. However, the orientations of active faults vary along the concave part of the Hellenic arc (e.g. Pliny and Strabo trenches) in accordance with subduction of remnants of old lithospheric slab (Taymaz *et al.* 1990, 1991). Hence, the Hellenic trench in the vicinity of Crete should be considered to be a seismogenic zone of considerable importance in the Mediterranean region (Guidoboni & Comastri 1997).

Figure 9 shows seismic activity of the Hellenic arc and surroundings. The rectangular box with a star indicates the approximate location of the 8 August 1303 Crete earthquake (M $\sim$ 8.0) and tsunami for which simulations are generated by applying an analogous representative earthquake derived from teleseismic P- and SH-wave modelling studies of current earthquakes during instrumental seismology (Taymaz *et al.* 1990; Yolsal & Taymaz 2004, 2005). A snapshot of the initial tsunami waves generated with the parameters given in Figure 9a is shown in Figure 9b. Letters and numbers refer to geographical locations where macroseismic observations were partly reported (Guidoboni & Comastri 2005*a*), and synthetic tsunami mareograms generated at pseudo tide-gauge locations, respectively, and details are displayed in Figure 10. The effects of tsunami waves are clearly visible when synthetic tsunami mareograms at 30, 40, 50, 60, 70 and 80 min are compared (Fig. 10a). The largest wave amplitudes are calculated at the eastern part of Crete (*c.* 7 m at location d). In contrast, the effects of coastal bathymetry and nearshore topography are obvious along the northern African coastline near Alexandria (Egypt), where the maximum wave height is *c.* 1.15 m, whereas at location *q* it is *c.* 85 cm (Fig. 10b). A further clear feature is the extent of the Nile Delta deposits on the Herodotus abyssal plain, which is a NE–SW-trending elongated depression zone bounded by the *c.* 3000 m isobath lying seaward of the northwestern part of the Nile cone; the basin plain stratigraphy consists of thick turbidite muds with pelagic interbeds (see Rothwell *et al.* 2000, figs 3 and 9b). We have synthetically generated many tsunami simulations with different source orientations and displacements, and have been satisfied with a maximum displacement of 8 m ($D_{\max}$), which is in agreement with many historical reports (Ambraseys *et al.* 1994; Guidoboni & Comastri 1997; Hamouda 2006). We consider that this is not too high given the magnitude of reported intensities and macroseismic observations (M $\sim$ 8.0). On the other hand, a maximum displacement of 5 m does not adequately explain the lack (or the excess) of tsunami wave heights at most of the northern Africa coastal plains, including Alexandria and Gaza. This also indicates the importance of source parameters, maximum displacement and of the coastal topography acting as a natural barrier to slow down tsunami sea-waves. It can also be argued that the geometry of the Hellenic arc and of the Pliny–Strabo trenches is responsible for the 1303 tsunamigenic earthquake. This is a rather difficult question to answer, but analogy with current seismicity and source rupture studies implies that a north- to NE-dipping plane with a significant amount of strike-slip component mechanism would better fit the regional and local seismotectonic and geodynamics setting (Taymaz *et al.* 1990; Yolsal & Taymaz 2004, 2005). Of course this is a still open question, but we should keep in mind that there is a dispute about the size and location of the 1303 earthquake in reported historical catalogues. Thus, in the present study we have adopted the best suited location in our simulation study (Fig. 9; Guidoboni & Comastri 1997, 2005*a*). Nevertheless, there are several tsunamis reported to have been caused by strong earthquakes in the vicinity of Rhodes to the east or NE of the island (Papadopoulos *et al.* 2007; Table A3). In contrast, there were other large earthquakes, on 26 June 1926 (M $\sim$ 7.5–8.0) and 25 April 1957 (M $\sim$ 7.2), near Rhodes that did not generate tsunamis; these are, of course, important for tsunami hazard assessment. Thus the question remains: why are some Rhodes earthquakes tsunamigenic and some others not?

Remote sensing and GIS contribution to tsunami risk detection

LANDSAT ETM and digital elevation model (DEM) data derived by the Shuttle Radar Topography Mission (SRTM) provide an excellent opportunity to detect traces of historical tsunami events. On a regional scale the areas of potential tsunami risk are determined by an integration of remote sensing, geological, seismotectonic and topographic data, and historical reports. Furthermore, remote sensing and geographical information system (GIS) methods have proven their utility for the detection of morphological traces of potential ancient tsunami flooding, and in some cases morphological traces of sea-waves as curvilinear scarps open to the sea side are clearly visible (Theilen-Willige 2006). Geomorphometric parameters (i.e. slope degree, minimum or maximum curvatures) provide information on the terrain morphology indicating geomorphological features that might be related to tsunami events. In addition, LANDSAT ETM data are also helpful for deriving information on near-surface water currents in coastal areas that might be useful to improve our understanding of the influence of coastal morphology on the streaming mechanism, which is of great importance for tsunami wave simulation. During the evaluation of the various remote sensing data it has become evident that slope failure caused by undercutting of slope profiles by flood waves is a widespread phenomenon in coastal areas of Eastern Mediterranean countries. One of the procedures to generate a tsunami hazard map could be a comparison between the morphological setting of regions historically affected by major tsunamis and of those affected by recent ones (e.g. the Sumatra earthquake and tsunami of 26 December 2004; Barber *et al.* 2005) and potential risk sites on the Eastern Mediterranean coasts. There are typical features observed in regions prone to catastrophic tsunami events, such as fan-shaped flat areas, drainage patterns, arc-shaped walls and scarps, to name a few. On the other hand, the remnants of tsunami floods can be summarized as irregular swamps, ponds and lagoons near the coast. The details of these issues will be discussed elsewhere.

Implications

There are about a dozen or so strong tsunami events in the Eastern Mediterranean, which reflects an apparent recurrence interval of about 150–200 years (Tables A1–A3), and they are associated mainly either with seismogenic zones of the Hellenic arc, the Gulf of Corinth, central Greece, Sea of Marmara, SW Black Sea, SW Cyprus, the Dead Sea fault and Levantine basin, or with the active volcanic complex of Thera (Santorini, Columbos) and seamounts of the Eastern Mediterranean Sea (Fig. 2). On the other hand, some of the damaging historical tsunamis (e.g. those of 1303 and 1481) in the eastern Hellenic arc also threatened the coastal plains of the Cyprus–Levantine and the Nile Delta regions, as confirmed by our simulations, and thus special care should be taken in the tsunami risk assessment of the region. In this study, numerical tsunami simulations with representative simple source models are presented to examine the probable effects of tsunamis originating in the Hellenic and Cyprus arcs. The likelihood of occurrence of such earthquakes and associated tsunamis is relatively low, but theoretical arrival times (i.e. computed travel times) and water surface elevation (wse) distributions are useful to evaluate the tsunami hazard in the region. In addition, it is obvious for us that tsunami simulations present only rough estimates of expected damage. However, it is essential systematically to conduct field and sedimentological studies to refine flooding estimates for historical and recent tsunamis, as discussed by Dominey-Howes *et al.* (2000). It is also vital to make use of remote sensing technologies embedded in a GIS information database as a complementary tool to existing tsunami hazard studies, to offer an independent approach and to provide a base for further field research. Furthermore, it is important to note that high-resolution bathymetry maps are a crucial component in tsunami wave simulations, and this aspect is rather poorly developed in the Eastern Mediterranean. Thus, future oceanographic and marine geophysical research should aim to improve the resolution of bathymetric maps, particularly for the details of the continental shelf and seamounts.

We thank Istanbul Technical University Research Fund (ITU-BAP), the Turkish National Scientific and Technological Foundation (TÜBİTAK-ÇAYDAG), the Turkish Academy of Sciences (TÜBA) in the framework for Young Scientist Award Program (TT-TÜBA-GEBİP/2001-2-17), and Alexander von Humboldt (AvH) Stiftung for financial support. We are also grateful to P. Wessel and W. H. F. Smith of Hawaii University–NOAA for providing GMT software (Wessel & Smith 1991, 1995). The authors thank N. Shuto, F. Imamura, E. Pelinovsky and A. Kurkin for their invaluable efforts and support for developments of the numerical codes TUNAMI-N2, AVI-NAMI and NAMI-DANCE.

Appendix A

In this Appendix historical tsunamis are summarized in Tables A1–A3 after reliable sources, and relevant references are provided for further sources of observations. UT, Universal Time; *I*, macroseismic intensity; Io, epicentral intensity; Me, equivalent magnitude value (calculated using the method of Gasperini *et al.* (1999) and Gasperini & Ferrari (2000).

Table A1. *Reported tsunamis during the 11th–15th centuries in the Eastern Mediterranean region (compiled after Guidoboni & Comastri 2005a)*

No.	Date	Region of destruction	Geomorphological observations
1	5 December 1033	Israeli–Palestinian (at night; 32°00′N, 35°12′E; Io = IX, Me = 6.0)	Tsunami and subsidence
	There was serious and widespread damage at Jerusalem: part of the city walls and some churches and convents were damaged, as well as the prayer niche (*mihrab*) in the mosque. This earthquake was felt from Egypt to the Negev desert, and from the mountains of Galilee to Syria in the north. The earthquake also had substantial environmental effects: the sources record a tsunami on the coast of Palestine, causing the water in the port of Acre to recede. Evidence of tsunami effects is confined to Acre, because it was a city able to produce written evidence, but it is reasonable to suppose that the tsunami affected the whole coast (Ergin *et al.* 1967; Taher 1979; Soloviev *et al.* 2000).		
2	12 March 1036 or 11 March 1037	Cilicia?–southern Turkey	Tsunami and landslides
	This earthquake is unknown to the seismic catalogue tradition. There is no record of damage, but the earthquake had substantial effects on the environment: mountains were severely shaken, and there were probably some landslides. It has also been reported that there was a strong tsunami in relation to this earthquake (Canard & Berberian 1973).		
3	18 March 1068	Aila (Elat), Israel (06:30 UT; 29° 33′N, 34°57′E; Io = IX, Me = 8.1)	Fissures, formation of new springs
4	29 May 1068	Ramla, Jerusalem (32°34′N, 35°17′E; Io = IX, Me = 6.0)	Tsunami, Euphrates overflowed
	This earthquake is to be located in the sparsely inhabited region between Aila and Taima, and caused environmental effects as the sea withdrew from the Mediterranean coast of Palestine, and then flowed back, engulfing many people. The River Euphrates overflowed its banks (Taher 1979; Ambraseys *et al.* 1994; Soloviev *et al.* 2000).		
5	20 May 1202	Western Syria–Lebanon (02:40 UT; 33°26′N, 35°43′E; Io = X, Me = 7.6)	Tsunami and landslides
	This is one of the strongest and the best documented seismic events in the Mediterranean area, the most advanced historical and seismological study of which has been provided by Ambraseys & Melville (1988). Gigantic waves rose up in the sea between Cyprus and the coast of Syria. The sea withdrew from the coast, ships were hurled onto the eastern coast of Cyprus, fish were thrown onto the shore, and lighthouses were severely damaged (Guidoboni & Traina 1996; Ambraseys & Jackson 1998; Ellenblum *et al.* 1998).		
6	11 May 1222	Cyprus (06:15 UT; 34°42′N, 32°48′E; Io = IX, Me = 6.0)	Tsunami, springs and lake formation
	The towns of Limasol and Nicosia were affected, and especially Paphos, where the castle collapsed and there were many victims. The earthquake was also felt in Egypt. Limasol and Paphos, which are situated on the south and west coast of the island, respectively, were struck by a strong tsunami. The harbour at Paphos was left completely without water (Ergin *et al.* 1967; Ambraseys *et al.* 1994).		

7	8 August 1303	Crete (Greece) (03:30 UT; 35°11′N, 25°38′E; Io = X, Me = 8.0)	Tsunami, landslides, flooding
	This earthquake was one of the largest events in the Mediterranean area, and is referred to in the seismological tradition. There was an enormous reported tsunami that struck Crete, the coast of Egypt and part of Palestine, and less serious effects were observed in the Adriatic (Ambraseys 1962; Taher 1979; Guidoboni & Comastri 1997; El-Sayed *et al.* 2000; Soloviev *et al.* 2000).		
8	18 October 1343	Sea of Marmara (Turkey) (16:15 UT; 41°03′N, 29°04′E; I = VIII)	Tsunami
	A large sea wave penetrated a long stretch of lowlying coast to a distance of about 1.8 km. Some boats were dragged from ports or other coastal places where they were lying, and left behind. After a considerable time, the sea receded, leaving mud and dead fish on land (Ergin *et al.* 1967; Taher 1979; Papazachos & Papazachou 1997; Soloviev *et al.* 2000; Ambraseys 2002*a*,*b*; Guidoboni & Comastri 2002).		
9	20 March 1389	Chios (Greece) (12:30 UT; 38°16′N, 26°31′E; Io = VIII–IX, Me = 5.8)	Tsunami
	A sea-wave penetrated as far as the market square in Chios. Those present fled in fright to a nearby hill (Ambraseys 1962; Ergin *et al.* 1967; Papazachos & Papazachou 1997; Soloviev *et al.* 2000).		
10	July 1402	Gulf of Corinth (Greece) (38°09′N, 22°20′E; Io = X, Me = 6.3)	Tsunami, landslides, fissures
	The earthquake was accompanied by a strong tsunami that struck both shores of the Gulf of Corinth. The tsunami was more violent on the northern shore: first the sea drew back about 970 m from the coast, and then it flowed back over the shore, penetrating more than 200 m, destroying the wheat crop. There was an increased flow of water from springs at Patras and Corinth (Ambraseys 1962; Ergin *et al.* 1967; Ambraseys & Melville 1995; Soloviev *et al.* 2000).		
11	29 December 1408	Western Syria and Cyprus (35°40′N, 36°10′E; Io = IX, Me = 6.0)	Tsunami, landslides, avalanches
	The earthquake was accompanied by a tsunami, perhaps in the stretch of a sea opposite or to the south of Mt. Cassius, but there is no specific information available about its location. However, it has been suggested that the 'surface-faulting' may have stretched for at least 20 km from Qusayr, either SW towards the coast, or southwards along one or more strands of the Dead Sea fault. The tsunami threw boats out of the sea onto the shore, but there was no other damage (Taher 1979; Ambraseys & Melville 1995; Ambraseys & Jackson 1998).		
12	19 December 1419 or 16 January 1420	İstanbul, NW Turkey	Tsunami?
	The earthquake was a damaging one, and there was probably a tsunami as well, although the sources simply refer to unusual tides (Taher 1979; Papazachos & Papazachou 1997; Ambraseys 2002*b*).		
13	3 May 1481 and 17–18–19 December 1481	Southern Aegean, Rhodes, Antalya	Tsunami

(*Continued*)

Table A1. Continued

No.	Date	Region of destruction	Geomorphological observations
	3 May 1481	06:30 UT:	Rhodes, 36°26′N, 28°13′E; I = V–VI
			Antalya, 36°53′N, 30°42′E
	3 October 1481		Rhodes, 36°26′N, 28°13′E; I = V–VI
	17 December 1481	22:00 UT:	Rhodes, 36°26′N, 28°13′E; I = V–VI
	18 December 1481	03:00 UT:	Rhodes, 36°26′N, 28°13′E; I = V–VI
	18 December 1481	05:15 UT:	Rhodes, 36°26′N, 28°13′E; I = VIII–IX
	19 December 1481	05:15 UT:	Rhodes, 36°26′N, 28°13′E; I = IV–V
	This earthquake series was very violent with many aftershocks, but did not cause extensive damage. However, the town of Rhodes was flooded by a tsunami that reached a height of about 3 m. Immediately afterwards, the sea flowed back and returned to its normal level. Some collapses were reported (Ambraseys 1962; Ben-Menahem 1979; Papazachos *et al.* 1986; Soloviev *et al.* 2000).		
14	1 July 1494	Crete (Greece) (10:10 UT; 35°12′N, 24°55′E; Io = VIII–IX)	Tsunami
	This earthquake was strong enough to create a local tsunami at Candia harbour. All the ships at anchor struck violently against each other to the extent that they seemed likely to break up (Ambraseys 1962; Soloviev *et al.* 2000).		

Table A2. *Reported tsunamis in the Levantine basin and surrounding regions from 1365* BC *to* AD *1900 (compiled after Sbeinati* et al. 2005)

No.	Date and magnitude	Observed intensity locations	Geomorphological observations	References
1	*c.* 1365 BC	VIII–IX (Ugharit) VII (Tyre)	Tsunami and fire	S48, S82, BM79
2	590 BC (M_L 6.8)	VII (Tyre)	Tsunami (Tyre and Lebanese coast)	BM79, PK81, S32
3	525 BC (M_L 7.5)	VIII–IX (Tyre, Sidon) III–IV (Cyclades and Euboea islands)	Tsunami (Bisri and Lebanese coast)	S32, BM79, PK81
4	148–130 BC	VII (Antioch)	Tsunami (Syrian coast)	S32, G94
5	92 BC (M_L 7.1)	III–IV (Syria, Egypt)	Tsunami (Levantine coast)	BM79, PK81
6	13 December 115 (M_L 7.4)	VII (Antioch) VI–VII (Mirana) V (Rhodes), Pitana	Tsunami (Antioch, Yavne, Caesaria, Palestine)	BM79, PK81
7	303–304 (M_L 7.1)	VIII (Sidon, Tyre) VII (Syria) III–IV (Al-Quds)	Tsunami (Caesaria in Palestine)	BM79, PK81, G94
8	348–349 (M_L 7.0)	VII (Beirut) VI (Arwad)	Tsunami (Beirut, Arwad)	S32, BM79, G94
9	9 July 551	IX–X (Beirut, Sur, Tripoli, Byblus, Al-Batron) VII–VIII (Sarfand, Sidon) III–IV (Arwad)	Tsunami (Lebanese coast) Landslide (near Al-Batron)	S32, PK81, D00
10	18 January 749	VII–IX (Mount Tabor) VIII (Baalbak, Nawa, Balqa) VII (Basra, Al-Quds, Tabaryya, Al-Ghouta, Manbej) VII–VIII (Beit Qubayeh) VI (Daraya) V–VI (Ariha)	Tsunami, liquefaction, landslides and surface faulting	BM79, R85, G94
11	5 April 991 (M_L 6.5)	VIII–IX (Baalbak) VII–VIII (Damascus) III–IV (Egypt)	Tsunami (Syria)	S32, G94

(*Continued*)

Table A2. Continued

No.	Date and magnitude	Observed intensity locations	Geomorphological observations	References
12	November 1114 (M_L 7.0)	VIII (Maskaneh) VII–VIII (Maraş, Samsat, Urfa) VII (Harran) V (Aleppo) IV (Antioch)	Tsunami (in Palestine) and landslides	BM79, PK81
13	29 June 1170 (M_L 7.9)	VII–VIII Damascus, Homs, Hama, Al-Sham, Lattakia, Baalbak, Shaizar) VII (Barin) VII–VIII (Aleppo) V (Iraq, Al Jazira, Al-Mousel)	Tsunami	AB89, PK81
14	20 May 1202	IX (Mount Lebanon, Baalbak, Tyre, Beit Jin) VIII (Nablus, Banyas, Al-Samyra, Damascus, Hauran, Hama, Tripoli)	Tsunami, landslides and aftershocks	AM88, A94, Table A1
15	8 August 1303	VII (Cairo, Alexandria, Damanhur, Safad) VI (Damascus, Hama) IV (Antioch, Tunus, Barqa, Morocco, Cyprus, Istanbul, Sicily)	Tsunami (Alexandria, Cairo-Damascus)	PK81, Table A1
16	20 February 1404	VI–VII (Qalaat) VIII (Bkas) VII–VIII (West of Aleppo, Qalaat Al-Marqeb) VII (Tripoli, Lattakia, Jableh)	Tsunami (Aleppo)	S32, AB89, AM95
17	29 December 1408	VIII–IX (Shugr, Bkas) VIII (Blatnes) VII (Lattakia, Jableh, Antioch) VI (Syria)	Tsunami	PT80, AM95, Table A1
18	29 September 1546	VI–VII (Nablus) V (Damascus) VI (Al-Quds, Yafa, Tripoli) V (Famagusta)	Tsunami	S32, PK81
19	21 July 1752 (M_L 7.0)	VII (Lattakia) V (Tripoli)	Tsunami (Syrian coast)	S32, BM79

20	30 October 1759 (M_L 6.5)	VIII (Al-Qunaytra) VII (Safad) VI (Acre, An-Nasra, Sidon, Saasaa) V (Damascus) IV (Aleppo, Al-Quds, Beirut, Antioch, Gaza, Cyprus)	Tsunami (Acre and Tripoli)	S32, BM79, AM89
21	25 November 1759 (M_L 6.8)	>VIII (Baalbak, Serghaya, Zabadani) VII (Ras Baalbak, Al-Qunaytra) VII–VIII (Damascus, Beirut, Sidon, Safad, Sur) VII (Tripoli, Acre) VI–VII (Homs, Hama, An-Nasra, Hosn Al-Akrad) V–VI (Lattakia, Al-Quds, Gaza)	Tsunami and faulting	BM79, AB89
22	13 August 1822 (M_L 7.1; M_S 7.4)	IX (Jisr Ash Shoughour, Quseir) VIII–IX (Aleppo, Darkoush) VIII (Antioch, Iskenderun, Idleb, Sarmeen, Kelless) VII–VIII (Armanaz, Sarmada) VII (Lattakia, Homs, Hama, Maraş, Ram Hamadan, Bennesh, Maarret Missrin) III (Damascus, Gaza, Al-Quds, Black Sea, Cyprus)	Tsunami and faulting (İskenderun, Beirut, Cyprus, Jerusalem)	BM79, PT80, A89
23	1 January 1837 (M_L 6.4)	VII–VIII (Safad, Nablus, Beit Lahm, Al-Khali) VII (Tabariya) VI–VII (Beirut) VI (Damascus)	Tsunami	S32, BM79
24	3 April 1872 (M_L 7.3; M_S 7.2)	VIII (Harem, Armanaz) VII–VIII (Buhyret Al-Amq, Antioch) VII (Aleppo, Suaidiya) VI–VII (Izaz, Idleb, Iskenderun) IV (Hama, Homs, Tripoli) III (Damascus, Beirut, Sidon, Diyarbakır, Egypt, Rhodes)	Tsunami	A89, AB89, PT80

References: A89, Ambraseys 1989; AB89, Ambraseys & Barazangi 1989; AF95, Ambraseys & Finkel 1995; AF93, Ambraseys & Finkel 1993; AM88, Ambraseys & Melville 1988; AM95, Ambraseys & Melville 1995; A94, Ambraseys *et al.* 1994; BM79, Ben-Menahem 1979; D00, Darawcheh *et al.* 2000; E67, Ergin *et al.* 1967; G94, Guidoboni *et al.* 1994; PK81, Plassard & Kogoj 1981; PT80, Poirier & Taher 1980; R85, Russel 1985; S82, Saadeh 1982; S05, Sbeinati *et al.* 2005; S48, Schaeffer 1948; S32, Sieberg 1932; M_L, local magnitude.

Table A3. *Reported tsunamis along the Hellenic arc and trench system (compiled after Papadopoulos* et al. *(2007) and Papadopoulos & Fokaefs (2005)*

No.	Date	Geographical region, Intensity, magnitude	Geomorphological observations
1	(222?) 227 BC	Rhodes, Tilos, Carian and Lycian towns (36°30′N, 27°48′E; Io: IX; Ms: 7.5)	Sea wave in Rhodes
	Sieberg (1932) stated that a seismic sea wave was associated with the earthquake but this is not justified by the available historical documentation Ambraseys 1962; Papadopoulos 2001).		
2	66	South of Crete (35°12′N, 23°30′E)	Crete
3	148	Rhodes, Dodecanese Islands (36°18′N, 28°36′E; Io: IX; Ms: 7.0)	Destructive sea inundation
	In the rhetorical speech 'Rodiakos' of Aristides Aelius (AD129–189), to the citizens of Rhodes, a strong tsunami caused by the earthquake is described (see passages 20–26): 'and I remember that in that fatal noon, when the calamity that happened to you started, when the sea was calm, . . . and all the earthquake force was directed against the city. . . Then the sea water retreated and the ports dried up. . . And the ports became as the dry ground. . . And everything happened at the same moment: the earthquake of the sea, the clouds of the roaring, the lamentations, the noise of the dead bodies, the ground subsidence. . . Everything had collapsed'. (Papadopoulos *et al.* 2007).		
4	262	South Asia Minor (36°30′N, 27°48′E)	Sea inundation
	Although there were a number of serious disasters in several parts of the Mediterranean, namely at Rome, and in Libya and Asia Minor (Ambraseys 1962; Antonopoulos 1980; Guidoboni *et al.* 1994), documents concerning earthquakes and associated sea disturbances remain rather uncertain.		
5	21 July 365	Crete (36°N, 23°E)	Crete
6	(554?) 556	Cos, Dodecanese islands (36°48′N, 27°18′E; Io: 10; Ms: 7.0)	Destructive earthquake and tsunami
	Agathias (536–582) reported that the island of Cos 'was shaken and only a very small part of it saved, while the rest collapsed . . . The sea rose up greatly and inundated the buildings along the coast and caused destruction to human beings and their property.' The exact date of the earthquake remains uncertain.		
7	8 August 1303	Crete and Dodecanese Islands (35°00′N, 27°00′E; Io: X; Ms: 8.0)	Destructive inundation
	According to many historical records (Ambraseys 1962; Ambraseys *et al.* 1994; Guidoboni & Comastri 1997), this earthquake was a very large tsunamigenic earthquake that occurred in the eastern segment of the Hellenic arc between Crete and Rhodes. It created a strong tsunami wave that affected Crete, Acre, Alexandria and Rhodes. Historical records indicated that the sea swept into Crete with such a force that it destroyed buildings and killed inhabitants; also, the Nile was flooded with great sound, throwing boats a bow-shot on land and smashing their anchors; then the water retreated leaving the boats on land (Antonopoulos 1980; El-Sayed *et al.* 2000; Papadopoulos *et al.* 2007). However, Evagelatou-Notara (1993) and Guidoboni & Comastri, (1997) did not support that Rhodes was damaged by this tsunami, and considered that three sediment layers found in Dalaman, SW Turkey, could be attributed to the 1303 tsunami (Papadopoulos *et al.* 2004). (See also Tables A1 and A2.)		

8	03 May 1481	Rhodes (at night; 36°30′N, 28°20′E; Io: VII; Ms: 6.5)	Damaging wave in Rhodes
	This earthquake caused a 3 m high tsunami that flooded the coast, and a ship that was moved ashore was smashed in a reef and sank with all its crew (Coronelli & Parisotti 1688; Tables A1 and A2). Ben-Menahem (1979) reported a tsunami as far away as the Levantine coasts, and Papadopoulos *et al.* (2004) suggested that radiocarbon dating of the medium tsunami sediment layer found at Dalaman indicates deposition in AD 1473 ± 46.		
9	1489	Dodecanese Islands, Antalya (36°36′N, 28°24′E)	Strong withdrawal
	This earthquake is questionable; a tsunami was described by Leonardo da Vinci to have occurred in 1489: 'In (fourteen hundred) and 89 there was an earthquake in the sea of Adalia (Antalya, south of Anatolia) near Rhodes, which opened the (floor of the) sea, and into this opening such a torrent water poured that for more than three hours the floor of the sea was uncovered by the reason of the water which was lost in it, and then it closed (the sea coming) its former level' (Baratta 1903; Ambraseys 1962; Antonopoulos 1980).		
10	1 July 1494	Crete	Tsunami
	See Tables A1 and A2 for details.		
11	April 1609	Rhodes, SE Aegean Sea (36°24′N, 28°20′E; Io: IX; Ms: 7.2)	Sea-waves in Rhodes, Dalaman
	This earthquake was very large and caused tsunami waves affecting Rhodes and the Eastern Mediterranean coasts. Ambraseys & Finkel (1995) compiled the historical catalogues and showed that: 'half of the town, including the castle, was ruined, and (an exaggerated figure of) over 10 000 people were reported drowned by a sea-wave... This appears to have been a great earthquake, felt also in various places in Egypt and the Syrian coast but further details are lacking.' Papadopoulos *et al.* (2004) did not find the 1609 tsunami in the Dalaman sedimentary stratigraphy.		
12	8 November 1612	North coasts of Crete (35°30′N, 25°12′E; Io: VIII; Ms: 7.0)	Damaging inundation
13	9 March 1630	Kythira, SE Aegean Sea	Tsunami
14	14 February 1672	North Aegean Sea, Cos (39°42′N, 26°00′E; Io: VIII; Ms: 6.8)	Sea-waves in Cos
	This strong earthquake was felt not only in Lesvos and Tenedos, and then NE Aegean Sea, but also as far away as Cos, in the SE Aegean Sea, where a sea wave is said to have followed the shock (Sieberg 1932; Montandon 1953; Ambraseys & Finkel 1995; Papadopoulos 2001; Papadopoulos *et al.* 2007).		
15	31 January 1741	Rhodes, SE Aegean Sea (to: UT01 (Universal Time) 15; 36°12′N, 28°30′E; Io: VIII; Ms: 7.3)	Very strong waves in Rhodes

(*Continued*)

Table A3. Continued

No.	Date	Geographical region, Intensity, magnitude	Geomorphological observations
	This large, destructive and tsunamigenic event occurred between Rhodes and Cyprus, where minarets fell and the church of Santa Sophia in Famagusta was damaged (Ambraseys *et al.* 1994). According to historical document (Ambraseys & Finkel 1995): 'As a result of the earthquake, the sea in Rhodes retreated and then flooded the coast 12 times with great violence, submerging the coast opposite the island and destroying five or six villages located a kilometre inland'. Also, the upper tsunami sediment layer discovered in Dalaman by Papadopoulos *et al.* (2004) could be attributed to the 1741 tsunami.		
16	14 March 1743	Antalya	Sea withdrawal
	Ambraseys & Finkel (1995) reviewed a report from Cyprus indicating the strong destructive earthquakes that shook the Antalya area to the east of Rhodes. In this report: 'I have been informed from Satalia (Antalya) that from the 8th to 20th of the month there were terrible earthquakes as a result of which the port dried up for some time, many houses collapsed as well as part of the walls at different places which fell on the consul's house, destroying it. Many villages were lost in these earthquakes and a mountain opposite that, which lies west of the islet of Rachat, sunk completely'.		
17	28 February 1851	South Asia Minor, Fethiye, Mugla, Rhodes (36°24′N, 28°42′E; Io: IX; Ms: 7.1)	Sea-wave in Fethiye
	Many historical documents indicate that this event was very strong and it caused destruction in Fethiye, SW Turkey and in Rhodes (Ambraseys 1962; Ambraseys *et al.* 1994; Papazachos & Papazachou 1997). It is supported that the coast was flooded about 0.6 m above the normal sea-water level at Fethiye (Perrey 1855; Ambraseys 1962; Antonopoulos 1980).		
18	3 April 1851	South Asia Minor, Fethiye Gulf (36°24′N, 28°42′E)	Sea-waves in Fethiye
	The tsunami intensity is proposed as three, and this event was possibly an aftershock of the 28 February 1851 earthquake (Ambraseys *et al.* 1994). Historical documents indicated about 1.8 m high inundation (height of the tsunami wave) in Fethiye region.		
19	23 May 1851	Rhodes, Dodecanese Islands (36°24′N, 28°42′E)	Sea-waves in Rhodes
	This was another aftershock of the 28 February 1851 earthquake, and reported inundation by tsunami waves in Rhodes and Chalki is doubtful.		

20	13 February 1855	South Asia Minor, Fethiye Gulf	Sea-waves in Fethiye
	Doubtful inundation in Fethiye was reported by Schmidt (1879) and Perrey (1855) for this earthquake.		
21	30 August 1926	Argolikos Gulf, SW Aegean Sea (36°30′N, 23°18′E)	Tsunami
22	9 February 1948	Karpathos, SE Aegean Sea (to UT 12:58:13; 35°30′N, 27°12′E; Io: IX; Ms: 7.1)	Damaging sea-waves in Karpathos
	As reported by Galanopoulos (1955, 1960), destruction was caused near Pigadia, Karpathos and 'a huge seismic sea-wave that penetrated inland 1 km'; the first motion of the sea was withdrawal (Papadopoulos *et al.* 2007). A few eyewitnesses reported that the first tsunami wave arrived about 5–10 min after the earthquake and many vessels were moved ashore and destroyed. Field surveys indicated that the wave height was about 2.5 m in Pigadia, penetration inland was about 250 m at its maximum to the west of Pigadia, and time of arrival after the earthquake occurrence ranged between 5 and 10 min (Papadopoulos *et al.* 2007).		
23	9 July 1956	Greek archipelago, South Aegean Sea Amorgos, Astypalea (to UT 03:11:40; 36°38′N, 25°58′E; Io: IX; Ms: 7.5)	Destructive large wave
	This event is one of the largest and best documented tsunamigenic earthquakes in the Aegean Sea during the 20th century (Stiros *et al.* 1994). It occurred near the SW coast of the island of Amorgos, killing 53 people, injuring 100 and destroying hundreds of houses. Tsunami waves were reported to be particularly high on the SE coast of Amorgos (*c.* 30 m) (Ambraseys 1962) and on the north coast of the island of Astypalaea. It was probably generated by either one or a series of submarine sediment slides down the slopes of the Amorgos–Astypalaea Trough (Perissoratis & Papadopoulos 1999; Dominey-Howes *et al.* 2000). The reported run-up elevations for the tsunami waves varied depending to local bathymetry and coastal configuration (Ambraseys 1962; Antonopoulos 1980; Papazachos *et al.* 1986).		
24	5 April 2000	Heraklion, North Crete	Tsunami
25	24 March 2002	City of Rhodes (36°27′N, 28°12′E; Ms: 7.5)	
	Local press reports indicated that sea waves along the coasts of Rhodes reached 3–4 m high and they overtopped an elevated wall, which protects the coastal street from sea-waves, and inundated a coastal segment as long as 2 km. The origin of the tsunami was thought to be aseismic submarine slides, because of the lack of an earthquake before or after the tsunami occurrence in the general region of Rhodes (Papadopoulos *et al.* 2007).		

References

ABE, K. 1979. Size of great earthquakes of 1837–1974 inferred from tsunami data. *Journal of Geophysical Research*, **84**, 1561–1568.

ABE, K. 1981. Physical size of tsunamigenic earthquakes of the northwestern Pacific. *Physics of the Earth and Planetary Interiors*, **27**, 194–205.

ABE, K. 1985. Quantification of major earthquake tsunamis of the Japan Sea. *Physics of the Earth and Planetary Interiors*, **38**, 214–223.

ABE, K. 1989. Quantification of tsunamigenic earthquakes by the Mt scale. *Tectonophysics*, **166**, 27–34.

AMBRASEYS, N. N. 1962. Data for the investigation of the seismic sea-waves in the Eastern Mediterranean. *Bulletin of the Seismological Society of America*, **52**, 895–913.

AMBRASEYS, N. N. 1989. Temporary seismic quiescence: SE Turkey *Geophysical Journal International*, **96**, 311–331.

AMBRASEYS, N. N. 2002*a*. The seismic activity of the Marmara Sea region over the last 2000 years. *Bulletin of the Seismological Society of America*, **92**, 1–18.

AMBRASEYS, N. N. 2002*b*. Seismic sea-waves in the Marmara Sea region during the last 20 centuries. *Journal of Seismology*, **6**, 571–578.

AMBRASEYS, N. N. & BARAZANGI, M. 1989. The 1759 earthquake in the Bekaa Valley: implications for earthquake hazard assessment in the Eastern Mediterranean region *Journal of Geophysical Research*, **94**, 4007–4013.

AMBRASEYS, N. N. & FINKEL, C. F. 1993. Material for the investigation of seismicity of the Eastern Mediterranean region during the period 1690–1710. *In*: STUCCHI, M. (ed.) *Materials of the CEC Project 'Review of Historical Seismicity in Europe'*, 1. CNR, Milan, 173–194.

AMBRASEYS, N. N. & FINKEL, C. F. 1995. *The seismicity of Turkey and adjacent areas: a historical review (1500–1800)*. Muhittin Salih EREN Publications, İstanbul.

AMBRASEYS, N. N. & JACKSON, J. A. 1998. Faulting associated with historical and recent earthquakes in the eastern Mediterranean region. *Geophysical Journal International*, **133**, 390–406.

AMBRASEYS, N. N. & MELVILLE, C. P. 1988. An analysis of the Eastern Mediterranean Earthquake of 20 May 1202, *In*: LEE, W. H., MEYERS, H. & SHIMAZAKI, K. (eds) *Historical Seismograms and Earthquakes of the World*. Academic Press, San Diego, CA, 181–200.

AMBRASEYS, N. N. & MELVILLE, C. P. 1995. Historical evidence of faulting in eastern Anatolia and northern Syria. *Annali di Geofisica*, **38**, 337–343.

AMBRASEYS, N. N., MELVILLE, C. P. & ADAMS, R. D. 1994. *The Seismicity of Egypt, Arabia and the Red Sea: a Historical Review*. Cambridge University Press, Cambridge.

ANTONOPOULOS, J. 1980. Data from investigation on seismic sea-waves events in the Eastern Mediterranean from the birth of Christ to 1980 AD (6 parts). *Annali di Geofisica*, **33**, 141–248.

BARATTA, M. 1903. Leonardo da Vinci ed i problemi della terra. *Biblioteca Vinciana*, **1**, 292–293.

BARBER, A. J., CROW, M. J. & MILSOM, J. S. (eds) 2005. *Sumatra: Geology, Resources and Tectonic Evolution*. Geological Society, London, Memoirs, **31**.

BEN-MENAHEM, A. 1979. Earthquake catalogue for the Middle East (92 B.C.–1980 A.D.). *Bolletino di Geofisica Teorica ed Applicata*, **21**, 245–310.

BOSCHI, E., TINTI, S., ARMIGLIATO, A. ET AL. 2005. A tsunami warning system for the Mediterranean: a utopia that could be implemented in a short time. EGU-2005 General Assembly, NH6.01 Tsunamis. *Geophysical Research Abstracts*, **7**, Sref-ID:1607-7962/gra/EGU05-A-09852.

CANARD, H. & BERBERIAN, M. 1973. *Aristakes de Lastivert. Recit des malheurs de la nation armenienne*. Editions de Byzantion, Brussels.

CORONELLI, P. & PARISOTTI, V. 1688. *Isola di Rodi geografica, storica, antica e moderna*. Venice.

DARAWCHEH, R., SBEINATI, M. R., MARGOTTINI, C. & PAOLINI, S. 2000. The 9 July 551 AD Beirut earthquake, Eastern Mediterranean region. *Journal of Earthquake Engineering*, **4**, 403–414.

DOMINEY-HOWES, D., CUNDY, A. & CROUDACE, I. 2000. High energy marine flood deposits on Astypalaea Island, Greece: possible evidence for the AD 1956 southern Aegean tsunami. *Marine Geology*, **163**, 303–315.

ELLENBLUM, R., MARCO, S., AGNON, A., ROCKWELL, T. & BOAS, A. 1998. Crusader castle torn apart by earthquake at dawn, 20 May 1202. *Geology*, **26**, 303–306.

EL-SAYED, A., ROMANELLI, F. & PANZA, G. 2000. Recent seismicity and realistic waveforms modeling to reduce the ambiguities about the 1303 seismic activity in Egypt. *Tectonophysics*, **328**, 341–357.

ERGIN, K., GÜÇLÜ, U. & UZ, Z. 1967. *A Catalogue of Earthquakes of Turkey and Surrounding Area (11 A.D. to 1964 A.D.)*. Istanbul Technical University Press, Istanbul.

EVAGELATOU-NOTARA, F. 1993. *Earthquakes in Byzantium from 13th to 15th century—a Historical Examination*, **24**. Parousia, Athens [in Greek with English summary].

FOKAEFS, A. & PAPADOPOULOS, G. A. 2006. Tsunami hazard in the Eastern Mediterranean: strong earthquakes and tsunamis in Cyprus and the Levantine Sea. *Natural Hazard*, doi:10.1007/s11069-006-9011-3.

GALANOPOULOS, A. G. 1955. The seismic geography of Greece. *Annales Géologiques des Pays Helléniques*, **6**, 83–121 [in Greek].

GALANOPOULOS, A. G. 1960. Tsunamis observed on the coasts of Greece from Antiquity to present time. *Annale di Géofisica*, **13**, 369–386.

GASPERINI, P. & FERRARI, G. 2000. Deriving numerical estimates from descriptive information: the computation earthquake parameters. *Annali di Geofisica*, **43**, 729–746.

GASPERINI, P., BERNARDINI, F., VALENSISE, G. & BOSCHI, E. 1999. Defining seismogenic sources from historical felt reports. *Bulletin of the Seismological Society of America*, **89**, 94–110.

GEBCO 1997. *General Bathymetric Chart of the Oceans, Digital Version*. Distributed on CD-ROM by the British Oceanographic Data Centre, Birkenhead.

GOTO, C., OGAWA, Y., SHUTO, N. & IMAMURA, F. 1997. Numerical method of tsunami simulation with the leap-frog scheme (IUGG/IOC Time Project). *IOC Manual, UNESCO*, **35**.

GUIDOBONI, E. & COMASTRI, A. 1997. The large earthquake of 8 August 1303 in Crete: seismic scenario and tsunami in the Mediterranean area. *Journal of Seismology*, **1**, 55–72.

GUIDOBONI, E. & COMASTRI, A. 2002. A 'belated' collapse and a false earthquake in Constantinople: 19 May 1346. *European Earthquake Engineering*, **3**, 22–26.

GUIDOBONI, E. & COMASTRI, A. 2005*a*. *Catalogue of earthquakes and tsunamis in the Mediterranean area from the 11th to the 15th century*. INGV-SGA, Bologna.

GUIDOBONI, E. & COMASTRI, A. 2005*b*. Two thousand years of earthquakes and tsunamis in the Aegean arc (from 5th BC to 15th century). *In*: TAYMAZ, T. (ed.) *International Symposium on the Geodynamics of Eastern Mediterranean: Active Tectonics of the Aegean Region*, 15–18, June 2005, Kadir Has University, İstanbul, Abstracts Volume, p. 242.

GUIDOBONI, E. & TRAINA, G. 1996. Earthquakes in medieval Sicily: a historical revision (7th–13th century). *Annali di Geofisica*, **39**, 1201–1225.

GUIDOBONI, E., COMASTRI, A. & TRAINA, G. 1994. *Catalogue of ancient earthquakes in the Mediterranean area up to the 10th century*. ING-SGA, Bologna.

HAMOUDA, A. Z. 2006. Numerical computations of 1303 tsunamigenic propagation towards Alexandria, Egyptian coast. *Journal of African Earth Sciences*, **44**, 37–44.

LE PICHON, X., LYBERIS, N. & ALVAREZ, F. 1984. Subsidence history of the North Aegean trough. *In*: DIXON, J. E. & ROBERTSON, A. H. F. (eds) *Geological Evolution of the Eastern Mediterranean*. Geological Society, London, Special Publications, **17**, 27–741.

MAKRIS, J. & STOBBE, C. 1984. Physical properties and state of the crust and upper mantle of the eastern Mediterranean Sea deduced from geophysical data. *Marine Geology*, **55**, 347–363.

MCCLUSKY, S., BALASSANIAN, S., BARKA, A. ET AL. 2000. Global positioning system constraints on plate kinematics and dynamics in the eastern Mediterranean and Caucasus. *Journal of Geophysical Research*, **105**, 5695–5719.

MCCLUSKY, S., REILINGER, R., MAHMOUD, S., BEN-SARI, D. & TEALEB, A. 2003. GPS constraints on Africa (Nubia) and Arabia plate motions. *Geophysical Journal International*, **155**, 126–138.

MINOURA, K., IMAMURA, F., KURAN, U., NAKAMURA, T., PAPADOPOULOS, G. A., TAKAHASHI, T. & YALÇINER, A. C. 2000. Discovery of Minoan tsunami deposits. *Geology*, **28**, 59–62.

MONTANDON, F. 1953. *Les tremblements de terre destructeurs en Europe, Catalogue par territoires seismique, de l'an 1000 à 1940*. L'Union internationale de Secours, Geneva.

MURTY, T. S. & LOOMIS, H. G. 1980. A new objective tsunami magnitude scale. *Marine Geodesy*, **4**, 267–282.

OKADA, Y. 1985. Surface deformation due to shear and tensile faults in a half-space. *Bulletin of the Seismological Society of America*, **75**, 1135–1154.

OKAL, E. A. 1988. Seismic parameters controlling far-field tsunami amplitudes: a review. *Natural Hazards*, **1**, 67–96.

PAPADOPOULOS, G. A. 2001. Tsunamis in the East Mediterranean: a catalogue for the area of Greece and adjacent seas. *In*: *Proceedings of Joint IOC–IUGG International Workshop, 'Tsunami Risk Assesment beyond 2000: Theory, Practice and Plans'*, 14–16 June 2000, Moscow, 34–43.

PAPADOPOULOS, G. A. & FOKAEFS, A. 2005. Strong tsunamis in the Mediterranean Sea: a re-evaluation. *ISET Journal of Earthquake Technology*, **42**, 159–170.

PAPADOPOULOS, G. A. & SATAKE, K. (eds) 2005. *Proceedings of the 22nd IUGG International Tsunami Symposium*, 27–29, June 2005. IUGG-UGGI, Chaina, Crete, Greece.

PAPADOPOULOS, G. A., IMAMURA, F., MINOURA, K., TAKAHASI, T., KURAN, U. & YALÇINER, A. C. 2004. Strong earthquakes and tsunamis in the East Hellenic Arc. *European Geosciences Union—Geophysical Research Abstracts*, **6**, 03212.

PAPADOPOULOS, G. A., DASKALAKI, E., FOKAEFS, A. & GIRALEAS, N. 2007. Tsunami hazard in the Eastern Mediterranean: strong earthquakes and tsunamis in the east Hellenic Arc and Trench system. *Natural Hazards and Earth System Sciences*, **7**, 57–64.

PAPAZACHOS, B. C. & PAPAZACHOU, C. 1997. *The Earthquakes of Greece*. Ziti Publications, Thessaloniki.

PAPAZACHOS, B. C., KOUTITAS, CH., HATZIDIMITRIOU, P. M. & PAPAIOANNOU, CH. A. 1986. Tsunami hazard in Greece and the surrounding area. *Annali di Geofisica*, **4B**, 79–90.

PAPAZACHOS, B. C., PAPADIMITRIOU, E. E., KIRATZI, A. A., PAPAZACHOS, C. B. & LOUVARI, E. K. 1998. Fault plane solutions in the Aegean Sea and the surrounding area and their implication. *Bollettino di Geofisica Teorica ed Applicata*, **39**, 199–218.

PAPAZACHOS, B. C., PAPAIOANNOU, C. A., PAPAZACHOS, C. B. & SAVVAIDIS, A. S. 1999. Rupture zones in the Aegean region. *Tectonophysics*, **308**, 205–221.

PELINOVSKY, E., KHARIF, C., RIABOV, I. & FRANCIUS, M. 2001. Study of tsunami propagation in the Ligurian Sea. *Natural Hazards and Earth System Sciences*, **1**, 195–201.

PERISSORATIS, C. & PAPADOPOULOS, G. C. B. 1999. Sediment instability and slumping in the southern Aegean Sea and the case history of the 1956 tsunami. *Marine Geology*, **161**, 287–305.

PERREY, A. 1855. Note sur les tremblements de terre en 1584, avec supplements pour les années antérieures 1852–1853. *Bulletin of the Royal Academy of Sciences, Brussels*, **22-I**, 526–572.

PLASSARD, J. & KOGOJ, B. 1981. *Seismicité du Liban: catalogue des seismes ressentis*, 3rd edn. Collection des Annales-Mémoires de l'Observatoire de Ksara, IV, Beirut.

POIRIER, J. & TAHER, M. 1980. Historical seismicity in the Near and Middle East, North Africa, and Spain from Arabic documents (VII–XVIIIth century). *Bulletin of the Seismological Society of America*, **70**, 2185–2201.

ROTHWELL, R. G., REEDER, M. S., ANASTASAKIS, G., STOW, D. A. V., THOMSON, J. & KAHLER, G. 2000. Low sea-level stand emplacement of megaturbidites in the western and eastern Mediterranean Sea. *Sedimentary Geology*, **135**, 75–88.

RUSSEL, K. W. 1985. The earthquake chronology of Palestine and northwest Arabia from the 2nd through the mid-8th century A.D. *Bulletin of the American School of Oriental Research*, **260**, 37–59.

SAADEH, S. 1982. *Ugharit, Mu'assaset Al-Fikr lil'abhath we Al-Nasher*, 1st edn. Beirut.

ŞAROĞLU, F., EMRE, Ö. & KUŞÇU, İ. 1992. *Active Fault Map of Turkey, 2 sheets*, Maden ve Tetkik Arama Enstitüsü, Ankara.

SBEINATI, M. R., DARAWCHEH, R. & MOUTY, M. 2005. The historical earthquakes of Syria: an analysis of large and moderate earthquakes from 1365 B.C. to 1900 A.D. *Annals of Geophysics*, **48**, 347–436.

SCHAEFFER, C. F. A. 1948. *Stratigraphie Comparée et Chronologie de l'Asie Occidentale (IIIe et IIe millenaires)*. Oxford University Press, Oxford.

SCHEFFERS, A. & KELLETAT, D. 2005. Tsunami relics on the coastal landscape west of Lisbon (Portugal). *Science of Tsunami Hazard*, **23**, 3.

SCHMIDT, J. 1879. *Studien über Erdbeben*. Carl Scholtze, Leipzig.

SHUTO, N., GOTO, C. & IMAMURA, F. 1990. Numerical simulation as a means of warning for near-field tsunami. *Coastal Engineering in Japan*, **33**, 773–793.

SIEBERG, A. 1932. Untersuchungen übar Erdbeben und Bruchscholenbau im Östlichen Mittelmeergebeit. *Denkschrifften der Medizinsch-Naturwissenschaft Gesellschaft zu Jena*, **18**, 161–273.

SMITH, W. H. F. & SANDWELL, D. T. 1997*a*. *Measured and estimated seafloor topography (version 4.2)*. World Data Centre-A for Marine Geology and Geophysics Research Publication **RP-1**.

SMITH, W. H. F. & SANDWELL, D. T. 1997*b*. Global seafloor topography from satellite altimetry and ship depth soundings. *Science*, **277**, 1957–1962.

SOLOVIEV, S. L., SOLOVIEVA, O. N., CHAN, N. G., KHEN, S. K. & SHCHETNIKOV, N. A. 2000. *Tsunamis in the Mediterranean Sea 2000 B.C.–2000 A.D.* Advances in Natural and Technological Hazard Research, **13**.

STIROS, S. C., MARANGOU, L. & ARNOLD, M. 1994. Quaternary uplift and tilting of Amorgos Island (southern Aegean) and the 1956 earthquake. *Earth and Planetary Science Letters*, **128**, 65–76.

TAHER, M. A. 1979. *Corpus des texts arabes relatifs aux tremblements de terre et autres catastrophes naturelles de la conquête arabe au XII H. /XVIII J.C.* Thèse de Doctorat d'Etat, Université Paris 1.

TAYMAZ, T., JACKSON, J. A. & WESTAWAY, R. 1990. Earthquake mechanisms in the Hellenic trench near Crete. *Geophysical Journal International*, **102**, 695–731.

TAYMAZ, T., JACKSON, J. A. & MCKENZIE, D. 1991. Active tectonics of the north and central Aegean Sea. *Geophysical Journal International*, **106**, 433–490.

TAYMAZ, T., WESTAWAY, R. & REILINGER, R. (eds) 2004. Active faulting and crustal deformation in the eastern Mediterranean region. *Tectonophysics*, (Special issue), **391**.

TAYMAZ, T., WRIGHT, T. J., YOLSAL, S., TAN, O., FIELDING, E. & SEYİTOĞLU, G. 2007. Source characteristics of the 6 June 2000 Orta–Çankırı (central Turkey) earthquake: a synthesis of seismological, geological and geodetic (InSAR) observations, and internal deformation of the Anatolian plate. *In*: TAYMAZ, T., YILMAZ, Y. & DILEK, Y. (eds) *The Geodynamics of the Aegean and Anatolia*. Geological Society, London, Special Publications, **291**, 259–290.

TEN VEEN, J. H., WOODSIDE, J. M., ZITTER, T. A. C., DUMONT, J. F., MASCLE, J. & VOLKONSKAIA, A. 2004. Neotectonic evolution of the Anaximander mountains at the junction of the Hellenic and Cyprus arcs. *Tectonophysics*, **391**, 35–65.

THEILEN-WILLIGE, B. 2006. Tsunami risk site detection in Greece based on remote sensing and GIS methods. *Science of Tsunami Hazard*, **24**, 35–48.

WESSEL, P. & SMITH, W. H. F. 1991. Free software helps map and display data. *EOS Transactions, American Geophysical Union*, **72**, 441, 445–446.

WESSEL, P. & SMITH, W. H. F. 1995. New version of the generic mapping tools released. *EOS Transactions, American Geophysical Union*, **76**, 329.

YALÇINER, A. C. & PELINOVSKY, E. 2007. A short cut numerical method for determination of periods of free oscillations for basins with irregular geometry and bathymetry. *Ocean Engineering*, doi:10.1016/j.oceaneng.2006.05.016, **34**, 747–757.

YALÇINER, A. C., ALPAR, B., ALTINOK, Y., ÖZBAY, I. & IMAMURA, F. 2002. Tsunamis in the Sea of Marmara: historical documents for the past, models for the future. *Marine Geology*, **190**, 445–463.

YALÇINER, A. C., PELINOVSKY, E., OKAL, E. & SYNOLAKIS, C. E. 2003. *Proceedings of the NATO Advanced Research Workshop on underwater ground failures on tsunami generation, modeling, risk and mitigation*, 23–26, May 2001. Intanbul. NATO Science Series.

YALÇINER, A. C., PELINOVSKY, E., TALIPOVA, T., KURKIN, A., KOZELKOV, A. & ZAITSEV, A. 2004. Tsunami in the Black Sea: comparison of the historical, instrumental and numerical data. *Journal of Geophysical Research*, **109**, C12023, doi:10.1029/2003JC002113.

YOLSAL, S. & TAYMAZ, T. 2004. Seismotectonics of the Cyprus Arc and Dead-Sea Fault Zone: Eastern Mediterranean. *EOS Transactions, American Geophysical Union*, **85**, Fall Meeting Supplements, Abstract T52B-06.

YOLSAL, S. & TAYMAZ, T. 2005. Potential source regions of earthquakes and tsunamis along the Hellenic and Cyprus arcs, eastern Mediterranean. *In*: TAYMAZ, T. (ed.) *International Symposium on the Geodynamics of Eastern Mediterranean: Active Tectonics of the Aegean Region*, 15–18, June 2005, Kadir Has University, İstanbul, Abstract Volume, Tübitak, İstanbul, Turkey, 240.

ZAHIBO, N., PELINOVSKY, E., YALÇINER, A. C., KURKIN, A., KOSELKOV, A. & ZAITSEV, A. 2003. The 1867 Virgin Island tsunami. *Natural Hazards and Earth System Sciences*, **3**, 367–376.

Late Quaternary sedimentation and tectonics in the submarine Şarköy Canyon, western Marmara Sea (Turkey)

M. ERGİN[1], E. ULUADAM[2], K. SARIKAVAK[3], Ş. KESKİN[4], E. GÖKAŞAN[5] & H. TUR[6]

[1]*Ankara University, Faculty of Engineering, Department of Geological Engineering/Geological Research Centre for Fluvial, Lacustrine and Marine Environments (AGDEJAM), Tandogan, Ankara 06100, Turkey (e-mail: ergin@eng.ankara.edu.tr)*

[2]*Ankara University, Faculty of Engineering, Department of Geological Engineering, Tandogan, Ankara 06100, Turkey*

[3]*General Directorate of Mineral Research and Exploration (MTA), Eskişehir Yolu, Balgat, Ankara, Turkey*

[4]*Niğde University, Faculty of Engineering, Department of Geological Engineering, Niğde, Turkey*

[5]*Yıldız Technical University, Natural Sciences Research Centre, Beşiktaş, 34349 İstanbul, Turkey*

[6]*İstanbul University, Department of Geophysics, Avcılar, İstanbul, Turkey*

Abstract: Influences of tectonics and late Quaternary sea-level changes on sedimentation in the submarine Şarköy Canyon, western Marmara Sea (Turkey) were investigated using a total of 37 seismic reflection profiles and 12 gravity sediment cores (with 63–435 cm thicknesses), which were collected at water depths ranging from 62 to 245 m. ^{14}C ages of base sections in three cores (11.585, 11.845 and 24.915 ka BP) and upward fining of grain size in the cores suggest that these sediments must have been deposited since the sea-level lowstand at about 12 ka BP, when the conditions in the Marmara Sea began to change from lacustrine to the present marine phase. With some exceptions, siliciclastic mud (silt + clay >90%) with low carbonate contents (<15% $CaCO_3$) is the dominant sediment type covering the floor of the canyon. The high organic carbon contents (1–2%) with slight downcore-increasing tendencies reflect higher primary organic productivities towards the early Holocene. Faults, sedimentation deformation structures, and submarine slides or slumps observed on seismic profiles, varying elevations of dated lowstand palaeoshores and low water contents (19–25%) of sediments at some sites together strongly indicate the important effect of neotectonics on sedimentation in this canyon. On the seismic profiles at least four stratigraphic units were recognized overlying the pre-Miocene basement, which indicate not only the effects of faulting and folding but also changing conditions and related depositional environments in and around the canyon. Geological evolution and thus the sea-floor morphology of the Şarköy Canyon is controlled by both regional Plio-Quaternary tectonics and global Quaternary sea-level changes.

The Marmara Sea constitutes an east–west-oriented elliptical and deep through, which consists of four basins (the Çınarcık, Kumburgaz, Central and Tekirdağ basins) and three ridges between them, with a narrow shelf on the north flank and a broader shelf on the south flank of the trough (Fig. 1). The basins are located on the western segment of the tectonically active, dextral North Anatolian Fault Zone (NAFZ; Fig. 2), which splits into three segments in the eastern part of the Marmara Sea. The northern segment of the NAFZ, which is called the Marmara Fault, is the most active rupture of the NAFZ at present (Taymaz *et al.* 1991; Smith *et al.* 1995; Gürbüz *et al.* 2000; Sato *et al.* 2004). The western extension of this fault, which is known as the Ganos Fault, passes through the deep Tekirdağ Basin delimiting the Şarköy Canyon from the north (Okay *et al.* 1999; İmren *et al.* 2001; Le Pichon *et al.* 2001; Gazioğlu *et al.* 2002; Yaltırak *et al.* 2002; Gökaşan *et al.* 2003).

The submarine Şarköy Canyon, also called the 'Dardanelles Canyon' (Ergin *et al.* 1991) or Çanakkale Underwater Valley (Yaltırak *et al.* 2002), is located at the northeastern exit of the Çanakkale Strait, the western edge of the southern Marmara Sea shelf (Fig. 2). It connects the deep Tekirdağ Basin in the NE with the shallow and narrow

From: TAYMAZ, T., YILMAZ, Y. & DILEK, Y. (eds) *The Geodynamics of the Aegean and Anatolia.* Geological Society, London, Special Publications, **291**, 231–257.
DOI: 10.1144/SP291.11 0305-8719/07/$15.00

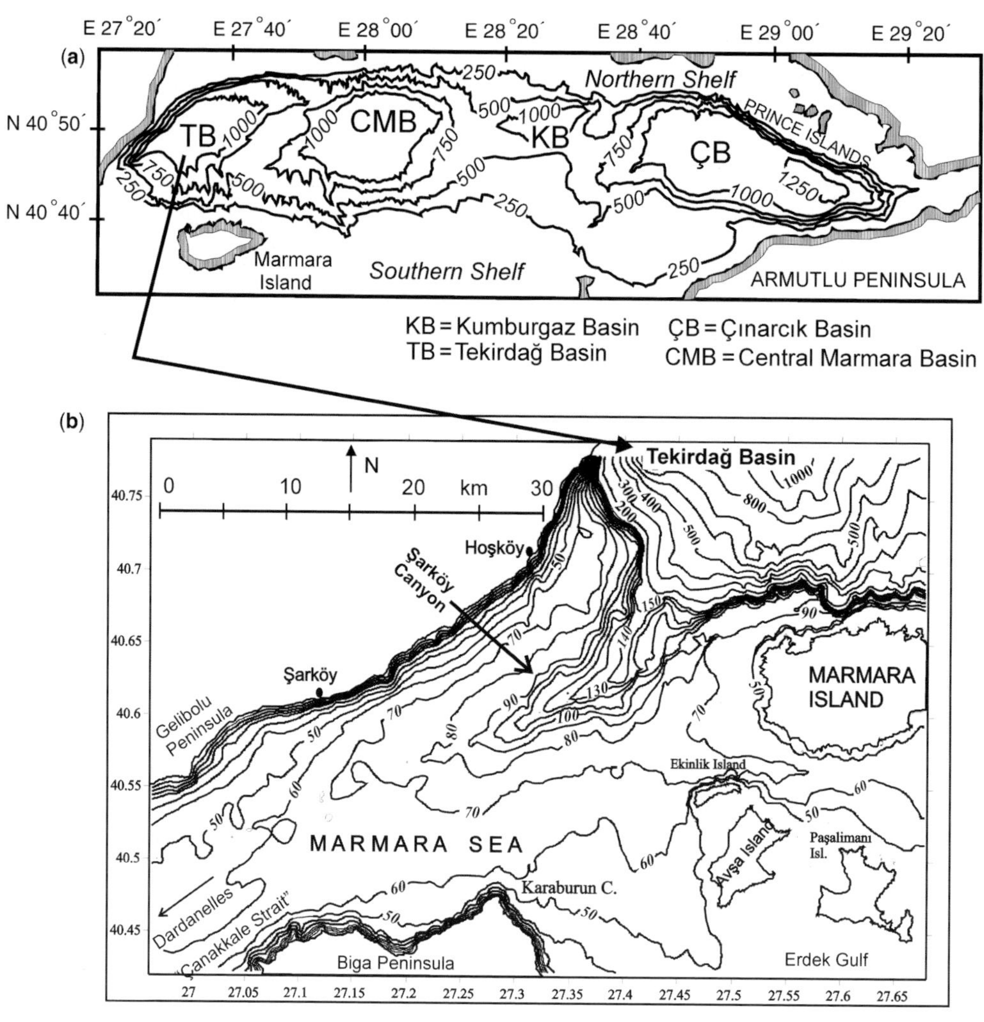

Fig. 1. (**a**) Multibeam bathymetric map of the Marmara Sea; (**b**) bathymetric map of the study area both compiled from SHOD (1983) and MTA (2004).

Çanakkale Strait in the SW (Fig. 2). Although the formation and evolution of this canyon seem to be controlled largely by tectonic activity along the western NAFZ, global climatic and associated sea-level changes in the Quaternary (Stanley & Blanpied 1980; Smith *et al.* 1995; Ergin *et al.* 1997; Aksu *et al.* 1999; Çağatay *et al.* 2000) must have further shaped the present morphology of this canyon.

In the Marmara Sea, available data have focused on seismic stratigraphy and neotectonics (Şengör *et al.* 1985; Wong *et al.* 1995; Saunders *et al.* 1998; Okay *et al.* 1999, 2004; Yaltırak *et al.* 2000; Le Pichon *et al.* 2001; Yaltırak 2002; Yaltırak & Alpar 2002; Gazioğlu *et al.* 2002; Demirbağ *et al.* 2003; Gökaşan *et al.* 2003; Seeber *et al.* 2004; Hall *et al.* 2005), onshore tectonic uplift (Yaltırak *et al.* 2002; Seeber *et al.* 2004) and active tectonics in relation to earthquakes (Taymaz *et al.* 1991, 2001, 2004; Erdik *et al.* 2004; Sato *et al.* 2004). Other seismic profiling and sediment core studies have been concerned with the Quaternary sea-level changes (Smith *et al.* 1995; Aksu *et al.* 1999, 2002; Ergin *et al.* 1999; Hiscott & Aksu 2002). Palaeogeographical reconstructions were attempted by Görür *et al.* (1997), Çağatay *et al.* (2000), Aksu *et al.* (2002) and Mudie *et al.* (2002). Very limited data have been presented on depositional, transportation and source characteristics of surface and core sediments in the southwestern Marmara Sea (Ergin *et al.* 1991, 1993, 1994, 1997, 1999; Bodur & Ergin 1994; Ergin & Bodur 1999).

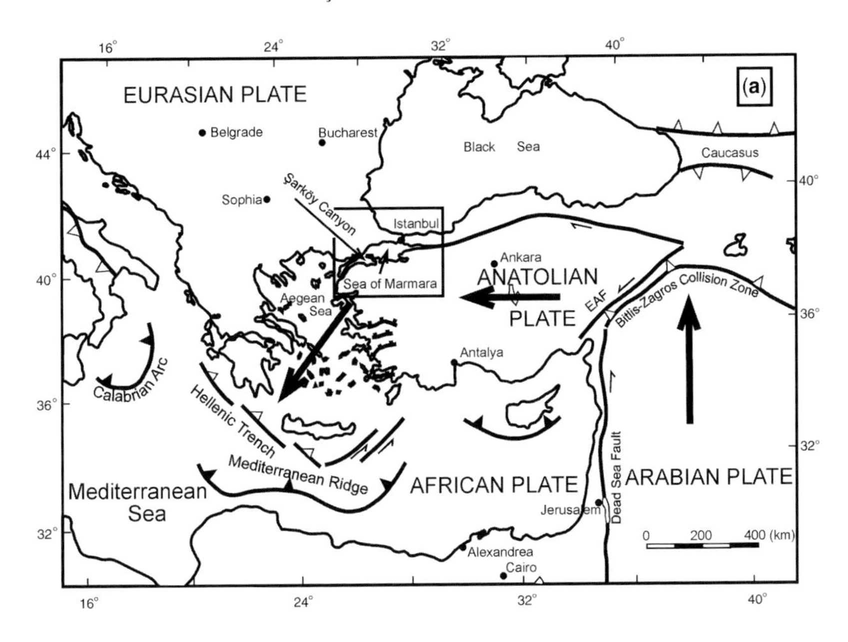

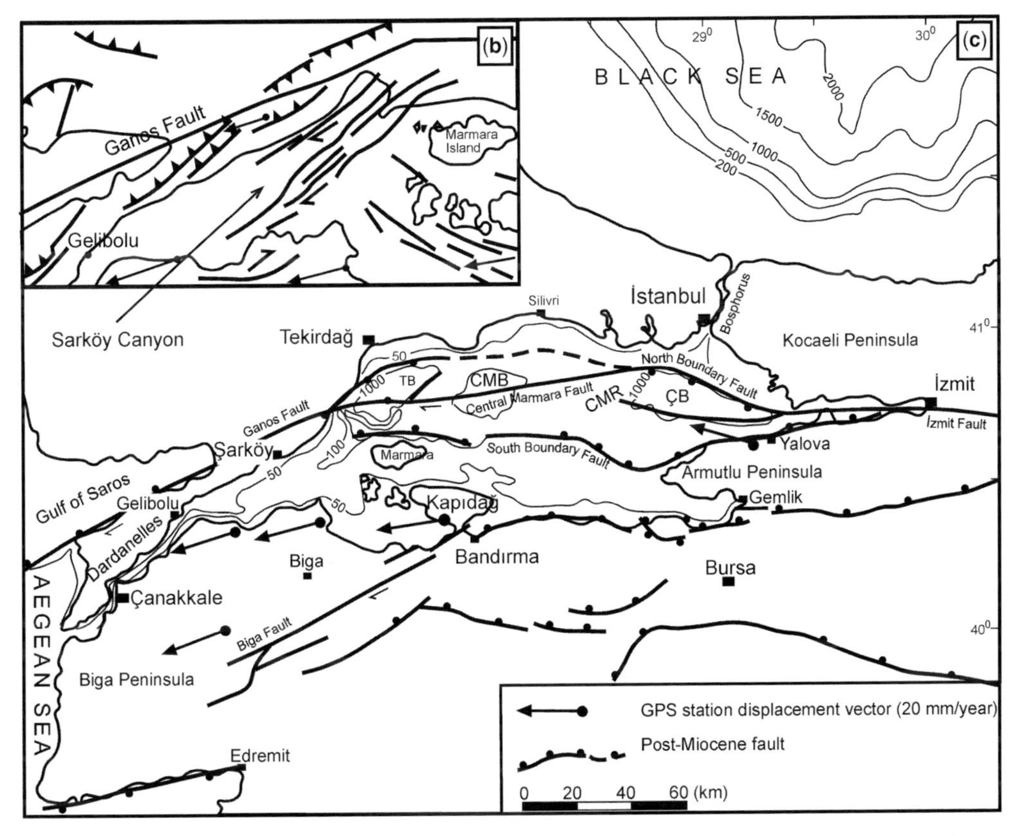

Fig. 2. Maps showing the study area in the Şarköy Canyon and its surroundings. (**a**) Simplified neotectonics of Turkey (Okay *et al.* 2004); (**b**) faults in and around the canyon (Yaltırak 2002); (**c**) Şarköy Canyon and its relation to major Marmara faults (compiled from various sources by Okay *et al.* 2004).

As mentioned above, there are many geological and geophysical investigations on the western Marmara Sea; however, almost none of them have addressed the tectonics and late Quaternary sedimentology of the Şarköy Canyon (Ergin *et al.* 2005). Thus, the present study is believed to provide important knowledge, which will help in understanding the role of this transitional valley in the late Quaternary with regard to sedimentation and neotectonics.

General setting

Morphological and geological setting of the Şarköy Canyon

In the western Marmara Sea and thus around the study area, the continental shelf break is normally at *c.* 90–104 m water depth (Smith *et al.* 1995; Gazioğlu *et al.* 2002), and farther to the NE there is a narrow continental slope (10 km wide) to the deep Tekirdağ Basin which has maximum water depths of between 1133 m (Gazioğlu *et al.* 2002) and 1151 m (Okay *et al.* 1999). Sea-floor morphology shows that the present Şarköy Canyon connects the Çanakkale Strait with the Tekirdağ Basin (Fig. 1). A NE–SW-oriented narrow channel at the Marmara Sea entrance of the Çanakkale Strait follows the Biga Peninsula coastline to meet the northeastern Aegean Sea. Morphologically, the Şarköy Canyon (Figs 1 and 2) can be classified, according to Shepard & Dill (1966) as both a U-type (towards the Çanakkale Strait) and a V-type (towards the Tekirdağ Basin) submarine valley. The Şarköy Canyon, with a SW–NE-extending axis that is about 50 km long, can be divided into an eastern sector (upper canyon) and a western sector (lower canyon). The upper canyon, with its head at about 55 m water depth, extends to about 95 m water depth at the start of the lower canyon, which continues, with a steeper dip, to its foot at about 150–200 m water depth (Figs 1 and 3).

The geology of the terrestrial areas around the canyon is apparently very complex, as a result of the influence of several tectonic events both in palaeotectonic and neotectonic periods. This area was in a zone of collision between the Sakarya Continent and İstanbul Zone during the closing of the Tethys (Şengör & Yilmaz 1981; Okay *et al.* 1994). After the collision period, this area also experienced the opening of the Marmara Basin, which possibly started in the Late Miocene–Pliocene (Saner 1985; Barka & Kadinsky-Cade 1988; Okay *et al.* 1999, 2000). Recent studies have revealed that this area has been in a collapse phase since the Quaternary, as a result of the evolution of the North Anatolian Fault Zone in the Marmara Sea (İmren *et al.* 2001; Le Pichon *et al.* 2001; Yaltırak *et al.* 2002; Demirbağ *et al.* 2003; Gökaşan *et al.* 2003; Rangin *et al.* 2004; Şengör *et al.* 2004). The multi-phase evolution of this area provides several geological units of different ages that crop out in the terrestrial areas around

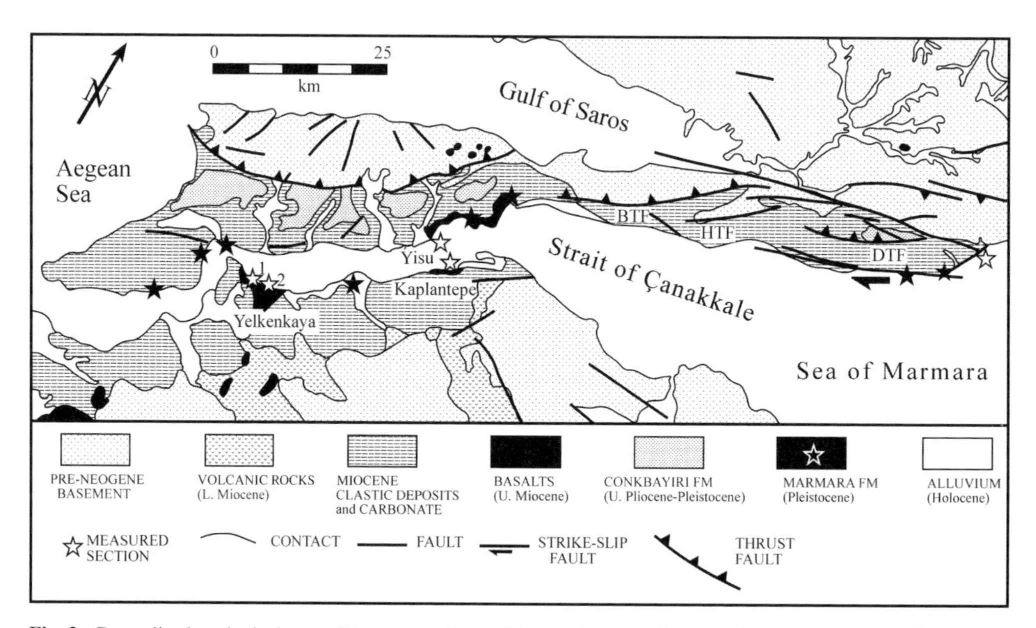

Fig. 3. Generalized geological map of the surroundings of the southwestern Marmara Sea (Yaltırak *et al.* 2000). BTF, Tahtatepe Thrust Fault; HTF, Helvatepe Thrust Fault; DTF, Dolucatepe Thrust Fault.

the canyon (Yaltırak *et al.* 2002; Fig. 3). These units may be divided into two main groups: pre-Miocene basement and Miocene–Quaternary deposits.

Pre-Miocene basement mainly consists of ophiolites, metamorphic rocks, volcanic rocks and clastic deposits of Triassic to Miocene age (Okay & Tansel 1992; Çağatay *et al.* 1998; Elmas & Meriç 1998; Tüysüz *et al.* 1998; Yaltırak *et al.* 1998, 2002; Fig. 3). On the other hand, in the coastal zones of the Çanakkale Strait on both the Biga and Gelibolu peninsulas, Late Miocene and younger deposits appear. These sediments consist of clastic deposits and limestone of fluvio-lacustrine, beach and shallow-marine environments (Çağatay *et al.* 1998; Elmas & Meriç 1998; Tüysüz *et al.* 1998; Yaltırak *et al.* 1998, 2002). Pliocene deposits overlie the Late Miocene sediments. Quaternary marine and fluvial terraces and alluvial deposits of the present rivers constitute the recent deposits around the study area.

Hydrography and river drainage regime

The water circulation and hydrography of the Marmara Sea is controlled by two-layer stratified flows. An upper layer circulation of Black Sea

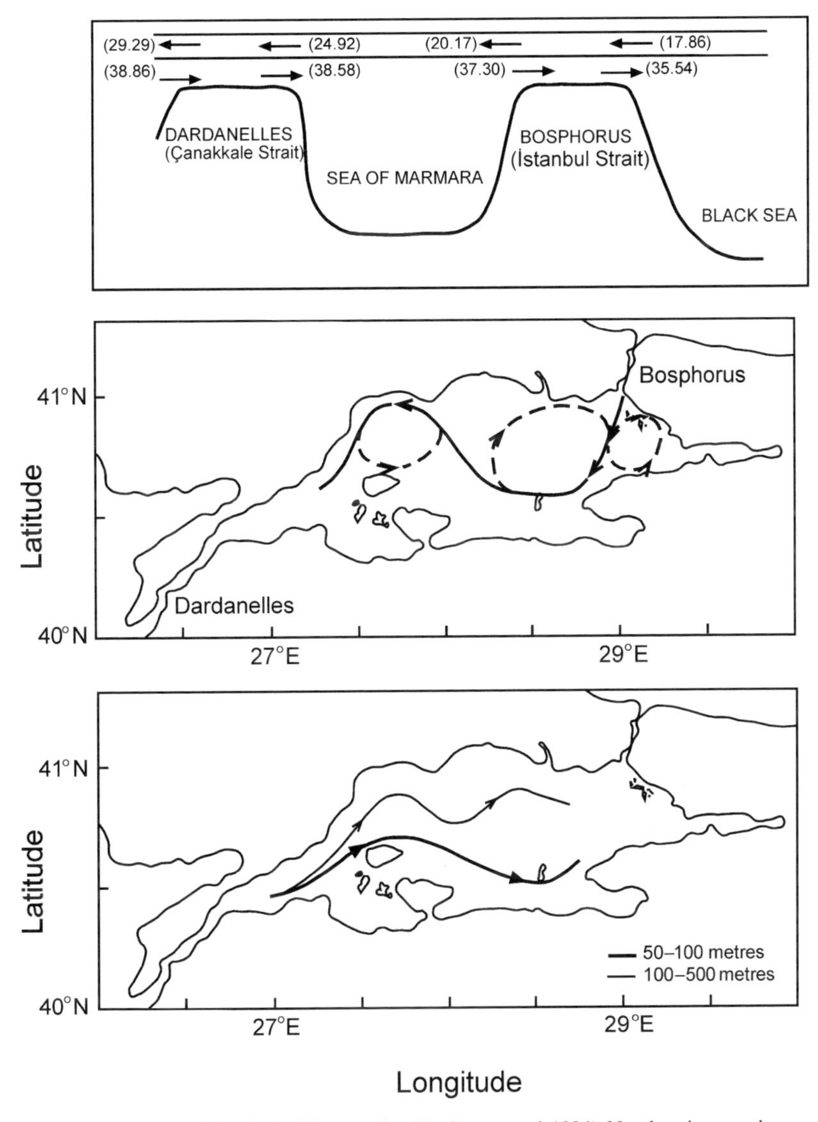

Fig. 4. Water exchange and circulation in the Marmara Sea (Beşiktepe *et al.* 1994). Numbers in parentheses are salinity values (in ppt).

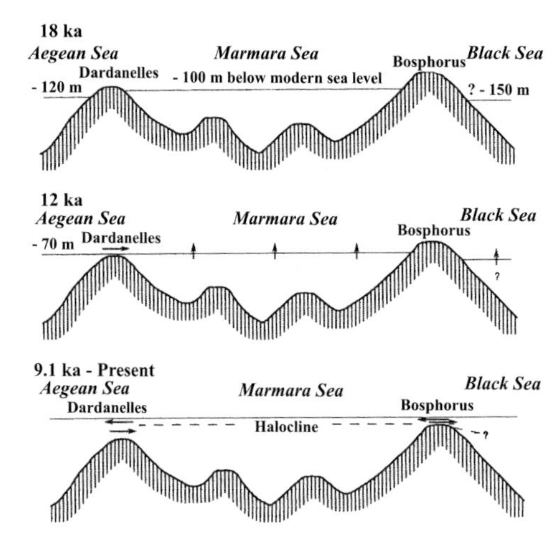

Fig. 5. Sea-level changes between the Aegean and Black Seas during the last 18 ka (modified from Kaminski *et al.* 2002).

origin (with salinities of about 20 ppt) enters the Marmara Sea and moves towards the Aegean Sea through the Çanakkale Strait ('Dardanelles') with basin-scale anti-cyclonic gyres (Ünlüata *et al.* 1990; Beşiktepe *et al.* 1994; Fig. 4). To compensate this, denser Mediterranean waters (with salinities of about 38.5 ppt) flow as a lower layer through the Çanakkale Strait and then towards the Marmara Sea and Black Sea (Fig. 4). The halocline boundary between these two opposite inflows in the Marmara Sea is generally located at 25 m water depth, but may be as shallow as 15–20 m in the Çanakkale Strait (Ünlüata *et al.* 1990). We may therefore infer that the Mediterranean waters inundating the Marmara Sea at *c.* 12–9 ka BP (Stanley & Blanpied 1980; Aksu *et al.* 1999; Çağatay *et al.* 2000; Hiscott & Aksu 2002; Kaminski *et al.* 2002; Major *et al.* 2002; Fig. 5) must have significantly influenced the morphology of the Şarköy Canyon and its tributaries when the Marmara Sea was a lake and its palaeoshorelines were located at the present upper and lower canyon boundary, at *c.* 100 m water depth.

Data from major rivers entering the Marmara Sea compiled from various sources (e.g. EİEİ 1984, 1993; Fig. 6) indicate that the southerly located rivers Kocasu, Gönen and Biga are of primary importance and introduce most of the terrigenous sediment into the Marmara Sea (Ergin *et al.* 1991; Smith *et al.* 1995; Aksu *et al.* 1999; Okay & Ergün 2005). These rivers discharge a total of *c.* 2.0×10^6, 0.08×10^6 and 0.1×10^6 tons a^{-1} suspended sediment, respectively. The terrigenous contribution from the northerly rivers or streams is of minor importance (Ergin *et al.* 1991; Smith *et al.* 1995; Aksu *et al.* 1999; Okay & Ergün 2005). However, suspended solid discharge into the Marmara Sea from the lower layer circulation through the Çanakkale Strait (0.9×10^6 tons a^{-1} (Baştürk *et al.* 1986; Özsoy *et al.* 1986) should also be considered, at least for sedimentation in the Şarköy Canyon.

Material and methods

In 2002, onboard R.V. *MTA-Sismik 1*, a total of 12 gravity sediment cores and five seismic reflection profiles were obtained in and around the Şarköy Canyon, western Marmara Sea (Fig. 7a). Additionally, 32 high-resolution seismic profiles were collected onboard TCG *Çubuklu* and TCG *Çeşme* of

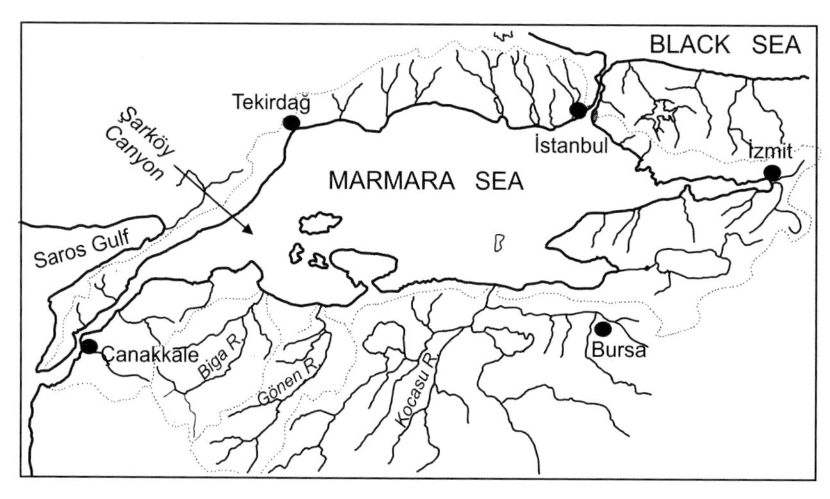

Fig. 6. Important rivers discharging into the Marmara Sea (EİEİ 1993).

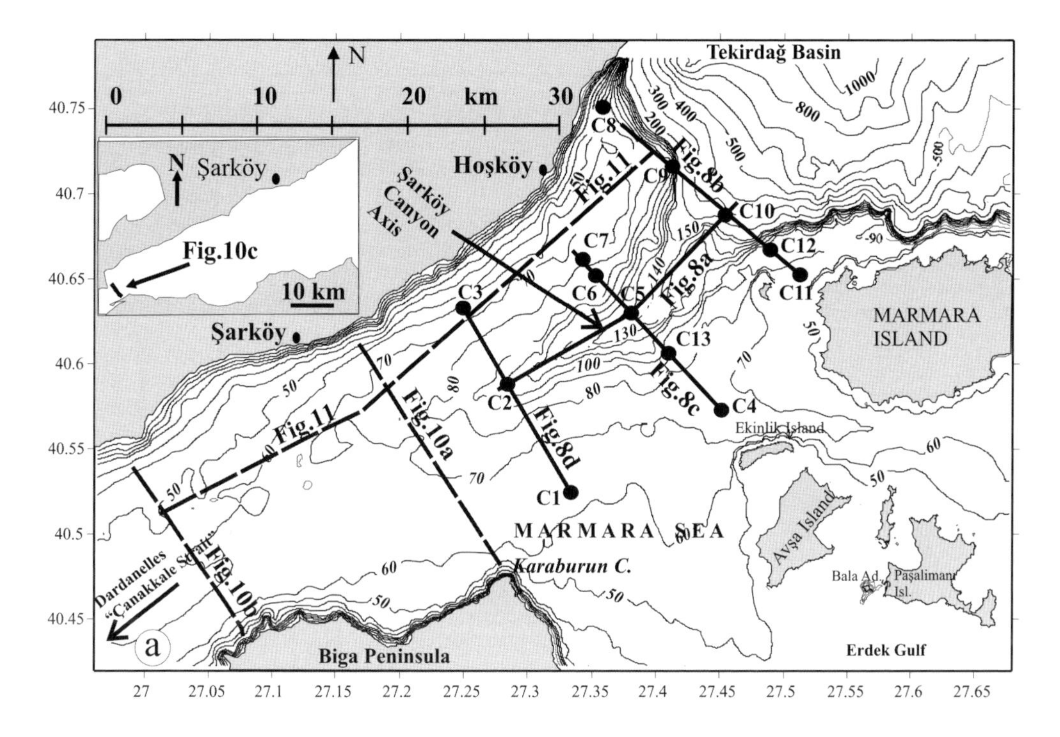

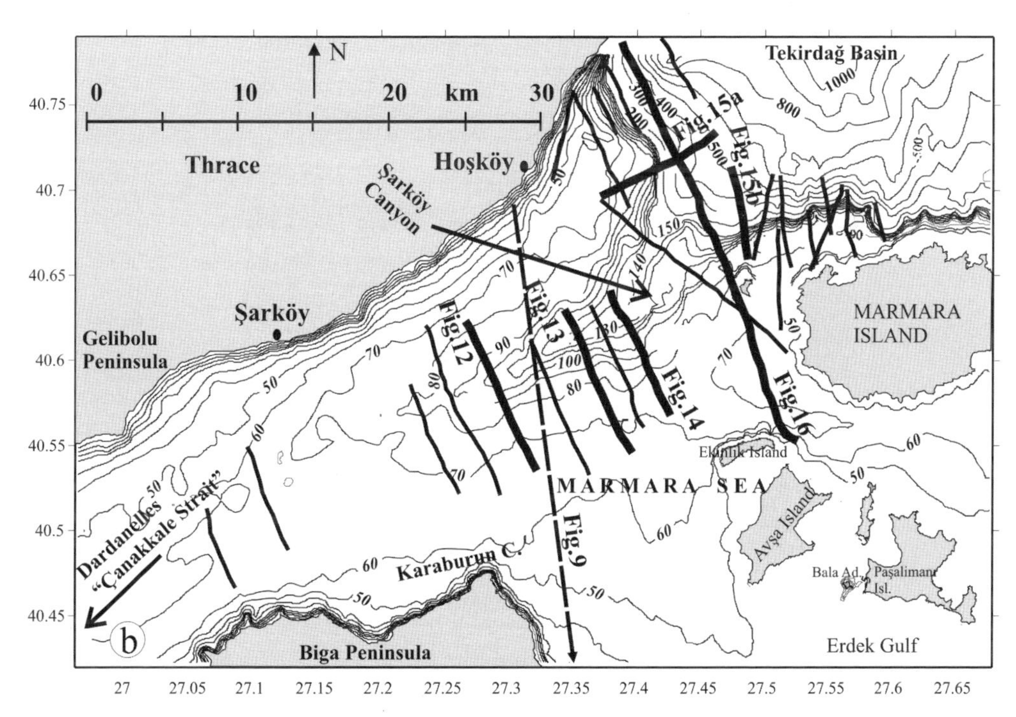

Fig. 7. (**a**) Sediment coring sites (C1–C13) and seismic reflection tracks (Fig. 8a–d) obtained in this study; (**b**) seismic profiles collected by Smith *et al.* (1995); shown on the bathymetric map of MTA (2004).

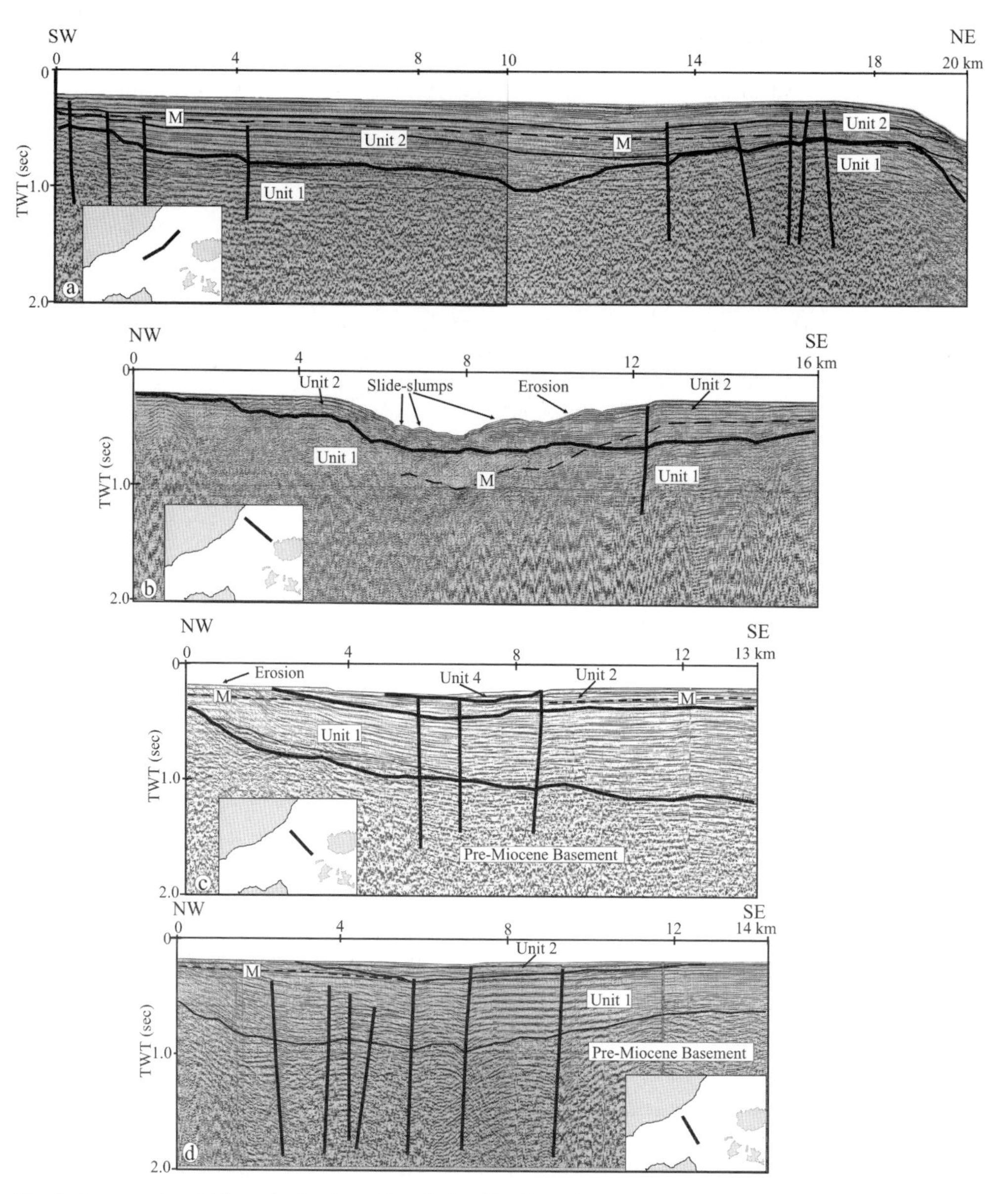

Fig. 8. Interpreted multichannel seismic profiles obtained in this study (see Fig. 7a for location). M, multiples; TWT, two-way travel time.

the Turkish Navy Department of Navigation, Hydrography and Oceanography in 1988 and in 2005, respectively (Fig. 7b). Some of the data obtained during the 1988 cruise were previously used by Smith *et al.* (1995; Fig. 7b) and the rest of the data are discussed here (Fig. 8). For the deep seismic profiling of *MTA Sismik 1*, a SYNTRAK-480MSTP recording system with a 24-channel streamer, and four airguns of 600 in^3 and 1000 p.s.i. was used. The high-resolution seismic data were collected in 1988, using a spark-array analogue seismic system comprising a 1 kJ seismic energy source, a transducer, a single-channel hydrophone streamer, and an analogue recorder run of 100–900 ms scan lengths. The vessel speed was about 4 knots. Positioning was achieved by means of a trisponder system. Seismic stratigraphic analysis methods described by Mitchum *et al.* (1977) and Brown & Fisher (1979) were used to interpret seismic reflections and related depositional characteristics.

Sediment cores with 63–435 cm recoveries were taken from water depths of between 62 and 245 m (Fig. 7a). The grain-size distribution in sediments was determined using sieving and Atterberg sedimentation techniques (after Folk 1980) whereby clay (<0.002 mm ø), silt (0.002–0.063 mm ø), and sand and gravel (>0.063 mm ø) fractions were separated. Water contents were calculated from measurements of wet and dry weights of about 20 cm^3 of sediment following the recommendation of the ODP (Ocean Drilling Program) Shipboard Measurements Panel (1991). No salinity correction was made. Total carbonate contents were determined by the gasometric–volumetric method of Loring & Rantala (1992), whereby dried and ground samples were reacted with 10% HCl solution. Organic carbon was measured using the modified Walkley–Black method (Gaudette *et al.* 1974). Selected sediment samples from the base or near-base sections of three cores were dated using the conventional radiocarbon method and the ages were calculated as ^{14}C years BP, corrected for ^{13}C at the Isotope Geochemistry Laboratory of the University of Arizona.

Seismic stratigraphy

Seismic data indicate that four units (Units 1–4) overlying the pre-Miocene basement exist along the Şarköy Canyon (Figs 8–15).

Pre-Miocene basement

This unit was observed only on seismic profiles off the Biga and Gelibolu peninsulas, west of the canyon (Figs 8c and 9). It is characterized by chaotic reflection configurations. This unit, on the basis of seismic and drill-hole data as well as onshore correlations presented by Okay *et al.* (1999), can be considered to represent pre-Pliocene (Eocene–Oligocene clastic sequence and Miocene strata with underlying crystalline rocks) basement.

Unit 1

Unit 1 (Figs 8–12) overlies the pre-Miocene basement and its reflection configuration pattern changes from chaotic to parallel. Reflections with parallel strata are generally observed as gently folded (Fig. 11). However, close to the fault zones, folds become tight and Unit 1 seems to be strongly deformed (inset of Fig. 10c, inset b of Fig. 11 and Fig. 12). Yaltırak *et al.* (2000), from the seismic data for the Çanakkale Strait, considered that the SE correspondence of this unit is the submarine extension of the Late Miocene deposits that crop widely out along the coastal zone of the Strait. The upper boundary of Unit 1 is represented by an erosional unconformity (Figs 10–12). This boundary surface is nearly exposed at the sea floor of the Marmara Sea exit of the Çanakkale Strait, and offshore along the Gelibolu Peninsula and Şarköy–Gaziköy coastline (Figs 8b, 9, 10 and 11). Thus, the NE shoulder of the Şarköy Canyon is mostly formed by this unit. On the other hand, the upper surface of Unit 1 becomes gradually deeper towards the canyon axis and it extends below the penetration depth of the seismic energy of the shallow data (Figs 8c, 9 and 12).

Unit 2

Unit 2 unconformably overlies the erosional upper surface of Unit 1. This unit has a continuous parallel reflection configuration (Figs 10–16). This unit corresponds to Unit C as previously interpreted by Smith *et al.* (1995). The reflection configuration of Unit 2 indicates that this unit must have been deposited under low-energy conditions of a marine or lake environment. Parallel reflections of this unit are folded where faults exist (inset of Fig. 10c, inset b of Fig. 11). Along the Gelibolu–Gaziköy coastal zone, the thickness of this unit is very limited (Fig. 11). However, along the axis of the Şarköy Canyon from the Çanakkale Strait to the Tekirdağ Basin and towards the SE of the canyon, thicknesses of this unit markedly increase. However, at the shelf edge, Unit 2 pinches out, possibly as a result of uplift of the base level. Some sub-units may be distinguished within Unit 2 by changes in the amplitudes of parallel reflections, as also reported by Smith *et al.* (1995). Nevertheless, no erosional unconformity was observed along these reflections, indicating that these boundaries do not imply any environmental change. At the shelf edge, parallel reflections of this unit were cut by the sea-floor surface (Fig. 8b), which could be the result of erosion caused by sediment transportation along the canyon on the southern slope of the Tekirdağ Basin (Fig. 2).

Unit 3

Unit 3 overlies Unit 2 and is characterized by its sigmoid–oblique progradational reflections, indicating sediments deposited under lowstand conditions in the Marmara Sea (Figs 12–15a). Smith *et al.* (1995) interpreted this unit as basinward-prograding clinoforms deposited during the last glacial maximum and early phase of deglaciation (stratigraphic unit B in their study). This unit is mostly observed along the southeastern side of the

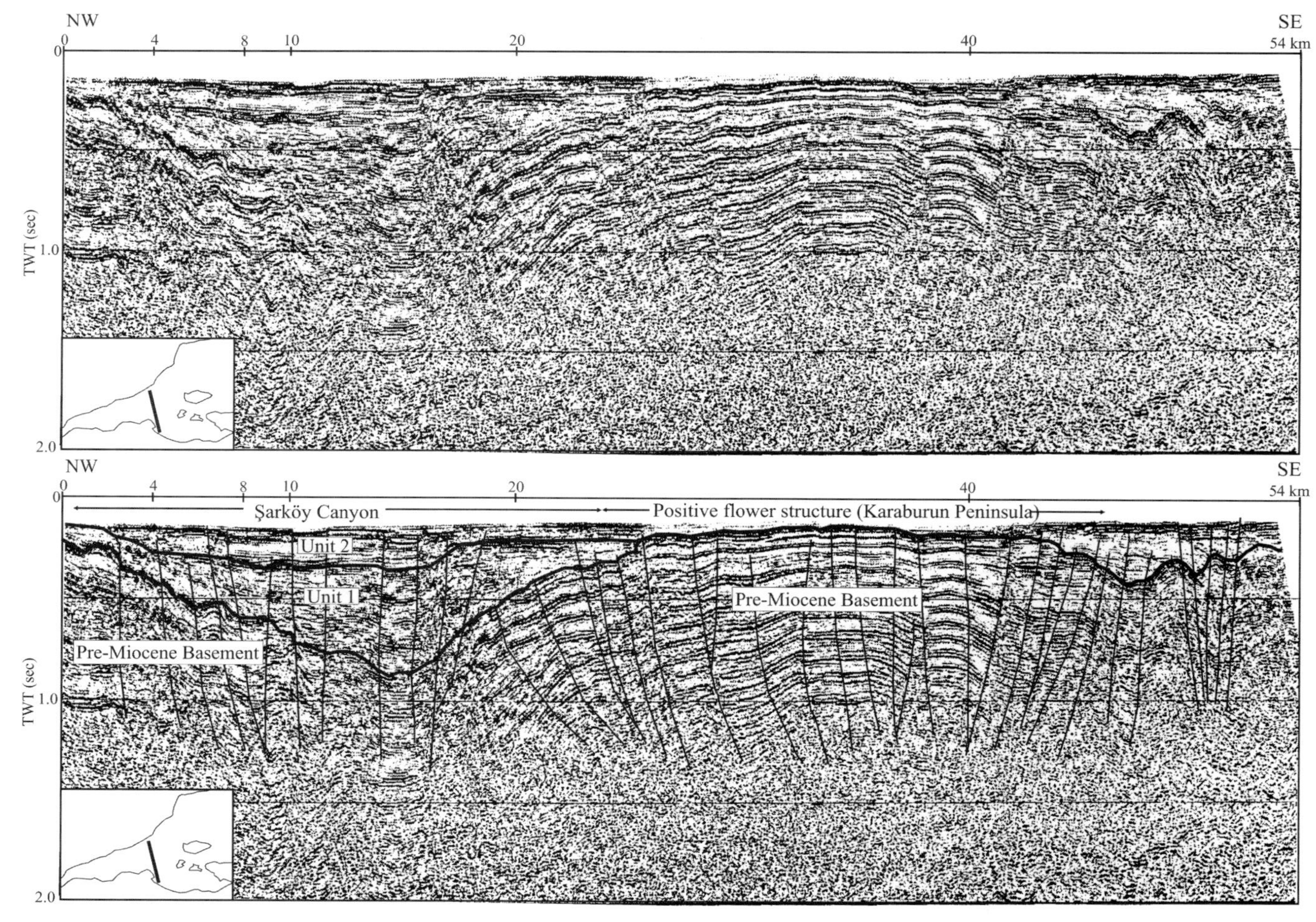

Fig. 9. Uninterpreted and interpreted multichannel seismic profile cross-cutting the Şarköy Canyon (modified from Demirbağ *et al.* 1998). (See Fig. 7b for location.)

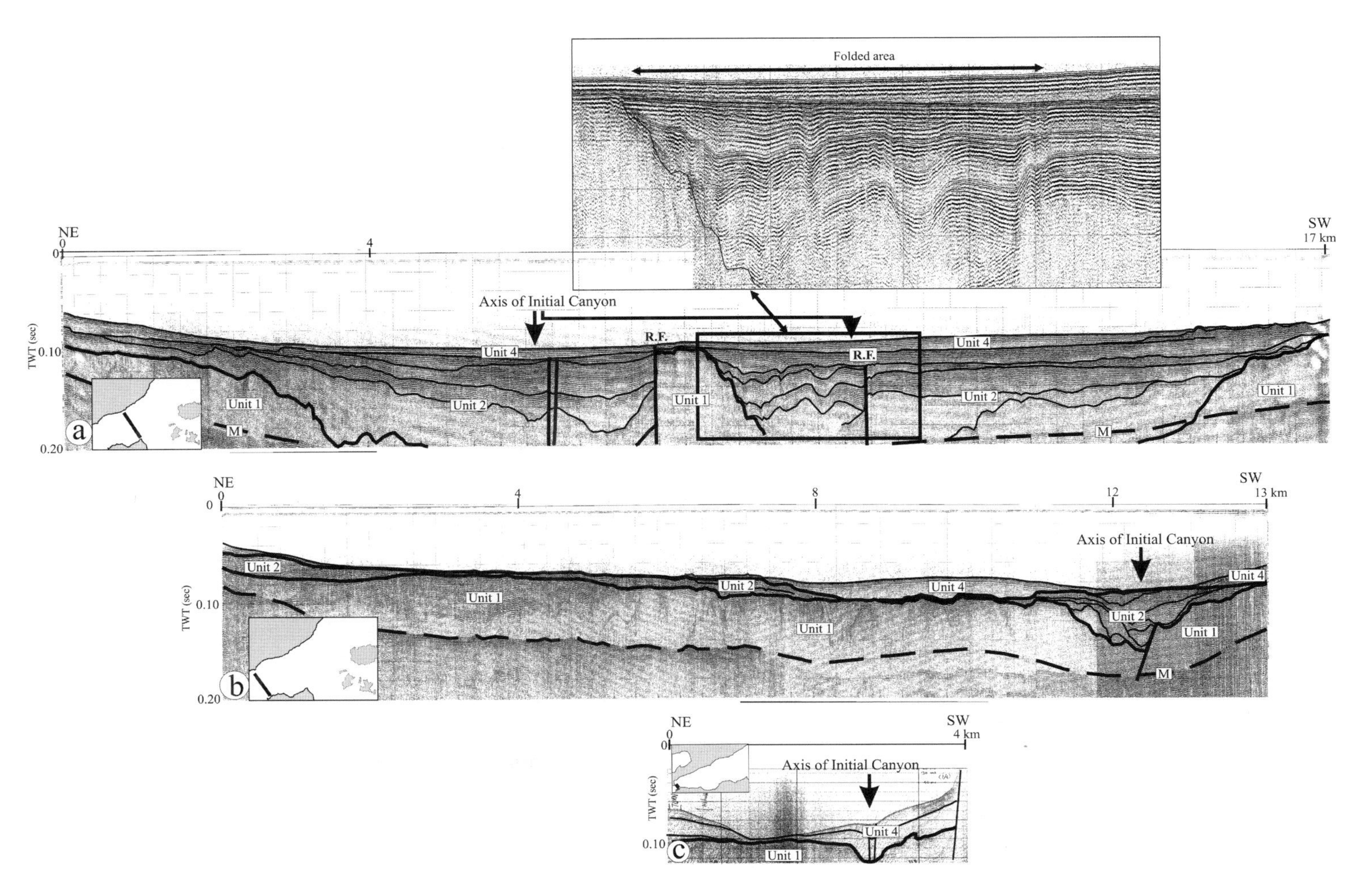

Fig. 10. Interpreted high-resolution seismic profiles cross-cutting the outer Çanakkale Strait at the Marmara Sea exit. (See Fig. 7a for location.) M, multiples; R.F., reverse fault.

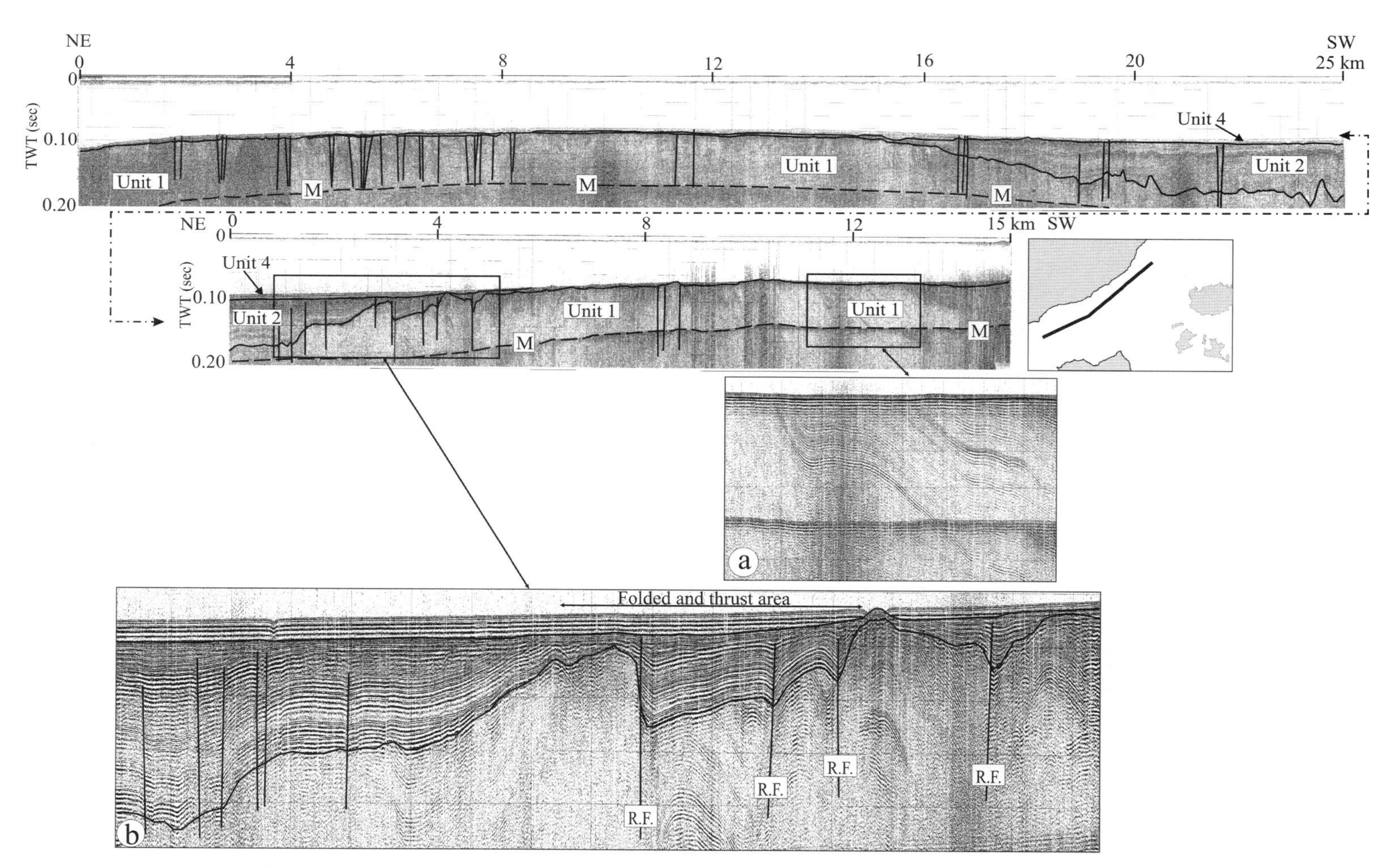

Fig. 11. Interpreted high-resolution seismic profiling along the NE–SW-directed wall of the Şarköy Canyon. (See Fig. 7a for location.) M, multiples; R.F., reverse fault.

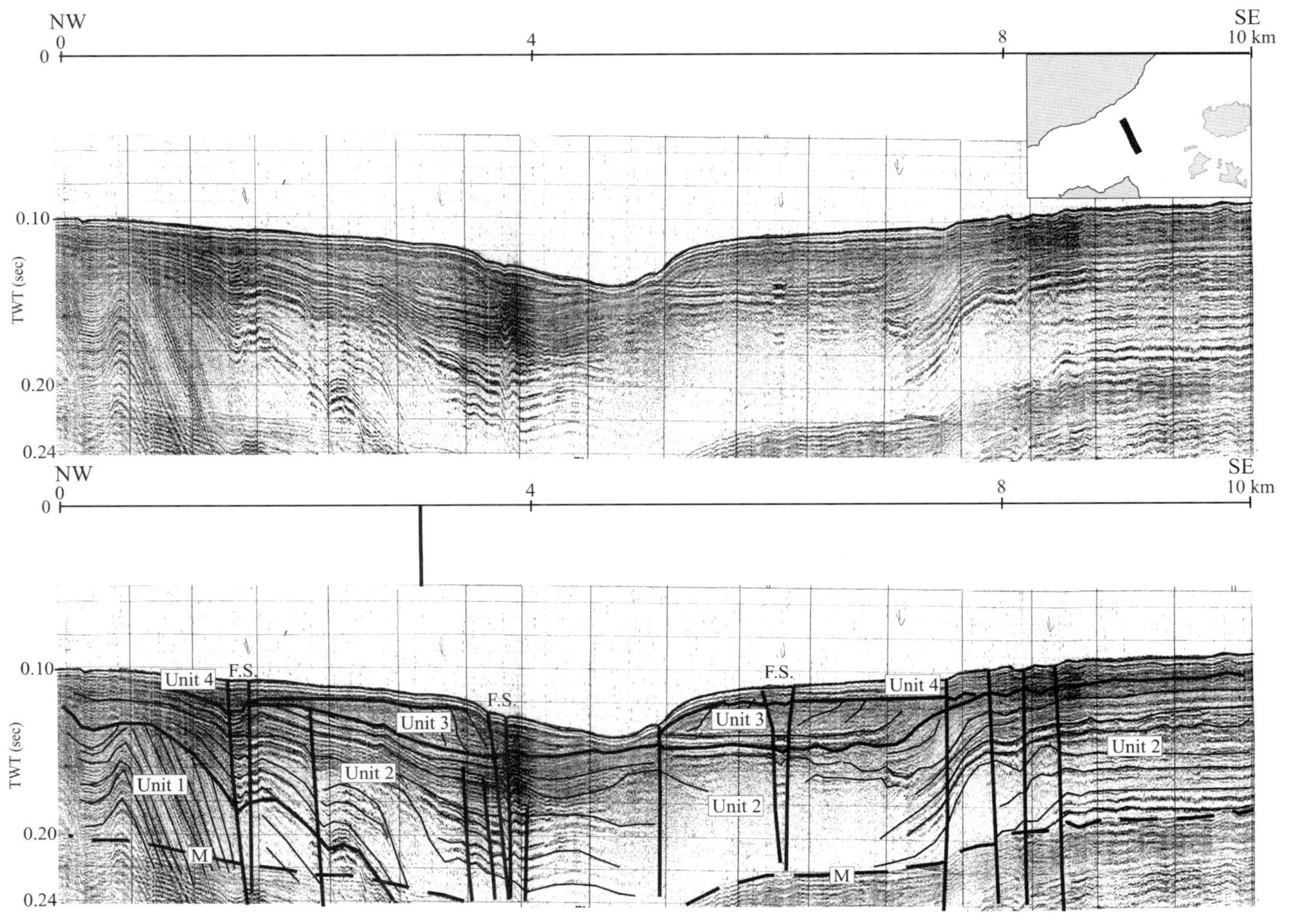

Fig. 12. Uninterpreted and interpreted high-resolution seismic profile across the head of the Şarköy Canyon. (See Fig. 7b for location.) M, multiples; F.S., flower structure.

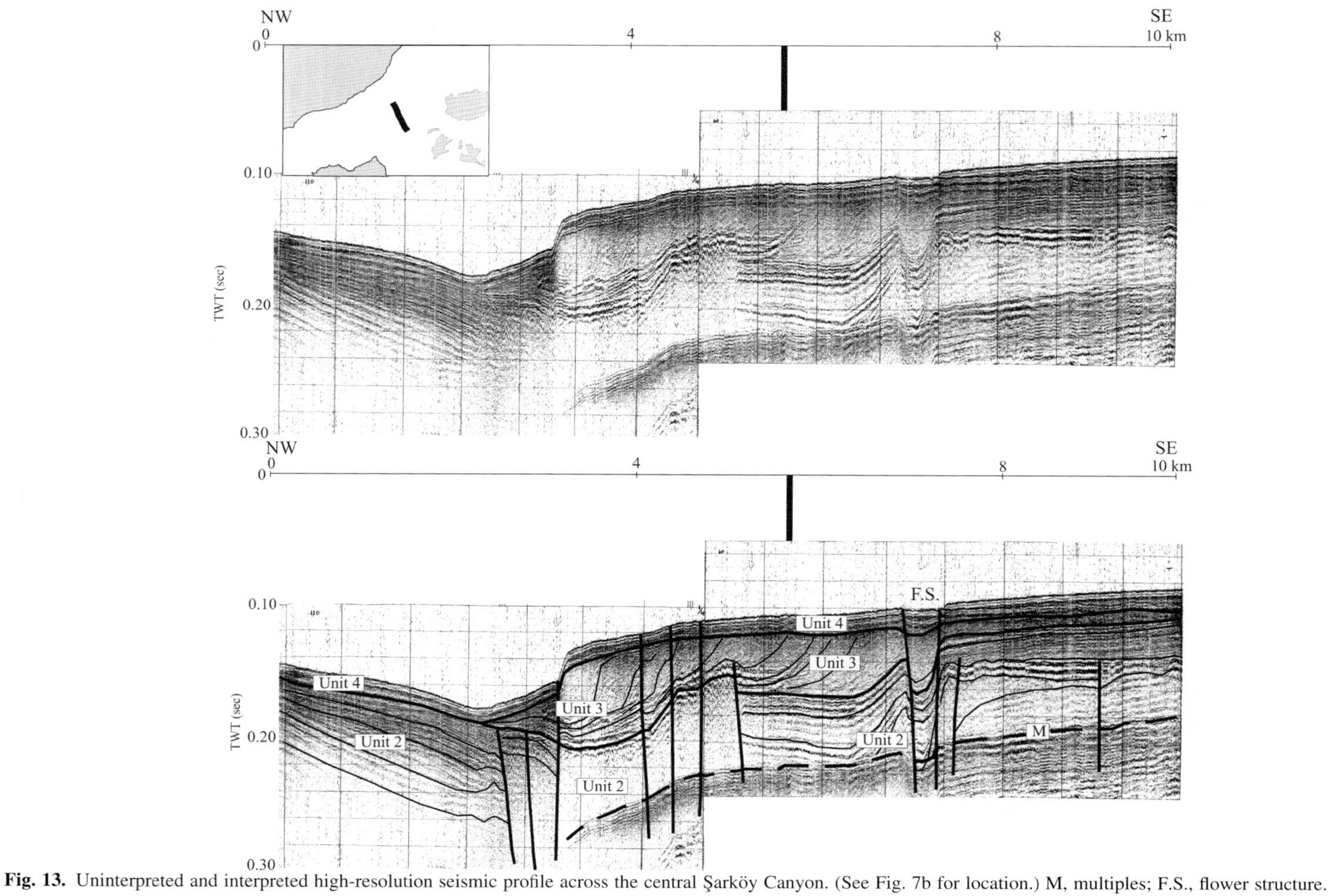

Fig. 13. Uninterpreted and interpreted high-resolution seismic profile across the central Şarköy Canyon. (See Fig. 7b for location.) M, multiples; F.S., flower structure.

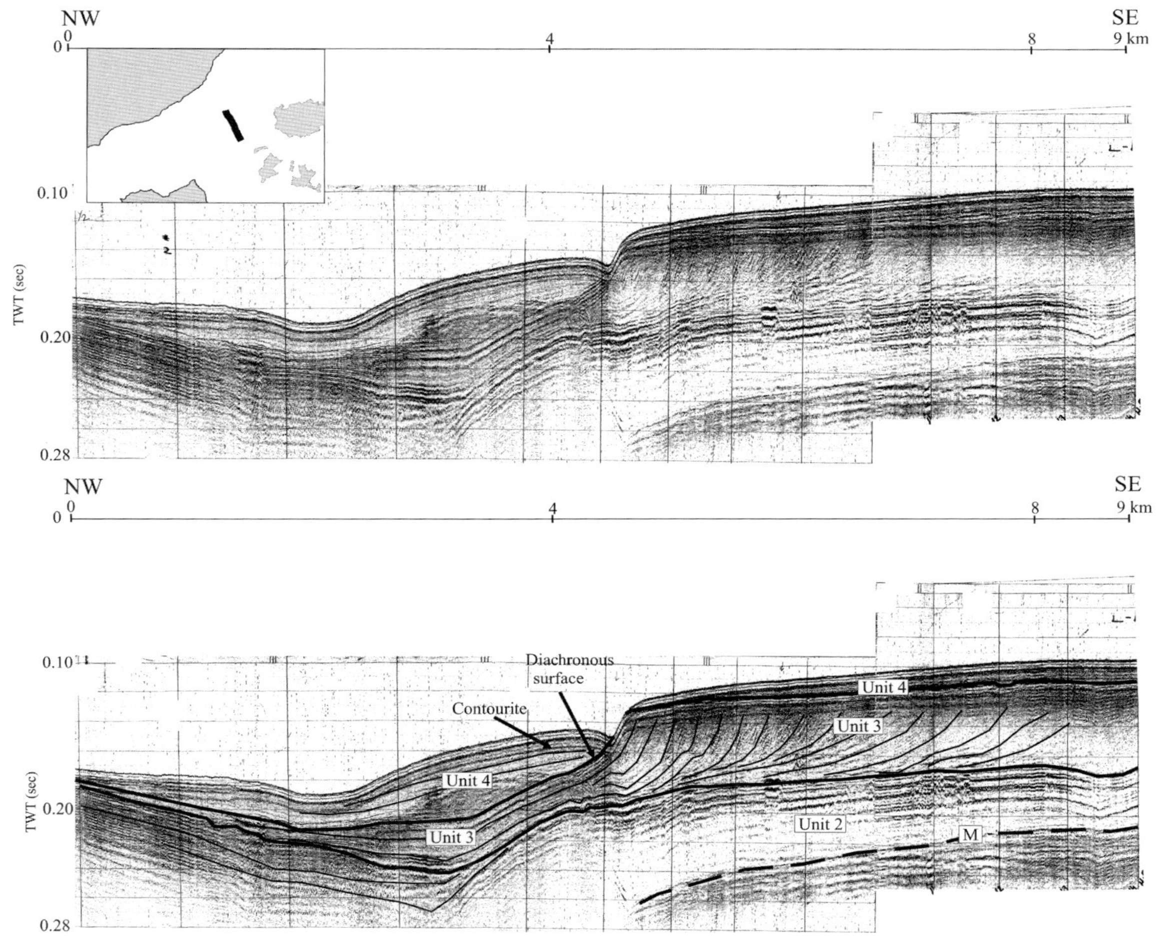

Fig. 14. Uninterpreted and interpreted high-resolution seismic profile across the central Şarköy Canyon. (See Fig. 7b for location.) M, multiples.

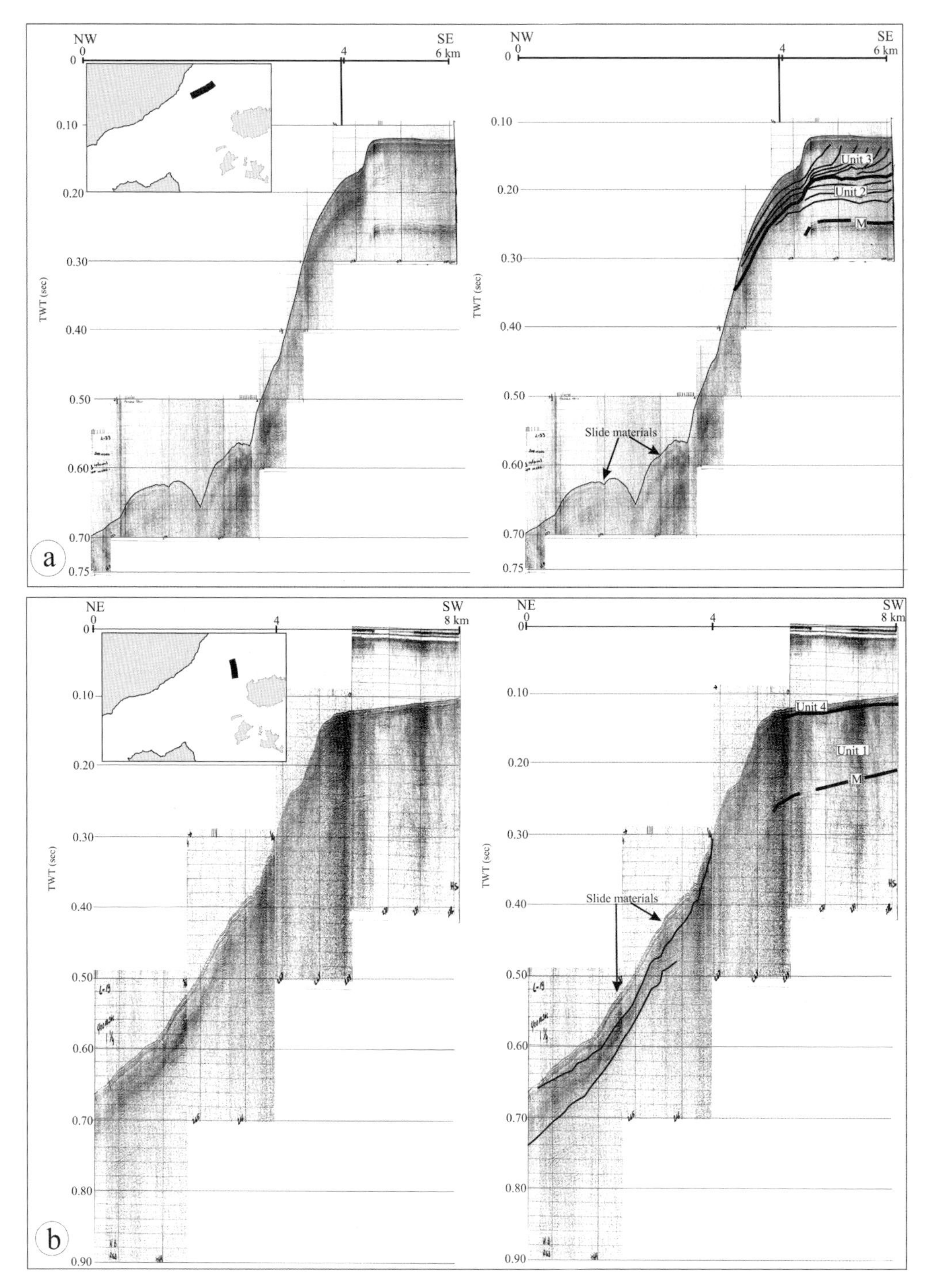

Fig. 15. Uninterpreted and interpreted high-resolution seismic profile across the foot of the Şarköy Canyon and the Tekirdağ Basin. (See Fig. 7b for location.) M, multiples.

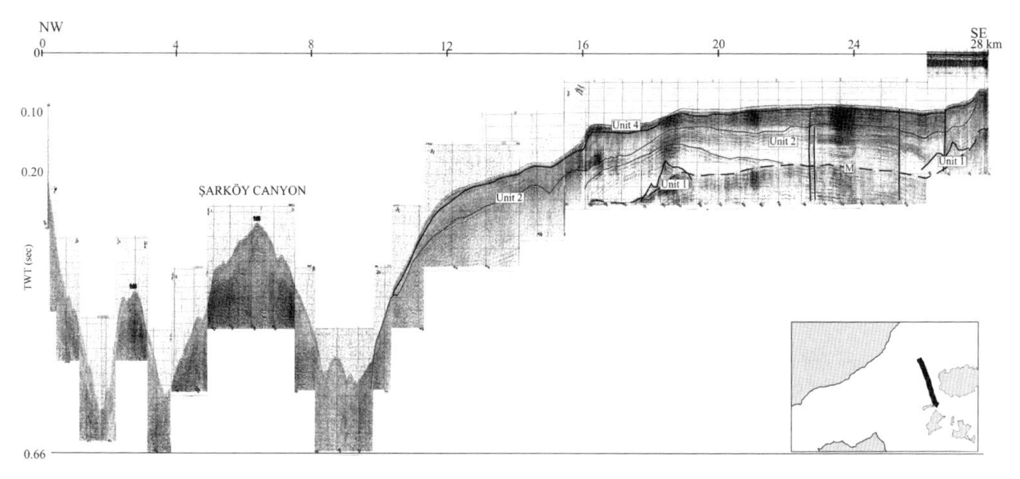

Fig. 16. Interpreted high-resolution seismic profile crossing the northern Şarköy Canyon wall towards the Tekirdağ Basin. (See Fig. 7b for location.) M, multiples.

canyon, and the progradation direction indicates that it was developed from SE to NW. Furthermore, this indicates that Unit 3 must have been deltaic deposits of the Biga and/or Gönen rivers, which probably developed during the last glacial period and early phase of the interglacial period.

Unit 4

Unit 4 contains the uppermost deposits in the study area. It has generally flat-lying parallel reflection configuration, indicating low-energy conditions similar to the present environmental conditions of the Marmara Sea (Figs 10–14). Thus, this unit is believed to represent recent deposits of the Marmara Sea. The reflections of this unit cover sigmoid–oblique reflections of Unit 2 without retrogradational deposits, which could imply a transgressive systems tract (Figs 12–14), resulted from a lowstand environment of the Marmara Sea followed by later highstand conditions during the last sea-level rise. On some seismic profiles, Unit 4 is observed to lie on the latest offlap of Unit 3, which forms the southeastern slope of the Şarköy Canyon (Fig. 14). This reflection configuration indicates that Unit 4 could probably form contourite-type deposition in some areas of the Şarköy Canyon. This type of deposition could also indicate sediment drift along the canyon under the present current regime.

^{14}C dates of the core sediments

Figure 17 shows three ^{14}C-dated sediment cores of this study. From section 270–280 cm of core C2 from the canyon head (at 99 m water depth), sediments were dated at 11.585 ± 0.142 ka BP. Sediments at 223–228 cm depth from coring site C10 from the lower canyon axis and foot (245 m water depth) are dated at 11.845 ± 0.137 ka BP. These two radiocarbon dates of the sediment cores correspond to Unit 4 on seismic profiles presented here, indicating deposition after the Mediterranean invasion of the Marmara Sea. At coring site C9, at

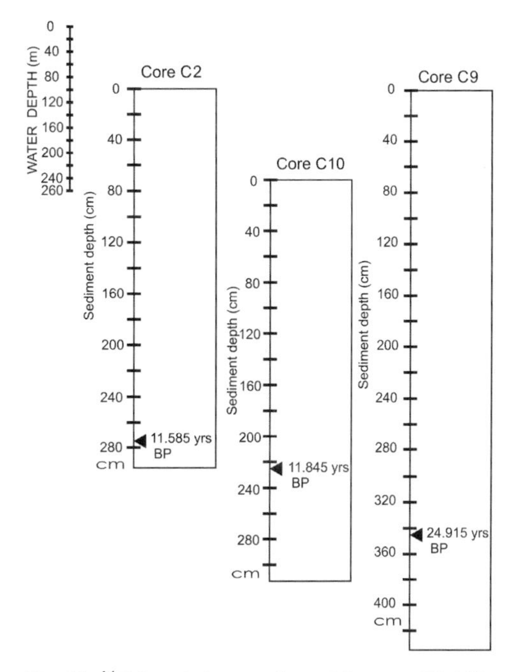

Fig. 17. ^{14}C dates in base sections of the cores C2, C10 and C9. (See Fig. 7a for locations.)

a water depth of 100 m from the northwestern canyon wall, sediments of between 338–353 cm core section were dated at 24.915 ± 0.800 ka BP. This date is comparable with the deposition of Unit 3 observed on seismic profiles, which probably represents prograded deposits of the lowstand marine environment. Available data from the literature (Aksu *et al.* 2002) revealed a ^{14}C date of 15.630 ± 0.100 ka BP, which was obtained from a core (MAR97-17, 175 cm sediment depth and 125 m water depth) located less than some 5 km from coring site C5 of this study (Fig. 7).

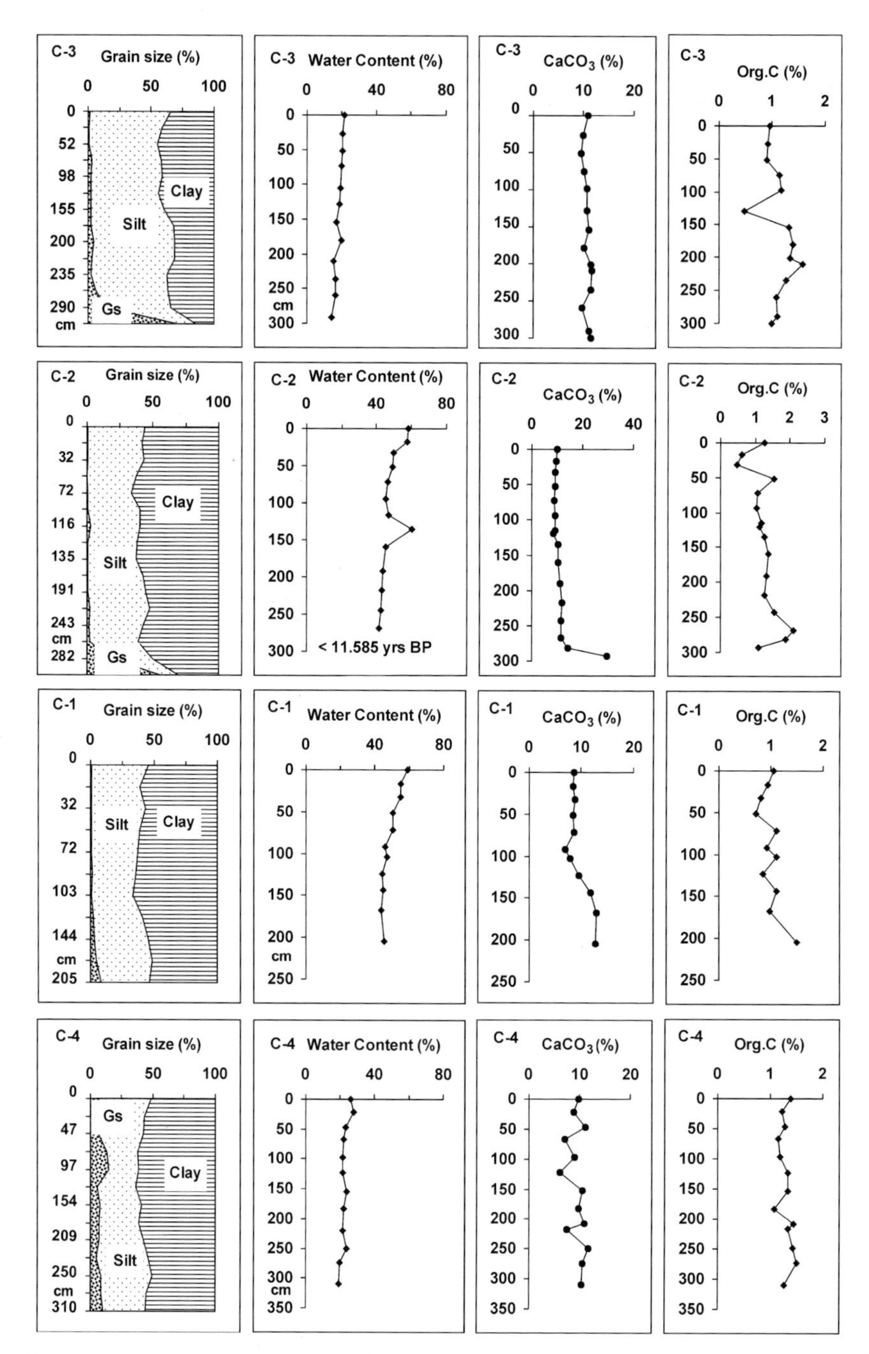

Fig. 18. Grain size, water content, carbonate distribution and organic carbon content in cores C3, C2, C1 and C4; Gs, gravel and sand.

Grain-size distribution

As shown (Figs 18–20), mud is the dominant sediment type in almost all cores, where silt and clay portions together make up to 99% of bulk sediment. Apart from some slight fluctuations, silt (13–68%) and clay (14–67%) contents do not change significantly with depth. Average silt contents ranged from 10 to 41% in the east, where the relatively broader shelf receives large amounts of terrigenous

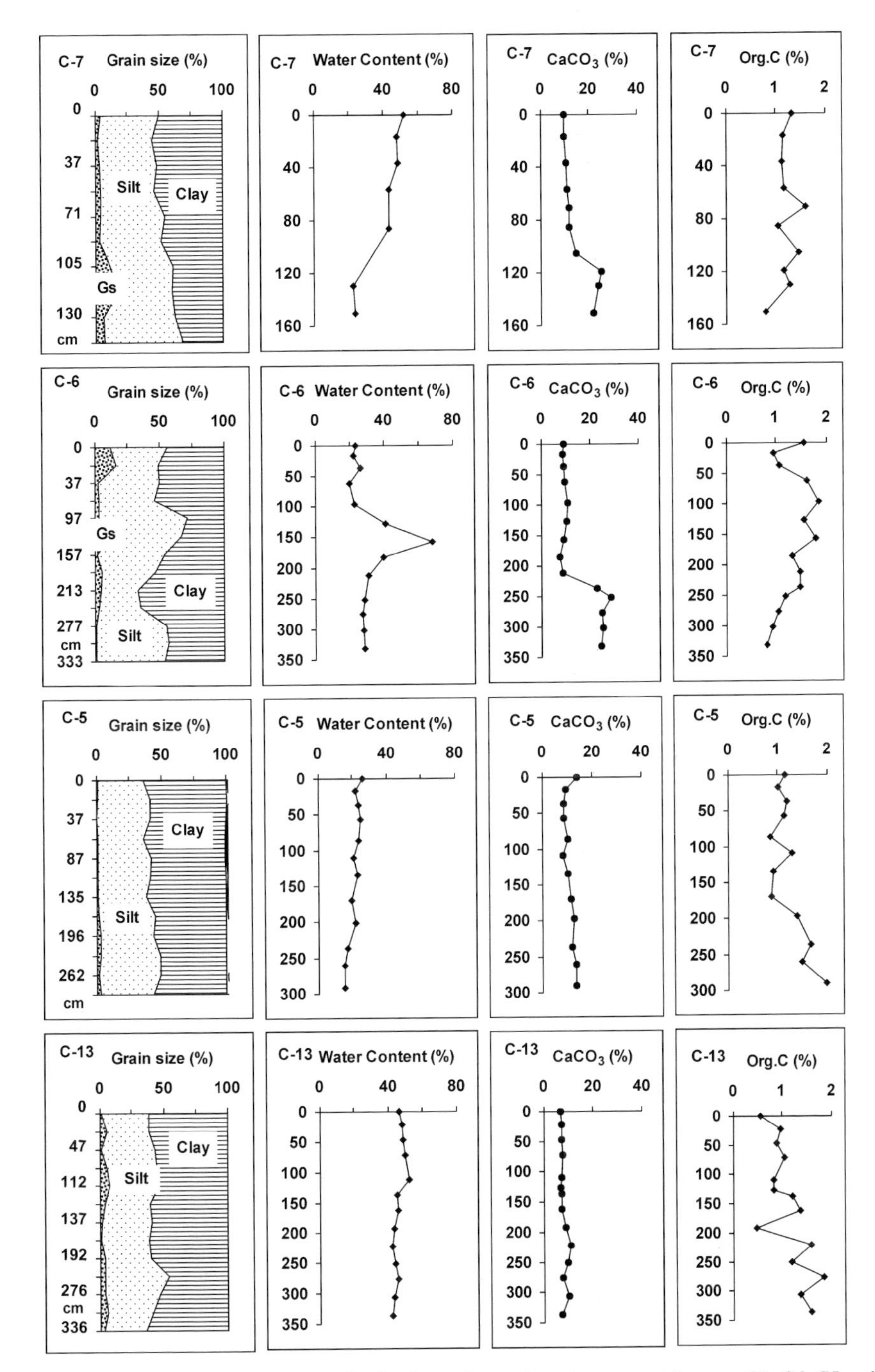

Fig. 19. Grain size, water content, carbonate distribution and organic carbon content in cores C7, C6, C5 and C13.

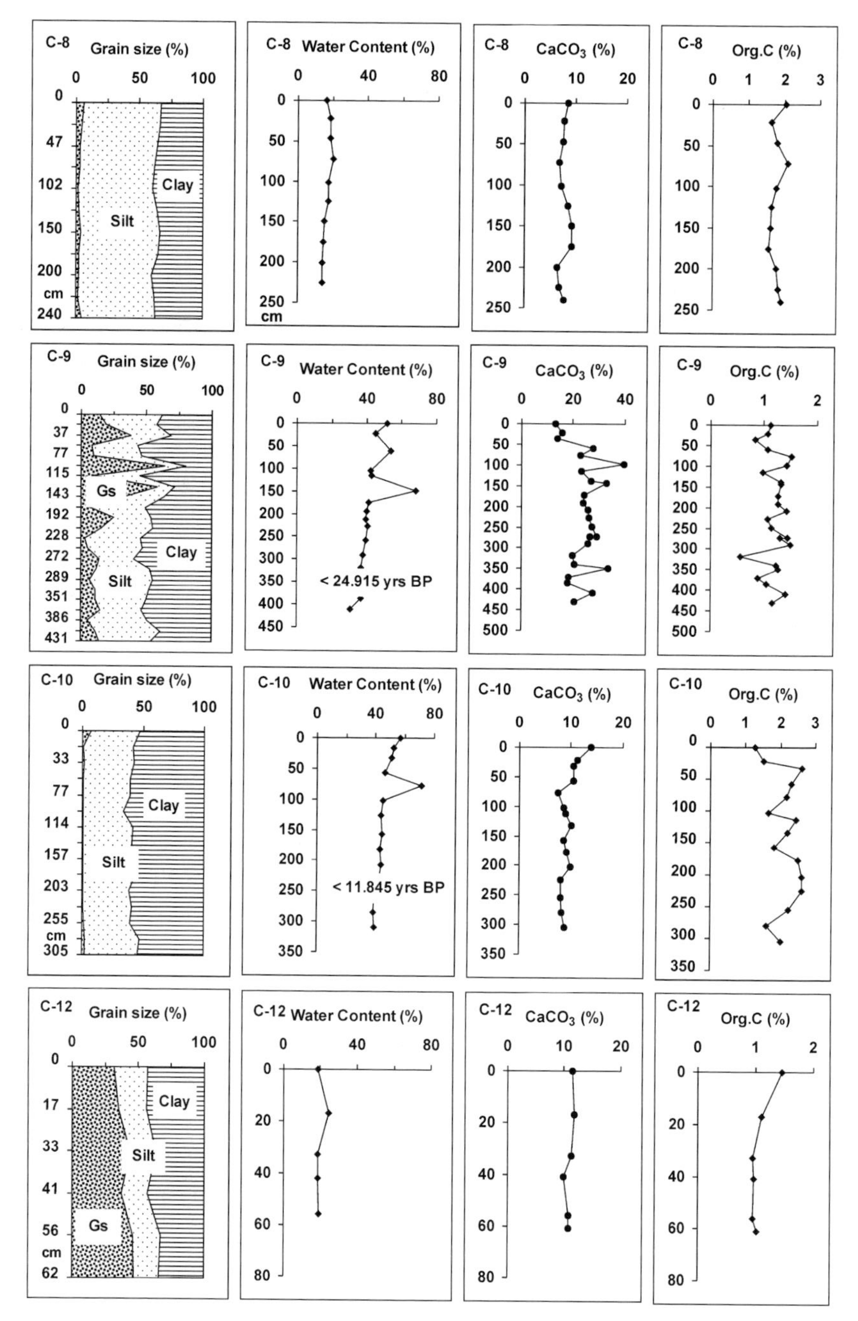

Fig. 20. Grain size, water content, carbonate distribution and organic carbon content in cores C8, C9, C10 and C12.

material from the Biga and Gönen rivers. In contrast, in the west of canyon, with no significant fluvial input and relatively narrower shelf, core sediments are marked by much higher silt contents (from 36 to 60%). Ergin & Bodur (1999) showed that silt/clay ratios in surface sediments ranged from 0.3–0.5 in the lower canyon (with lower-energy conditions) to 1.3 in the upper canyon (with higher-energy conditions) as a result of changing hydrography in this region.

In most cores, sand and gravel contents were 1–9%. Exceptionally, sediments from cores C9 and

C12 from both the easterly and westerly down-canyon walls contained 3–65 and 32–47% sand and gravel (Fig. 20), respectively. Another station (C4) located close to easterly islands also had slightly higher sand and gravel percentages (5–18%; Fig. 18).

Core C2 from the canyon head at 99 m water depth and core C3 from the southwestern wall of the canyon (62 m water depth) displayed upcore-fining and downcore-coarsening sediment texture (Fig. 18). For example, from the top to 263 cm depth in core C3, sediments contained 1–5.7% sand and gravel, which abruptly increased to 70.9% at 297–303 cm. Similarly, in core C2, sediments between 0 and 271 cm depth contained 0.1–3% sand and gravel, and below this depth this value reached 55% at 289–297 cm depth (Fig. 18). In particular, the ^{14}C age of the coarser-grained near-base section of core C2 is determined as 11.585 years BP (Fig. 17).

Total carbonate distribution

The percentages of total carbonate contents in most sediment cores varied between 8 and 15%, with no significant downcore changes (Figs 18–20). However, sediments in lower sections of cores C1, C6 and C7 are marked by slightly higher carbonate contents (7–28%) whereas the high mud percentages of sediments remained nearly the same (Figs 18 and 19). Carbonate contents fluctuating over a wide range of values between 13 and 39% (average 24%) were observed in core C9, where sand and gravel were also high (Fig. 20). The coarser-grained bottom section of core C2 displayed the highest carbonate content (29%, Fig. 18), which is due to the presence of various benthic organism remains.

Organic carbon distribution

Organic carbon contents in the sediment cores generally range between 0.47 and 2.1% (on average 1–2%; Figs 18–20), although higher values (2.61%) were obtained in core C10. Except for two cores, C6 and C7 (Fig. 19), sediments from other sites showed significantly increasing organic carbon contents down the cores (Figs 18–20). Nevertheless, as in core C2 (Fig. 18), downcore increases of organic carbon contents (from <0.47% at 30–35 cm to 2.1% at 275 cm) could result, at least in part, from increased sedimentation rates (i.e. earthquake-related sediment mass flows) and thus dilution of organic matter in sediments from early Holocene time. However, this needs to be confirmed by analytical data, which are at present lacking.

Water content distribution

The water contents in the core sediments ranged from 11 to 72%, being relatively low in cores C3, C4, C5, C8 and C12 and high in cores C1, C2, C6, C7, C9, C10 and C13 (Figs 18–20). Decreasing water contents with depth in cores are normally explained in terms of the physical diagenetic process of compaction with sediment burial (Füchtbauer & Müller 1977). The lower water content values in core C12 (19–25%; Fig. 20) are probably due to the higher sand and gravel contents of the sediments.

Results and implications

Seismic stratigraphy and tectonics

According to most of the recent studies that post-dated the great earthquake in the Marmara Sea (which occurred on 17 August 1999), it is suggested that the origin and present tectonics of the Marmara Sea basin can be explained by different phases of the evolution of this basin (İmren *et al.* 2001; Le Pichon *et al.* 2001; Gazioğlu *et al.* 2002; Yaltırak *et al.* 2002; Demirbağ *et al.* 2003; Gökaşan *et al.* 2003; Rangin *et al.* 2004; Şengör *et al.* 2004). Also, these studies further indicate that the origin of the Marmara Sea Basin was controlled by an earlier occurrence of extensional tectonics of the principal deformation zone of the NAFZ in the Marmara Sea. On the other hand, transpression structures observed in deep seismic data from the Marmara Sea were interpreted as the product of active tectonics indicating the collapse phase of the Marmara Sea evolution. Similar active transpressional structures were also observed in deep seismic data from offshore of the northeastern coastline of the Aegean Sea (Boztepe-Güney *et al.* 2001; Ocakoğlu *et al.* 2004, 2005). Marine terraces uplifted 50 m since 225 Ma along the western coast of the Marmara Sea and the Çanakkale Strait prove the activity of the transpressional structures in the Marmara Sea. Evidence of this evolution is also present in the deposits of the Şarköy Canyon.

Four seismic units and pre-Miocene basement underlying the units are observed in the Şarköy Canyon (Figs 8–15). Pre-Miocene basement is accepted as the previous formation for origination of the Marmara Sea Basin (Şengör *et al.* 1985, 2004; Barka & Kadinsky-Cade 1988). Thus, the upper surface morphology of this unit is important to determine the evolution of the Marmara Sea and the Şarköy Canyon. In the study area, the morphology of upper surface of the pre-Miocene basement is clearly seen on the seismic profile of Figure 9, which crosscuts the Karaburun Peninsula

(Cape) and Şarköy Canyon. Between 15 and 45 km on the seismic profile, a ridge-like formation, on the submarine extension of the Karaburun Cape, is observed (Fig. 9). The basement upper surface gradually becomes shallower towards the northwestern end of the profile (Fig. 9). This morphology shows a syncline-type structure where the profile crosscuts the Şarköy Canyon on the NW side of the ridge. Gökaşan *et al.* (2003) suggested that uplift on the pre-Miocene basement has been controlled by a large positive flower structure, which also forms the Karaburun Peninsula (Cape) on land (Fig. 9). Those researchers further argued that the Şarköy Canyon was formed by subsidence in front of this transpressional structure (Fig. 9).

The reflection configuration of deposits that filled the canyon (Units 1 and 2) seems similar to that of the Plio-Quaternary and older deposits of the Tekirdağ Basin (Okay *et al.* 1999, 2004). This may indicate that Units 1 and 2 originally constituted the southern extension of deposits of the Tekirdağ Basin. If this is the case, Units 1 and 2 should have been deposited during the opening phase of the Marmara Sea Basin; thus, these must be pre-tectonic sediments that preceded transpressional motion indicating the collapse period of the basin. High-resolution seismic data show the folded parallel strata of Units 1 and 2 without thickness variations of the sediment layers towards the fold axis or limbs (Figs 10–14). This indicates that the initial base level of these deposits should have been horizontal, similar to the basin floor in the Marmara Sea; thus, deposition of Units 1 and 2 was prior to transpressional tectonics in the Marmara Sea.

On the other hand, Units 3 and 4 are deposits related to the latest phase of evolution of the Şarköy Canyon. Unit 3 indicates a delta deposition developed from SE to NW during the lowstand environment. ^{14}C dating on samples from this unit indicates that it was deposited during the last glacial maximum in the Marmara Sea. Deposition of Unit 4 followed Unit 3, indicating a highstand environment. Seismic stratigraphic interpretation and ^{14}C dating of sediment samples show that these deposits imply the present environment in the Marmara Sea, which has existed since the end of the last glacial period.

It is shown that the study area has been strongly influenced by tectonic activity, which is reflected in the sedimentary units by faulting and folding. Vertical faults are generally observed on seismic profiles (Figs 8–13). Some of these faults affect the whole sediment column and the sea floor, indicating that they are active. Some faults coalescence at depth and form flower structures (Figs 9, 12 and 13). The lack of vertical displacements on sedimentary layers along these faults indicates that the slip motion is in the third dimension. This evidence shows that these faults may be strike-slip faults. Because the core of the largest flower structure in the study area is uplifted, some of these strike-slip faults also have a compressional component (Fig. 9). Transpressional effects in the study area are observed better in high-resolution seismic data. Gently folded parallel reflections of Units 1 and 2 are tightly folded along some zones on profiles (inset of Fig. 10c; compare insets a and b of Fig. 11), indicating that these sediments have been affected along these zones by transpressional tectonics. These layers are delimited by some vertical faults, which are probably reverse faults (Figs 10 and 11). The lack of thickness variations of sediment layers of Units 1 and 2 towards the fold axis or limbs, and of reverse faults, indicates that Units 1 and 2 are pre-tectonic deposits that preceded transpressional tectonics. On the other hand, Unit 4 covers the folded layers, and therefore consists of post-tectonic deposits. In the study area, some vertical displacements are also interpreted as evidence of normal faults (Fig. 8c). These faults should have caused subsidence of the Şarköy Canyon.

Seismic data also indicate several slides or slumps, and secondary canyons at the shelf edge and slope (Figs 8b, 15 and 16). The largest mass movement, which was determined as a mudflow by Gazioğlu *et al.* (2002), is at the SW corner of the Tekirdağ Basin (Fig. 1b). Bathymetric data indicate that the mass of this landslide almost completely filled the NE extension of the Şarköy Canyon along the southern slope of the Tekirdağ Basin (Fig. 2b). This material also delimits the secondary canyons on the slope (Figs 2b and 16). These landslides must be related to active tectonism in and around the Şarköy Canyon.

Evolution of the Şarköy Canyon

The results of seismic stratigraphic interpretations indicate that the Şarköy Canyon was initially a flat-lying basin possibly forming the southern extension of the Tekirdağ Basin during the initial phase of Marmara Sea formation. In this period, Late Miocene and Plio-Quaternary aged Units 1 and 2 were deposited as basinal deposits with horizontal parallel reflections. During the collapse phase of the Marmara Sea, the basin was compressed in a NW–SE orientation (Kahle *et al.* 1998; Yaltırak *et al.* 1998, 2000; Gökaşan *et al.* 2003) possibly as a result of the southerly curvature of the NAFZ through the study area and westerly movement of Anatolia (e.g. Yılmaz *et al.* 2000). During this phase, a positive flower structure formed the Karaburun Peninsula (Cape) as a ridge. On the northwestern side of this ridge, pre-Miocene basement started to be shortened and subsided, and Miocene and younger deposits on the basement were folded. The subsided region formed the Şarköy Canyon. After its formation, the Şarköy Canyon has experienced both depositional and erosional events as a result of the tectonic activity and sea-level fluctuations. In the last glacial maximum and the early phase of deglaciation, delta development occurred (Unit 3). This delta formation is completely preserved on the sea floor except for a very thin drape of current deposition. Offlaps of this delta are still forming the present southeastern slope of the canyon. This may indicate that the last sea-level rise has been very fast. It may have been caused by blocking and sudden release of the raised Mediterranean waters by the Çanakkale–Gelibolu Ridge. In the present stage of this evolution, sediment drifting and contourite-type deposition have occurred as a result of currents along the present canyon system. In addition, at the shelf edge and the slope of the Tekirdağ Basin, a secondary canyon development and submarine landslides give the present shape of the Şarköy Canyon. In particular, the large mass movement at the SE corner of the slope has strongly affected the present morphology of the canyon at the slope.

The geological evolution of the Şarköy Canyon, based on available data, seems to be controlled by both the Plio-Quaternary tectonics of the Çanakkale Strait (Elmas & Meriç 1998; Yaltırak *et al.* 2000, 2002; Yaltırak & Alpar 2002) and Tekirdağ Basin (Wong *et al.* 1995; Okay *et al.* 1999, 2004; Le Pichon *et al.* 2001; Demirbağ *et al.* 2003; Gökaşan *et al.* 2003; Seeber *et al.* 2004) and Quaternary sea-level changes between the Black Sea and Aegean or Mediterranean Sea (Stanley & Blanpied 1980; Smith *et al.* 1995; Ergin *et al.* 1997; Görür *et al.* 1997; Aksu *et al.* 1999; Çağatay *et al.* 2000; Hiscott & Aksu 2002; Kaminski *et al.* 2002; Fig. 3). According to Yaltırak *et al.* (1998, 2000, 2002) and Gökaşan *et al.* (2003), the Western Marmara Sea including the Çanakkale Strait resulted from regional uplift of the Gelibolu and Biga Peninsulas as a consequence of compressional events generated by strike-slip faulting activity of the Ganos Fault (a northern strand of the North Anatolian Fault) and escape tectonics of the Anatolian Block. Tectonic activity in the region continues at present, mainly in the form of earthquakes (Taymaz *et al.* 1991, 2001, 2004; Erdik *et al.* 2004; Sato *et al.* 2004) and coastal uplift (Yaltırak *et al.* 2002). For example, the western Marmara shelf (including the Çanakkale Strait) has been undergoing tectonic uplift at an average rate of 0.40 mm a^{-1} since 225 ka (Yaltırak *et al.* 2002).

Sedimentary characteristics

Radiocarbon ages obtained from the two cores (*c.* 11–12 ka BP) suggest that sediments in this study must have accumulated since pre-early Holocene time, following the Flandrian transgression. At that time (near the Pleistocene–Holocene boundary), the Marmara Sea was a lake with palaeoshorelines at present-day 90–100 m water depth, before the Mediterranean incursion commenced into the Marmara Sea through the Dardanelles or Çanakkale Strait (Stanley & Blanpied 1980; Aksu *et al.* 1999, 2002; Çağatay *et al.* 2000; Caner & Algan 2002; Hiscott & Aksu 2002; Kaminski *et al.* 2002; Major *et al.* 2002; Tolun *et al.* 2002). The age of the base of core site C9 (*c.* 25 ka BP) is known as a period of regression during the late glacial maximum, just before the sea level reached its lowest position (−110 to −120 m) at *c.* 18 ka (Aksu *et al.* 2002; Caner & Algan 2002; Kaminski *et al.* 2002; Mudie *et al.* 2002; Perissoratis & Conispoliatis 2003).

Low-calcareous (8–15%) mud (rich in siliciclastic silt and clay) is the dominant sediment type covering the sea floor of the Şarköy Canyon. This type of sediment in the western Marmara Sea is also known as 'mud drape' (Aksu *et al.* 1999; Hiscott & Aksu 2002), 'Unit 1' (Çağatay *et al.* 2000), 'Unit A' (Smith *et al.* 1995), and Unit 4 of the present study. It corresponds in age to the Holocene and must have deposited since about 12 ka BP. Based on high-resolution shallow-seismic records, thicknesses of this Holocene mud drape in some localities in the Şarköy Canyon varied between 0 and 18 m (Smith *et al.* 1995; Aksu *et al.* 1999), being lower in the SW and higher in the NE (Aksu *et al.* 1999) probably because of proximity to the Biga and Gönen rivers.

The presence of relatively higher sand and gravel percentages at core sites C9 and C12, where present-day shelf edges occur at water depths of 100 and 92 m, closely corresponds to palaeoshorelines during the last sea-level lowstand (Skene *et al.* 1998). This is further confirmed by the changes in grain size in sediment cores, where the transition from a coarser-grained bottom section to a finer-grained upper section reflects that this boundary site was once possibly a palaeoshore and thus subjected to higher wave energies during sea-level lowstands, when the Marmara Sea was a lake (Stanley & Blanpied 1980; Çağatay *et al.* 2000; Aksu *et al.* 2002; Hiscott & Aksu 2002; Kaminski *et al.* 2002). The presence of coarser-grained and palaeoshore sediments underlying transgressive shelf mud of Holocene age at the head of the Şarköy Canyon was also previously reported (Ergin *et al.* 1999).

Available sedimentary carbonate data from the western Marmara Sea (Ergin *et al.* 1991, 1993, 1997; Bodur & Ergin 1994), reveal similar results to those presented here. Benthic and planktonic foraminiferal groups are abundant in the western

Marmara sediments (Aksu *et al.* 2002; Kaminski *et al.* 2002), and probably constitute the majority of the carbonate contents. Highest carbonate contents of biogenic origin were found in the bottom section of core C2 (dated at 11.585 ka BP; Fig. 17), which apparently accumulated during sea-level lowstands during the early Holocene. Such 'bioherm' accumulations in the Marmara Sea were also observed on seismic profiles (Aksu *et al.* 1999; Hiscott & Aksu 2002) and in sediments from other studies (Ergin *et al.* 1997, 1999; Hiscott & Aksu 2002).

The organic carbon contents of sediments from the Şarköy Canyon, in general, compared with those from other marine Mediterranean regions (<1%; Emelyanov & Shimkus 1986) seem to be rather high because of the relatively high primary organic productivity in the Marmara Sea (Ergin *et al.* 1993, 1994). The overall high organic carbon percentages in sediments from the Marmara Sea, however, could also be derived from anthropogenic and terrigenous sources around the Marmara region, as well as from Black Sea inflows (Ergin *et al.* 1993). The slight increases in organic carbon contents in the topmost sections suggest recent deposition of organic matter before burial and diagenetic decomposition. Recent studies (Çağatay *et al.* 2000; Aksu *et al.* 2002; Caner & Algan 2002; Tolun *et al.* 2002; Sperling *et al.* 2003) have shown that organic carbon enrichment in late Quaternary sediments from the Marmara Sea can be related indirectly to climatic changes. For example, around 4750–3200 years and 11–6 ka BP, large amounts of fresh waters from the Black Sea flowed into the Marmara Sea; this resulted in water stratification and thus deposition of organic-rich sediments or 'sapropel'. It is therefore reasonable to expect that upcore-decreasing organic carbon contents in the cores of this study would partly reflect the early to middle Holocene water circulation in this sea.

Although sediments displayed similar composition (i.e. mud, organic carbon, carbonate), cores C1, C2, C10, C7 and C13 are characterized by generally higher water contents (40–49%) whereas sediments in cores C6, C5, C8 and C3 are marked by lower water contents (16–31%) (Figs 18–21). The only reasonable explanation for this seems to be the effects of regional neotectonics, unless other data exist to make this hypothesis questionable.

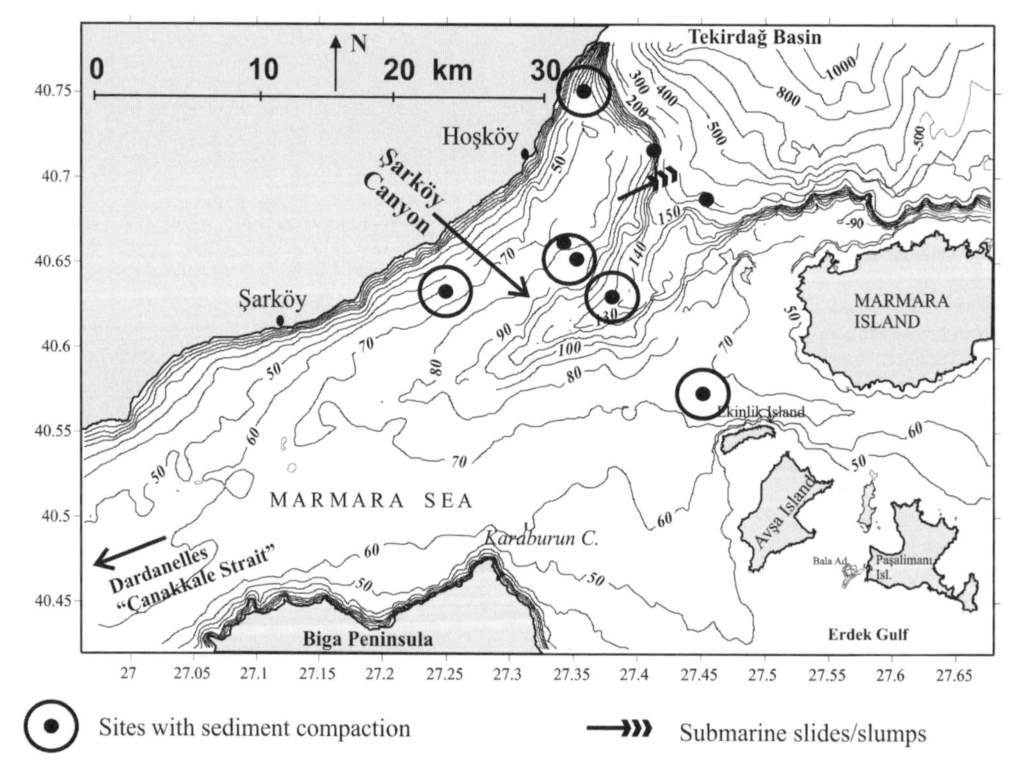

Fig. 21. Bathymetric map of the study area (based on SHOD 1983 and MTA 2004) showing the sites of sediment compaction as determined on the basis of lower water contents and submarine slides or slumps.

We acknowledge the Turkish Scientific and Technical Research Council (TÜBİTAK) for funding this work within the projects YDABAG-102Y113 and ÇAYDAG-104Y024. We also thank the officers and crew as well as scientists and technicians onboard the R.V. *MTA Sismik-1* of the General Directorate of the Mineral Research and Exploration, and onboard the TCG *Çeşme* and onboard TCG *Çubuklu* of the Turkish Navy, Department of Navigation, Hydrography and Oceanography. The raw seismic data were processed using appropriate packages in the Geophysical Department of Faculty of Mines, the İstanbul Technical University. Detailed reviews by E. Demirbağ and an anonymous reviewer helped significantly to improve the final paper. We are grateful to T. Taymaz for his editorial help and patience.

References

AKSU, A. E., HISCOTT, R. N. & YAŞAR, D. 1999. Oscillating Quaternary water levels of the Marmara Sea and vigorous outflow into the Aegean Sea from the Marmara Sea-Black Sea drainage corridor. *Marine Geology*, **153**, 275–302.

AKSU, A. E., YALTIRAK, C. & HISCOTT, R. N. 2002. Quaternary palaeoclimatic-palaeoceanographic and tectonic evolution of the Marmara Sea and environs. *Marine Geology*, **190**, 9–18.

BARKA, A. A. & KADINSKY-CADE, K. 1988. Strike-slip fault geometry in Turkey and its influence on earthquake activity. *Tectonics*, **7**, 663–684.

BAŞTÜRK, Ö., SAYDAM, A. C., SALIHOĞLU, İ. & YILMAZ, A. 1986. *Oceanography of the Turkish Straits, Vol. III: Health of the Straits, II: Chemical and Environmental Aspects of the Sea of Marmara.* Institute of Marine Sciences, METU, Erdemli-İçel, Turkey, First Annual Report.

BEŞIKTEPE, Ş. T., SUR, H. I., ÖZSOY, E., LATIF, M. A., OĞUZ, T. & ÜNLÜATA, A. 1994. The circulation and hydrography of the Marmara Sea. *Progress in Oceanography*, **34**, 285–334.

BODUR, M. N. & ERGIN, M. 1994. Geochemical characteristics of the Late-Holocene sediments from the Sea of Marmara. *Chemical Geology*, **115**, 73–101.

BOZTEPE-GÜNEY, A., YILMAZ, Y., DEMIRBAĞ, E., ECEVITOĞLU, B., ARZUMAN, S. & KUŞÇU, İ. 2001. Reflection seismic study across the continental shelf of Baba Burnu promontory of Biga Peninsula, northwest Turkey. *Marine Geology*, **176**, 75–85.

BROWN, L. F., JR & FISHER, W. L. 1979. *Seismic Stratigraphic Interpretation and Petroleum Exploration.* AAPG Continuing Education Course Note Series, **16**.

CANER, H. & ALGAN, O. 2002. Palynology of sapropelic layers from the Marmara Sea. *Marine Geology*, **190**, 35–46.

ÇAĞATAY, N., GÖRÜR, N., ALPAR, B. *ET AL.* 1998. Geological evolution of the Gulf of Saros, NE Aegean Sea. *Geo-Marine Letters*, **18**, 1–9.

ÇAĞATAY, M. N., GÖRÜR, N., ALGAN, O. *ET AL.* 2000. Last Glacial–Holocene palaeoceanography of the Sea of Marmara: timing of connections with the Mediterranean and the Black Seas. *Marine Geology*, **167**, 191–206.

DEMIRBAĞ, E., GÖKAŞAN, E., KURT, H. & TEPE, C. M. 1998. [On the origin of the formation of the NE Dardanelles.] Workshop IV, Abstracts, Tubitak National Marine Geology and Geophysics Program, 1–3. (In Turkish)

DEMIRBAĞ, E., RANGIN, C., LE PICHON, X. & ŞENGÖR, A. M. C. 2003. Investigation of the tectonics of the Main Marmara Fault by means of deep-towed seismic data. *Tectonophysics*, **361**, 1–19.

EIEI (ELEKTRIK İŞLERI ETÜT İDARESI) 1984. *1981 Water Year Discharges.* Elektrik İsleri Etüd İdaresi Genel Direktörlüğü, Ankara.

EIEI (ELEKTRIK İŞLERI ETÜT İDARESI) 1993. *Sediment data and sediment transport amount for surface waters in Turkey.* Elektrik İsleri Etüd İdaresi Genel Direktörlüğü, Ankara, **93–59**.

ELMAS, A. & MERİÇ, E. 1998. The seaway connection between the Sea of Marmara and Mediterranean: tectonic development of the Dardanelles. *International Geology Review*, **40**, 144–163.

EMELYANOV, E. V. & SHIMKUS, K. M. 1986. *Geochemistry and Sedimentology of the Mediterranean Sea.* Reidel, Dordrecht.

ERDIK, M., DEMIRCIOĞLU, M., SESETYAN, K., DURUKAL, E. & SIYAHI, B. 2004. Earthquake hazard in Marmara Region, Turkey. *Soil Dynamics and Earthquake Engineering*, **24**, 605–631.

ERGIN, M. & BODUR, M. N. 1999. Silt/clay fractionation in surficial marine sediments: implication for water movement and sediment transport paths in a semi-enclosed and two-layered flow system (northeastern Mediterranean Sea). *Geo-Marine Letters*, **18**, 225–233.

ERGIN, M., BODUR, M. N. & EDIGER, V. 1991. Distribution of surficial shelf sediments in the northeastern and southwestern parts of the Sea of Marmara: strait and canyon regimes of the Dardanelles and Bosporus. *Marine Geology*, **96**, 313–340.

ERGIN, M., BODUR, M. N., EDIGER, V. & YILMAZ, A. 1993. Organic carbon distribution in the surface sediments of the Sea of Marmara and its control by the inflows from adjacent water masses. *Marine Chemistry*, **41**, 311–326.

ERGIN, M., BODUR, M. N., YILDIZ, M., EDIGER, D., EDIGER, V. & YÜCSOY, F. 1994. Sedimentation rates in the Sea of Marmara: a comparison of results based on organic carbon-primary productivity and Pb-210 dating. *Continental Shelf Research*, **14**, 1371–1387.

ERGIN, M., KAZANCI, N., VAROL, B., İLERI, Ö. & KARADENIZLI, L. 1997. Sea-level changes and related depositional environments on the southern Marmara Shelf. *Marine Geology*, **140**, 391–403.

ERGIN, M., KAPUR, S., KARAKAŞ, Z., AKÇA, E., KANGAL, Ö. & KESKIN, Ş. 1999. Grain size and clay mineralogy of Late Quaternary sediments on a tectonically active shelf, the southern Sea of Marmara: clues to hydrographic, tectonic and climatic evolution. *Geological Journal*, **34**, 199–210.

ERGIN, M., SARIKAVAK, K., KESKIN, Ş., HAKYEMEZ, Y. & ULUADAM, E. 2005. Late Quaternary sea-level changes in tectonics and sedimentation in a submarine canyon off Hoşköy, SW Marmara Sea (Turkey). *In: International Symposium on the Geodynamics of*

Eastern Mediterranean: Active Tectonics of the Aegean Region, 15–18 June 2005, Abstracts. Kadir Has University, İstanbul.

FOLK, L. 1980. *Petrology of Sedimentary Rocks*. Hemphill, Tulsa, OK.

FÜCHTBAUER, H. & MÜLLER, G. 1977. *Sedimente and Sedimentgesteine*. Schweizerbart, Stuttgart.

GAUDETTE, H. E., FLIGHT, W. R., TONER, L. & FOLGER, D. V. 1974. An inexpensive titration method for determination of organic carbon in recent sediment. *Journal of Sedimentary Petrology*, **44**, 249–253.

GAZIOĞLU, C., GÖKAŞAN, E., ALGAN, O., YÜCEL, Z., TOK, B. & DOĞAN, E. 2002. Morphologic features of the Marmara Sea from multi-beam data. *Marine Geology*, **190**, 397–420.

GÖKAŞAN, E., USTAÖMER, T., GAZIOĞLU, C. *ET AL.* 2003. Morpho-tectonic evolution of the Marmara Sea inferred from multi-beam bathymetric and seismic data. *Geo-Marine Letters*, **23**, 19–33.

GÖRÜR, N., ÇAĞATAY, M. N., SAKINÇ, M., SÜMENGEN, M., ŞENTÜRK, K., YALTIRAK, C. & TCHAPALYGA, A. 1997. Origin of the Sea of Marmara as deduced from Neogene to Quaternary paleogeographic evolution of its frame. *International Geology Review*, **39**, 342–352.

GÜRBÜZ, C., AKTAR, M., EYIDOĞAN, H. *ET AL.* 2000. The seismotectonics of the Marmara Region (Turkey): results from a microseismic experiment. *Tectonophysics*, **316**, 1–17.

HALL, J., AKSU, A. E. & YALTIRAK, C. 2005. Miocene to Recent tectonic evolution of the Eastern Mediterranean. *Marine Geology* (special issue), **221**, 1–440.

HISCOTT, R. N. & AKSU, A. E. 2002. Late Quaternary history of the Marmara Sea and Black Sea from high-resolution seismic and gravity-core studies. *Marine Geology*, **190**, 261–282.

İMREN, C., LE PICHON, X., RANGIN, C., DEMIRBAĞ, E., ECEVITOĞLU, B. & GÖRÜR, N. 2001. The North Anatolian Fault within the Sea of Marmara: a new evaluation based on multichannel seismic and multibeam data. *Earth and Planetary Science Letters*, **186**, 143–158.

KAHLE, H. G., STRAUB, C., REILINGER, R. *ET AL.* 1998. The strain rate field in the eastern Mediterranean region, estimated by repeated GPS measurements. *Tectonophysics*, **294**, 237–252.

KAMINSKI, M. A., AKSU, A., BOX, M., HISCOTT, R. N., FILIPESCU, S. & AL-SALAMEEN, M. 2002. Late Glacial to Holocene benthic foraminifera in the Marmara Sea: implications for Black Sea–Mediterranean Sea connections following the last deglacation. *Marine Geology*, **190**, 165–202.

LE PICHON, X., ŞENGÖR, A. M. C., DEMIRBAĞ, E. *ET AL.* 2001. The active main Marmara fault. *Earth and Planetary Science Letters*, **192**, 595–616.

LORING, D. H. & RANTALA, R. T. T. 1992. Manual for the geochemical analyses of marine sediments and suspended particulate matter. *Earth-Science Review*, **32**, 235–283.

MAJOR, C., RYAN, W., LERICOLAIS, G. & HAJDAS, I. 2002. Constraints on Black Sea outflow to the Sea of Marmara during the last glacial–interglacial transition. *Marine Geology*, **190**, 19–34.

MITCHUM,R. M. JR., VAIL, P. R. & SANGREE, J. B. 1977. Seismic stratigraphy and global changes of sea level, part 6: stratigraphic interpretation of seismic reflection patterns in depositional sequences. *In*: PAYTON, C. E. (ed.) *Seismic Stratigraphy—Applications to Hydrocarbon Exploration*. American Association of Petroleum Geologists, Memoirs, **26**, 117–133.

MTA, 2004. *Bathymetric Chart of the Sea of Marmara, scale: 1/250 000*. General Directorate of Mineral Research and Exploration, Ankara.

MUDIE, P. J., ROCHON, A. & AKSU, A. E. 2002. Pollen stratigraphy of Late Quaternary cores from Marmara Sea: land–sea correlation and paleoclimatic history. *Marine Geology*, **190**, 233–260.

OCAKOĞLU, N., DEMIRBAĞ, E. & KUŞÇU, İ. 2004. Neotectonic structures in the area offshore of Alaçati, Doğanbey and Kuşadası (eastern Turkey): evidence of strike-slip faulting in the Aegean extensional province. *Tectonophysics*, **391**, 67–83.

OCAKOĞLU, N., DEMIRBAĞ, E. & KUŞÇU, İ. 2005. Neotectonic structures in the Gulf of İzmir and surrounding regions (western Turkey): evidence of transpressional faulting in the Aegean extensional regime. *Marine Geology*, **219**, 155–171.

ODP SHIPBOARD MEASUREMENTS PANEL 1991. *Recommended methods for the discrete measurement of index properties on the* JOIDES Resolution*: water content, bulk density, and grain density*. Ocean Drilling Program College Station, TX.

OKAY, A. İ. & TANSEL, İ. 1992. New data on the upper age of the Intra-Pontid Ocean from north of Şarköy (Thrace). *MTA Bulletin*, **114**, 23–26.

OKAY, A. I., SENGÖR, A. M. C. & GÖRÜR, N. 1994. Kinematic history of the opening of the Black Sea and its effects on surrounding regions. *Geology*, **22**, 267–270.

OKAY, A. I., DEMIRBAĞ, E., KURT, H., OKAY, N. & KUŞÇU, İ. 1999. An active, deep marine strike-slip basin along the North Anatolian Fault, Turkey. *Tectonics*, **18**, 129–147.

OKAY, A. I., KAŞLILAR-ÖZCAN, A., İMREN, C., BOZTEPE-GÜNEY, A., DEMIRBAĞ, E. & KUŞÇU, İ. 2000. Active faults and evolving strike-slip basins in the Marmara Sea, northwest Turkey: a multichannel seismic reflection study. *Tectonophysics*, **321**, 189–218.

OKAY, A. I., TÜYSÜZ, O. & KAYA, Ş. 2004. From transpression to transtension: changes in morphology and structure around a bend on the North Anatolian Fault in the Marmara region. *Tectonophysics*, **391**, 259–282.

OKAY, N. & ERGÜN, B. 2005. Source of the basinal sediments in the Marmara Sea investigated using heavy minerals in the modern beach sediments. *Marine Geology*, **216**, 1–15.

ÖZSOY, E., OĞUZ, T., LATIF, M. A. & ÜNLÜATA, Ü. 1986. *Oceanography of the Turkish Straits—Volume I, Physical Oceanography of the Turkish Straits*. Institute of Marine Sciences, METU, Erdemli, İçel, Turkey, First Annual Report.

PERISSORATIS, C. & CONISPOLIATIS, N. 2003. The impacts of sea-level changes during latest Pleistocene and Holocene times on the morphology of the Ionian and Aegean seas (SE Alpine Europe). *Marine Geology*, **196**, 145–156.

RANGIN, C., LE PICHON, X., DEMIRBAĞ, E. & İMREN, C. 2004. Strain localization in the Sea of Marmara: propagation of the North Anatolian Fault in a now inactive pull-apart. *Tectonics*, **23**, TC2014, doi:10.1029/2002TC001437.

SANER, S. 1985. Saros Körfezi dolayının çökelme istifleri ve tektonik yerleşimi, Kuzeydoğu Ege Denizi, Türkiye. *TJK Bülteni*, **28**, 1–18.

SATO, T., KASAHARA, J., TAYMAZ, T., ITO, M., KAMIMURA, A., HAYAKAWA, T. & TAN, O. 2004. A study of microearthquake seismicity and focal mechanisms within the Sea of Marmara (NW Turkey) using ocean bottom seismometers (OBSs). *Tectonophysics*, **391**, 303–314.

SAUNDERS, P., PRIESTLEY, K. & TAYMAZ, T. 1998. Variations in crustal structure beneath western Turkey. *Geophysical Journal International*, **134**, 373–389.

SEEBER, L., EMRE, O., CORNIER, M. H. *ET AL*. 2004. Uplift and subsidence from oblique slip: the Ganos–Marmara bend of the North Anatolian Transform, western Turkey. *Tectonophysics*, **391**, 239–258.

ŞENGÖR, A. M. C. & YILMAZ, Y. 1981. Tethyan evolution of Turkey: a plate tectonic approach. *Tectonophysics*, **75**, 181–241.

ŞENGÖR, A. M. C., GÖRÜR, N. & ŞAROĞLU, F. 1985. Strike-slip faulting and related basin formation zones of tectonic escape: Turkey as a case study. *In*: BIDDLE, K. T. & CHRISTIE-BLICK, N. (eds) *Strike-Slip Deformation, Basin Formation and Sedimentation*. Society of Economic Paleontologists and Mineralogists, Special Publications, **37**, 227–264.

ŞENGÖR, A. M. C., TÜYSÜZ, O., İMREN, C. *ET AL*. 2004. The North Anatolian Fault. A new look. *Annual Review of Earth and Planetary Sciences*, **33**, 1–75

SHEPARD, F. P. & DILL, R. F. 1966. *Submarine Canyons and other Sea Valleys*. Rand McNally, Chicago, IL.

SHOD, 1983. *Bathymetric chart of the Marmara Sea between Hoşköy and Gelibolu. Scale 1:75 000. Map No. 295 (INT 3752)*. Turkish Navy, Department of Navigation, Hydrography and Oceanography, İstanbul.

SKENE, K. I., PIPER, D. J. W., AKSU, A. E. & SYVITSKI, J. P. M. 1998. Evaluation of the global oxygen isotope curve as a proxy for Quaternary sea level by modeling of delta progradation. *Journal of Sedimentary Research*, **68**, 1077–1092.

SMITH, A. D., TAYMAZ, T., OKTAY, F. Y. *ET AL*. 1995. High-resolution seismic profiling in the Sea of Marmara (NW Turkey): Late Quaternary sedimentation and sea-level changes. *Geological Society of America Bulletin*, **107**, 923–936.

SPERLING, M., SCHMIEDL, G., HEMLEBEN, CH., EMEIS, K. C., ERLENKEUSER, H. & GROOTES, P. M. 2003. Black Sea impact on the formation of eastern Mediterranean sapropel S1? Evidence from the Marmara Sea. *Palaeogeography, Palaeoclimatology, Palaeoecology*, **190**, 9–21.

STANLEY, D. J. & BLANPIED, C. 1980. Late Quaternary water exchange between the eastern Mediterranean and the Black Sea. *Nature*, **285**, 537–541.

TAYMAZ, T., JACKSON, J. A. & MCKENZIE, D. 1991. Active tectonics of the north and central Aegean Sea. *Geophysical Journal International*, **106**, 433–490.

TAYMAZ, T., KASAHARA, J., HIRN, A. & SATO, T. 2001. Investigations of micro-earthquake activity within the Sea of Marmara and surrounding regions by using ocean bottom seismometers (OBS) and land seismographs: initial results. *In*: TAYMAZ, T. (ed.) *Symposia on Seismotectonics of the North-Western Anatolia—Aegean and Recent Turkish Earthquakes*. ATLAS & ITU Faculty of Mines, İstanbul, 42–51.

TAYMAZ, T., WESTAWAY, R. & REILINGER, R. (eds) 2004. Active Faulting and Crustal Deformation in the Eastern Mediterranean Region. *Tectonophysics*, (special issue) **391**.

TOLUN, L., ÇAĞATAY, N. & CARRIGAN, W. J. 2002. Organic geochemistry and origin of Late Glacial–Holocene sapropelic layers and associated sediments in Marmara Sea. *Marine Geology*, **190**, 47–60.

TÜYSÜZ, O., BARKA, A. A. & YIĞITBAŞ, E. 1998. Geology of the Saros Graben: its implications on the evolution of the North Anatolian Fault in the Ganos–Saros Region, NW Turkey. *Tectonophysics*, **293**, 105–126.

ÜNLÜATA, Ü., OĞUZ, T., LATIF, M. A. & ÖZSOY, E. 1990. On the physical oceanography of the Turkish Straits. *In*: PRATT, L. J. (ed.), *The Physical Oceanography of Sea Straits*. NATO Advanced Study Institutes Series. Series C, **318**, 25–60.

WONG, H. K., LUDMANN, T., ULUĞ, A. & GÖRÜR, N. 1995. The Sea of Marmara: a plate boundary sea in an escape tectonic regime. *Tectonophysics*, **244**, 231–250.

YALTIRAK, C. 2002. Tectonic evolution of the Marmara Sea and its surroundings. *Marine Geology*, **190**, 493–529.

YALTIRAK, C. & ALPAR, B. 2002. Kinematics and evolution of the northern branch of the North Anatolian Fault (Ganos Fault) between the Sea of Marmara and the Gulf of Saros. *Marine Geology*, **190**, 352–366.

YALTIRAK, C., ALPAR, B. & YÜCE, H. 1998. Tectonic elements controlling the evolution of the Gulf of Saros (northeastern Aegean Sea, Turkey). *Tectonophysics*, **300**, 227–248.

YALTIRAK, C., ALPAR, B., SAKINÇ, M. & YÜCE, H. 2000. Origin of the Strait of Çanakkale (Dardanelles): regional tectonics and the Mediterranean–Marmara incursion. *Marine Geology*, **164**, 139–156.

YALTIRAK, C., SAKINÇ, M., AKSU, A. E., HISCOTT, R. N., GALLEB, B. & ÜLGEN, U. B. 2002. Late Pleistocene uplift history along the southwestern Marmara Sea determined from raised coastal deposits and global sea-level variations. *Marine Geology*, **190**, 283–305.

YILMAZ, Y., GENÇ, Ş. C., GÜRER, F. *ET AL*. 2000. When did the western Anatolian grabens begin to develop? *In*: BOZKURT, E., WINCHESTER, J. A. & PIPER, J. D. A. (eds) *Tectonics and Magmatism in Turkey and the Surrounding Area*. Geological Society, London, Special Publications, **173**, 131–162.

Source characteristics of the 6 June 2000 Orta–Çankırı (central Turkey) earthquake: a synthesis of seismological, geological and geodetic (InSAR) observations, and internal deformation of the Anatolian plate

T. TAYMAZ[1], T. J. WRIGHT[2], S. YOLSAL[1], O. TAN[3], E. FIELDING[4] & G. SEYİTOĞLU[5]

[1]*Department of Geophysics, İstanbul Technical University, Maslak-TR 34469, İstanbul, Turkey (e-mail: taymaz@itu.edu.tr)*

[2]*Department of Earth Sciences, University of Oxford, Parks Road, Oxford OX1 3PR, UK*

[3]*TÜBİTAK-Marmara Research Centre, 41470 Gebze, Kocaeli, Turkey*

[4]*Jet Propulsion Laboratory, 4800 Oak Grove Drive, Caltech, Pasadena, CA 91109, USA*

[5]*Department of Geological Engineering, Ankara University, 06100 Ankara, Turkey*

Abstract: This paper is concerned with the seismotectonics of the North Anatolian Fault in the vicinity of the Orta–Çankırı region, and consists of a study of a moderate-sized (Mw = 6. 0) earthquake that occurred on 6 June 2000. The instrumental epicentre of this earthquake is far from the North Anatolian Fault Zone (NAFZ), and rapid focal mechanism solutions of USGS–NEIC and Harvard-CMT also demonstrate that this earthquake is not directly related to the right-lateral movement of the North Anatolian Fault. This earthquake is the only instrumentally recorded event of magnitude (Mw) >5.5 since 1900 between Ankara and Çankırı, and therefore provides valuable data to improve our understanding of the neotectonic framework of NW central Anatolia. Field observations carried out in the vicinity of Orta town and neighbouring villages immediately after the earthquake indicated no apparent surface rupture, but the reported damage was most intense in the villages to the SW of Orta. We used teleseismic long-period P- and SH-body waveforms and first-motion polarities of P-waves, broadband P-waves, and InSAR data to determine the source parameters of the 6 June 2000 (Orta–Çankırı, t_o = 02:41:53.2, Mw = 6. 0) earthquake. We compared the shapes and amplitudes of long-period P- and SH-waveforms recorded by GDSN stations in the distance range 30–90°, for which signal amplitudes were large enough, with synthetic waveforms. The best-fitting fault-plane solution of the Orta–Çankırı earthquake shows normal faulting with a left-lateral component with no apparent surface rupture in the vicinity of the epicentre. The source parameters and uncertainties of this earthquake were: Nodal Plane 1: strike 2° ± 5°, dip 46° ± 5°, rake –29° ± 5°; Nodal Plane 2: strike 113°, dip 70°, rake –132°; principal axes: P = 338° (48°), T = 232° (14°), B = 131° (39°); focal depth 8 ± 2 km (though this does not include uncertainty related to velocity structure), and seismic moment $M_o = (140–185) \times 10^{16}$ N m. Furthermore, analysis of a coseismic interferogram also allows the source mechanism and location of the earthquake to be determined. The InSAR data suggest that the north–south fault plane (Nodal Plane 1 above) was the one that ruptured during the earthquake. The InSAR mechanism is in good agreement with the minimum misfit solution of P- and SH-waveforms. Although the magnitude of slip was poorly constrained, trade-off with the depth range of faulting accurred such that solutions with a large depth range had small values of slip and vice versa. The misfit was small and the geodetic moment constant for fault slips greater than *c.* 1 m. The 6 June 2000 Orta–Çankırı earthquake occurred close to a restraining bend in the east–west-striking rightlateral strike-slip fault that moved in the much larger earthquake of 13 August 1951 (Ms = 6.7). The faulting in this anomalous earthquake could be related to the local geometry of the main strike-slip system, and may not be a reliable guide to the regional strain field in NW central Turkey. We tentatively suggest that one possible explanation for the occurrence of the 6 June 2000 Orta–Çankırı earthquake could be localized clockwise rotations as a result of shear of the lower crust and lithosphere.

The epicentre of the 6 June 2000 Orta earthquake (Mw = 6.0) is far from the North Anatolian Fault Zone (Figs 1 and 2). Rapid focal mechanism solutions of USGS–NEIC and Harvard-CMT also demonstrate that this earthquake is not directly related to the right-lateral movement of the North Anatolian Fault (Fig. 2; Taymaz *et al.* 2002*a*, *b*). Field observations carried out in Orta town and

From: TAYMAZ, T., YILMAZ, Y. & DILEK, Y. (eds) *The Geodynamics of the Aegean and Anatolia.* Geological Society, London, Special Publications, **291**, 259–290.
DOI: 10.1144/SP291.12 0305-8719/07/$15.00

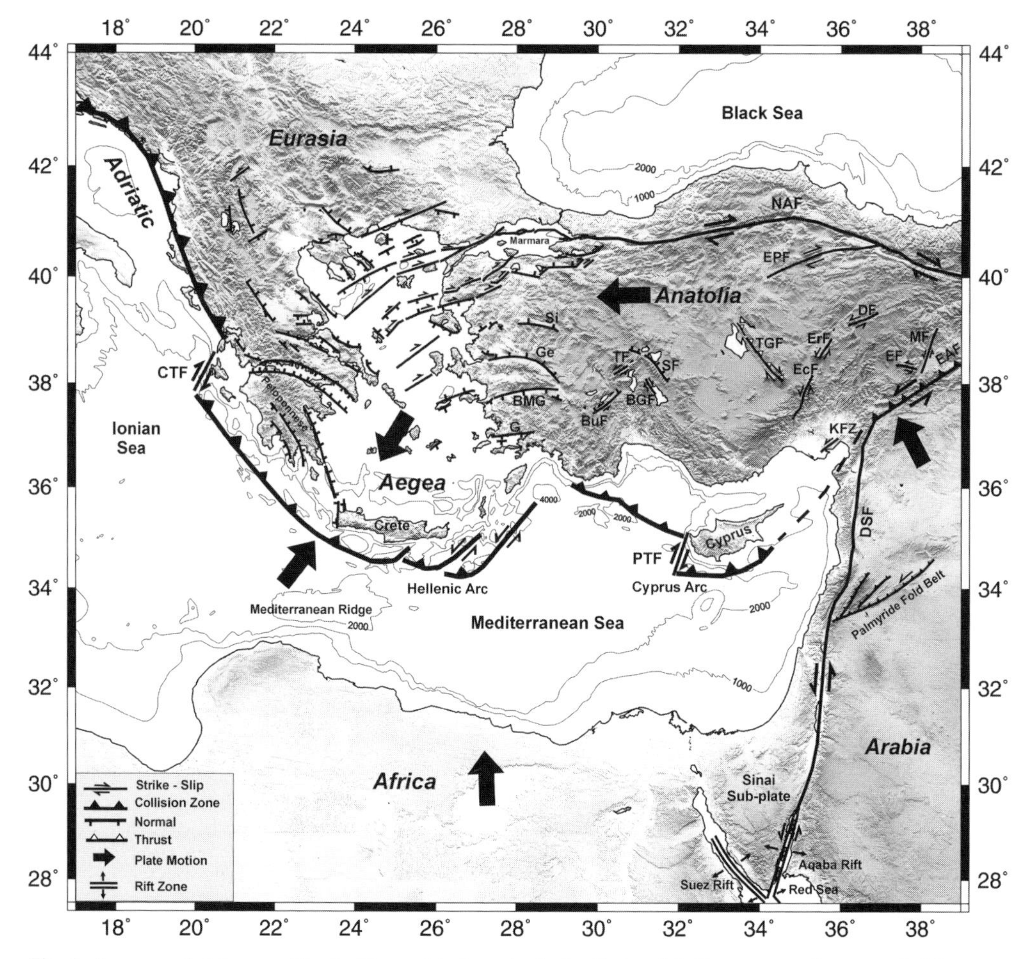

Fig. 1. Summary sketch map of the faulting and bathymetry in the Eastern Mediterranean region, compiled from our observations and those of Le Pichon *et al.* (1984, 2001), Şengör *et al.* (1985), Mercier *et al.* (1989), Taymaz *et al.* (1990, 1991*a*, *b*, 2002*a*, *b*), Şaroğlu *et al.* (1992), Taymaz & Price (1992), Jackson (1994), Price & Scott (1994), Alsdorf *et al.* (1995), Papazachos *et al.* (1998), Kurt *et al.* (1999, 2000), McClusky *et al.* (2000), Taymaz & Tan (2001), and Demirbağ *et al.* (2003). NAF, North Anatolian Fault; EAF, East Anatolian Fault; DSF, Dead Sea Fault; EPF, Ezinepazarı Fault; PTF, Paphos Transform Fault; CTF, Cephalonia Transform Fault; G, Gökova Fault; BMG, Büyük Menderes Graben; Ge, Gediz Graben; Si, Simav Graben; BuF, Burdur Fault; BGF, Beysehir Gölü Fault; TF, Tatarlı Fault; SF, Sultandağ Fault; TGF, Tuz Gölü Fault; EcF, Ecemiş Fault; ErF, Erciyes Fault; DF, Deliler Fault; EF, Elbistan Fault; MF, Malatya Fault; KFZ, Karataş–Osmaniye Fault Zone.

the neighbouring villages immediately after the earthquake indicated no apparent surface rupture, but the damage was most intense in the villages to the SW of Orta. To explain the cause of this earthquake the regional neotectonic framework should be taken into account. There are three main neotectonic elements in NW central Anatolia that correspond to the regional earthquake epicentre distribution ($M > 2$) between 1964 and 2000 (Fig. 2). The first is the well-known North Anatolian Fault Zone (NAFZ), which has the ability to generate earthquakes with magnitudes >5. The second main tectonic element is the Kırıkkale–Erbaa Fault Zone (KEFZ), which is responsible for the 10 June 1985, 14 February 1992 and 14 August 1996 earthquakes. The third tectonic element is a NNE–SSW-trending pinched crustal wedge between Ankara and Çankırı (see below). Its neotectonic importance was not accurately recognized until recently (Seyitoğlu *et al.* 2000, 2001). The closer distribution of epicentre locations (Fig. 2) along the line of Ankara and Çankırı corresponds to the NNE–SSW-trending İzmir–Ankara suture zone.

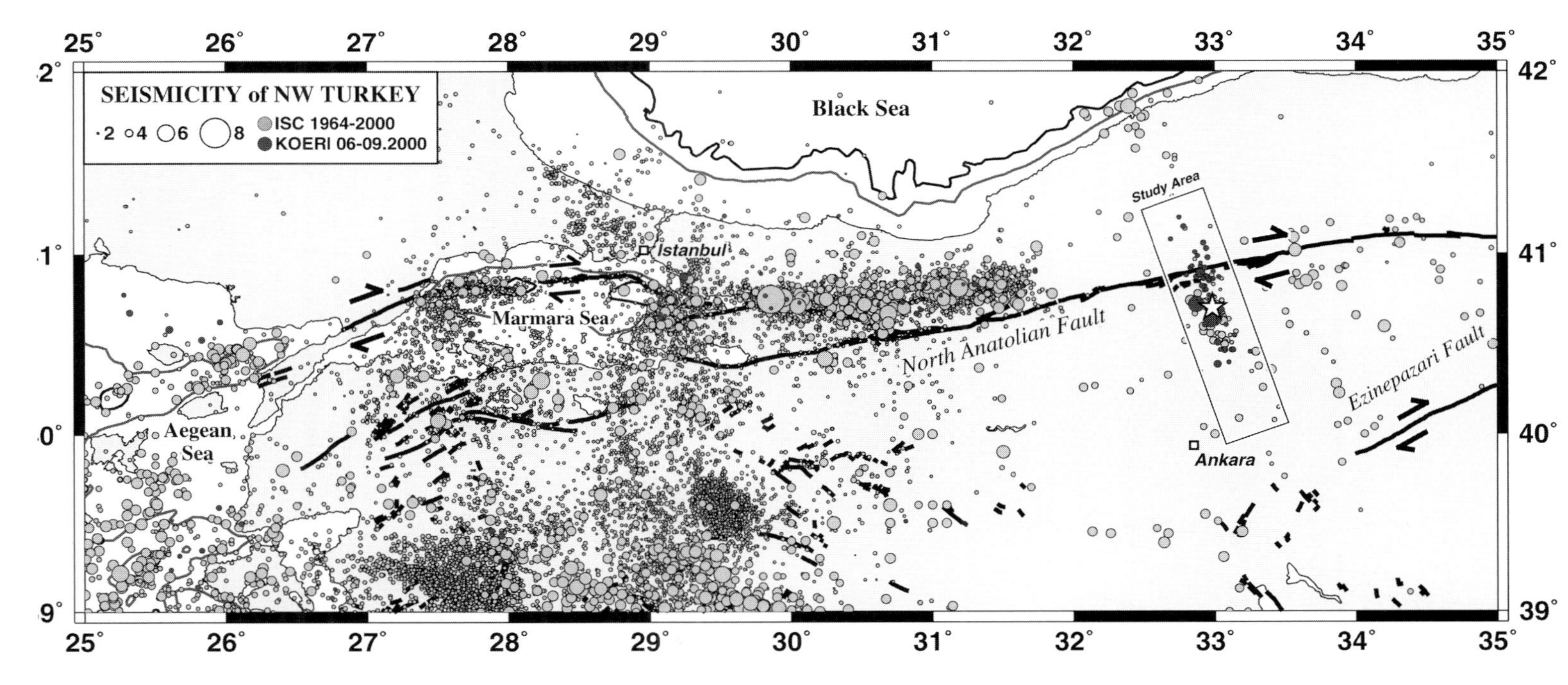

Fig. 2. Seismicity of NW Turkey and the character of the North Anatolian Fault (NAF; after Şaroğlu *et al.* 1992) reported by ISC during 1964–2000 for M >2 superimposed on a shaded relief map derived from the GTOPO-30 Global Topography Data from USGS. Bathymetry data are from Smith & Sandwell (1997). Geographical location of the study area is outlined by a rectangular box. The anomalous seismic activity (Kandilli Observatory) orthogonal to the trace of the NAF should be noted. Star indicates the instrumental epicentre of 6 June 2000 Orta–Çankırı earthquake reported by the USGS.

The 6 June 2000 Orta earthquake is the only instrumentally recorded event that has magnitude >5 since 1900 between Ankara and Çankırı, and therefore provides valuable data on the neotectonic framework of NW central Anatolia. In this paper, we first introduce the overall regional tectonic framework and then present seismological and InSAR investigations of the 6 June 2000 Orta–Çankırı earthquake. Finally, its implications on the internal deformation of the Anatolian plate will be discussed.

Neotectonic framework of NW central Anatolia

This section generally focuses on the less well-known reactivated section of the Izmir–Ankara suture zone between Ankara and Çankırı rather than relatively well-known structures such as the North Anatolia Fault (Ambraseys 1970; Ambraseys & Jackson 1998; McKenzie 1969, 1970; Nowroozi 1972; Şengör 1979; Stein *et al.* 1997; Taymaz *et al.* 2002*a*, *b*, 2004) and its splay, the Kırıkkale–Erbaa Fault (Figs 3 and 4; Polat 1988).

The remnant of the Neo-Tethyan ocean known as the Izmir–Ankara–Erzincan suture trends approximately east–west to the west of Ankara. Further to the east, it turns nearly 90° and has a NNE trend towards Çankırı (Figs 3 and 4). Although synorogenic basin development related to the closure of the northern branch of the Neo-Tethyan ocean during Cretaceous to Eocene time has been discussed (Şengör & Yılmaz 1981), there are few studies on the detailed post-collisional evolution of NW central Turkey. Okay & Tüysüz (1999) have recently studied this suture of northern Turkey and reported that the Tethyan subduction–accretion complexes along the İzmir–Ankara–Erzincan suture around Çankırı form a 5–10 km wide tectonic belt, which circles and radially thrusts the Eocene–Miocene sedimentary rocks of the Çankırı Basin, resulting in a large loop of the suture (Fig. 3). Furthermore, the Senonian andesitic volcanism observed in the northern parts of the Sakarya zone is thought to be related to northward subduction of the along İzmir–Ankara–Erzincan ocean. Palaeomagnetic data within the Sakarya Zone also indicate that it was close to the Laurasian margin during Liassic and Late Cretaceous time (Sarıbudak, 1989; Channel *et al.* 1996; Erdoğan *et al.* 1996). Although there is a considerable number of studies on the synorogenic basin development related to the closure of the northern branch of the Neo-Tethyan ocean during Cretaceous to Eocene time (e. g. see Görür *et al.* 1998, for a summary) there is only a limited number of field-oriented studies on the post-collisional evolution of NW central Turkey.

Koçyiğit (1991*a*, *b*, 1992) and Koçyiğit *et al.* (1995) proposed that intracontinental convergence related to the closure of the Neo-Tethyan ocean continued until the Late Pliocene, and called this the 'Ankara Orogenic Phase'. This suggestion was based on the south-vergent thrusting of the east–west-trending Izmir–Ankara suture zone over Neogene sedimentary units towards the east and west in the western margin of the Çankırı basin. Koçyiğit *et al.* (1995) also claimed that the sedimentary units were deposited in thrust-related basins. After the Late Pliocene, intracontinental convergence gave way to an extensional regime as indicated by the NE–SW-trending normal faults with related horizontal Pliocene deposits SW Çankırı (Koçyiğit *et al.* 1995).

However, Seyitoğlu *et al.* (1997) demonstrated that south-vergent thrusting of the suture zone does not exist NW of Ankara and argued that inter-continental convergence must have been ceased before Miocene times because Miocene to Pliocene geochemical evolution of the Galatia volcanic series (Wilson *et al.* 1997; Gürsoy *et al.* 1999) shows lithospheric thinning rather than thickening in the region. Alternatively, an extensional regime as a result of orogenic collapse has been proposed during Early Miocene–Pliocene time with emphasis on a post-Pliocene NAFZ effect on the region (Seyitoğlu *et al.* 1997). In addition, recent work (Seyitoğlu *et al.* 2000) has shown that the NNE–SSW-trending section of the İzmir–Ankara suture zone between Ankara and Çankırı has been reactivated as a east-vergent tectonic sliver that fragments and deforms Miocene–early Pliocene basin deposits. Thrust-related younger clastic deposits unconformably overlie the older deformed sediments in front of the tectonic sliver SE of Çankırı (Fig. 4c; Polat 1988). This latest east-vergent thrusting of the NNE–SSW-trending sector of the Izmir–Ankara suture is related to the NW–SE contraction caused by the movements of the right-lateral North Anatolian Fault and its splay, the Kırıkkale–Erbaa Fault Zone (Fig. 4c). Between these two strike-slip fault zones internal deformation of the Anatolian plate was mostly taken up by the weaker zones of the Neo-Tethyan orogeny.

The overall neotectonic framework of north central Anatolia is controlled mainly by splay faults of the NAFZ. The splay faults from SE to NW are the Almus Fault Zone (Bozkurt & Koçyiğit 1996), Ezinepazarı–Sungurlu Fault Zone, Kızılırmak Fault Zone and Laçin Fault Zone. They bifurcate from the NAF, and divide the Anatolian Block into east–west-trending wedge-like blocks that are deforming internally and rotating in a counter-clockwise sense (Tatar *et al.* 1995; Bozkurt & Koçyiğit 1996; Piper *et al.* 1996; Kaymakçı 2000). Chorowicz *et al.* (1999)

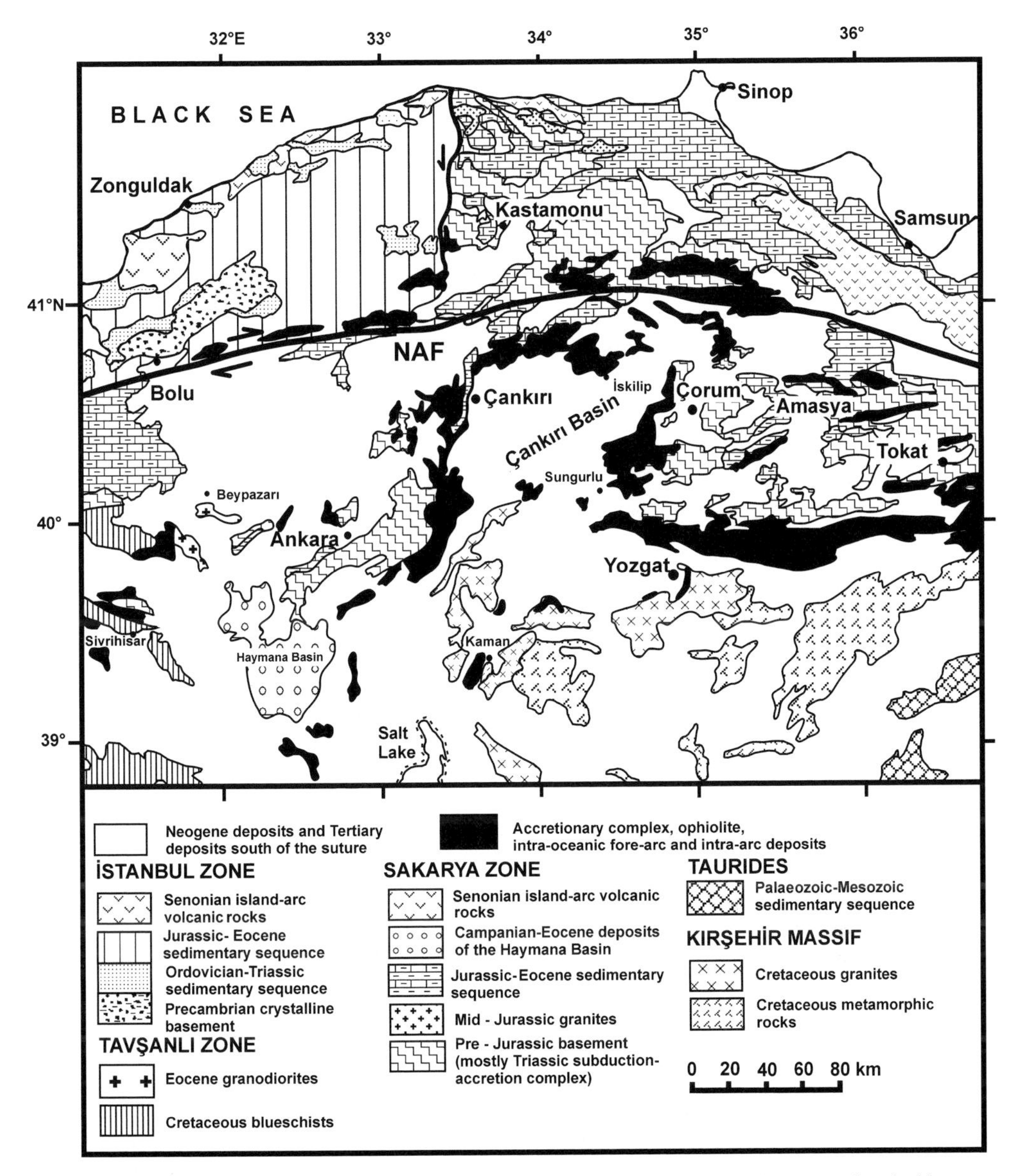

Fig. 3. Summary geological map of Çankırı Basin, central Turkey, and surrounding regions (reproduced with permission from Okay & Tüysüz (1999)).

interpreted these splays in a different way, as an element of extensional escape wedges, and suggested that the Anatolian Block is pulled by the Hellenic trench rather than pushing from the east because of collision of the Arabian and Eurasian plates.

Apart from these splays, nearly north–south-trending tectonic lines play an important role in the neotectonic framework of the region. However, there are varying descriptions and interpretations of these lines. These wedges are dissected by north–south-trending faults, which further complicate the deformation styles of the Anatolian Block in this region. The most important of these faults are the Eldivan Fault Zone (Kaymakçı 2000; Kaymakçı *et al.* 2003) and the Dodurga Fault Zone (Koçyiğit *et al.* 2001).

Seyitoğlu *et al.* (2000, 2001) described a NNE–SSW-trending east-vergent pinched crustal wedge between Ankara and Çankırı. The western margin

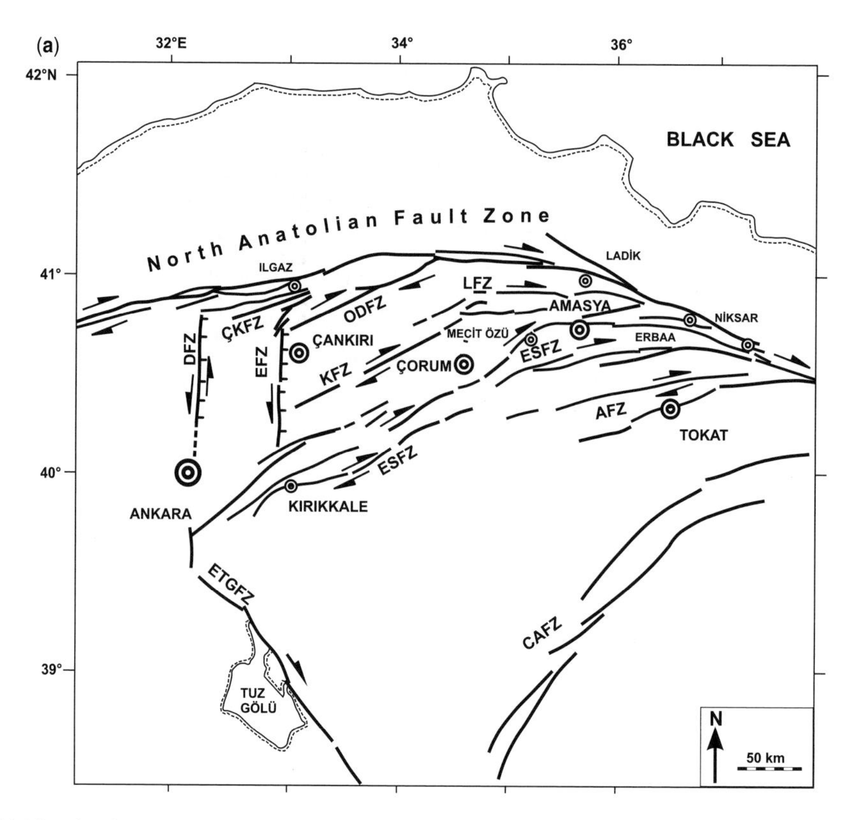

Fig. 4. (**a**) Map showing major structural elements of central Anatolia. Bold lines are strike-slip faults with arrows showing relative motion (after Bozkurt 2001). (**b**) Summary tectonic map of NW central Anatolia compiled from our observations and those of Şaroğlu *et al.* (1992), Gökten *et al.* (1996), Emre *et al.* (2000), and Koçyiğit *et al.* (2001). Main faults are marked with continuous lines. (**c**) Simplified main neotectonic elements of NW central Anatolia, North Anatolian Fault Zone (NAFZ), Kırıkkale–Erbaa Fault Zone (KEFZ), and the cause of the east-vergent tectonic sliver between Ankara and Çankırı. NW–SE-trending greatest principal stress (σ_1) is created by the NAFZ and KEFZ and activates the east-vergent tectonic sliver. 1, Neo-Tethyan suture zone; 2, Galatia volcanic complex (after Seyitoğlu *et al.* 2000).

of this wedge is limited by west-dipping normal faults but its eastern margin is a thrust that controls the accumulation of Late Pliocene–Pleistocene clastic deposits that unconformably overlie deformed Neogene successions in the western Çankırı basin (Fig. 3; Sen *et al.* 1998; Nemec & Kazancı 1999). Seyitoğlu *et al.* proposed that the thrusting and the related deformations in this region are linked with the post-Late Pliocene neotectonic pinched crustal wedge rather than intra-continental convergence (i.e. the Ankara orogenic phase). This wedge is developed as a result of the NW–SE contraction caused by the movements of the right-lateral NAFZ and its splay, the Kırıkkale–Erbaa Fault Zone (Fig. 4c).

In the same area, Kaymakçı (2000) re-mapped double-vergent thrusting towards the east and west after Akyürek *et al.* (1980). These inverted structures were interpreted as a consequence of the NNE–SSW-trending sinistral strike-slip Eldivan Fault Zone and they suggest that this fault zone was created by a major principal stress (σ_1) oriented NW–SE owing to transcurrent tectonics (Fig. 4c). Adıyaman *et al.* (2001) named the same fault zone the Korgun Fault, and considered that it plays an important role in creating the extensional escape wedges that moved to the SW during Early–Middle Miocene times and to WSW during Late Pliocene–Plio-Quaternary times (Fig. 4). West of the Eldivan or Korgun Fault, the north–south-trending Dodurga Fault has been mapped and nominated as a cause of the Orta earthquake (Fig. 4, Emre *et al.* 2000; Koçyiğit *et al.* 2001). The Dodurga Fault Zone (DFZ) bifurcates from the Çerkeş–Kurşunlu segment of the NAFZ around Çerkeş.

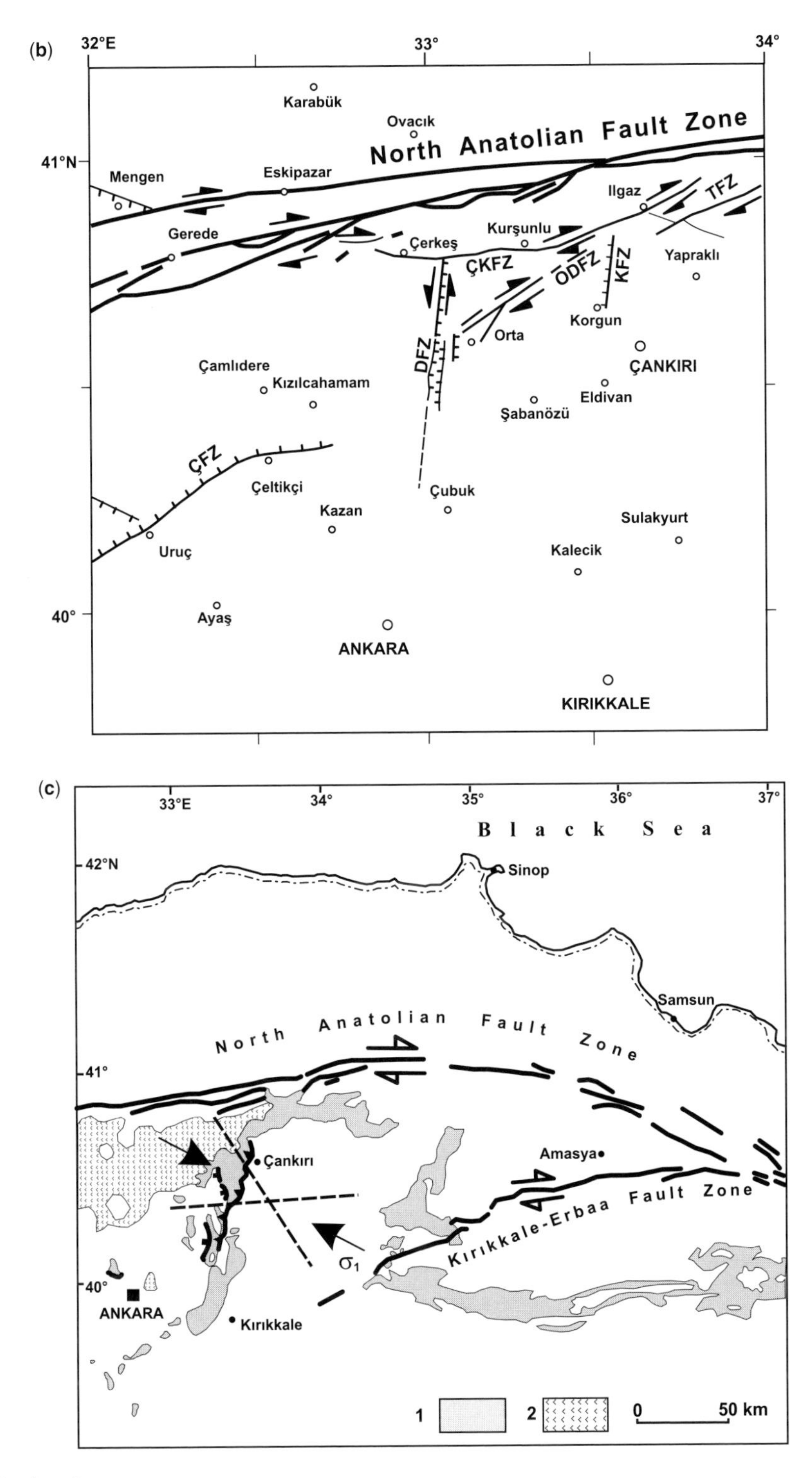

Fig. 4. *Continued.*

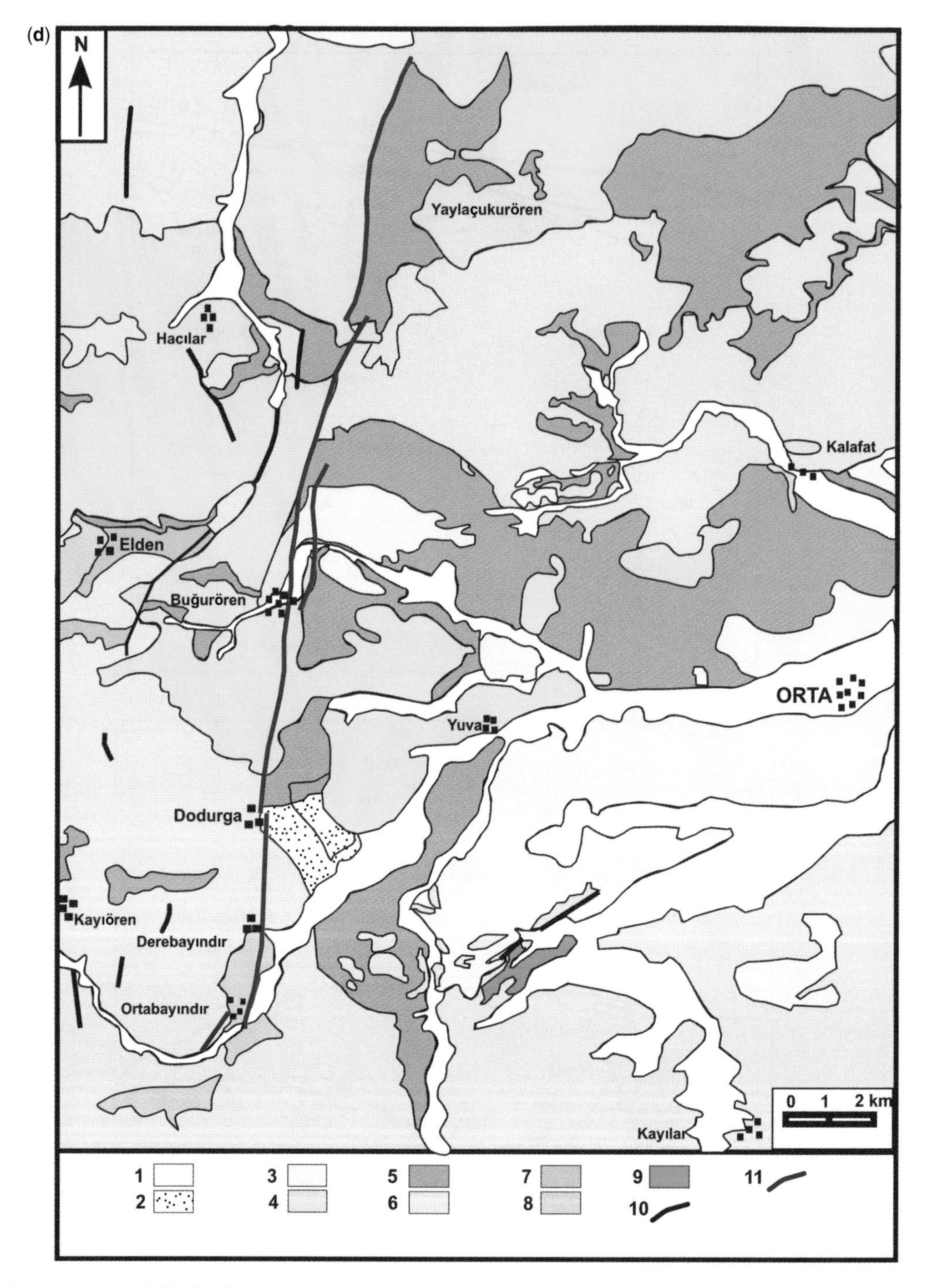

Fig. 4. *Continued.* (**d**) Simplified geological map of western part of Orta (Çankırı) and surrounding regions (modified after Türkecan *et al.* 1991; Emre *et al.* 2000): 1, alluvium (Quaternary); 2, alluvial fan (Quaternary); 3, conglomerate, sandstone, mudstone and limestone (Pliocene); 4, basalt (Pliocene); 5, conglomerate, sandstone, mudstone, limestone, and evaporite (Miocene); 6, andesite, basalt, rhyolite, dacite and pyroclastic deposits (Miocene); 7, andesite, basalt, dacite and pyroclastic deposits (Eocene); 8, sandstone, mudstone, limestone olistostrome (Late Cretaceous); 9, metadetrital deposits (Triassic), 10, faults; 11, Dodurga Fault.

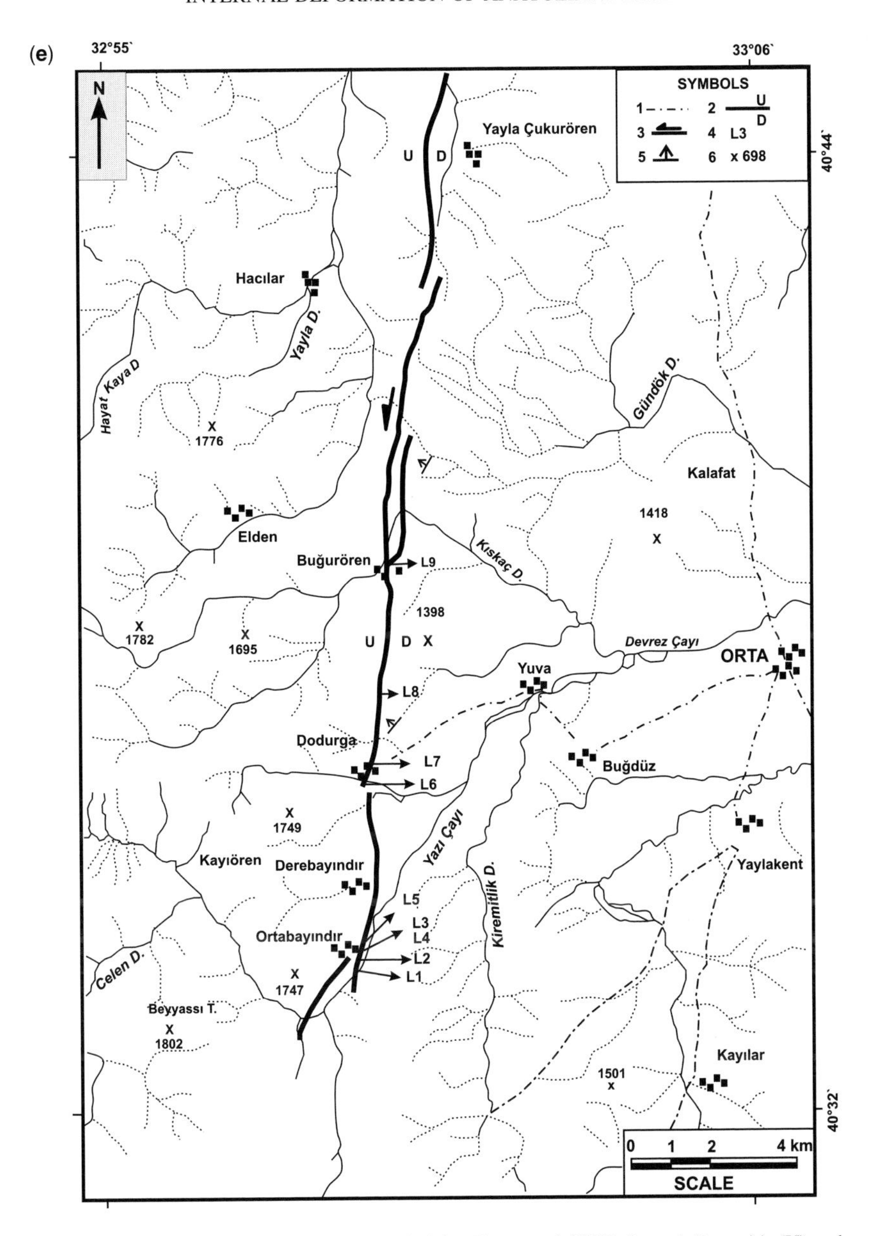

Fig. 4. *Continued.* (**e**) Detailed map of the Dodurga Fault (after Emre *et al.* 2000): 1, road; 2, up side (U) or down side (D); 3, left-lateral offset; 4, localities of observation sites; 5, direction of tilt; 6, elevation marks. AFZ, Almus Fault Zone; ÇFZ, Çeltikçi Fault Zone; ÇKFZ, Çerkeş–Kurşunlu Fault Zone; DFZ, Dodurga Fault Zone; EFZ, Eldivan Fault Zone; ESFZ, Ezinepazarı–Sungurlu Fault Zone; ETGFZ, Ezinepazarı–Tuz Gölü Fault Zone; ODFZ, Orta–Devrez Fault Zone; KFZ, Korgun Fault Zone; LFZ, Laçin Fault Zone; TFZ, Tosya Fault Zone; CAFZ, Central Anatolian Fault Zone.

The DFZ is an approximately north–south-trending fault zone of about 4–7 km width and 36 km length. It is composed of a number of parallel to obliquely oriented faults with considerable amounts of normal-slip component. Along the DFZ a number of strike-slip-induced morphotectonic features are present. These are the Yalaközü pull-apart basin, the Yalakçukurören pull-apart basin, alluvial fans, landslides and a number of historical ruins indirectly indicating historical activity of the DFZ (Fig. 4d and e; Emre *et al.* 2000).

Adıyaman *et al.* (2001) reported that the age of the same fault, named the Orta Fault, is Middle Miocene or older because it is covered by middle Miocene sediments of the Çerkeş–Kurşunlu–Ilgaz basin. Structural analysis of this fault (Adıyaman *et al.* 2001) has shown general west- to NW-dipping fault surfaces with left-lateral slip in the Early–Middle Miocene and dominantly normal slip in the Late Miocene–Plio-Quaternary (Fig. 4b and c). Kaymakçı *et al.* (2003) reported that the Eldivan Fault Zone (EFZ) is a sinistral strike-slip fault zone with a reverse component, and has resulted from reactivation of a Neotethyan suture zone that dips NW, and the reverse nature of the EFZ is due to the palaeotectonic nature of the fault zone. Palaeostress patterns constructed by Kaymakçı *et al.* (2000, 2003) indicate that the orientation of major principal stress (σ_1) is consistently NW–SE whereas the intermediate stress is vertical, indicating strike-slip deformation in the region.

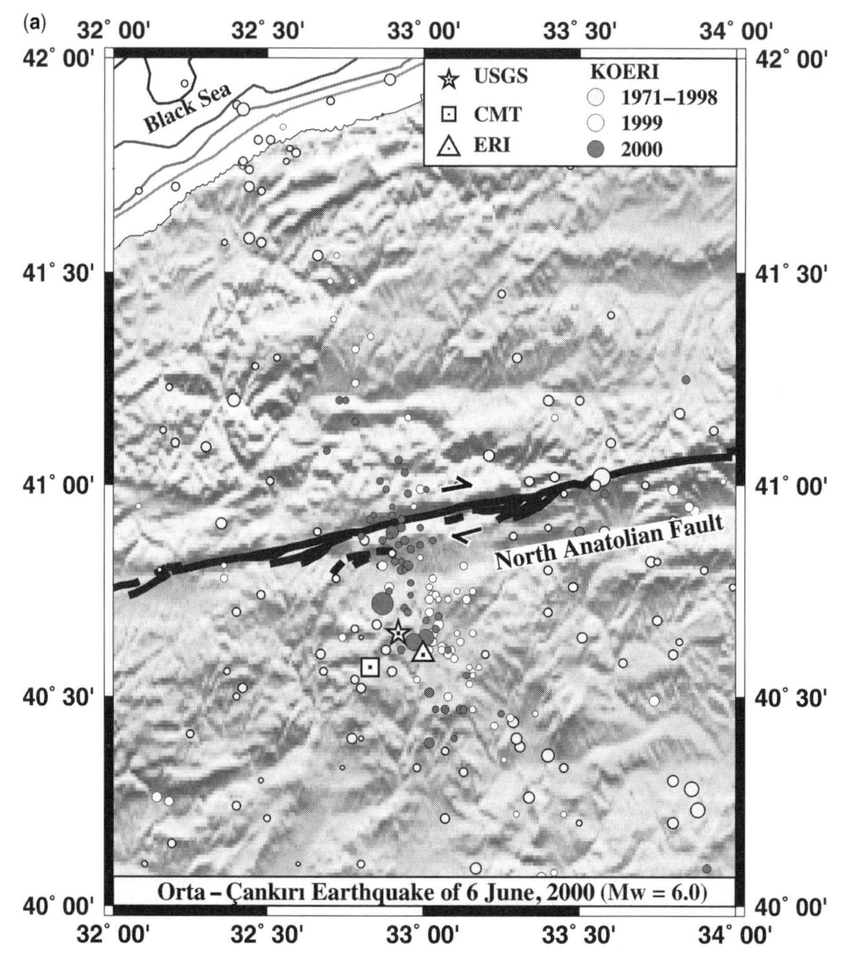

Fig. 5. (**a**) Regional seismicity in the central segment of North Anatolian Fault (NAF; mapped by Şaroğlu *et al.* 1987) during 1971–2000 for M >2, with aftershock data after Kandilli Observatory. Star, square and triangle indicate the instrumental epicentres of the 6 June 2000 Orta–Çankırı earthquake reported by the USGS, Harvard-CMT and ERI, respectively. GTOPO-30 global topography data are from the USGS and re-sampled at 0.1 min. Bathymetry data are from Smith & Sandwell (1997). (**b**) Lower hemisphere projections of the focal mechanisms corresponding to the minimum misfit solutions of earthquakes studied here and by earlier workers (see Table 1 for details). Compressional quadrants are shaded. Numbers in dilatational quadrants identify the focal depths obtained from the inversion.

Earthquake source parameters from inversion of teleseismic body-waveforms

Data reduction

We used both P- and SH-waveforms and first-motion polarities of P-waves to constrain earthquake source parameters. The approach we followed is that described by Taymaz *et al.* (1990, 1991*a*, *b*) to study earthquakes in the Hellenic trench, the Aegean region and on the East Anatolian Fault Zone (EAFZ), respectively. We compared the shapes and amplitudes of long-period P- and SH-waveforms recorded by GDSN stations in the distance range 30–90° with synthetic waveforms. To determine source parameters we used the McCaffrey & Abers (1988) version of Nábělek's (1984) inversion procedure, which minimizes, in a weighted least-squares sense, the misfit between observed and synthetic seismograms (McCaffrey & Nábělek 1987; Nelson *et al.* 1987; Fredrich *et al.* 1988). Seismograms are generated by combining direct (P or S) and reflected (pP and sP, or sS) phases from a point source embedded in a given velocity structure. Receiver structures are assumed to be homogeneous half-spaces. Amplitudes are adjusted for geometric spreading, and for attenuation using Futterman's (1962) operator, with $t^* = 1$s for P and $t^* = 4$ s for SH. As explained by Fredrich *et al.* (1988), uncertainties in t^* affect mainly source duration and seismic moment, rather than source orientation or centroid depth. Seismograms were weighted according to the azimuthal distribution of stations, such that stations clustered together were given smaller weights than those of isolated stations (McCaffrey & Abers 1988). The inversion routine then adjusts the strike, dip, rake, centroid depth and source time

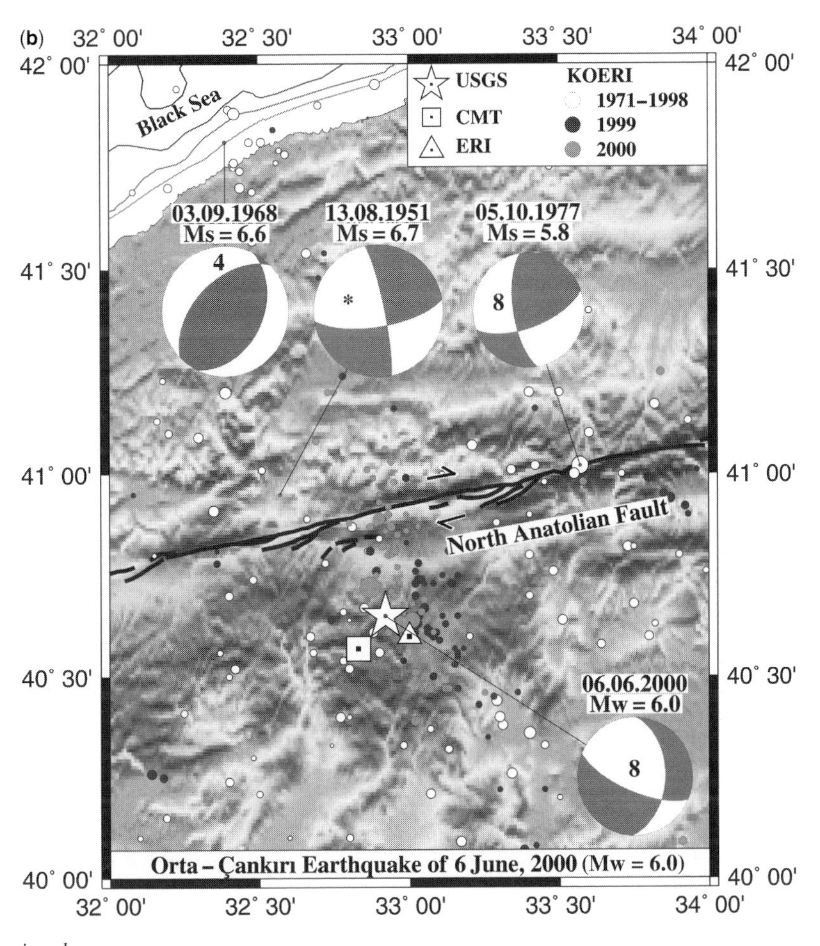

Fig. 5. *Continued.*

function, which is described by a series of overlapping isosceles triangles (Nábělek 1984) whose number and duration we selected.

Our experience with the inversion routine was very similar to that of Nelson *et al.* (1987), McCaffrey (1988), Fredrich *et al.* (1988) and Molnar & Lyon-Caen (1989). We found that a point source, in which all slip occurs at the same point (the centroid) in space but not in time, was a good approximation; that is, we saw no indication of systematic azimuthal variations in waveforms that might be associated with rupture propagation. However, the waveforms show evidence of multiple ruptures that we attempted to match by later sub-events that are discussed in greater detail below. The focal sphere was generally covered by observations in all quadrants, although with more stations to the north than the south, and we found that estimates of the strike, dip, rake and centroid depth were relatively independent of each other. Thus if one parameter was fixed at a value within a few degrees or kilometres of its value yielded by the minimum misfit of observed and synthetic seismograms, the inversion routine usually returned values for the other parameters that were close to those of the minimum misfit solution. The strikes and dips of nodal planes were consistent, within a few degrees, with virtually all first-motion polarities (Figs 5–8). The estimate of seismic moment clearly depended on the duration of the source time function, and to some extent on centroid depth and velocity structure. As our main interest is in source orientation and depth, we did not concern ourselves much with uncertainties in seismic moment, which in most cases is probably about 30%. We estimated the lengths of the time functions by increasing the number of isosceles triangles until the amplitudes of the later ones became insignificant. The seismogram lengths we selected for inversion were sufficient to include the reflected phases pP, sP and sS. We examined the P-waves for PcP arrivals, where they were anticipated within the selected window, but this phase was never of significant amplitude. ScS presented a greater problem, and we generally truncated our inversion window for SH-waves before the ScS arrival. The source velocity structures we used to calculate the synthetic seismograms are those reported by Jeffreys & Bullen (1940) and the IASPE91 (Lee 1991) Earth models. We assumed the centroid was in a layer with P velocity 6.5 km s^{-1}, which is a likely lower crustal velocity (Makris & Stobbe 1984) and appropriate for calculating the take-off angles of ray paths leaving the source. Uncertainty in the average velocity above the source leads directly to an uncertainty in centroid depth, which we estimate to be about ± 2 km for the range of depths involved in this study.

Uncertainties in source parameters

Having found a set of acceptable source parameters, we followed the procedure described by McCaffrey & Nábělek (1987), Nelson *et al.* (1987), Fredrich *et al.* (1988), Taymaz *et al.* (1990, 1991*a*, *b*), and Taymaz & Price (1992), in which the inversion routine is used to carry out experiments to test how well individual source parameters are resolved. We investigated one parameter at a time by fixing it at a series of values to either side of its value yielded by the minimum misfit solution, and allowing the other parameters to be found by the inversion routine. We then visually examined the quality of fit between observed and synthetic seismograms to see whether it had deteriorated from the minimum misfit solution. In this way we were able to estimate the uncertainty in strike, dip, rake and depth for each event. In common with the researchers cited above, we believe that this procedure gives a more realistic quantification of likely errors than the formal errors derived from the covariance matrix of the solution. We found that changing the depth of this earthquake by more than 2–3 km produced a noticeable degradation in the fit of waveforms, and take this to be a realistic estimate of the uncertainty in focal depth. This estimate, which is listed in Table 1, does not include the uncertainty related to the unknown average velocity above the source, discussed above. These tests give us some confidence

Fig. 6. This (and subsequent similar figures) shows the radiation patterns and synthetic waveforms for the minimum misfit solution returned by the inversion routine, as well as the observed waveforms. Continuous lines are observed waveforms, and the inversion window is identified by a short vertical bar. Dashed lines indicate synthetic waveforms. For the purposes of display, waveform amplitudes have been normalized to that of an instrument with a gain of 6000 at a distance of 60°. The station code is shown to the left of each waveform, together with an upper-case letter that identifies its position on the focal sphere and a lower-case letter that identifies the type of instrument (d, GDSN long-period). The vertical bar beneath the focal spheres shows the scale in microns, with the lower-case letter identifying the instrument type as before. The source time function is shown in the middle of the figure, and beneath it is the time scale used for the waveforms. Focal spheres are shown with P and SH nodal planes, in lower hemisphere projection. Station positions are indicated by letter, and are arranged alphabetically clockwise, starting from north. ●, ○, P- and T-axes, respectively. Beneath the header at the top of the figure, which shows the date, body-wave and surface-wave magnitudes, are given the strike, dip, rake, centroid depth and seismic moment (in units of $\times 10^{18}$ N m) of the minimum misfit solution obtained from the inversion of teleseismic long-period P and S body waveforms.

6 JUNE 2000 — ORTA–ÇANKIRI (Mw = 6.0)

A: NP1: 113 / 70 / -132 NP2: 2 / 46 / -29 h: 8 km Mo: 124.1E16 Nm
B: NP1: 90 / 86 / 49 NP2: 355 / 41 / 174 h: 7 km Mo: 61.68E16 Nm

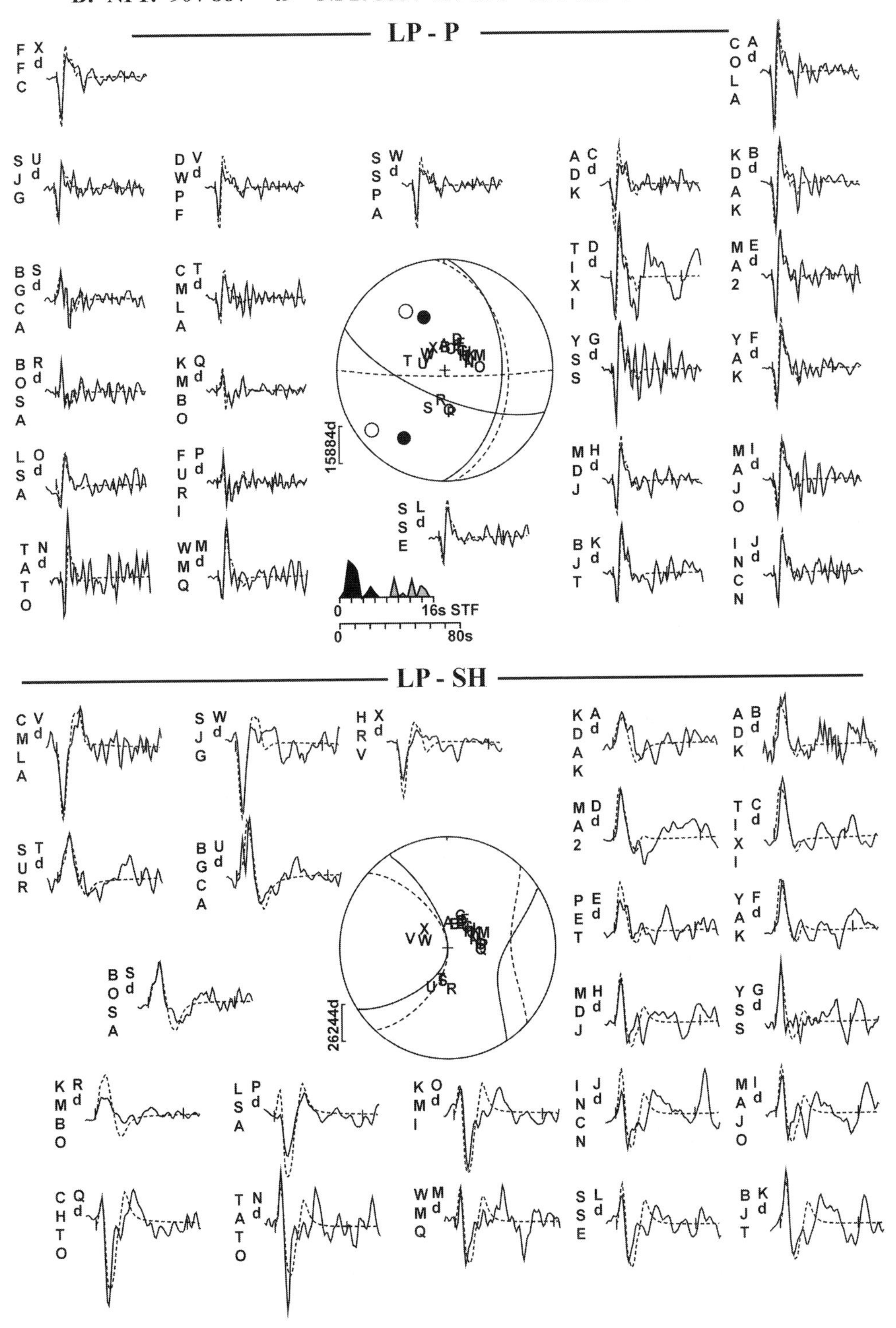

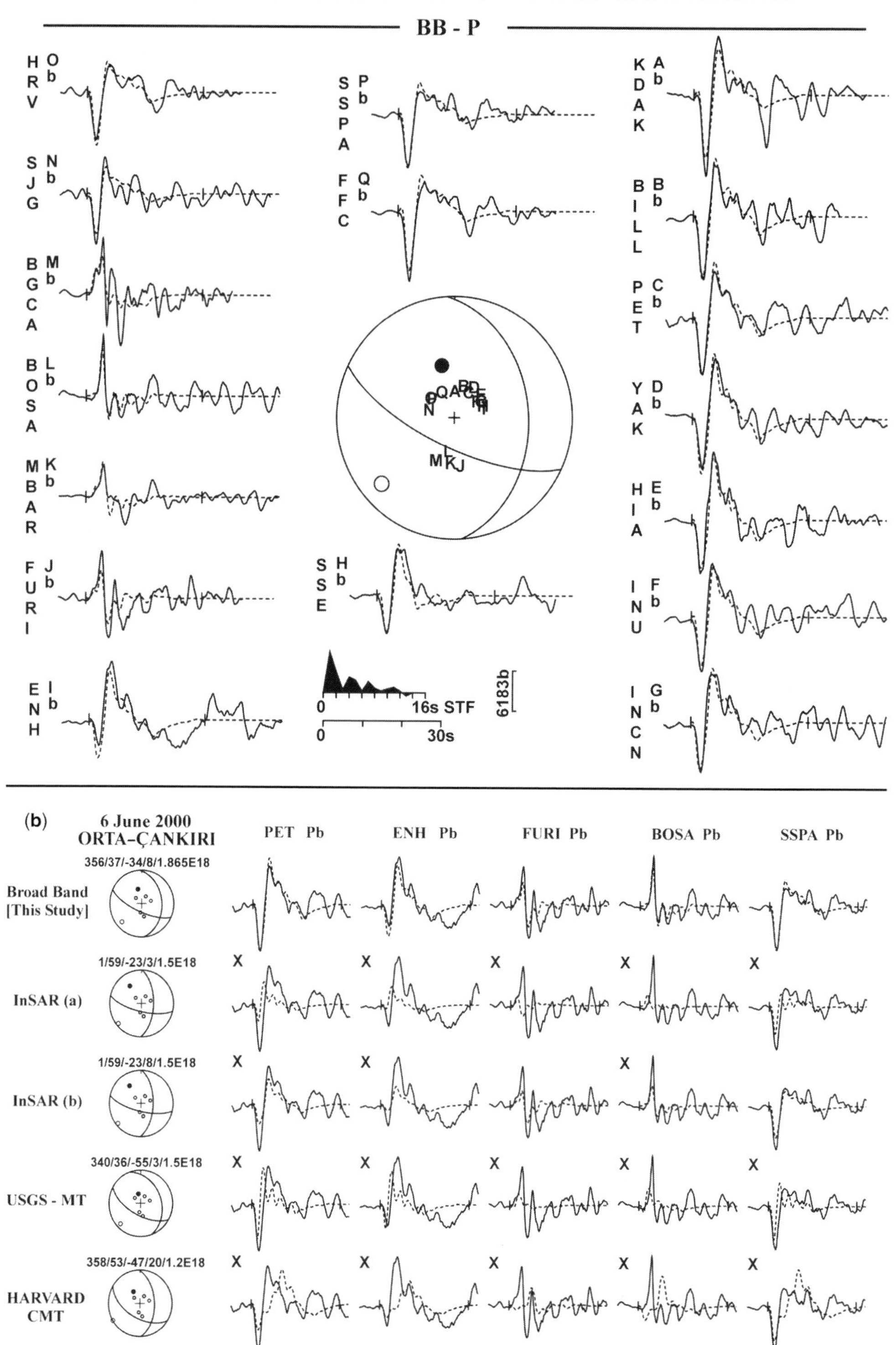
(a)
6 JUNE 2000 — ORTA–ÇANKIRI (Mw = 6.0)
NP1: 115 / 70 / -122 NP2: 356 / 37 / -34 h: 8 km Mo: 186.5E16 Nm
BB - P
HRV O b
SJG N b
BGCA M b
BOSA L b
MBAR K b
FURI J b
ENH I b
SSPA P b
FFC Q b
SSE H b
KDAK A b
BILL B b
PET C b
YAK D b
HIA E b
INU F b
INCN G b
16s STF
30s
6183b
(b)
6 June 2000
ORTA–ÇANKIRI
PET Pb
ENH Pb
FURI Pb
BOSA Pb
SSPA Pb
356/37/-34/8/1.865E18
Broad Band [This Study]
1/59/-23/3/1.5E18
InSAR (a)
1/59/-23/8/1.5E18
InSAR (b)
340/36/-55/3/1.5E18
USGS - MT
358/53/-47/20/1.2E18
HARVARD CMT

that there is no significant trade-off between source parameters for this event. The greatest change occurs to the value of seismic moment, which varies by 20%.

The pattern of focal mechanisms

Aftershock distributions

Figure 5 shows the background seismicity and aftershock distributions of the 6 June 2000 Orta–Çankırı sequence reported by KOERI, using P-wave arrival times picked at regional stations by station operators. The aftershocks, which occur in a broad zone elongated north–south to NW–SE, are subparallel to the trend of the major faults in the region (Fig. 4d and e, Emre *et al.* 2000), but lie to the east of the macroseismic epicentre for the main shock. The USGS, Harvard-CMT, ERI and KOERI locations of the Orta earthquake lie about 10–15 km west of the macroseismic epicentre. It is likely that the locations for earthquakes the size of the Orta earthquake may be as much as 15–20 km in error (Taymaz *et al.* 1991; Taymaz & Price 1992). The influence of station distribution is also significant for the mislocation vectors, which are similar to that for J–B times, as reported by Kennett & Engdahl (1991). Their study showed a systematic shift of ISC locations in the Aegean.

Focal mechanisms in this area for earthquakes larger than Ms = 5.8 since 1960 were determined using the same inversion algorithm and procedure (Fig. 5b). Two types of mechanism are common: strike-slip faulting with roughly east–west nodal planes and normal faulting with nodal planes trending NNW–SSE. Most of the strike-slip solutions are near the NAFZ, whereas the normal faulting solution of the Orta–Çankırı earthquake is on the nearby secondary faults. The one obvious anomaly is the high-angle reverse faulting solution of the 3 September 1968 Bartın earthquake (Fig. 5b and Table 1) in NW Turkey, which shows shortening in a roughly NW–SE direction. All of solutions are well constrained by first-motion polarities and waveforms (Özay 1996; Tan 1996).

The 6 June 2000 Orta–Çankırı earthquake was the largest instrumentally recorded event occurred we were able to study (Table 1; Taymaz & Tan 2001; Taymaz *et al.* 2002*a*), with a moment of $M_o = 1.85 \times 10^{18}$ N m. It occurred close to a restraining bend in the north–south-striking right-lateral strike-slip fault that moved in the much larger earthquake of 13 August 1951 (McKenzie 1972). It is further obvious for small earthquakes, particularly aftershocks, to have mechanisms incompatible with a uniform regional strain field (e.g. Richens *et al.* 1987), but unusual to see this effect with earthquakes large enough to study teleseismically. Hence, the faulting in this anomalous earthquake could be related to the local geometry of the main strike-slip system, and may not be a reliable guide to the regional strain field in central Turkey.

Teleseismic body waveforms

Long-period P, SH and broadband P waveforms at teleseismic distances were clearly recorded for this earthquake. Thus in Figures 6–8 we have compared the observed body-wave seismograms with synthetics generated following the procedure described by Taymaz *et al.* (1991*a*, *b*) and Taymaz & Price (1992), among many others.

Long-period P and SH body waveforms

The minimum misfit solution for the earthquake of 6 June 2000 is shown in Figures 6–8. The nodal planes of this solution are compatible with virtually all first-motion P polarities, shown in Figure 8b. The P and SH pulses are simple at all azimuths and characteristic of normal faulting with a shallow focal depth. There is good coverage of the focal sphere at all azimuths, for both first-motion and long-period P, SH and broadband P waveform data. We carried out many experiments as described by McCaffrey & Nábělek (1987),

Fig. 7. (**a**) Minimum misfit solution for the broadband P waveforms. For the purposes of display, waveform amplitudes have been normalized to that of an instrument with a gain of 3000 at a distance of 70°. The remainder of the display convention is as in Figure 6. (**b**) A selection of waveforms from a run of the inversion program. The top row shows waveforms from the minimum misfit solution. The stations are identified at the top of each column, with type of waveform marked by P and followed by the instrument type (b, broadband). At the start of each row the P focal sphere is shown for the focal parameters represented by the five numbers (strike, dip, rake, depth and moment), showing the positions on the focal spheres of the stations chosen. The convention for the waveforms is as in Figures 6 and 7, but here the large × shows matches of observed to synthetic waveforms that are worse than in the minimum misfit solution. We compare selected waveforms from misfit solution (a) with those generated by sources with orientations of the InSAR obtained in the present study, and reported USGS-MT and Harvard-CMT solutions. In the following rows the strike, dip, rake, depth and moment were fixed at the values of InSAR, USGS-MT and Harvard-CMT modelling results (see Table 2).

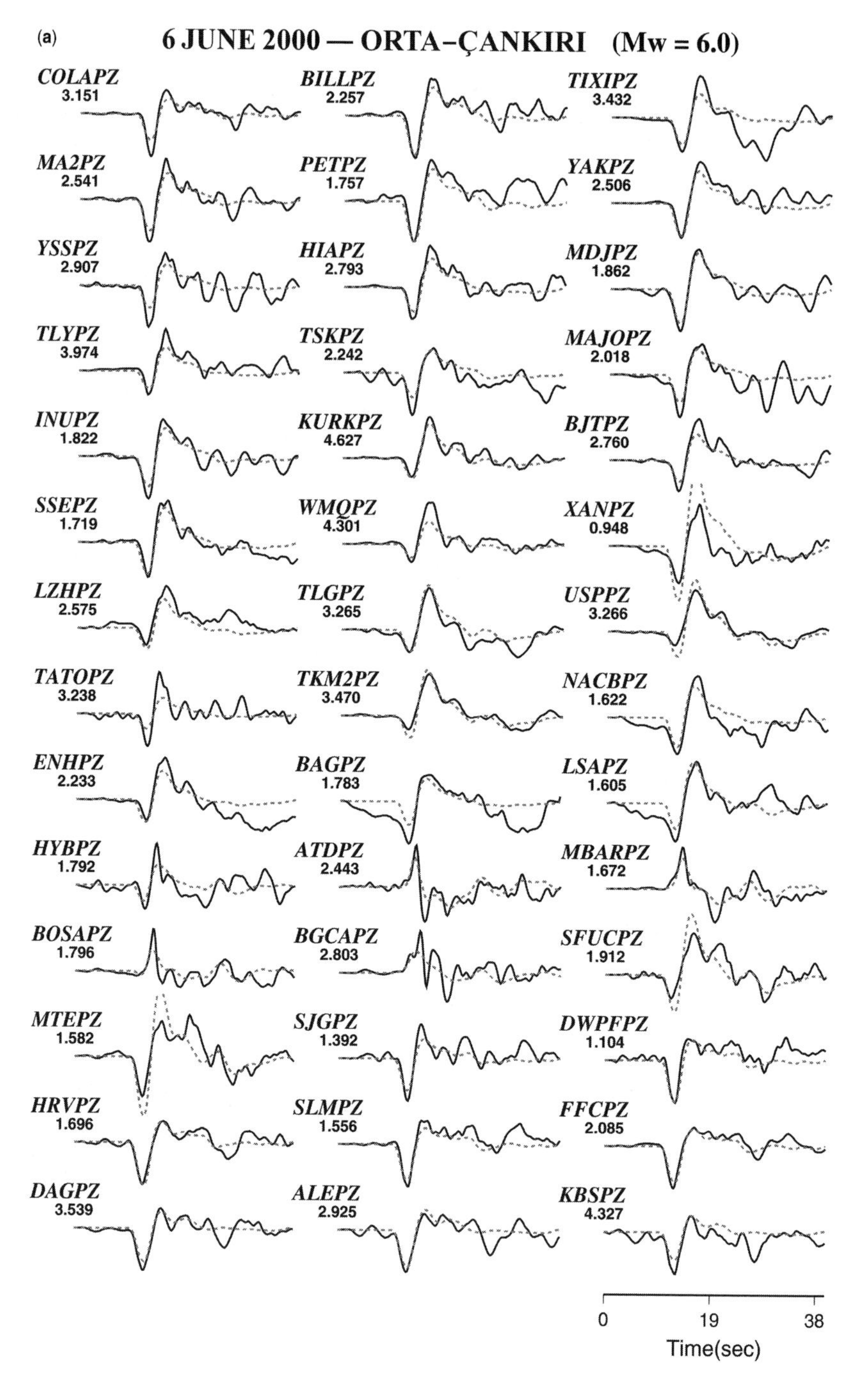

Fig. 8. (**a**) Comparison of the observed broadband P waveforms (continuous lines) with the synthetic waveforms (dashed lines) used in rupture history and slip distribution analyses. The numbers below the station code indicate maximum amplitude, and the time scale is shown below the figure. (**b**) Lower hemisphere equal area projections of the first-motion polarity data. Station positions of the focal sphere have been plotted using the same velocity below source (6.5 km s^{-1}) that was used in our waveform inversion procedure. ●, compressional first motions, ○, dilatational; all were read on long-period, short-period and broadband instruments of the GDSN, when available. Nodal planes are those of the minimum misfit solutions. Beneath the selected waveforms, marked with station codes, at the bottom of the figure is given the event's header (month, date, year, geographical location), with both of the nodal planes that illustrate the strike, dip and rake of the minimum solution. ■, □, P- and T-axis, respectively.

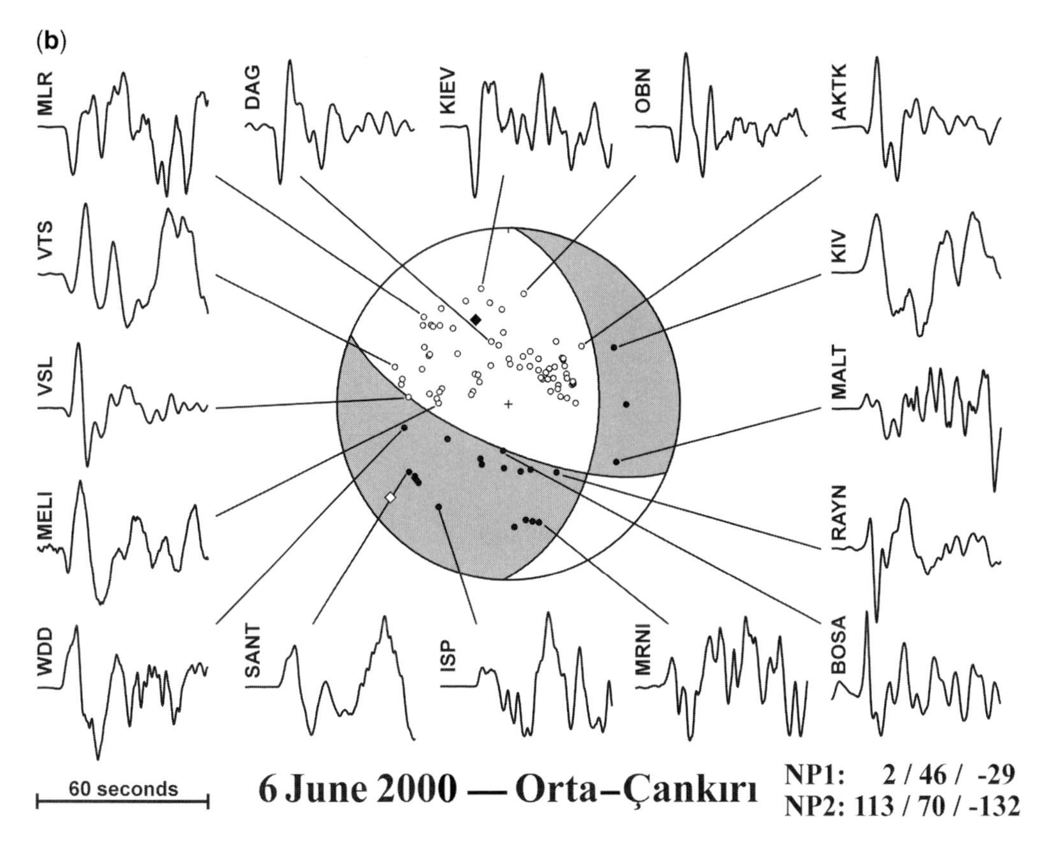

Fig. 8. *Continued.*

Nelson *et al.* (1987), Taymaz *et al.* (1990, 1991*a*, *b*), Taymaz & Price (1992) and Taymaz (1993), in which the inversion routine is used to carry out experiments to test how well individual source parameters resolved, and based on these, estimate the source parameters and uncertainties of the 6 June 2000 Orta–Çankırı earthquake to be: Nodal Plane 1: strike $2 \pm 5^\circ$, dip $46^\circ \pm 5^\circ$; rake $-29 \pm 5^\circ$; Nodal Plane 2: strike 113°, dip 70°; rake -132°; principle axes: P = 338 (48); T = 232 (14); B = 131 (39); depth 8 ± 2 km (although this does not include uncertainty related to velocity structure), and seismic moment (M_o) in the range of $(140–185) \times 10^{16}$ N m.

In Figure 7b, we compare, at selected stations, the waveforms from the minimum misfit solution with those from some of the other reported solutions. The strike, dip, rake, depth and moment were fixed using the values of InSAR, USGS-MT and Harvard-CMT reported results (see Table 2). The minimum misfit solution obtained from joint inversion of long-period P, SH and broadband P waveform data is clearly much better constrained than that of InSAR modelling results, which produces a poor fit (marked with $\times$) of the seismograms at several stations (Fig. 7b, rows 2 and 3). In rows 2 and 3, InSAR parameters (see Table 2) are used, and the only difference is the focal depth, which was fixed at values of 3 and 8 km to test the lower and upper limits, respectively. The InSAR solution obtained in the present study differs from the body-wave modelling results by 5° in strike, 22° in dip, and 11° in rake. The differences in dip, rake and depth are marginally outside the acceptable errors of body-wave solutions (Tables 1 and 2). In rows 4 and 5 of Figure 7b we fixed the source orientation using USGS-MT and Harvard-CMT reported solutions. The fit of waveforms is noticeably worse than in the minimum misfit solution, and the polarities are incorrect (Fig. 7b).

Rupture history and slip distribution inversion of broadband P body waveforms

In recent years there has been significant progress in our understanding of the nature of earthquakes and Earth structure, mainly as a result of our improved ability to interpret broadband seismograms

Table 1. *Source parameters of earthquakes shown in Figures 4–6*

Origin time, t_0 (GMT)	Latitude (°N)	Longitude (°E)	M_w	Nodal Plane 1			Nodal Plane 2			Focal depth (km)	Seismic moment ($\times 10^{16}$ Nm)	STF duration $\Delta\tau_{95}$(s)
				Strike (°)	Dip (°)	Rake (°)	Strike (°)	Dip (°)	Rake (°)			
13 August 1951 Çerkeş (McKenzie 1972)												
18:33:30.0	40.95	32.57	6.7	81	70	−172	348	82	−20	–	–	–
3 September 1968 Bartın (Tan 1996)												
08:19:52.0	41.81	32.39	6.3	26	40	75	255	52	102	4	400	*c.* 14
5 October 1977 Kurşunlu (Özay 1996)												
05:34:43.0	41.02	33.57	5.8	70	65	155	171	67	27	8	58	*c.* 6
6 June 2000 Orta–Çankırı (this study)												
02:41:53.2	40.70	32.98	6.0									
InSAR				1	59	−23	–	–	–	3–8	150	–
Broadband P-waves				356	37	−34	115	70	−122	8	186	*c.* 15
Long-period P- and SH-waves and rupture or slip distribution				2	46	−29	113	70	−132	6–8	140–185	*c.* 16
							Fault rupture area: *c.* 42 km^2; stress drop: 128 bar					
							Maximum and average slip: *c.* 231 cm, *c.* 111 cm					

The detailed source parameters of the Orta–Çankırı earthquake are obtained from inversion of teleseismic P and SH body waveforms and InSAR data with rupture history (slip distribution). STF, Source Time Function (in seconds).

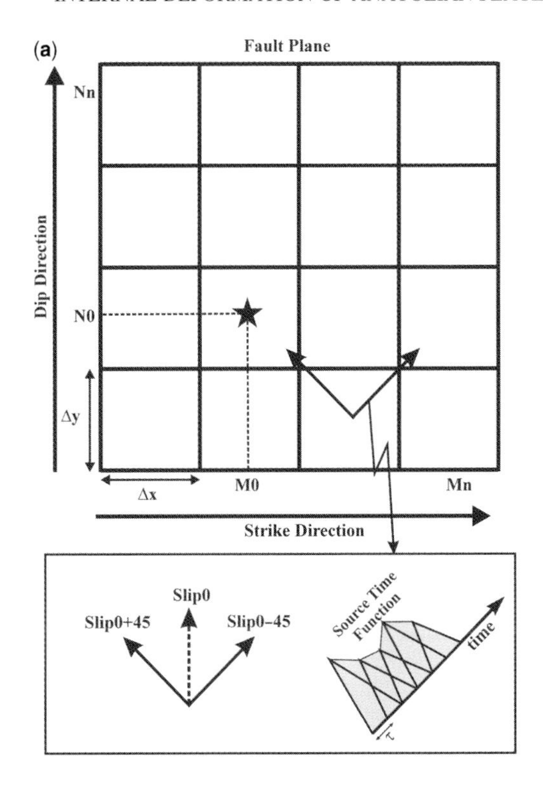

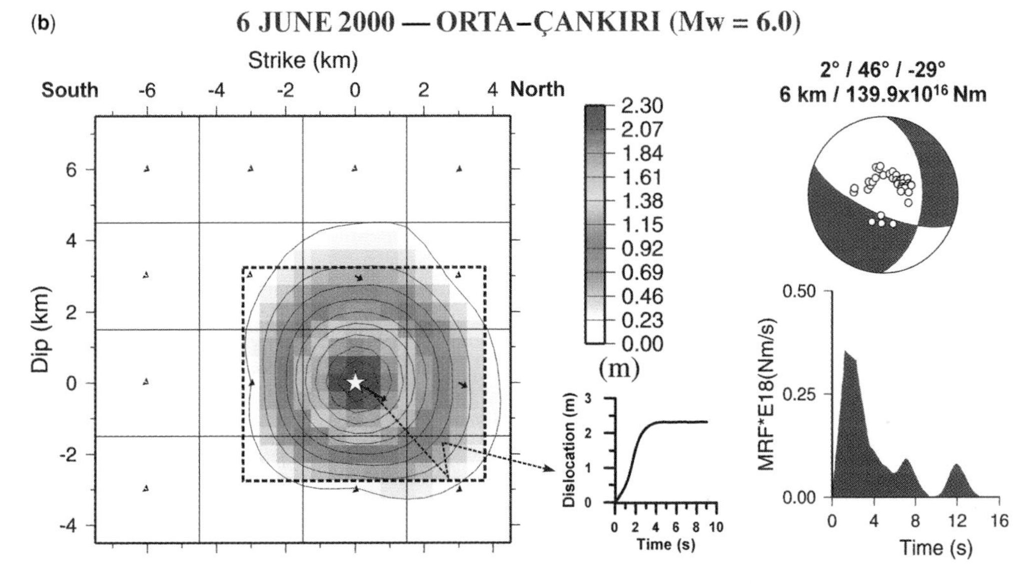

Fig. 9. (**a**) Schematic diagram summarizing the source expression during the rupture process. The fault plane is divided into sub-faults whose numbers (Nn), and dimensions ($\Delta x - \Delta y$) are predefined along the strike and dip on the fault plane. Also shown are the two components of slip vector with rakes (slip 0 ± 45), and parameterized moment-rate (source-time) function, which is described by the amplitudes of a series of overlapping isosceles triangles (Nábělek 1984; Yagi & Kikuchi 2000) whose numbers and duration were determined during the inversion procedure. (**b**) The rupture history obtained from our joint inversion of teleseismic P and SH long-period and broadband body waveforms. Earthquake focal mechanism, total moment rate function (source-time function), and distribution of coseismic slip are shown. The star indicates the location of the rupture initiation (initial break) located at a depth of about 6 km (0 km×0 km). The slip vectors and the distribution of slip magnitudes are also presented.

Table 2. *Source parameters of the Orta–Çankırı earthquake from InSAR and seismology*

	HRV CMT	USGS MT	InSAR
Scarp latitude (°N)	40.57	40.62	40.63 ± 0.01*
Scarp longitude (°E)	32.83	32.97	32.96 ± 0.01*
Length (km)	–	–	9 ± 1
$M_0 (10^{16}$ N m)	120	150	150 ± 40†
Slip (m)	–	–	1(fixed)‡
Strike§	358°/121°	340°/122°	$1 \pm 7°$
Dip§	53°/54°	36°/69°	$59 \pm 10°$
Rake§	−47°/−132°	−55°/−106°	$-23 \pm 8°$
Depth (km)	20.5	3	3.2 ± 0.6–7.7 ± 1.9‡
l.o.s. offset (mm)	–	–	-13.6 ± 6

InSAR data are from the present study. Error bounds of 1σ are given for the InSAR fault parameters.
*Location of surface scrap, projected up-dip from the centroid location.
†Assuming Lamé elastic constants $\mu = 3.23 \times 10^{10}$ Pa and $\lambda = 3.23 \times 10^{10}$ Pa.
‡Slip trades-off with depth extent of faulting. If a 1 m slip is chosen then the best-fit depth range is 3.2–7.7 km.
§The two numbers for the seismological solutions are the two nodal planes of the focal mechanisms reported.
l.o.s., line of sight.

recorded globally after major earthquakes. Kikuchi & Kanamori (1991) developed a procedure to determine the detailed source rupture processes, which provide valuable information on the nature of Earth dynamics (plate interactions, etc.), and this has recently been improved by Yoshida *et al.* (1996), Yagi & Kikuchi (2000) and Yagi (2002), among many others. The results contain valuable information, which is also related to strong ground motion experienced at the vicinity of faults. Yagi & Kikuchi (2000) expressed the rupture process as a spatio-temporal slip distribution on the fault plane (see Fig. 9 for details).

We followed the procedure developed by Yagi and Kikuchi (2000), and retrieved 42 teleseismic broadband P- body-wave data that were band-pass filtered between 0.01 and 1 Hz, and converted into ground displacement with a sampling time of 0.2 s (Fig. 8a). We have further introduced time corrections using the standard Jeffreys & Bullen (1940) crustal structure so that the observed P arrivals coincide with the theoretical arrival time of P-waves. We determined the spatio-temporal distribution of fault slip by applying a multi-time window inversion to the data obtained. Then, we resolved their relative weight so that the standard deviations of the teleseismic P-wave is about 10% of their individual maximum amplitudes by analysing the quality of observed records and background noise level. We assumed that faulting occurs on a single fault plane and that the slip angle is unchanged during the rupture. There is good coverage of the focal sphere at all azimuths, for both waveform (Fig. 8a) and first-motion (Fig. 8b) data. The waveform inversion provides us with direct information about the extent of the coseismic rupture area. Thus, to obtain the detailed rupture history, first we selected a fault area of 12 km × 12 km to obtain a rough estimate of the rupture area (Fig. 8a and b), which we divided into 4 × 4 sub-faults, each with an area of 3 km × 3 km. The source-time (slip-rate) function of each sub-fault is expanded in a series of 10 triangle functions with a rise time (τ), of 1.2 s. A rupture front velocity (*Vr*) is set at 2.8 km s^{-1} (for further information, see Hartzell & Heaton 1983; Lay & Wallace 1995; Ide & Takeo 1997; Imanishi *et al.* 2004), which gives the start time of the base function at each sub-fault. The inversion results are given in Figures 8 and 9, and Table 1. Figure 9b shows the final dislocation distribution. The source rupture process is rather simple and characteristic of a shallow crustal normal faulting.

We estimate the source parameters and uncertainties of the 6 June 2000 Orta–Çankırı earthquake from rupture history analyses to be: Nodal Plane 1: strike $2 \pm 5°$, dip $46° \pm 5°$; rake $-29 \pm 5°$; Nodal Plane 2: strike 113°, dip 70°; rake −132°; depth 6 ± 2 km (although this does not include uncertainty related to velocity structure, and is nearly constant along the fault strike), and seismic moment, $M_o = 140 \times 10^{16}$ N m (Mw 6.0). There is only one asperity recovered on the fault plane and the rupture is localized and estimated to propagate evenly in NE–SW direction with a slip vector of 203°. The largest slip of 2.31 m occurred at a depth of 6 km in the first 3 s of source rupture history (Fig. 9b). We have estimated the effective rupture area, stress drop ($\Delta\sigma = 2.5 \times M_o/S^{1.5}$, where S is the area of the maximum slip of 2.31 m, $6 \times 7\ km^2$), maximum slip and average slip to be *c.* 42 km^2, 128 bar, 231 cm and 111 cm, respectively. The total source duration is *c.* 16 s.

Earthquake source parameters from InSAR

In recent years, synthetic aperture radar interferometry (InSAR) has been established as a valuable technique with which to measure the surface deformation caused by earthquakes (Massonet *et al.* 1993; Zebker *et al.* 1994; Feigl *et al.* 1995; Wright 2002). By differencing the phase of the ground returns from an area, illuminated on repeated occasions by a SAR carried on an orbiting satellite, detailed maps of crustal deformation can be obtained with a spatial resolution of a few tens of metres and a measurement precision of a few millimetres. Further details on the technique have been given in reviews by Massonet & Feigl (1998) and Bürgmann *et al.* (2000). We use data from the ERS-2 SAR, which has a wavelength of 56 mm (C-band). Each interference 'fringe' corresponds to 28 mm of range change (the component of displacement in the satellite line of sight, 23° from the vertical at the scene centre).

InSAR data

Only limited SAR data were available for the Orta–Çankırı earthquake (Taymaz *et al.* 2002*a*). The best coseismic pair had an altitude of ambiguity of *c.* 270 m and a temporal separation of 10 months, only 5 days of which were after the earthquake (ERS-2, track 479, frame 2786, orbits 22878 (5 September 1999) and 26886 (11 June 2000)). The interferogram was processed using the ROI_pac software, and a 3 arcsecond (*c.* 90 m) digital elevation model (DEM) from the US Department of Defense was used to make the topographic correction. The altitude of ambiguity is the magnitude of topographic error that will cause a single interference fringe of erroneous phase. In this case, the DEM is thought to have height errors of less than 50 m, corresponding to a phase error of *c.* 1.2 radians, or an error in range change of *c.* 5 mm. A nonlinear, power-spectrum filter (Goldstein & Werner 1998) was applied to the interferogram.

The resultant interferogram (Fig. 10) is mostly coherent and shows concentric fringes in the expected epicentral are. The fringe pattern is asymmetrical, with a pear-shaped pattern of about five concentric fringes of range (distance to satellite) increase in the south, and about two fringes of range decrease in the north. The interferogram's noise, mostly arising from changes in atmospheric conditions, can be assessed by examining the phase signal away from the epicentral area, where we do not expect any deformation. The range changes have an r.m.s. error of 14 mm. We also investigate spatial correlations of the InSAR noise, by determining the 1D covariance function (e.g. Hanssen 2001). The e-folding length scale of the InSAR noise for this interferogram is *c.* 6 km.

Inversion procedure

We use a hybrid Monte-Carlo, downhill simplex inversion procedure (Wright *et al.* 1999, 2001*a, b*) to determine the best-fitting model parameters. This procedure minimizes the misfit between our interferometric measurements of range change, sampled at discrete locations using the Quadtree Algorithm (e.g. Jónsson *et al.* 2002), and those predicted by a simple elastic dislocation model (Okada 1985). We solved for the nine parameters required to describe a single rectangular fault with uniform slip (strike, dip, rake, slip, latitude and longitude, fault length, minimum and maximum depth), as well as a line-of-sight offset and gradients in the x- and y-directions to account for orbital errors in the interferogram.

A posteriori errors to the best-fit fault parameters (Tables 1 and 2) were determined using a Monte-Carlo simulation technique (T.J. Wright pers. comm.). In this technique, 100 simulations of the InSAR noise, based on the 1D covariance function, are created. These are added to the original range change observations and each noisy dataset is inverted. The distribution of inverted parameters gives their error. The method also allows the trade-offs between parameters to be examined (Fig. 11).

Inversion results

The best-fit inversion solution comprises a nearly north–south fault plane, dipping *c.* 60° to the east, and with a combination of normal and left-lateral slip. Within error, this solution is in good agreement with the body-wave solution. The best-fit fault parameters were used to construct a model interferogram (Fig. 10c). This simple model reproduces the interferogram very well, as is evident by the lack of systematic misfit in the residual interferogram (Fig. 10d). The r.m.s. misfit is 8 mm, comparable with the level of atmospheric noise in the interferogram. Further complications to the model are therefore unnecessary. The geodetic moment of $(150 \pm 40) \times 10^{16}$ N m is also in good agreement with the estimate from body-wave modelling (Figs 6–9). One difficulty was that the fault slip trades-off strongly with the fault width for this inversion, such that large slips on narrow faults fit the data as well as lower slips on wider faults. To overcome this, we fixed the fault slip at 1 m, consistent with standard stress drops and slip to length ratios (e.g. Wells & Coppersmith 1994). This gives us a reasonable depth range of 3.2–7.7 km for this earthquake.

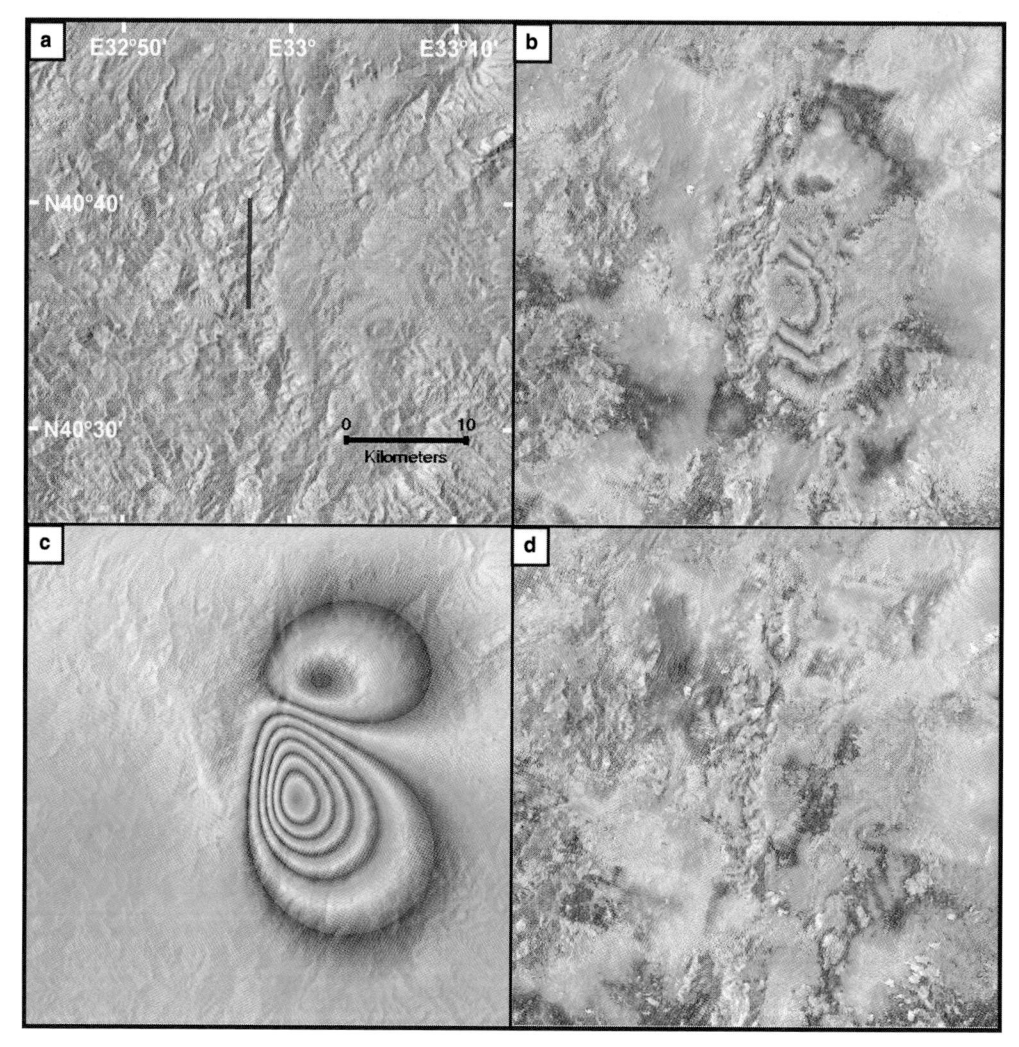

Fig. 10. Interferogram and model of the Orta–Çankırı earthquake. (**a**) SAR amplitude image, showing the location of the model surface rupture as a red line; (**b**) coseismic interferogram with a complete colour cycle of red through yellow to blue and back to red indicating an increase in range of 28 mm. Areas with no colour correspond to those where phase could not be unwrapped; (**c**) best-fit single-fault model, determined by inversion; (**d**) residual interferogram, obtained by subtracting (c) from (b).

We also investigated solutions that used the alternative (ESE–WNW-striking) nodal plane. Inversions in which the fault strike is held fixed at 113° (the strike determined from body-wave modelling) yield solutions that dip to the south at 70°, with a fault rake of –155°. The r.m.s. residual is 11 mm, larger than the 8 mm of the solution with a north–south-striking fault plane. Hence the north–south nodal plane is marginally preferred from the InSAR data alone. One surprising result of the inversion solution is that the best-fit slip was 7.7 ± 6 m. When combined with a very narrow best-fit depth range (4.8–5.3 km) this would give the earthquake an unusual aspect ratio, and abnormally high slip for an Mw = 6.0 earthquake (e.g. Wells & Coppersmith 1994). To determine if these values are well constrained by the InSAR data, we carried out a series of inversions, each time fixing the slip to a different value but solving for the best-fit depth range, with the fault geometry and rake held fixed. The result showed that these parameters trade-off against each other, such that for large values of slip, the depth range is small. For any value of slip equal to or greater than *c*. 1 m, the misfit and geodetic moment are approximately constant, but both increase sharply for lower

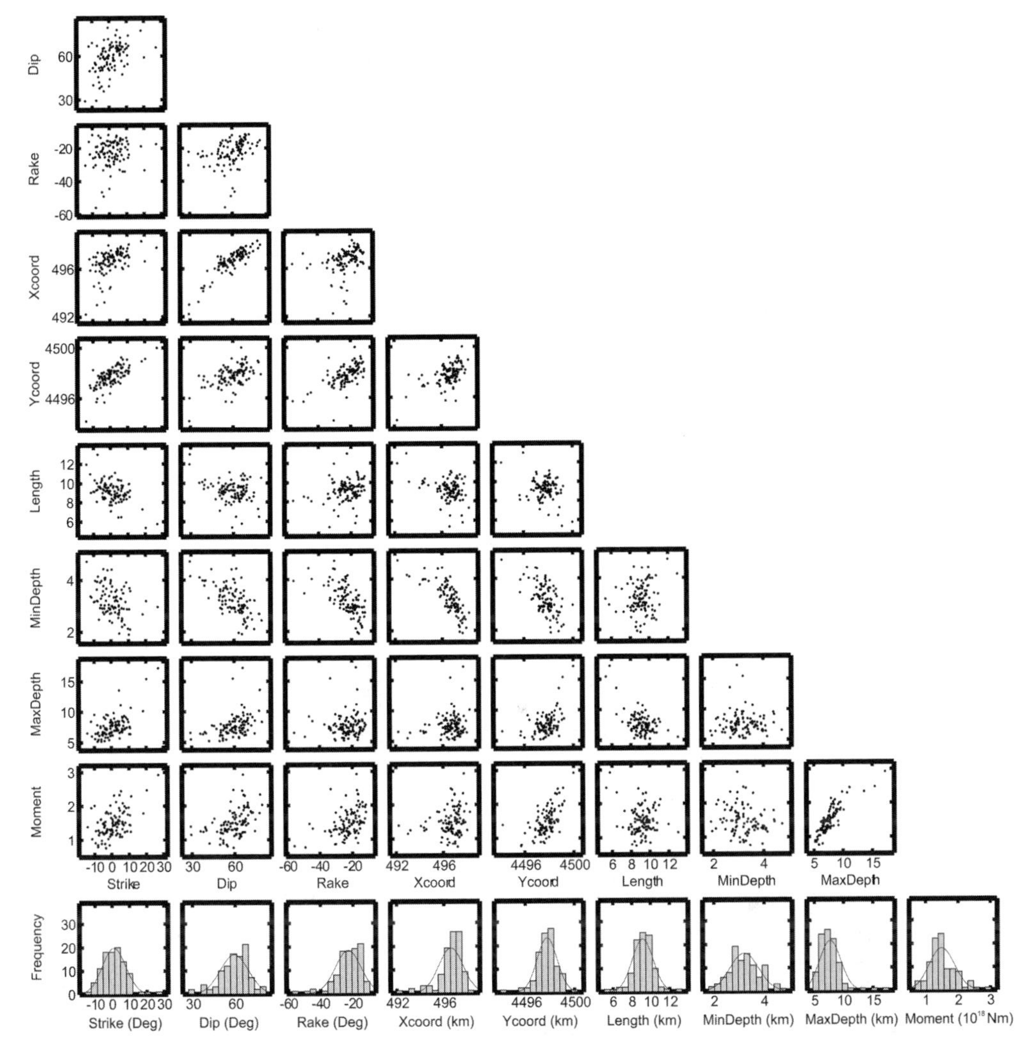

Fig. 11. Parameter trade-offs determined for the Orta–Çankırı earthquake using InSAR data. Each of the 100 dots in each figure is the solution of an inversion of the original InSAR data to which synthetic noise has been added. Histograms summarize the spread of solutions for each fault parameter, with the curves representing the Gaussian distributions with means and standard deviations as given in Table 2.

values of slip. The depth range found for a 1 m slip is more reasonable (3.6–7.4 km) and, given no additional constraints, these values are preferred. For very large values of slip, the depth range of the faulting becomes very small, and centred on a depth of 5 km.

Remote sensing data

One important factor that must be accounted for in seismotectonic studies is the site response caused by the surface conditions, the site effect. Earthquake damage may vary locally, being a function of the type of structures in the subsurface and/or soil mechanical ground conditions, such as faults and fractures, lithology or groundwater table. It has further been observed by macroseismic studies of topographic effects that in valleys and depressions damage intensity is higher because of higher earthquake vibration. These factors vary from earthquake to earthquake. Other influential factors are source distance and depth, azimuthal variation of source radiation, anelastic absorption and focusing effects of geological structures. Fault zones could cause constructive interference of multiple reflections

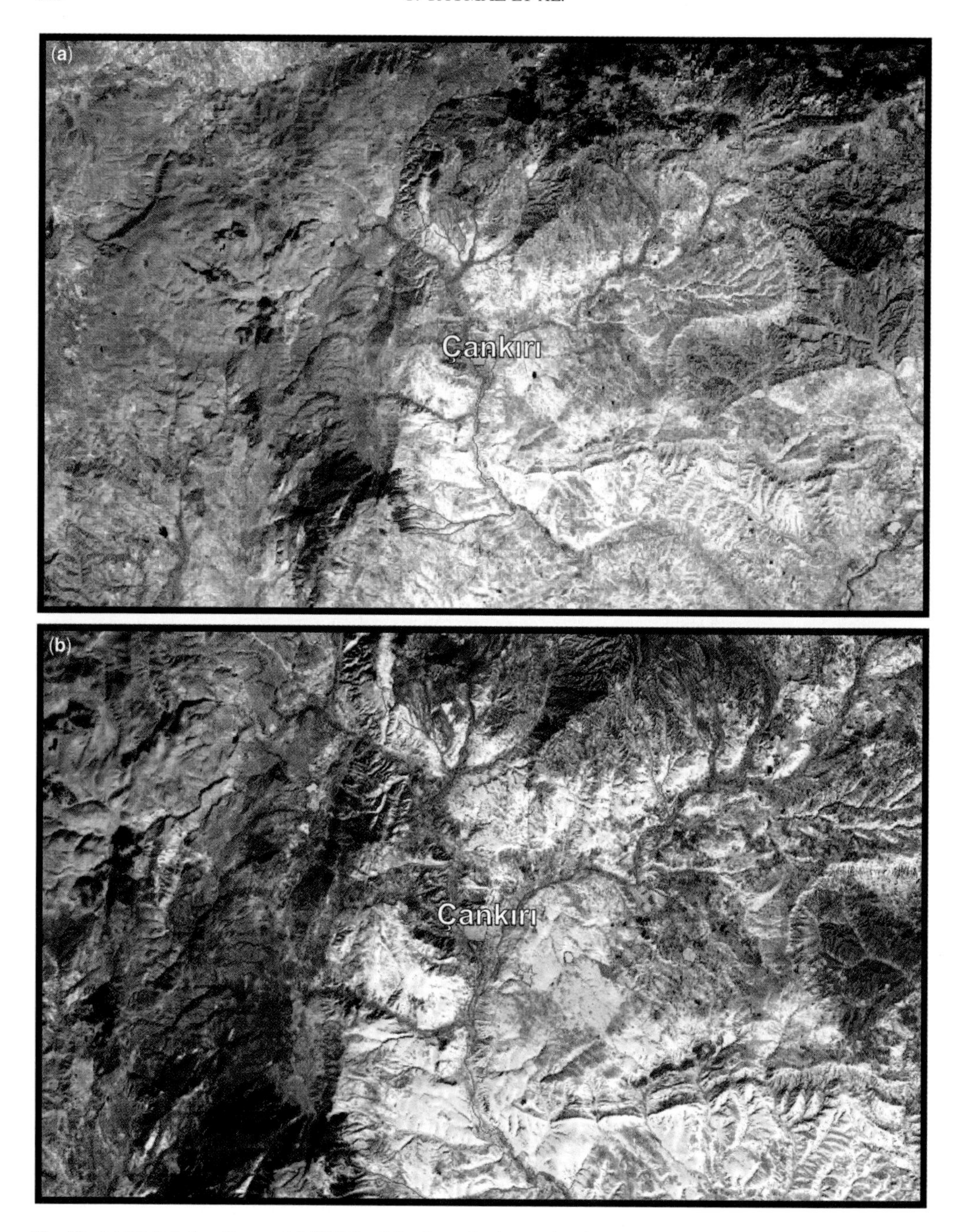

Fig. 12. (**a**) Digital elevation model (DEM) of the Orta–Çankırı region. LANDSAT ETM scene with HSV bands 4, 5, 3 and 8 is used. (**b**) LANDSAT ETM HSV image: bands 6, 5, 3 and 8 are used, and image resolution is 15 m. (**c**) DEM of North Anatolian Fault (NAF) at the Orta–Çankırı epicentre region.

of seismic waves at the boundaries between fault zones and surrounding rocks. Fault segments, their bends and intersection are the localized regions of stress concentration and seismic shock amplification. Intersecting fault zones could cause constructive interference of multiple reflections of seismic waves at the boundaries between fault zones and surrounding rocks. The

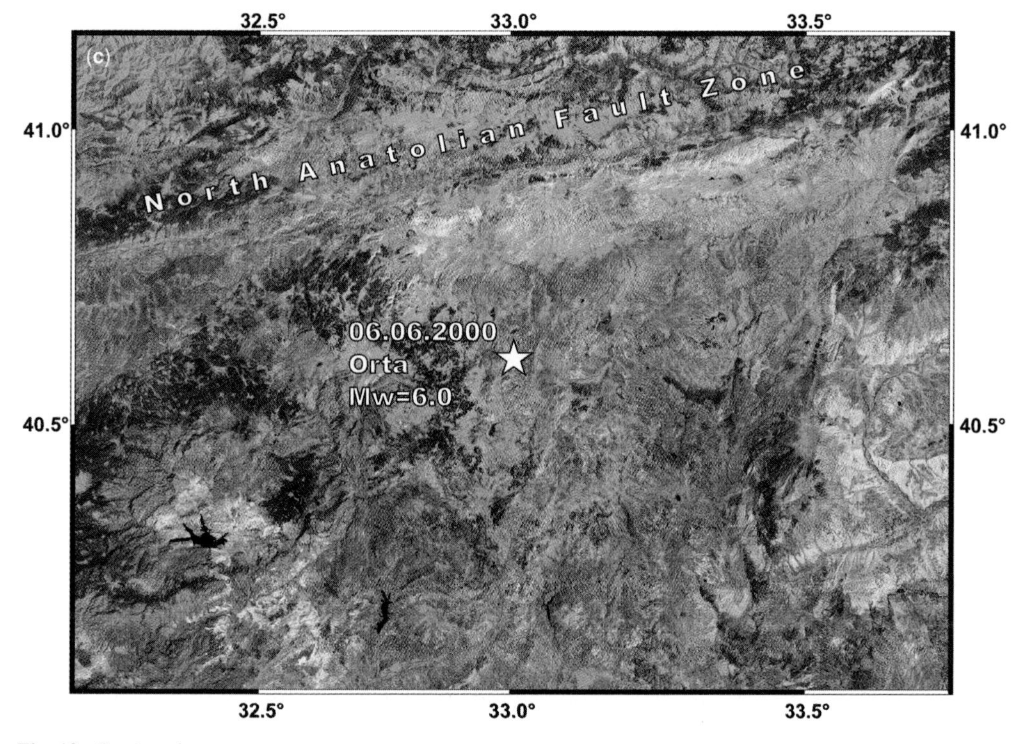

Fig. 12. *Continued.*

highest risk must be anticipated at the junctions of differently oriented ruptures, especially where one intersects the other. Compact fault zones consisting of distinct segments can be considered to be more dangerous in terms of seismic risk than those where active ruptures are scattered over a larger area. Those regions can be considered as being more exposed to earthquake shock as a result of amplification of guided seismic waves along crossing fault zones and soil amplification.

Therefore special attention is focused on precise mapping of traces of faults on satellite images, predominantly for areas with distinct expressed lineaments, as well as those with intersecting or overlapping lineaments or with unconsolidated sedimentary cover. Lineament analysis based on satellite images can help to delineate local fracture systems and faults that might influence seismic-wave propagation and influence the intensity of seismic shock. For example, merging lineament maps with isoseismal maps contributes to a better knowledge of subsurface structural influence on seismic shock intensity and on potentially earthquake-induced secondary effects such as landslides or soil liquefaction. Landslides cover a wide range including rock-fall, rockslide, debris slide and earth flow, and landslide risk high especially in steep slope areas. Landslides triggered by seismic shock have been documented in many parts of Turkey with existing slope instabilities. Subsurface fracture and fault patterns influence the shape and dimension of landslides to a large extent. Lineament analysis therefore provides important clues for delineation of areas prone to landslides, and especially a more precise localization of areas with a relatively high risk of slope failure.

Approach

For the present study LANDSAT 7 ETM images from northern central Turkey were obtained through the German Science Foundation (DFG, Bonn), the German Aerospace Centre (DLR, Oberpfaffenhofen) and EUROIMAGE (Rome). The LANDSAT ETM scene, LANDSAT 7 ETM, 29 May 1999, Track 177, Frame 32, was provided by EURIMAGE and DLR as Level 1G product (Fig. 12). This image product is radiometrically and geometrically corrected including output map projection and image orientation (UTM, WGS84), and resampled by using nearest neighbourhood algorithm. The LANDSAT data comprise six bands of multi-spectral data in the visible and IR portion of the spectrum. The image has a geometric resolution (30 m) that provides sufficient level of

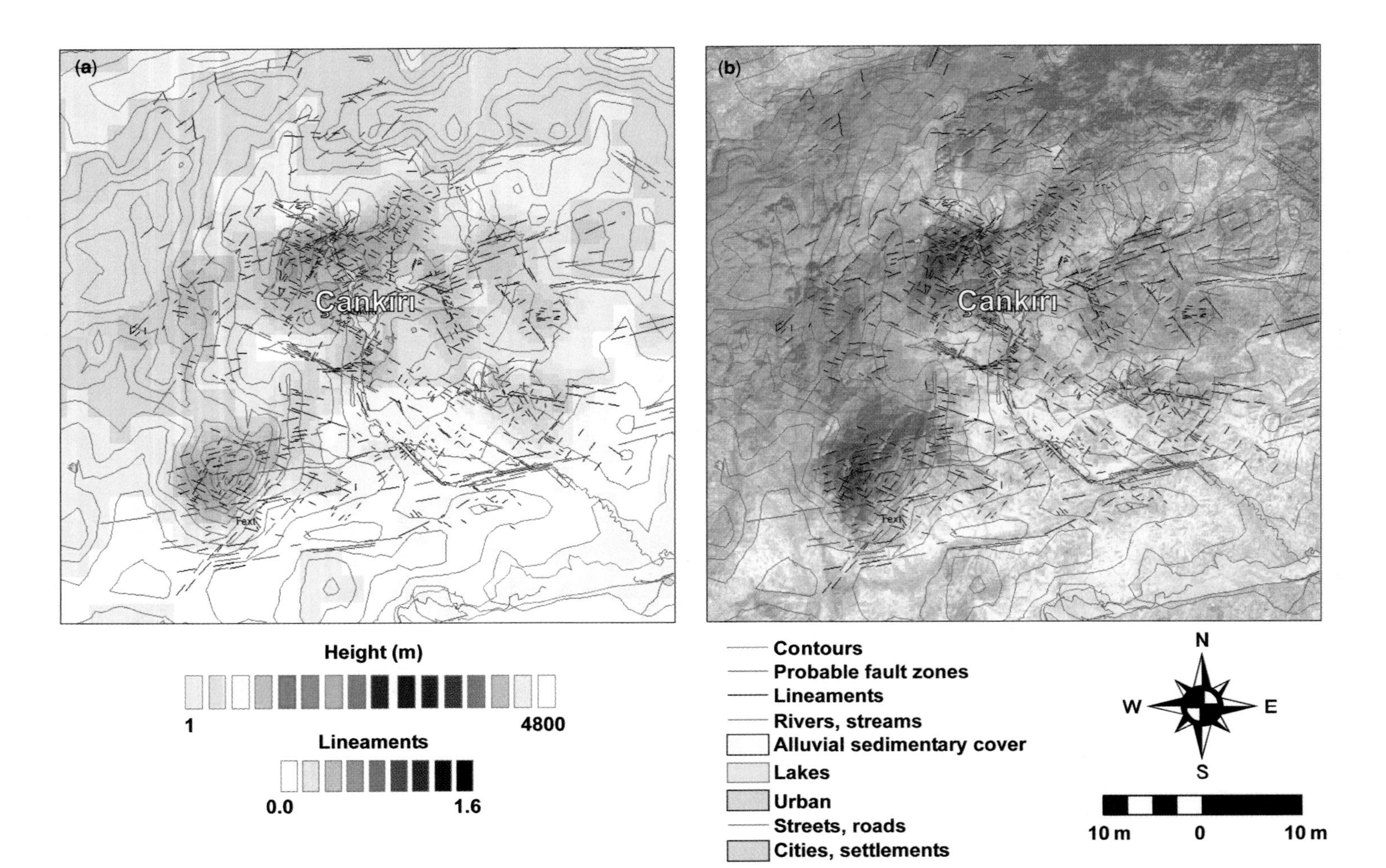

Fig. 13. (**a**) Merged topographic information and structural evaluations based on the LANDSAT ETM image. (**b**) Overlay of LANDSAT ETM, topographic and structural evaluation data.

detail for the purpose of application and maintains a sufficiently large scene size to give the 'overall picture'. The corresponding panchromatic band 8 (15 m) co-registered to the multi-spectral data was used to add crispness and detail to the multi-spectral data. The thermal band has 60 m resolution. The LANDSAT track is 183 km wide.

To enhance the LANDSAT ETM data digital image processing was carried out. Various image sharpening tools (ENVI Software/CREASO) were tested to find the best-suited combinations. The bands 5, 4, 3 (RGB) and 6, 5, 3 (RGB) and 8 were selected to transform to an HSV high-resolution (15 m) image product as shown in Figures 12 and 13. The various datasets (LANDSAT ETM data, and topographic, geological and geophysical data from the study area) were integrated into a GIS using the software ArcView GIS 3.2 with the extensions Spatial Analyst and 3D-Analyst and ArcGIS 8.2 of ESRI to obtain a better understanding of processes influencing the damage intensity of stronger earthquakes.

Lineament analysis

Based on the LANDSAT ETM data for the Orta–Çankırı region, lineament analysis has been carried out. Many of the known faults and fracture zones are traced as linear features and linear arrangements of pixels, especially on amplifications of the satellite images up to a scale of 1:50 000. Linear features are clearly detectable because of the linear arrangements of pixels with the same tone, and/or of linear topographic features such as scarps or drainage pattern. Special attention is focused on areas with distinct expressed lineaments, as well as on those with intersecting lineaments or unconsolidated sedimentary covers. The available data then were merged in a GIS to demonstrate the topographic situation together with tectonic information. The maps show clearly morphological depressions and areas with higher densities of lineaments where stronger ground shaking can be expected. Liquefaction can occur within fluvial sediments (see Fig. 13 for details).

Implications

The rupture process of the 6 June 2000 Orta–Çankırı earthquake (Mw = 6.0) is deduced from the joint inversion of the teleseismic P and SH body waveforms, and separately from InSAR data. We have been able to discriminate between the two possible nodal planes of the fault mechanism parameters. The fault length and rake are also well determined, showing that the earthquake was caused by a mixture of left-lateral strike-slip and normal slip on a north–south-striking fault plane that dips to the east. The fault rake is found to be $-29 \pm 5°$, indicating that the earthquake had a larger component of left-lateral slip than was previously reported by Harvard-CMT and USGS solutions. The magnitude of slip distribution and the depth range of the faulting is well constrained, and the geodetic moment is found to be within the acceptable range of the seismic moment obtained from the teleseismic body-wave inversion results.

The fault plane responsible for the Orta–Çankırı earthquake is at a high angle to that of the North Anatolian Fault, and because of large component of left-lateral slip confirmed by the joint inversion of the teleseismic P and SH body waveforms and InSAR solution, the earthquake slip vector is also oriented at a high angle with respect to slip on the North Anatolian Fault (Fig. 14). However, the horizontal projections of the P- and T-axes of the focal mechanism for the Orta–Çankırı earthquake are similar to those of focal mechanisms summarized in Table 1, suggesting that the same stress regime could be responsible for the earthquakes concerned. One possible interpretation is that right-lateral shear stresses form a shear zone at depth that causes a rotational torque to be applied. This would manifest itself seismically in left-lateral slip on roughly north–south-striking fault planes. In this model, the misalignment between the earthquake slip vectors would be caused by rotation (Fig. 15). Iio *et al.* (2002) also suggested that the observed aftershocks of the 17 August 1999 Gölcük–İzmit earthquake indicate afterslips, which could concentrate larger stress on the source region. Thus, this stress concentration may have triggered the 12 November 1999 Düzce earthquake in conjunction with the stress change caused by the main shock of the Gölcük–İzmit earthquake (e.g. Parsons *et al.* 2000). Similarly uncharacteristic seismic activity

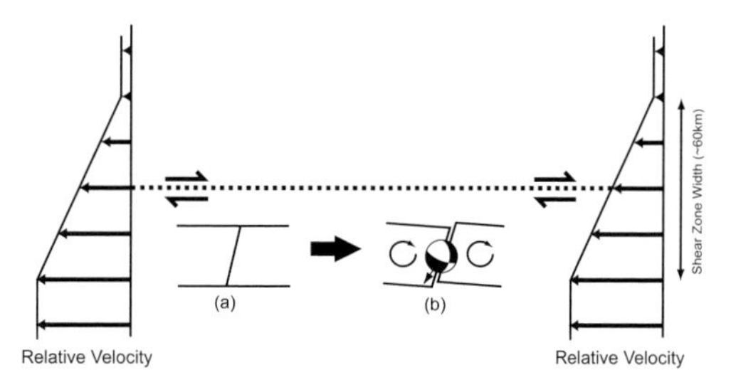

Fig. 14. Schematic diagram (map view) illustrating a possible cause of the Orta–Çankırı earthquake. The velocity arrows indicate the deformation at depth in a 60 km wide shear zone with a strike-slip fault at the centre. The shear zone causes rotational torque to be applied to the upper layer, which could result in an earthquake with the mechanism and slip vector as indicated. (**a**) is the starting condition, and (**b**) is the situation after one earthquake cycle.

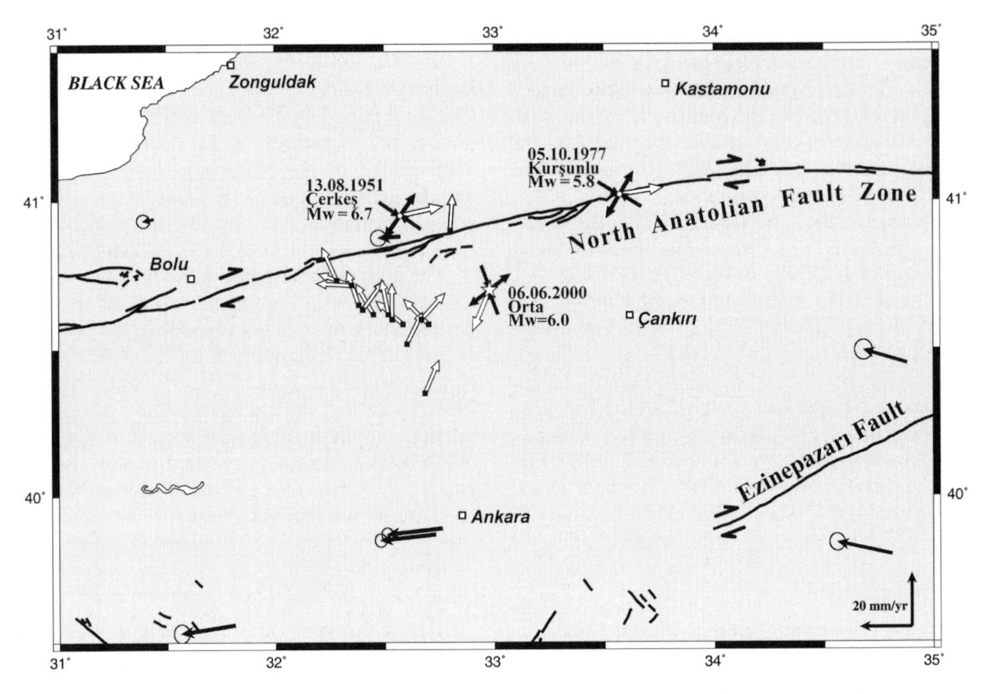

Fig. 15. Summary map showing the horizontal projections of the P- and T-axes and the slip vectors of the Orta–Çankırı earthquake inferred in the present study, and those of nearby earthquakes already reported, shown by bold filled arrows and large open arrows, respectively (see Table 1). Locations of the mapped strands of the North Anatolian Fault (Şaroğlu *et al.* 1992) are shown by bold continuous lines. The small open arrows are palaeomagnetic declinations reported by Platzman *et al.* (1994), obtained from volcanic rocks dated to a Miocene age (7.5–22 Ma). Star indicates the instrumental epicentre of the 6 June 2000 Orta–Çankırı earthquake reported by ISC. The bold filled arrows with error ellipses are interseismic velocities relative to Eurasia measured by GPS (McClusky *et al.* 2000).

is observed in the vicinity of Orta town orthogonal to the known geometry of the North Anatolian Fault Zone (see Fig. 2). Furthermore, Platzman *et al.* (1994) carried out a palaeomagnetic survey on a transect across the North Anatolian Fault close to the Orta–Çankırı earthquake (Fig. 15).

In general, there is no systematic clockwise rotation, but some sites near the Orta–Çankırı earthquake do show clockwise rotations of up to 43°. In addition, it may be concluded that Anatolia has deformed as a fundamentally integral plate subjected to discrete intervals of rotation that would be consistent with the 'instantaneous' solution derived from GPS observations. There are conflicting views on the neotectonic crustal deformation in this region. Some researchers have proposed that the region is deformed by differential extrusion and rotation of crustal blocks on a regional scale. Thus, Gürsoy *et al.* (1999) considered that rotations recognized in Eocene units are comparable over a large area of central Anatolia (see Tatar *et al.* 1995) and north of the NAFZ (Sarıbudak 1989; Piper *et al.* 1996, 2001), and they prove to be similar to rotations recognized in younger rocks influenced only by the neo-tectonic regime. Piper *et al.* (2001) considered that central Anatolia shows widespread regional post-Eocene anticlockwise rotations of *c.* 30°. Hence, they argued that rotations are present on either side of the NAFZ, and are therefore older than the establishment of the NAFZ in mid-Pliocene time. Piper *et al.* (2001) further advocated that the character of lithosphere deformation in the region is in conflict with a progressive decline in rotation away from the fault break predicted by the thin viscous sheet model of continental deformation (England & McKenzie 1982). They concluded that the brittle upper crust is detached from the lower crust, undergoes continuum deformation by creep, and the fault blocks are therefore interpreted to be pinned rather than free-floating. It is difficult to compare palaeomagnetic rotations with GPS results, as the former reflect variations on a regional scale with rates of differential rotation many times larger than those of the second. We can then argue that the time period currently investigated by GPS data is too short to yield rotation rates representative of the long-term crustal deformation in Anatolia. In conclusion, these observations suggest that although there is no general clockwise rotation, localized rotations do occur, and these may have been responsible for the Orta–Çankırı earthquake of 6 June 2000.

We thank personnel and station operators at WWSSN, GDSN and IRIS Data Centre for making waveform data available for this study. We are grateful to W. Lee of IASPEI for providing a complimentary copy of IASPEI-Software Library Volume 3: SYN4. We are especially indebted to P. Zwick and R. McCaffrey for helping us on daily matters encountered in use of SYN3, SYN4 and MT5 packages. This research could not have been done without their help. We appreciate Y. Yagi's generous and unselfish offer to let us use his rupture history and slip distribution code. We are grateful to the German Science Foundation (DFG, Bonn), the German Aerospace Centre (DLR, Oberpfaffenhofen) and EUROIMAGE (Rome) for freely providing LANDSAT 7 ETM images from northern central Turkey (Figs 12 and 13), and we thank B. Theilen-Willige for co-operation and discussions. We are also grateful to P. Wessel and W.H.F Smith of Hawaii University–NOAA for providing GMT software (Wessel & Smith 1991, 1995). This research has been partly funded by İstanbul Technical University Research Fund (ITU-BAP), TÜBİTAK, and the Turkish Academy of Sciences (TÜBA), in the framework of the Young Scientist Award Programme (TT/TÜBAGEBİP/2001-2-17). T.J.W. is an NERC (UK) Research Fellow. We also thank B. Parsons, K. Feigl, J. Woodhouse (discussions), ESA (data) and JPL/Caltech (ROI_pac).

References

ADIYAMAN, Ö., CHOROWICZ, J., ARNAUD, O. N., GÜNDOĞDU, M. N. & GOURGAUD, A. 2001. Late Cenozoic tectonics and volcanism along the North Anatolian Fault: new structural and geochemical data. *Tectonophysics*, **338**, 135–165.

AKYÜREK, B., BILGINER, E., ÇATAL, E., DAGER, Z., SOYSAL, Y. & SUNU, O. 1980. *The Geology of Eldivan–Sabanozu (Çankırr) and Hasayaz–Çandır (Kalecik–Ankara) and surrounding regions*. MTA Report 6741 [in Turkish].

ALSDORF, D., BARAZANGI, M., LITAK, R., SEBER, D., SAWAF, T. & AL-SAAD, D. 1995. The intraplate Euphrates fault system–Palmyrides mountain belt junction and relationship to Arabian plate boundary. *Annali di Geofisica*, **38**, 385–397.

AMBRASEYS, N. N. 1970. Some characteristic features of the Anatolian fault zone. *Tectonophysics*, **9**, 143–165.

AMBRASEYS, N. N. & JACKSON, J. A. 1998. Faulting associated with historical and recent earthquakes in the Eastern Mediterranean region. *Geophysical Journal International*, **133**, 390–406.

BOZKURT, E. 2001. Neotectonics of Turkey—a synthesis. *Geodinamica Acta*, **14**, 3–30.

BOZKURT, E. & KOÇYIĞIT, A. 1996. The Kazova basin: an active negative flower structure on the Almus Fault Zone, a splay fault system of the North Anatolian Fault Zone, Turkey. *Tectonophysics*, **265**, 239–254.

BÜRGMANN, R., ROSEN, P. & FIELDING, E. 2000. Synthetic aperture radar interferometry to measure Earth's surface topography and its deformation. *Annual Review of Earth and Planetary Sciences*, **28**, 169–209.

CHANNEL, J. E. T., TÜYSÜZ, O., BEKTAŞ, O. & ŞENGÖR, A. M. C. 1996. Jurassic–Cretaceous paleomagnetism and paleogeography of the Pontides (Turkey). *Tectonics*, **15**, 201–212.

CHOROWICZ, J., DHONT, D. & GÜNDOĞDU, N. 1999. Neotectonics in the eastern North Anatolian fault region (Turkey) advocates crustal extension: mapping from SAR ERS imagery and Digital Elevation Model. *Journal of Structural Geology*, **21**, 511–532.

DEMIRBAĞ, E., RANGIN, C., LE PICHON, X. & ŞENGÖR, A. M. C. 2003. Investigation of the tectonics of the Main Marmara Fault by means of deep-towed seismic data. *Tectonophysics*, **361**, 1–69.

EMRE, Ö., DUMAN, T. Y., DOGAN, A. & ÖZALP, S. 2000. *Evaluation Report of 6 June 2000 Orta (Çankırı-Turkey) Earthquake*. MTA Geology Department, Ankara [in Turkish].

ENGLAND, P. & MCKENZIE, D. P. 1982. A thin viscous sheet model for continental deformation. *Geophysical Journal of the Royal Astronomical Society*, **70**, 295–321.

ERDOĞAN, B., AKAY, E. & UĞUR, M. S. 1996. Geology of the Yozgat region and evolution of the collisional Çankırı basin. *International Geology Review*, **38**, 788–806.

FEIGL, K. L., SERGENT, A. & JACQ, D. 1995. Estimation of an earthquake focal mechanism from a satellite radar interferogram; application to the December 4, 1992 Landers aftershock. *Geophysical Research Letters*, **22**, 1037–1040.

FREDRICH, J., MCCAFFREY, R. & DENHAM, D. 1988. Source parameters of seven large Australian earthquakes determined by body waveforms inversion. *Geophysical Journal*, **95**, 1–13.

FUTTERMAN, W. I. 1962. Dispersive body waves. *Journal of Geophysical Research*, **67**, 5279–5291.

GÖKTEN, E., ÖZAKSOY, V. & KARAKUŞ, K. 1996. Tertiary volcanic and tectonic evolution of the Ayaş–Güdül–Çeltikçi region, Turkey. *International Geology Review*, **38**, 926–934.

GOLDSTEIN, R. M. & WERNER, C. L. 1998. Radar interferogram filtering for geophysical applications. *Geophysical Research Letters*, **25**, 4035–4038.

GÖRÜR, N., TÜYSÜZ, O. & ŞENGÖR, A. M. C. 1998. Tectonic evolution of the Central Anatolian basins. *International Geology Review*, **40**, 831–850.

GÜRSOY, H., PIPER, J. D. A. & TATAR, O. 1999. Palaeomagnetic study of the Galatean Volcanic Province, north–central Turkey: Neogene deformation at the northern border of the Anatolian Block. *Geological Journal*, **34**, 7–23.

HANSSEN, R. F. 2001. *Radar Interferometry: Data Interpretation and Error Analysis*. Kluwer, Dordrecht.

HARTZELL, S. H. & HEATON, T. H. 1983. Inversion of strong ground motion and teleseismic waveform data for fault rupture history of the 1979 Imperial Valley, California, earthquake. *Bulletin of the Seismological Society of America*, **73**, 1553–1583.

IDE, S. & TAKEO, M. 1997. Determination of constitutive relations of fault slip based on seismic wave analysis. *Journal of Geophysical Research*, **102**, 27379–27391.

IMANASHI, K., TAKEO, M., ITO, H. *ET AL.* 2004. Source parameters and rupture velocities of microearthquakes in western Nagano, Japan, determined using stopping phases. *Bulletin of the Seismological Society of America*, **94**, 1762–1780, doi:10.1785/012003085.

IIO, Y., HORIUCHI, S., BARIS, S. *ET AL.* 2002. Aftershock distribution in the eastern part of the aftershock region of the 1999 Izmit, Turkey, earthquake. *Bulletin of the Seismological Society of America*, **92**, 411–417.

JACKSON, J. 1994. Active tectonics of the Aegean region. *Annual Review of Earth and Planetary Sciences*, **22**, 239–271.

JEFFREYS, S. & BULLEN, K. 1940. *Seismological Tables*. British Association for the Advancement of Science, Gray–Milne Trust, London.

JÓNSSON, S., ZEBKER, H., SEGALL, P. & AMELUNG, F. 2002. Fault slip distribution of the 1999 Mw 7.1 Hector Mine earthquake, California, estimated from satellite radar and GPS measurements. *Bulletin of the Seismological Society of America*, **92**, 1377–1389.

KAYMAKÇI, N. 2000. *Tectono-stratigraphical evolution of the Çankırı Basin (Central Anatolia)*. PhD thesis, Universiteit Utrecht.

KAYMAKÇI, N., WHITE, S. H., VANDIJK, P. M. *ET AL.* 2000. Palaeostress inversion in a multiphase deformed area: kinematic and structural evolution of the Çankırı Basin (central Turkey), Part 1—northern area. *In*: BOZKURT, E., WINCHESTER, J. A. & PIPER, J. D. A. (eds) *Tectonics and Magmatism in Turkey and the Surrounding Area.* Geological Society, London, Special Publications, **173**, 295–323.

KAYMAKÇI, N., WHITE, S. H. & VANDIJK, P. M. 2003. Kinematic and structural development of the Çankırı Basin (Central Anatolia, Turkey): a palaeostress inversion study. *Tectonophysics*, **364**, 85–113.

KENNETT, B. L. N. & ENGDAHL, E. R. 1991. Traveltimes for global earthquake location and phase identification. *Geophysical Journal International*, **105**, 429–465.

KIKUCHI, M. & KANAMORI, H. 1991. Inversion of complex body waves—III. *Bulletin of the Seismological Society of America*, **81**, 2335–2350.

KOÇYIĞIT, A. 1991*a*. An example of an accretionary forearc basin from northern central Anatolia and its implications for the history of subduction of Neo-Tethys in Turkey. *Geological Society of America Bulletin*, **103**, 22–36.

KOÇYIĞIT, A. 1991*b*. Changing stress orientation in progressive intracontinental deformation as indicated by the neotectonics of the Ankara region (NW Central Anatolia). *Turkish Association of Petroleum Geologists Bulletin*, **3**, 43–55.

KOÇYIĞIT, A. 1992. Southward-vergent imbricate thrust zone in Yuvaköy: a record of the latest compressional event related to the collisional tectonic regime in Ankara–Erzincan Suture Zone. *Turkish Association of Petroleum Geologists Bulletin*, **4**, 111–118.

KOÇYIĞIT, A., TÜRKMENOĞLU, A., BEYHAN, A., KAYMAKÇI, N. & AKYOL, E. 1995. Post-collisional tectonics of Eskisehir–Ankara–Çankırı segment of Izmir–Ankara–Erzincan suture zone: Ankara Orogenic Phase. *Turkish Association of Petroleum Geologists Bulletin*, **6**, 69–86.

KOÇYIĞIT, A., ROJAY, B., CIHAN, M. & ÖZACAR, A. 2001. The June 6, 2000, Orta (Çankırı, Turkey) earthquake: sourced from a new antithetic sinistral strike-slip structure of the North Anatolian Fault System, the Dodurga Fault Zone. *Turkish Journal of Earth Sciences*, **10**, 69–82.

KURT, H., DEMIRBAĞ, E. & KUŞÇU, I. 1999. Investigation of the submarine active tectonism in the Gulf of Gökova, Southwest Anatolia–Southeast Aegean Sea, by multi-channel seismic reflection data. *Tectonophysics*, **305**, 477–496.

KURT, H., DEMIRBAĞ, E. & KUŞÇU, İ. 2000. Active submarine tectonism and formation of the Gulf of Saros; Northeast Aegean Sea; inferred from multi-channel seismic reflection data. *Marine Geology*, **165**, 13–26.

LAY, T. & WALLACE, T. C. 1995. *Modern Global Seismology*. Academic Press, San Diego, CA.

LEE, W. H. K. (ed.) 1991. *IASPEI Software Library Volume 3, Digital Seismogram Analysis and Waveform Inversion*. IASPEI Working Group on Personal Computers, Menlo Park, CA.

LE PICHON, X., LYBERIS, N. & ALVAREZ, F. 1984. Subsidence history of the North Aegean trough. *In*: DIXON, J. E. & ROBERTSON, A. H. F. (eds) *The Geological Evolution of the Eastern Mediterranean*. Geological Society, London, Special Publications, **17**, 27–741.

LE PICHON, X., ŞENGÖR, A. M. C., DEMIRBAĞ, E. *ET AL.* 2001. The active Main Marmara Fault. *Earth and Planetary Science Letters*, **192**, 595–616.

MAKRIS, J. & STOBBE, C. 1984. Physical properties and state of the crust and upper mantle of the Eastern Mediterranean Sea deduced from geophysical data. *Marine Geology*, **55**, 347–363.

MASSONET, D. & FEIGL, K. L. 1998. Radar interferometry and its application to changes in the Earth's surface. *Review of Geophysics*, **36**, 441–500.

MASSONET, D., ROSSI, M., CARMONA, C., ADRAGNA, F., PELTZER, G., FEIGL, K. & RABAUTE, T. 1993. The displacement field of the Landers earthquake mapped by radar interferometry. *Nature*, **364**, 138–142.

MCCAFFREY, R. 1988. Active tectonics of the Eastern Sunda and Banda Arcs. *Journal of Geophysical Research*, **93**, 15163–15182.

MCCAFFREY, R. & ABERS, G. 1988. *SYN3: a program for inversion of teleseismic body wave forms on microcomputers*. US Air Force Geophysics Laboratory Technical Report, **AFGL-TR-88-0099**.

MCCAFFREY, R. & NÁBĚLEK, J. 1987. Earthquakes, gravity, and the origin of the Bali Basin: an example of a nascent continental fold-and-thrust belt. *Journal of Geophysical Research*, **92**, 441–460.

MCCLUSKY, S., BALASSANIAN, S., BARKA, A. *ET AL.* 2000. Global positioning system constraints on plate kinematics and dynamics in the eastern Mediterranean and Caucasus. *Journal of Geophysical Research*, **105**, 5695–5719.

MCKENZIE, D. 1969. The relation between fault plane solutions for earthquakes and the directions of the principal stresses. *Bulletin of the Seismological Society of America*, **59**, 591–601.

McKenzie, D. 1970. Plate tectonics of the Mediterranean region. *Nature*, **226**, 239–243.
McKenzie, D. 1972. Active tectonics of the Mediterranean region. *Geophysical Journal of the Royal Astronomical Society*, **30**, 109–185.
Mercier, J. L., Sorel, D., Vergely, P. & Simeakis, K. 1989. Extensional tectonic regimes in the Aegean basins during the Cenozoic. *Basin Research*, **2**, 49–71.
Molnar, P. & Lyon-Caen, H. 1989. Fault plane solutions of earthquakes and active tectonics of the Tibetan Plateau and its margins. *Geophysical Journal International*, **99**, 123–153.
Nábělek, J. L. 1984. *Determination of earthquake source parameters from inversion of body waves*. PhD thesis, Massachusetts Institute of Technology, Cambridge, MA.
Nelson, M. R., McCaffrey, R. & Molnar, P. 1987. Source parameters for 11 earthquakes in the Tien Shan, central Asia, determined by P and SH waveform inversion. *Journal of Geophysical Research*, **92**, 12629–12648.
Nemec, W. & Kazanci, N. 1999. Quaternary colluvium in west–central Anatolia: sedimentary facies and paleoclimatic significance. *Sedimentology*, **46**, 139–170.
Nowroozi, A. A. 1972. Focal mechanism of earthquakes in Persia, Turkey, west Pakistan and Afghanistan and plate tectonics of the Middle East. *Bulletin of the Seismological Society of America*, **62**, 823–850.
Okada, Y. 1985. Surface deformation due to shear and tensile faults in a half-space. *Bulletin of the Seismological Society of America*, **75**, 1135–1154.
Okay, A. I. & Tüysüz, O. 1999. Tethyan sutures of northern Turkey. *In*: Durand, B., Jolivet, L., Horvath, F. & Serrane, M. (eds) *The Mediterranean Basins: Tertiary Extension within the Alpine Orogen*. Geological Society, London, Special Publications, **156**, 475–515.
Özay, İ. 1996. *Source parameters of 1977 October 5 Kurşunlu (NW Turkey) earthquake*. BSc Thesis, İstanbul Technical University.
Papazachos, B. C., Papadimitriou, E. E., Kiratzi, A. A., Papazachos, C. B. & Louvari, E. K. 1998. Fault plane solutions in the Aegean Sea and the surrounding area and their implication. *Bolletino di Geofisica Teorica ed Applicata*, **39**, 199–218.
Parsons, T., Toda, S., Stein, R., Barka, A. & Dieterich, J. 2000. Heightened odds of large earthquakes near Istanbul: an interaction-based probability calculation. *Science*, **288**, 661–665.
Piper, J. D. A., Moore, J., Tatar, O., Gürsoy, H. & Park, G. 1996. Palaeomagnetic study of crustal deformation across an intracontinental transform: the North Anatolian Fault Zone in Northern Turkey. *In*: Morris, A. & Tarling, D. H. (eds) *Palaeomagnetism of the Mediterranean Regions*. Geological Society, London, Special Publications, **105**, 191–203.
Piper, J. D. A., Gürsoy, H. & Tatar, O. 2001. Palaeomagnetic analysis of neotectonic crustal deformation in Turkey. *In*: Taymaz, T. (ed.) *Symposia on Seismotectonics of the North-Western Anatolia–Aegean and Recent Turkish Earthquakes, Extended Abstracts*. Istanbul Technical University, Istanbul, 4–13.
Platzman, E. S., Platt, J. P., Tapirdamaz, C., Sanver, M. & Rundle, C. C. 1994. Why are there no clockwise rotaions along the North Anatolian Fault Zone? *Journal of Geophysical Research*, **99**, 21705–21715.
Polat, A. 1988. Evidence for paleo- to neotectonic transition period in the Büyük Polat-Yarımsöğüt region and the origin of the Kırıkkale–Erbaa Fault Zone. *Turkish Association of Petroleum Geologists Bulletin*, 127–140.
Price, S. P. & Scott, B. 1994. Fault-block rotations at the edge of a zone of continental extension; southwest Turkey. *Journal of Structural Geology*, **16**, 381–392.
Richens, W. D., Pechmann, J. C., Smith, R. B., Langer, C. J., Goter, S. K., Zollweg, J. E. & King, J. J. 1987. The Borah Peak, Idaho, earthquake and its aftershocks. *Bulletin of the Seismological Society of America*, **77**, 694.
Saribudak, M. 1989. A paleomagnetic approach to the origin of the Black Sea. *Geophysical Journal International*, **99**, 247–251.
Şaroğlu, F., Emre, Ö. & Kuşçu, İ. 1992. *Active Fault Map of Turkey, 2 sheets*. Maden ve Tetkik Arama Enstitüsü, Ankara.
Şen, Ş., Seyitoğlu, G., Karadenizli, L., Kazanci, N., Varol, B. & Araz, H. 1998. Mammalian biochronology of Neogene deposits and its correlation with the lithostratigraphy in the Çankırı–Çorum basin, central Anatolia, Turkey. *Eclogae Geologicae Helvetiae*, **91**, 307–320.
Şengör, A. M. C. 1979. The north Anatolian transform fault: its age, offset and tectonic significance. *Journal of the Geological Society, London*, **136**, 269–282.
Şengör, A. M. C. & Yilmaz, Y. 1981. Tethyan evolution of Turkey: a plate tectonic approach. *Tectonophysics*, **75**, 181–241.
Şengör, A. M. C., Görür, N. & Şaroğlu, F. 1985. Strike-slip faulting and related basin formation in zones of tectonic escape: Turkey as a case study. *In*: Biddle, K. T. & Christie-Blick, N. (eds) *Strike-Slip Deformation, Basin Formation, and Sedimentation*. Society of Economic Paleontologists and Mineralogists, Special Publications, **37**, 227–264.
Seyitoğlu, G., Kazanci, N., Fodor, L., Karakuş, K., Araz, H. & Karadenizli, L. 1997. Does continuous compressive tectonic regime exist during late Paleogene to late Neogene in NW central Anatolia, Turkey? *Turkish Journal of Earth Sciences*, **6**, 77–83.
Seyitoğlu, G., Kazanci, N., Karadenizli, L., Şen, S., Varol, B. & Karabiyikoğlu, T. 2000. Normal fault induced rockfall avalanche deposits in the NW of Çankırı basin (central Turkey) and their implications for the post-collisional tectonic evolution of the Neo-Tethyan suture zone. *Terra Nova*, **12**, 245–251.
Seyitoğlu, G., Kazanci, N., Karadenizli, L., Şen, S. & Varol, B. 2001. A Neotectonic pinched crustal wedge in the west of Cankırı basin accommodating the intemal deformation of Anatolian Plate. *Fourth International Turkish Geology Symposium, Abstracts*. Çukurova University, Turkey, p. 103.

SMITH, W. H. F. & SANDWELL, D. T. 1997. Global seafloor topography from satellite altimetry and ship depth soundings. *Science*, **277**, 1957–1962.

STEIN, R. S., BARKA, A. A. & DIETERICH, J. D. 1997. Progressive failure on the North Anatolian fault since 1939 by earthquake stress triggering. *Geophysical Journal International*, **128**, 594–604.

TAN, O. 1996. *Source parameters of 1968 September 3 Bartın (NW Turkey) earthquake.* BSc thesis, İstanbul Technical University.

TATAR, O., PIPER, J. D. A., PARK, R. G. & GÜRSOY, H. 1995. Palaeomagnetic study of block rotations in the Niksar overlap region of the North Anatolian Fault Zone, central Turkey. *Tectonophysics*, **244**, 251–256.

TAYMAZ, T. 1990. *Earthquake source parameters in the Eastern Mediterranean region.* PhD thesis, University of Cambridge.

TAYMAZ, T. 1993. The source parameters of Çubukdağ (Western Turkey) earthquake of 11 October 1986. *Geophysical Journal International*, **113**, 260–267.

TAYMAZ, T. & PRICE, S. 1992. The 1971 May 12 Burdur earthquake sequence, SW Turkey: a synthesis of seismological and geological observations. *Geophysical Journal International*, **108**, 589–603.

TAYMAZ, T. & TAN, O. 2001. Source parameters of June 6, 2000 Orta–Çankırı (Mw = 6.0) and December 15, 2000 Sultandağ–Akşehir (Mw = 6.0) earthquakes obtained from inversion of teleseismic P- and SH waveforms. *In*: TAYMAZ, T. (ed.) *Symposia on Seismotectonics of the North-Western Anatolia–Aegean and Recent Turkish Earthquakes, Extended Abstracts.* Istanbul Technical University, Istanbul, 96–107.

TAYMAZ, T., JACKSON, J. A. & WESTAWAY, R. 1990. Earthquake mechanisms in the Hellenic Trench near Crete. *Geophysical Journal International*, **102**, 695–731.

TAYMAZ, T., JACKSON, J. A. & MCKENZIE, D. 1991*a*. Active tectonics of the North and Central Aegean Sea. *Geophysical Journal International*, **106**, 433–490.

TAYMAZ, T., EYIDOĞAN, H. & JACKSON, J. 1991*b*. Source parameters of large earthquakes in the East Anatolian Fault Zone (Turkey). *Geophysical Journal International*, **106**, 537–550.

TAYMAZ, T., WRIGHT, T., TAN, O., FIELDING, E. & SEYITOĞLU, G. 2002*a*. Source characteristics of Mw = 6.0 June 6, Orta–Çankırı (Central Turkey) earthquake: a synthesis of seismological, geological and geodetic (InSAR) observations, and internal deformation of Anatolian plate. *In*: *1st International Symposium of Istanbul Technical University the Faculty of Mines on Earth Sciences and Engineering, Abstracts Book.* Istanbul Technical University, Istanbul, 63.

TAYMAZ, T., TAN, O., ÖZALAYBEY, S. & KARABULUT, H. 2002*b*. Source characteristics of February 3, 2002 Çay–Sultandağı earthquake (Mw = 6.5) sequence in SW Turkey: a synthesis of seismological observations of body-waveforms, strong motions, and aftershock seismicity survey data. *In*: *1st International Symposium of Istanbul Technical University the Faculty of Mines on Earth Sciences and Engineering, Abstracts Book.* Istanbul Technical University, Istanbul, 60, Turkey.

TAYMAZ, T., WESTAWAY, R. & REILINGER, R. (eds) 2004. Active faulting and crustal deformation in the Eastern Mediterranean region. *Tectonophysics* (special issue), **391**,

TÜRKECAN, A., HEPŞEN, N., PAPAK, İ. *ET AL.* 1991. *Geology and petrology of volcanic rocks of Seben–Gerede (Bolu)–Güdül–Beypazarı (Ankara and Çerkeş–Orta–Kurşunlu, Çankırı and surrounding regions.* MTA Report 9193 [in Turkish].

WELLS, D. L. & COPPERSMITH, K. 1994. New empirical relationships among magnitude, rupture length, rupture width, rupture area, and surface displacement. *Bulletin of the Seismological Society of America*, **84**, 974–002.

WESSEL, P. & SMITH, W. H. F., 1991. Free software helps map and display data. *EOS Transactions, American Geophysical Union*, **72**, 441, 445–446.

WESSEL, P. & SMITH, W. H. F. 1995. New version of the generic mapping tools released. *EOS Transactions, American Geophysical Union*, **76**, 329.

WILSON, M., TANKUT, A. & GÜLEÇ, N. 1997. Tertiary volcanism of the Galatia province, north-west central Anatolia, Turkey. *Lithos*, **42**, 105–121.

WRIGHT, T. J. 2002. Remote monitoring of the earthquake cycle using satellite radar interferometry. *Philosophical Transactions of the Royal Society of London, Series A*, **360**, 2873–2888.

WRIGHT, T. J., PARSONS, B., JACKSON, J., HAYNES, M., FIELDING, E., ENGLAND, P. & CLARKE, P. 1999. Source parameters of the 1 October 1995 Dinar (Turkey) earthquake from SAR interferometry and seismic body-wave modelling. *Earth and Planetary Science Letters*, **172**, 23–37.

WRIGHT, T. J., PARSONS, B. & FIELDING, E. 2001*a*. The 1999 Turkish earthquakes: source parameters from InSAR and observations of triggered slip. *In*: TAYMAZ, T. (ed.) *Symposia on Seismotectonics of the North-Western Anatolia-Aegean and Recent Turkish Earthquakes, Extended Abstracts.* Istanbul Technical University, Istanbul, 54–71.

WRIGHT, T. J., FIELDING, E. J. & PARSONS, B. 2001*b*. Triggered slip: observations of the 17 August 1999 İzmit (Turkey) earthquake using radar interferometry. *Geophysical Research Letters*, **28**, 1079–1082.

YAGI, Y. 2000. Source process of large earthquakes (aseismic and coseismic slip). PhD thesis, University of Tokyo.

YAGI, Y. & KIKUCHI, M. 2000. Source rupture process of the Kocaeli, Turkey, earthquake of August 17, 1999, obtained by joint inversion of near-field data and teleseismic data. *Geophysical Research Letters*, **27**, 1969–1972.

YOSHIDA, S., KOKETSU, K., SHIBAZAKI, B., SAGIYA, T., KATO, T. & YOSHIDA, Y. 1996. Joint inversion of near- and far-field waveforms and geodetic data for rupture processes of the 1995. Kobe earthquakes. *Journal of Physics of the Earth*, **44**, 437–457.

ZEBKER, H., ROSEN, P., GOLDSTEIN, R., GABRIEL, R. & WERNER, C. 1994. On the derivation of coseismic displacement fields using differential radar interferometry: the Landers earthquake. *Journal of Geophysical Research*, **99**, 19617–19634.

The magnetism in tectonically controlled travertine as a palaeoseismological tool: examples from the Sıcak Çermik geothermal field, central Turkey

H. GÜRSOY[1], B. L. MESCI[1], J. D. A. PIPER[2], O. TATAR[1] & C. J. DAVIES[2]

[1]*Department of Geology, Cumhuriyet University, 58140 Sivas, Turkey (e-mail: gursoy@cumhuriyet.edu.tr)*

[2]*Geomagnetism Laboratory, Department of Earth and Ocean Sciences, University of Liverpool, Liverpool L69 7ZE, UK*

Abstract: In regions of neotectonic activity geothermal waters flow into extensional fissures and deposit successive layers of carbonate as fissure travertine incorporating small amounts of ferromagnetic grains. The same waters spill out onto the surface to deposit bedded travertine, which may also incorporate wind-blown dust with a ferromagnetic component. Travertine deposits are linked to earthquake activity because geothermal reservoirs are reset and activated by earthquake fracturing but tend to become sealed by deposition of carbonate between events. A weak ferromagnetism records the ambient field at the time of deposition and sequential deposition can identify cycles of secular variation of the geomagnetic field to provide a means of estimating the rate of travertine growth. The palaeomagnetic record in three travertine fissures from the Sıcak Çermik geothermal field in central Anatolia dated to between 100 and 360 ka by U–Th determinations has been examined to relate the geomagnetic signature to earthquake-induced layering. Sequential sampling from the margins (earliest deposition) to the centres (last deposition) identifies directional migrations reminiscent of geomagnetic secular variation. On the assumption that these cycles record time periods of 1–2 ka, the number of travertine layers identifies resetting of the geothermal system by earthquakes every 50–100 years. Travertine precipitation occurs at rates of 0.1–0.3 mm a^{-1} on each side of the extensional fissures and at a rate an order higher than for bedded travertine on the surface. Earthquakes of magnitude $M \leq 4$ occur too frequently in the Sivas Basin to have any apparent influence on travertine deposition but earthquakes with M in the range 4.5–5.5 occur with a frequency compatible with the travertine layering, and it appears to be events of this order that are recorded by sequential travertine deposition. Two signatures of much larger earthquakes on a 1–10 ka time scale are also present in the travertine deposition: (1) the incidental emplacement of massive travertine or fracturing of earlier travertine without destruction of the fissure as a site of travertine emplacement; (2) termination of the fissure as a site of deposition with transfer of the geothermal activity to a new fracture. The presence of some 25 fractures in the *c.* 300 ka Sıcak Çermik field growing at rates of 0.1–0.6 mm a^{-1} suggests that the type (2) signature may be achieved by an M *c.* 7.5 event approximately every 10 ka.

The record of past earthquakes beyond historical and archaeological time scales relies on the identification of catastrophic events in the geological record. Movements along fault planes are the most direct signatures, with slickenlines indicating the latest sense of motion and offsets providing the cumulative movement of many successive events. Another effect is the liquefaction of incompetent sedimentary layers; this may provide evidence for ground shaking by earthquake events but cannot be used to constrain the events within a narrow time frame. The past directional behaviour of the Earth's magnetic field expressed in the short term (10^3–10^5 years) as palaeosecular variation (PSV) and occasional excursions, and in the long term (10^5–10^6 years) by reversals of polarity, can provide a time scale for ancient earthquakes provided that these same events are expressed in a rock succession with recoverable primary magnetism. The long-term changes in polarity may be linked to the Geomagnetic Polarity Time Scale (GPTS), whereas the archaeomagnetic and lake sediment records covering the last 10 ka years (Thompson & Oldfield 1986) suggest that PSV is closely constrained to the Earth's rotation with complete cycles taking 1000–2000 years to complete (Butler 1992).

Tectonically controlled travertine

The sedimentary deposit that most closely identifies earthquake events is tectonically controlled travertine. Travertine is inorganic carbonate deposition

From: TAYMAZ, T., YILMAZ, Y. & DILEK, Y. (eds) *The Geodynamics of the Aegean and Anatolia.* Geological Society, London, Special Publications, **291**, 291–305.
DOI: 10.1144/SP291.13 0305-8719/07/$15.00

and occurs whenever the partial pressure of CO_2 is reduced in waters draining hinterland including limestone and dolomites. In regions of neotectonism the groundwater is frequently at elevated temperatures, thus enhancing solution of country rock. These waters circulate to the surface via extensional fissures that become progressively filled by layers of travertine as they open (Hancock *et al.* 1999). Such fissures are typically banded because the hydrothermal system is episodically reset, and the flow stimulated, by earthquake activity. The incremental growth on each side of the fissure axis is then especially useful, because the layers are often variegated and can be correlated with one another to evaluate temporal trends in the geomagnetic record. The fissures feed layered travertine at the surface, where stratified deposits are produced by outflow onto the surface (Fig. 1). When travertine flow is tectonically controlled the stratified deposits at the surface may also be distinctively layered and alternate between slowly deposited friable, open-textured bands (tufa) and more massive, rapidly deposited bands. The friable bands are the product of waning hydrothermal activity as the conduits are progressively sealed by precipitation, whereas the massive bands result from stimulated flow after the reservoir has been fractured and reset by an earthquake event. Hence in regions of neotectonic activity travertine preserves a layered signature recording the past earthquake activity.

In this paper we report a palaeomagnetic study of fissure fill travertine from a geothermal field in central Turkey with the main aim of identifying migrations of the palaeofield direction to correlate with cycles of secular variation. As the travertine deposits also contain a history of layering, with each layer resulting from an earthquake event, we have aimed to evaluate the number of layers within secular variation cycles to estimate the long-term earthquake frequency. We also note two longer-term signatures in the travertine deposition, which appear to define the incidence of much larger but infrequent earthquake events.

Geological framework

The travertine examined in this study is derived from the geothermal region of Sıcak Çermik (39.9°N, 36.9°E) 25 km west of the city of Sivas in central Turkey where hot springs permeate Late Miocene–Early Pliocene sediments (Ayaz & Gökçe 1998). The carbonate source is probably marble in underlying metamorphic rocks lying >100 m below the sediments. The regional tectonic regime is dominated by NNW–SSE compression (σ_1) resulting from the continuing northward movement of Afro-Arabia into the Anatolides, an assemblage of accreted terranes between the Pontide orogen at the Eurasian margin in the north and the Tauride orogen in the south, although regional variations are complex, because of the westward extrusion of fault blocks within Anatolia (Tatar *et al.* 1996; Gürsoy *et al.* 1997; Piper *et al.* 2007). The axis of minimum principal stress (σ_3) is a tension in this region, and the sinuous pattern of fissure development (Fig. 2) suggests that it results from a torsional stress regime between two major NE–SW-trending tear faults so that only the currently active fissures are oriented NNW–SSE perpendicular to σ_3 (Mesci 2004). Quarrying of the mounds of layered travertine has excavated the sampled fissures and U–Th dating indicates that geothermal activity has occurred here, at least episodically, over the past 300–400 ka. The fissure fills consist of a succession of pale, straight, sinuous or undulating carbonate bands interrupted by darker bands of varying shades of brown; the more prominent examples of the latter can be reliably matched on either side of the fissure axes. The studied fissures are either completely sealed or are defined by a small gap showing that infilling by travertine occurred at approximately the same rate as the fissures were opened by the regional extension.

Palaeomagnetic study

Three travertine fissures at localities denoted 1, 2 and 3 in Figure 2 were drilled by coring outwards from the axes so that full thicknesses of travertine growth were sampled by successions of 2.4 cm diameter cores. At localities 1 and 2 suitable exposures allowed mirror sections to be sampled outwards on either side of the fissure axes. Orientations were by both Sun and magnetic compasses. Much of the layered travertine in this region comprises friable tufa and it was difficult to core coherent samples; isolated samples were obtained near the sampled fissures and at two sites (4 and 5) in more massive layered travertine, although the latter yielded few coherent data and are not further discussed here. Also, for comparative work, we have sampled 24 successive compact travertine layers from a travertine mound in a smaller geothermal field at Ortaköy near the town of Şarkışla 25 km SW of the Sıcak Çermik geothermal field.

The magnetizations in cores cut from field samples were determined by nitrogen SQUID (FIT) magnetometer following measurement of magnetic susceptibility using a Bartington Bridge. Progressive demagnetization was applied to resolve components of magnetization using either the alternating field (a.f.) method or thermal demagnetization. The preferred method of data analysis used orthogonal projections of the remanence vector onto horizontal and vertical planes accompanied by recognition of

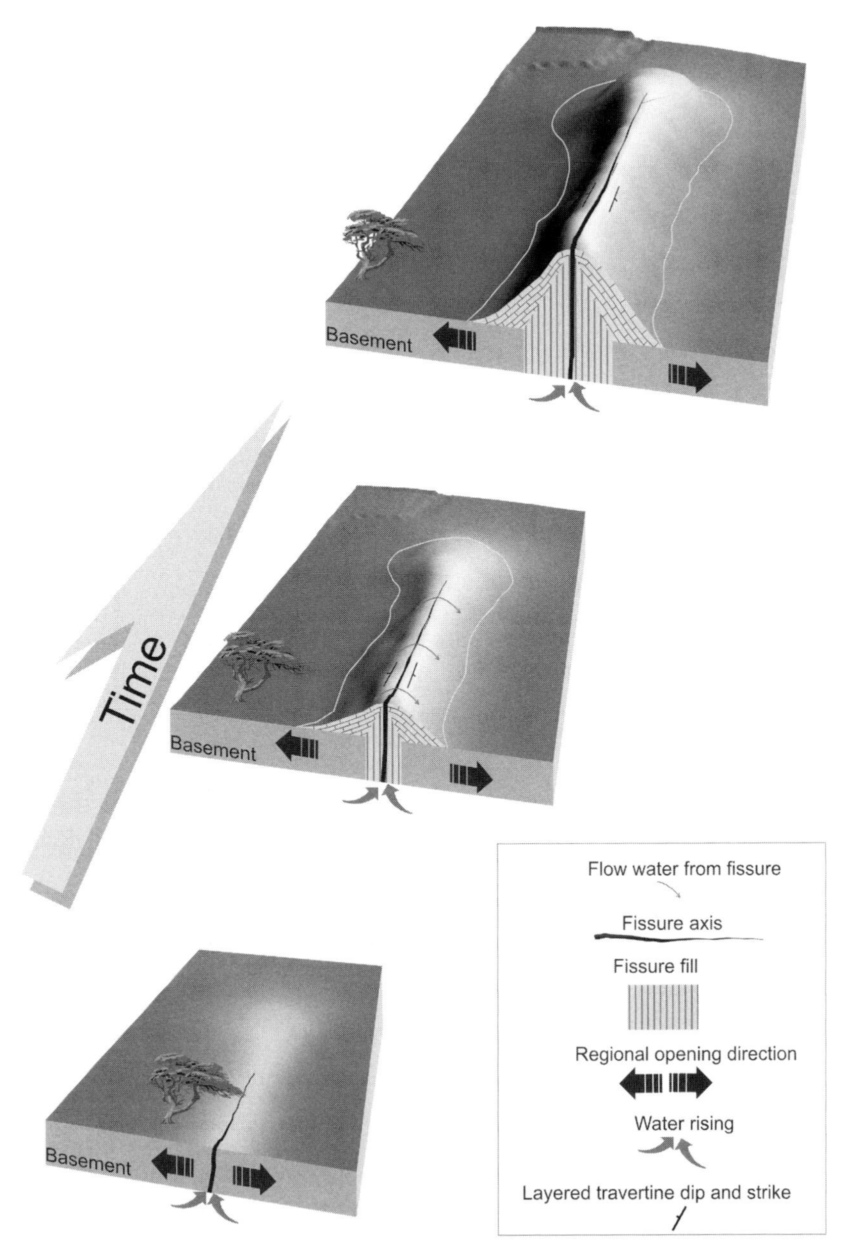

Fig. 1. The mode of formation of a travertine mound above an extensional fissure (after Mesci 2004).

components by eye and calculation of equivalent directions of magnetization by principal component analysis. Thermal demagnetization could be applied only to 350 °C because of disintegration of the carbonate host at higher temperatures, and directions of magnetization were determined by end-point analysis in samples showing stable behaviour where both methods of treatment produced minimal demagnetization trajectories.

Fissure locality 1

Fissure 1 is a 2 m wide example yielding U–Th ages of $120\,000 \pm 5$ years at the axis and $177\,000 + 14/-13$ years at the margin (Mesci 2004). Figure 3 illustrates the sample distribution, comprising sites of 23 cores to the right of the axis (Site 1, west) and 26 cores to the left (Site 2, east). Thermomagnetic spectra show weak,

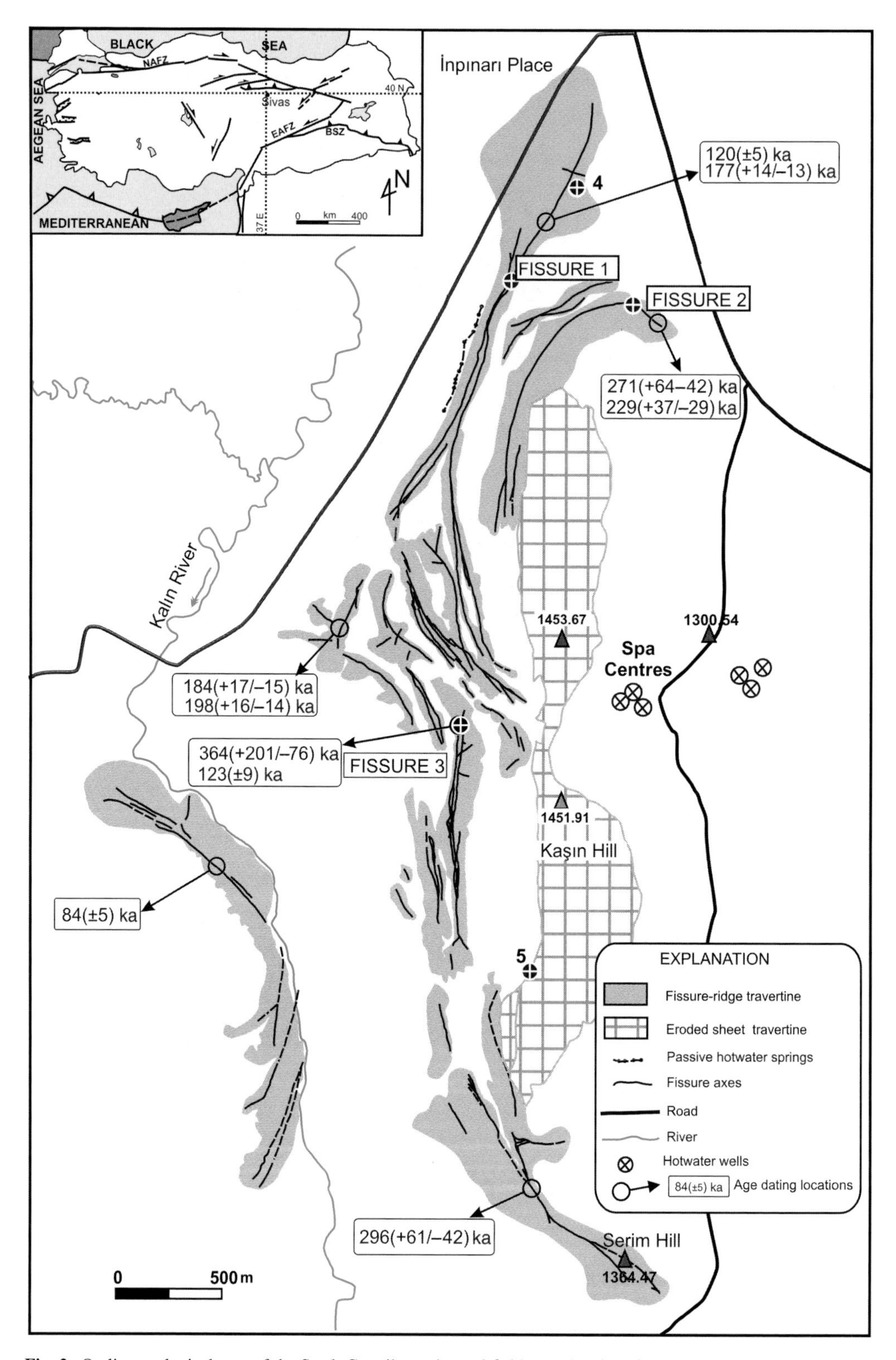

Fig. 2. Outline geological map of the Sıcak Çermik geothermal field near the city of Sivas, central Turkey. The locations of the three fissure travertines sampled for this study are shown, with two additional sites in bedded travertine (4 and 5). NAFZ, North Anatolian Fault Zone; EAFZ, East Anatolian Fault Zone; BSZ, Bitlis Suture Zone.

convex and asymptotic curves obeying the Curie Law; this indicates that saturation magnetization, M_s, is dominated by paramagnetism. Low-Ti magnetite is produced by alteration during heating, possibly from breakdown of siderite or reduction of limonite, and is reflected in large increase in M_s (Fig. 4). Acquisition of isothermal remanent magnetization (IRM) shows continuing rise of curves to the limits of laboratory fields (2.7 T) and identifies much of the ferromagnetism in these rocks as due to goethite or hematite (Fig. 4). The travertine has a weak remanence that is measurable following a.f. demagnetization in steps ranging from 2.5–20 mT to peak fields of 130 mT. Examples of orthogonal projections are illustrated in Figure 5. Components defining directions with mean angular deviations of $\leq 25^\circ$ were accepted and successive mean directions of magnetization were calculated from threefold running means from successive cores to evaluate directional trends across the fissure.

Magnetic declinations on the left- and right-hand sides of the fissure show similar variation, with a swing of the palaeofield initially to the east, then to the west, and finally back to the east (Fig. 6). The inclination records also correspond very closely out to sample 12, but beyond this point they diverge, with site 1 becoming steeper and site 2 shallower than the mean field inclination in this region. The variation in declination is of higher frequency than the variation in inclination; this same contrast is present, for example, in the historical PSV record in Europe (e.g. Marton 1996). The mean direction yielded by the running means of the samples from site 1 on the west side of the fissure axis is $D/I = 348.7/58.7^\circ$ ($N = 18$, $R = 17.21$, $\alpha_{95} = 7.6^\circ$, $k = 21.5$) and from site 2 on the east side of the axis is $D/I = 345.2/56.0^\circ$ ($N = 18$, $R = 16.93$, $\alpha_{95} = 8.9^\circ$, $k = 15.9$). The inclinations are not significantly shallower than the inclination predicted from the geocentric axial dipole field at the sample locality ($I = 59^\circ$). The declinations, however, appear to be rotated anticlockwise from north, consistent with the interpretation that the fissure system in this geothermal field is undergoing torsional rotation between bounding ENE–WSW strike-slip faults as a consequence of the tectonic escape of terranes in central Anatolia (Mesci 2004). Beyond sample 18 (Fig. 3) it is not possible to correlate bands on either side with confidence and beyond core 21 in site 1 the layering is truncated at the base of the outcrop, with older layers apparently recording a separate period of growth. We infer that a rare larger earthquake event occurred at this point, which dislocated the fissure but was not of sufficient magnitude to destroy the fissure as a conduit for water flow.

The time sequence of mean directions of magnetization show a net clockwise migration (Fig. 7), the anticipated response of the magnetic vector to the westward motion of a magnetic dipole in the Earth's core and consistent with well-known

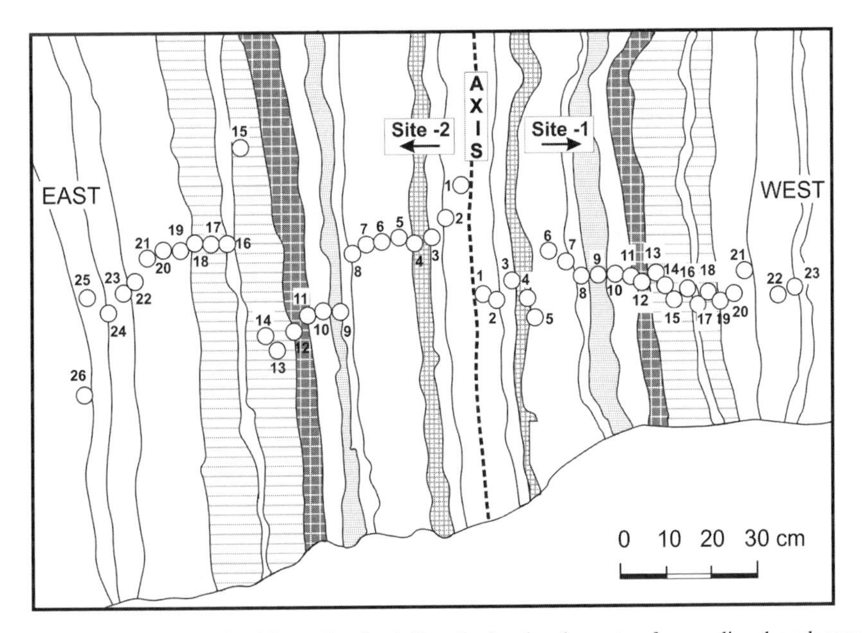

Fig. 3. Drawing from a photograph of fissure 1 at Sıcak Çermik, showing the system for sampling the palaeomagnetic cores across a fissure travertine.

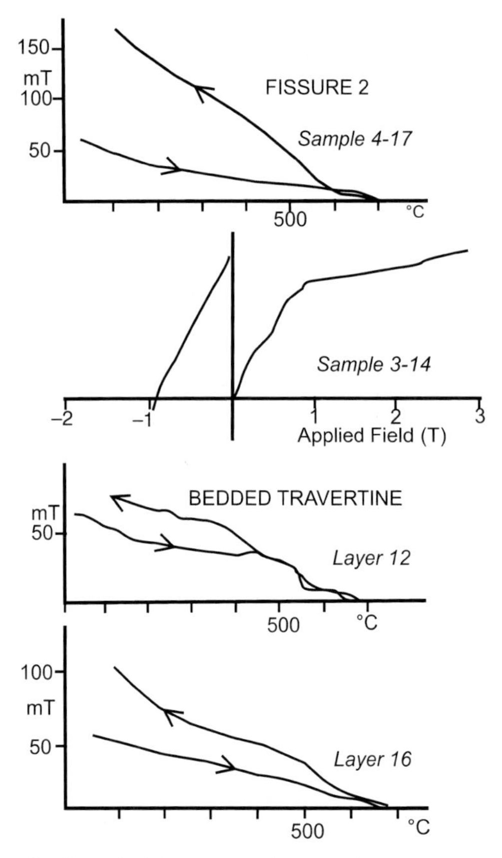

Fig. 4. Typical thermomagnetic (saturation magnetization (M_s) v. temperature) determinations and IRM acquisition curves for samples of fissure and bedded travertine.

European observatory data. Some 60 cm of travertine growth at this locality thus appear to be just short of recording a complete cycle of secular variation. Analogies with the historical field and the lake sediment record (Butler 1992) suggest that some 2000 years of deposition could be represented, and suggest a rate of carbonate incrementation of *c.* 0.3 mm a^{-1}. Approximately 50 layers are discernible within this thickness and suggest that earthquakes large and/or close enough to reset the reservoir and stimulate water flow occurred roughly every 40 years. At this locality the record of PSV thus suggests a fissure lifespan much shorter than the 177–120 ka interval suggested by the U–Th determinations; we address this discrepancy below in the context of the data from fissures 2 and 3.

Fissure locality 2

This is a *c.* 3 m wide fissure dated by U–Th at 271 + 64/−42 ka at the axis and 299 + 37/−29 ka at the margin (Mesci 2004). The palaeomagnetic sample comprises 49 cores from the east side (site 3) and 43 cores from the west side (site 4). The cores were subjected to a.f. demagnetization incremented in 5 mT steps to 35 mT and then in steps of 50, 75, 100 and 140 mT. All samples were characterized by high magnetic coercivities but this treatment was usually able to subtract sufficient remanence to demonstrate convergent vectors and identify single-component behaviours. Component directions fall into two groups: the majority of cores show directions close to the present field direction with small consecutive directional change between adjacent cores. The remainder show directions diverging widely from the recent field direction although approximately consistent between adjacent cores until there is a shift back to the predicted field direction. The outer parts of this fissure (core 28 and beyond) shows thin parallel layers of travertine that can be readily correlated on either side. Comparable swings in declination and inclination can be identified in this part of the fissure (Fig. 8) and suggest that one full cycle of PSV is recorded within this initial (outer) sequence

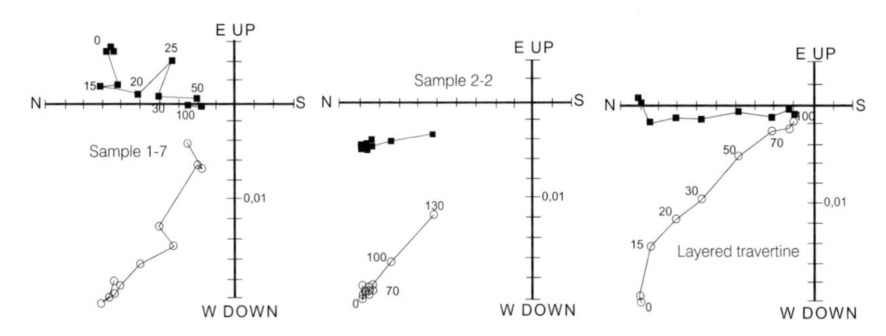

Fig. 5. Typical orthogonal projections of the magnetization vector during a.f. demagnetization from fissure travertine at sites 1 and 2 in fissure 1 and an example of demagnetization of bedded travertine bordering this fissure. Projections onto the horizontal plane are shown as filled symbols and projections onto the vertical plane are shown as open symbols. The intensity values are $\times 10^{-5}$ A m^2 kg^{-1}.

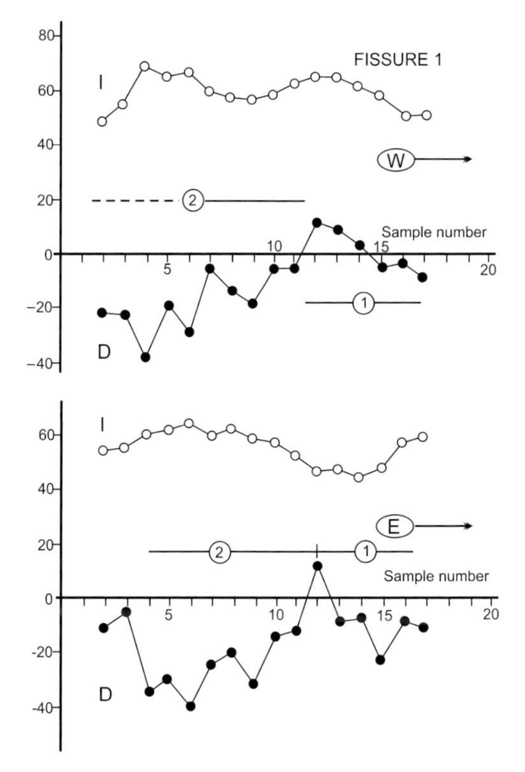

Fig. 6. Variation in the declination and inclination of the magnetization on each side of the axis of fissure 1.

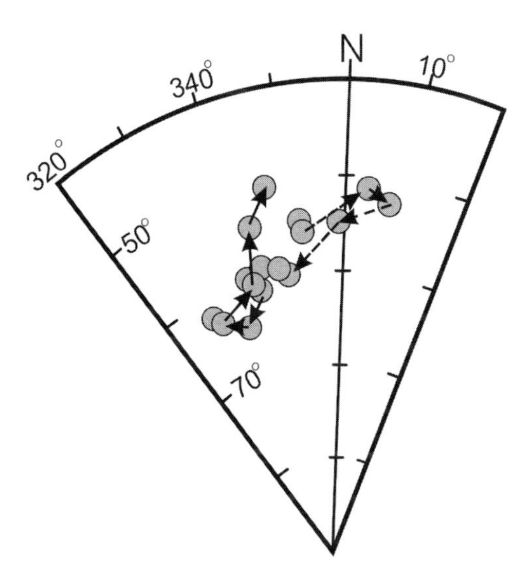

Fig. 7. Directional migration of the geomagnetic field recorded by successive travertine growth in fissure 1.

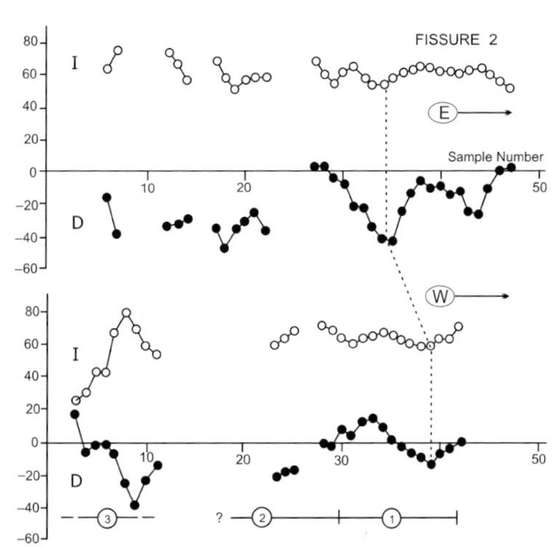

Fig. 8. Variation of the magnetic declination and inclination recorded by travertine growth across fissure 2.

of travertine deposition. Only fragments of a subsequent cycle were recorded although the last pulse of deposition recorded approximately half a cycle of PSV before the fissure became extinct. The initial phase of layered travertine (about 1.5 cycles of PSV) records 31 layers of travertine whereas the last (half-cycle) records eight layers. In each case the deposition of one travertine layer in *c.* 100 years is indicated and a rate of travertine growth of *c.* 0.2 mm a^{-1} is implied.

Thirty-eight of 47 samples from site 3 having directions comparable with the recent field yield a mean direction of $D/I = 339.8/60.3^\circ$ ($R = 36.99$, $k = 36.6$, $\alpha_{95} = 3.9^\circ$) whereas 28 of 42 samples from site 4 have a mean of $D/I = 356.8/60.3^\circ$, $R = 26.85$, $k = 23.4$, and $\alpha_{95} = 5.8^\circ$). Thus the mean field inclination recorded here is not significantly different from the geocentric dipole value of 59°. The declination, however, is rotated anticlockwise by a similar amount to the declination in fissure 1.

The incomplete recovery of an interpretable field direction correlates with the presence of travertine banding with contorted shapes and with massive (unbanded) travertine within the inner part of the fissure (Fig. 9), which has only incidentally recorded the predicted field. It appears that precipitation processes within the fissure during turbulent flow or incidental catastrophic events prevented recording of the ambient field direction; hence this kind of travertine would appear to be unsuitable for investigation of PSV. Because of the insertion of this massive travertine into this fissure we are unable to estimate the duration of activity but on the assumption that it spanned about three cycles of PSV, it could be of the order

Fig. 9. Photograph of fissure 2 (eastern side) showing the distribution of palaeomagnetic samples and three types of fissure travertine: uniformly layered travertine, convolute layered travertine and massive (non-layered) travertine.

of 5000 years, which is within the margin of error of the U–Th age determinations from the margin and axis noted above. Between 60 and 65 bands are discernible (uncertainties arise as a result of local truncation) between the axis and the outer core, suggesting that resetting of the reservoir by earthquake activity occurred approximately every 80 years. The average carbonate incrementation of the *c.* 145 cm of travertine sampled on each side of the fissure is *c.* 0.3 mm a^{-1}; however, this value is biased by the massive travertine layer and a more realistic estimate for the rate of growth of the banded travertine is the value of 0.2 mm a^{-1} noted above.

Fissure locality 3

This is a wide fissure dated by U–Th analysis at 123 ± 9 ka at the axis and $364 + 201/-76$ ka at the margins (Mesci 2004). It is the widest fissure investigated here, although it was cored on one side only so we are unable to compare growth on either side of the axis.

The cores in this example were treated by thermal demagnetization to determine whether the ferromagnet goethite, presumed to be a component of the brown colour of much of the travertine, is a significant remanence carrier. Thermal demagnetization could then surmount the difficulty that coercivities are very high in most of the travertine samples and component trajectories are correspondingly difficult to define. Because goethite has a Curie point of only *c.* 125 °C, steps of only 25 °C were used commencing at 50 °C and continuing to 150 °C. In the event, only marginal reduction of the remanent intensity of 5–10% had been achieved by this temperature, thus showing that hematite rather than goethite must be the carrier of the high-coercivity remanence identified by a.f. demagnetization. The thermal demagnetization was continued in 50 °C steps until the carbonate began to disintegrate at *c.* 350 °C, and this achieved variable reduction in natural remanent magnetization, which often allowed convergent vectors to be recognized and defined.

To illustrate the effect of calculating successive three-point means to determine palaeofield migration across a fissure we show both the raw and smoothed data for declination in Figure 10 and for inclination in Figure 11. Continuing to adopt the criteria that a swing in declination and/or inclination is significant only if it is recognized in more than one successive sample, we identify five complete cycles of PSV across this fissure. These are evident in both the raw and smoothed data (Fig. 10), with the chief uncertainty being

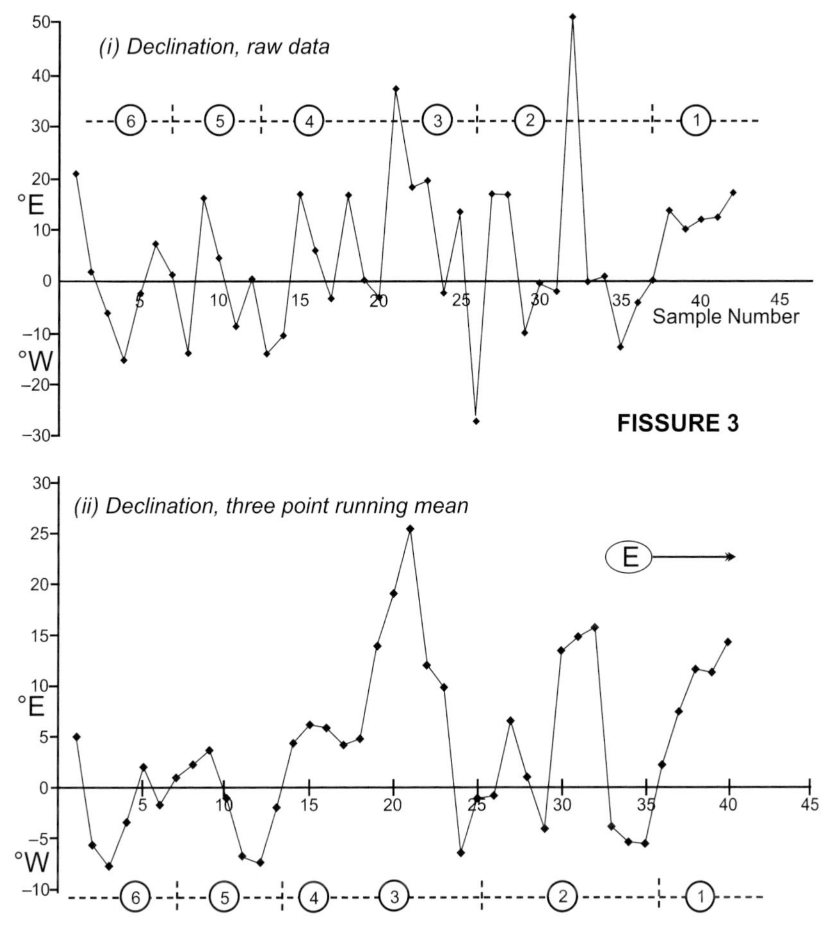

Fig. 10. Variation of the magnetic declination recorded by travertine growth across fissure 3. Both raw data and Fisherian means derived from applying a three-point filter are shown.

within cycles 3 and 4, where three counter-clockwise swings in declination are observed although the complementary clockwise swing is defined with confidence only in the earlier part of cycle 3. As in fissure 1, declination changes are higher in frequency and magnitude than inclination changes. Approximately 85 layers are discernible across this fissure as distinct white or brown bands, and indicate an earthquake event large enough to reset the reservoir approximately every 120 years. The rate of travertine growth appears to have been *c*. 0.1 mm a^{-1}.

Bedded travertine (Ortaköy geothermal field)

A comparative sample of layered travertine has been collected from the Ortaköy geothermal field located SW of Sıcak Çermik and comprises 24 successive layers derived from a single travertine mound. The basal layer overlies volcanoclastic sediments and a 2 m thick basal layer of massive and brecciated travertine. Succeeding layers range from 14 to 50 cm in thickness and make up a total thickness of 7.9 m to the summit of the mound. This deposit is undated, although geothermal activity is present nearby and it is likely to be no more that a few tens of thousands of years old. Two samples were drilled from each layer and used to derive layer means summarized in Table 1.

Mean declinations and inclinations are summarized as a log in Figure 12 and compared with magnetic susceptibility through the same section. All but two layers (8 and 20) yielded measurable and stable components of magnetization, although only one core at layers 1 and 20 yielded useful data. Layer mean directions summarized in Table 1 show typical normal polarity fields with declinations ranging from 324°E to 9°E and

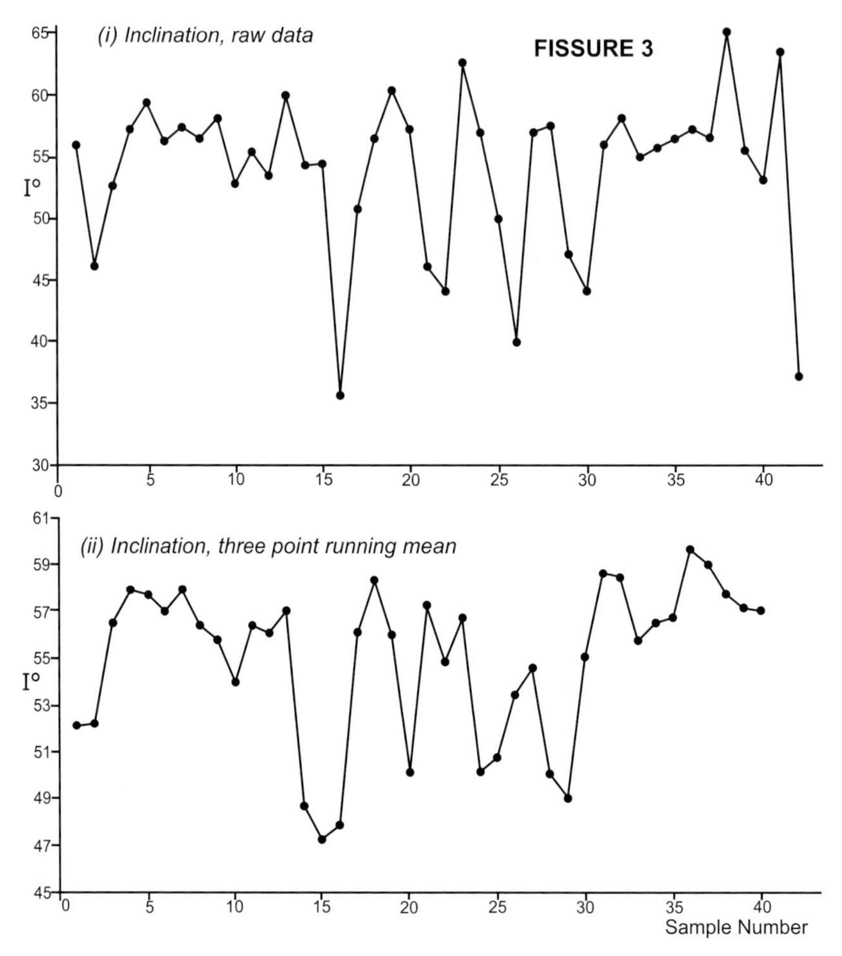

Fig. 11. Variation of the magnetic inclination recorded by travertine growth across fissure 3. Raw data and Fisherian means derived from applying a three-point filter are shown for comparison.

inclinations ranging from 34° to 66°. The mean direction derived from 20 layer means is $D/I = 353/47°$ ($\alpha_{95} = 5.2°$). Thus the significance of the direction of magnetization in this bedded travertine contrasts with the results from the fissure travertines: the declination barely differs from the mean ambient field direction within confidence limits although a small anticlockwise rotation consistent with the pattern of declinations across central Anatolia is possible (Gürsoy *et al.* 1997). In contrast, the inclination is notably shallower than the inclination (58.5°) predicted at this latitude (39.4°N) from a geocentric axial dipole source. Because no inclination shallowing is observed in the fissure travertine, this suggests that a depositional mechanism is responsible for this effect in bedded travertine; a likely mechanism is the settling of elongate and platy shaped ferromagnetic grains (Butler 1992) as they settle through the water films to become incorporated into the precipitating carbonate.

Earthquake frequency here has not been high enough to reset the hydrothermal system and deposit sufficient layers to record cycles of secular variation with complete confidence (although multiple layers between 2 and 3 were too thin to sample) and there are two possible interpretations of these data. First, the surface bedded travertine with an open porous texture is likely to be more susceptible to diagenesis than the compact and impervious fissure travertine and the magnetic record may be more susceptible to alteration by continuing fluid drainage. In these circumstances no consistent record of PSV might be recorded in layered travertine. Second, however, it is possible that detrital ferromagnetic grains could be fixed in the host carbonate and still be essentially impervious to significant later change. Thus if we exclude declination

Table 1. *Summary of palaeomagnetic results from bedded travertine deposit at Ortaköy, SW of Sivas, central Turkey*

Layer number	*D*	*I*	*N/R*	α_{95}
1	(348.5)	(49.5)	1/–	
2	5.8	41.1	2/2.00	6.7
3	348.6	50.3	2/2.00	2.2
4	354.6	44.7	2/2.00	10.6
5	359.6	51.6	2/2.00	7.8
6	22.4	53.4	3/2.99	8.3
7	1.6	47.7	2/1.98	32.2
8	–	–	–	–
9	6.2	47.2	2/1.98	35.2
10	338.0	34.7	2/2.00	9.9
11	346.4	36.2	4/3.96	10.8
12	223.9	33.8	2/1.99	24.5
13	2.6	36.4	2/2.00	3.5
14	327.9	51.8	3/2.94	22.1
15	338.9	40.7	3/2.97	14.6
16	358.8	39.6	2/2.00	13.1
17	349.0	48.1	2/1.99	22.8
18	6.2	48.3	2/1.99	19.7
19	8.9	59.3	2/1.99	24.6
20	–	–	–	–
21	(347.2)	(65.6)	1/–	
22	329.1	51.7	3/2.96	16.9
23	358.0	53.5	3/3.00	4.5
24	352.7	49.1	2/2.00	7.8
Mean result (20 layers)	353.4	46.9	20/19.50	5.2

Pole position is 77°N, 248°E ($dp/dm = 4.3/6.7°$).

and inclination swings defined by single layers only, it appears that two complete cycles of PSV are recorded by the log in Figure 12 so that some 2000–4000 years of deposition are represented here, with each layer (and equivalent earthquake cycle) recording intervals of the order of 100–200 years. Deposition rates of the travertine would then have been 2–4 mm a^{-1}. This is about an order higher than the rate of deposition of the layered fissure travertine at Sıcak Çermik and is consistently explained by the following: (1) the carbonate-charged water is preferentially blown forcibly out of the fissure by activity of the steam; (2) the largest reduction in PCO_2 takes place at the surface; (3) it is here that the water moves away only slowly and is subject to strong evaporation.

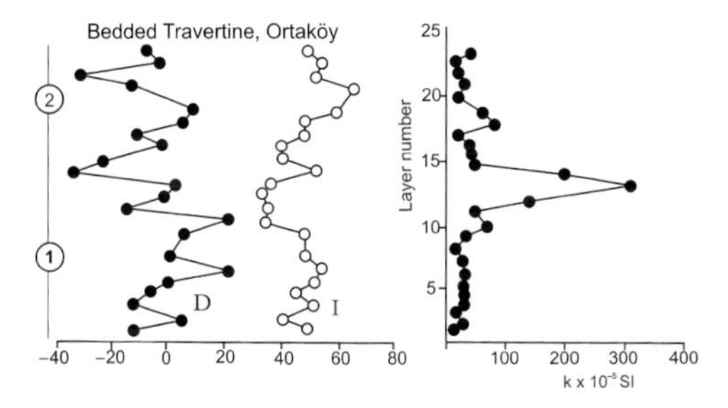

Fig. 12. Variation of magnetic declination and inclination derived from palaeomagnetic study of 24 successive layers through bedded travertine in the Ortaköy geothermal field. Also shown is the variation of magnetic susceptibility through this succession.

Of especial interest is the large increase in magnetic susceptibility observed between layers 11 and 15 (Fig. 11). The dramatic input of magnetic material at this point could have several possible explanations including: (1) the signature of a volcanic event; (2) a change in groundwater chemistry; (3) a change in the principal wind direction carrying enhanced magnetic detritus into the region. Because the enhanced susceptibility is recognized through three successive layers the time period involved is likely to have been several hundreds of years, and explanations (2) or (3) are therefore more likely. Volcanic outcrops are extensively exposed in the immediate vicinity of Ortaköy and are a possible source of wind-borne magnetite. Volcanic rocks around the town of Ortaköy were regarded as Pliocene–Pleistocene by Parlak *et al.* (2001) although K–Ar age dating of basalts by Tatar *et al.* (2004) has yielded a somewhat older age of 12.96 ± 0.10 Ma.

Table 2. *Historical earthquakes affecting central Anatolia*

Date	Magnitude	Number of deaths	Affected area	Origin
240	?	?	Kayseri–Sivas	CAFZ?
1205	?	?	Kayseri	CAFZ?
1268	?	15000	Erzincan–Erzurum	CAFZ?
1458	?	30000	Erzincan–Erzurum	CAFZ?
17.08.1668	?	?	Amasya–Tokat–Sivas	CAFZ?
11.01.1695	?	?	Sivas–Ordu	CAFZ?
06.09.1704	?	?	Kayseri	CAFZ?
1714	?	?	Kayseri	CAFZ?
09.05.1717	?	8000	Kayseri	CAFZ?
16.09.1754	?	?	SE Sivas	CAFZ?
14.03.1779	?	?	Divriği–Malatya	Malatya Fault?
28.05.1789	?	?	Divriği–Elazığ	Malatya Fault?
18.07.1794	?	?	Central Anatolia	NAFZ
02.06.1859	6.4	15000	Erzurum	?
10.1891	6 (I_o)	?	Sivas and Hafik	?
1893	6 (I_o)	?	Sivas and Zara	?

After data compiled from Ergin *et al.* (1967), Ambrasseys & Finkel (1995) and Boğaziçi University Kandilli Observatory and Earthquake Research Institute, Istanbul.
CAFZ, Central Anatolian Fault Zone; NAFZ, North Anatolian Fault Zone.

Table 3. *Earthquakes with magnitude >3 (M_s) that occurred between 1900 and 2005 in and around the study area*

Date	Time (GMT)	Latitude (°N)	Longitude (°E)	Depth (km)	Magnitude (M_s)
25.12.2005	03:50	39.64	37.33	8	3.3
12.10.2005	04:12	39.72	37.43	3	3.1
27.09.2005	05:24	39.70	37.30	25	3.1
06.08.2005	18:34	39.68	37.47	20	3.3
14.12.2004	22:49	39.77	36.75	5	3.2
14.12.2004	22:24	39.77	36.74	5	4.1
09.03.2004	12:39	39.43	37.32	3	3.2
02.02.2004	02:08	39.51	37.01	5	3.4
08.01.2004	22:28	39.39	37.01	13	3.1
26.11.2003	01:17	39.48	36.67	5	3.0
09.08.2003	15:07	39.76	37.44	5	3.4
18.06.2002	08:08	39.70	37.45	5	4.0
11.06.1999	05:44	39.53	36.83	9	4.0
11.06.1999	05:25	39.49	36.67	10	4.8
15.12.1998	20:15	38.96	35.81	8	4.1
14.12.1998	13:06	38.95	35.72	5	4.6
14.12.1998	12:44	38.96	35.91	6	4.3
31.07.1995	03:26	39.51	36.48	0	4.5
23.01.1985	01:23	39.11	35.94	33	4.6
31.08.1960	22:11	39.09	35.98	70	4.7
12.03.1960	21:25	39.40	36.40	0	4.5
07.06.1940	19:49	40.06	37.82	10	4.6
27.12.1939	02:48	39.99	38.14	50	5.5
28.06.1929	22:18	40.20	37.90	0	4.5
18.05.1929	06:37	40.20	37.90	10	6.1
10.02.1909	19:49	40.00	38.00	0	5.7
09.02.1909	14:38	40.00	38.00	0	5.8
09.02.1909	11:24	40.00	38.00	60	6.3

After Boğaziçi University Kandilli Observatory and Earthquake Research Institute, Istanbul.

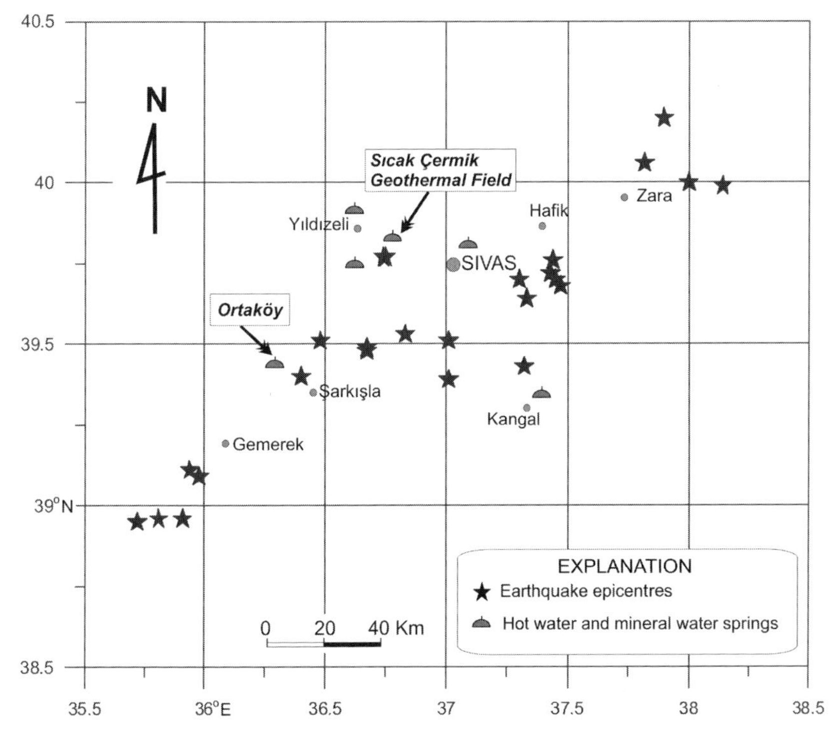

Fig. 13. The region of the Sivas Basin bordering the Sıcak Çermik geothermal field, showing epicentres of $M \geq 4$ 20th century earthquakes.

Discussion

Although the U–Th dating results are able to show that geothermal activity in the Sıcak Çermik field has spanned the last 300–400 ka, they are unable to constrain the intervals of activity within individual fissures with much precision. There are some 25 fissures in this geothermal field (Fig. 2) and, on the assumption that fluid outflow was concentrated on one fissure at a time, they may have been active for *c.* 12 ka on average, although this estimate will be a minimum because more than one fissure might have been active at any one time. The signature of PSV within the three fissures studied here implies activity over significantly shorter intervals than this. However, this discrepancy is more apparent than real because the full history of fissure growth is recorded only at the base of each fissure mound (Fig. 1); as the travertine pile builds up the fissure advances through successive layers so that the highest parts record only the last phase of activity of the fissure. Hence the interval of PSV recorded within a fissure will be controlled by the depth of exposure and the level of sampling.

Historical seismic activity in the Sivas Basin has been included in the assessments of a number of researchers (e.g. Gençoğlu *et al.* 1990; Ambraseys & Jackson 1998; Table 2). Damaging earthquakes were also noted in Sivas in 1695 and in the Kayseri district to the SW in *c.* 1745 (Ambraseys & Finkel 1995). A greater frequency of smaller earthquakes is evident from the high density of normal faults cutting Quaternary alluvial deposits bordering the adjoining Kızılırmak River (Gürsoy *et al.* 1992). Earthquake epicentres highlight a NE–SW trend of major faults delineating block boundaries and recording predominantly strike-slip motions resulting from the southwestward extrusion of terranes in central Anatolia by tectonic escape (Fig. 12). Most recently, Mesci (2004) has compiled a record of observatory recorded earthquakes and identified more than 50 earthquakes of $M \leq 4$ between 1995 and 2002 (Table 3; Fig. 13). Clearly, earthquakes of this magnitude occur much too frequently to have significantly influenced travertine formation. Twelve earthquakes with magnitudes between 4 and 4.8 occurred between 1929 and 2005 and five earthquakes with magnitudes between 5.5 and 6.3 occurred between 1909

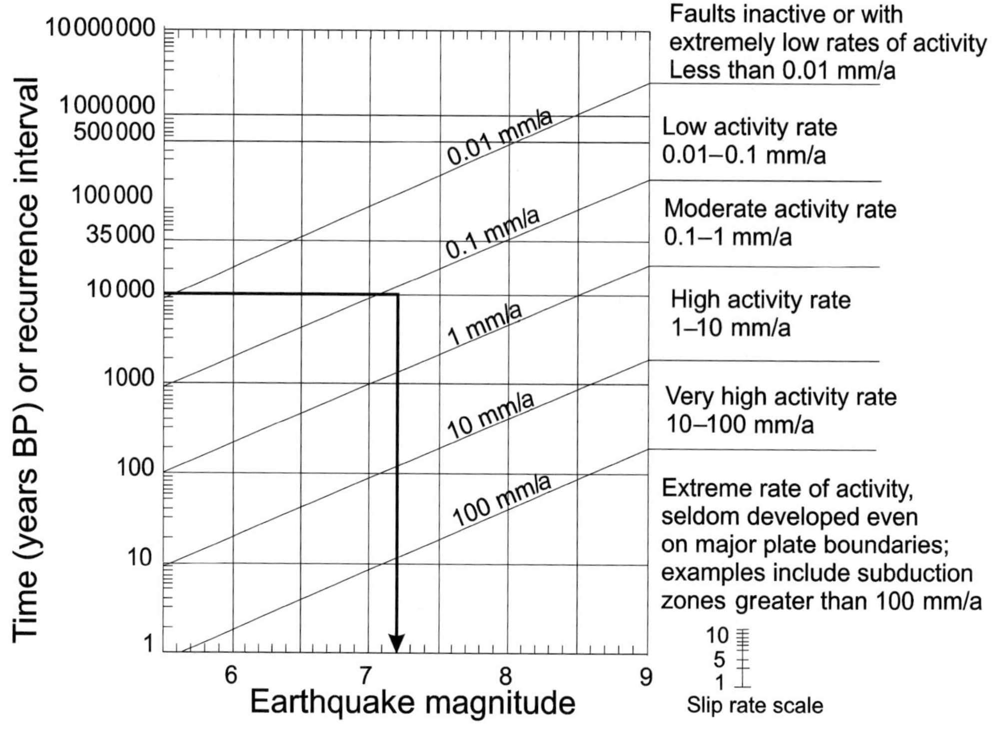

Fig. 14. The relationship between earthquake magnitude and fault offset used to estimate the recurrence time of earthquakes in western North America. After Schwartz & Coppersmith (1984) and Keller & Pinter (1996).

and 1939. Because only a few of these events would have occurred sufficiently close to influence the geothermal field, it seems likely that events in the magnitude range *c*. 4.5–5.5 are those capable of resetting the reservoir and instigating a revitalized water flow.

There are two further signatures within the record of travertine deposition in this geothermal field for the incidence of less frequent but much larger earthquakes. The first is the incidental emplacement of massive travertine without destroying the fissure as a site of travertine emplacement, such as that observed in the central part of the record at fissure 2, and perhaps the truncation of the outer record of the travertine in fissure 1 noted above. The second is the termination of the fissure as a site of deposition with transfer of the geothermal activity to a new fracture. A tentative estimate of the earthquake magnitude and frequency responsible for the second and most catastrophic type of earthquake event can be made from the relationship between earthquake magnitude and fault offset used to estimate the recurrence time of earthquakes in western North America (Schwartz & Coppersmith 1984). This approach is valid provided that repeated earthquakes on the fault segment occur with approximately the same magnitudes and amounts of displacement. The earthquake model assumes that the slip rate of the fault, the magnitude of the earthquake and the displacement per event are related in a systematic way. The model of Schwartz & Coppersmith relates the average recurrence time to the ratio of the displacement per event and the slip rate, on the assumption that an $M = 7$ earthquake is characterized by a 1 m displacement, an $M = 8$ event by a 5 m displacement, etc. The relationships for this example are shown in Figure 14 after Keller & Pinter (1996). In the Sıcak Çermik field the palaeomagnetic results show that the fissures have grown at rates of 0.1–0.3 mm a^{-1} on either side and have active durations of a few thousand years. As noted above, our samples will have embraced a significant part, but not the whole, of the history of each fissure, so an average lifespan of *c*. 10 ka is a reasonable estimate. Entering these values into the empirical relationship of Keller & Pinter (1996) suggests that an $M = 7.5$ event every *c*. 10 ka could record the extreme of earthquake activity in this region and be responsible for shifting the geothermal activity to a new fracture.

Conclusions

This investigation has been mainly exploratory in nature and designed to determine whether travertine, the natural deposit most closely linked to the regional earthquake signature, can yield a palaeomagnetic record of value to palaeoseismological investigation. In the event, it has been found that both fissure and bedded travertine yield an accurate record of the palaeomagnetic field direction and illustrate progressive changes in direction compatible with the signature of PSV. The record in bedded travertine is less clearly related to PSV than the record in fissure travertine and may be influenced by diagenesis; the value of bedded material for palaeoseismological investigation thus requires more investigation. The temporal link to earthquake frequency is based on the assumption that cycles of PSV were similar to those resolved from archaeological and historical records during the past few thousand years. By counting the number of travertine layers within a cycle of PSV we have a means of estimating the frequency of earthquakes capable of resetting the reservoir. The rate of travertine growth is also determined, as well as the nature of PSV. The typical magnitudes of the earthquake events responsible for influencing the travertine deposition can then be estimated from the historical records.

Two larger but more incidental signatures of earthquake activity are recorded in travertine deposits and suggested by the occasional emplacement of massive travertine and by the initiation of new travertine fissures. Travertine can thus provide an indication of bimodality in earthquake magnitude or frequency and indicate sourcing from two or more levels and stress regimes within the crust.

We are grateful to the British Council, NATO (Grant No. CLG-EST-977055) and TÜBİTAK (Grant YDABAG 101Y023) for supporting palaeomagnetic and neotectonic investigations in Turkey and for promoting the link between the Department of Geology at the Cumhuriyet University and the Geomagnetism Laboratory at Liverpool. We are grateful to E. Bozkurt and C. Yaltırak for reviews and for advice that improved the manuscript, and to T. Taymaz for editorial help.

References

AMBRASEYS, N. N. & FINKEL, C. F. 1995. *Seismicity of Turkey and Adjacent Areas: an Historical Catalogue of Earthquakes 1500–1799.* Eren, Istanbul.

AMBRASEYS, N. N. & JACKSON, J. A. 1998. Faulting associated with historical and recent earthquakes in the Eastern Mediterranean region. *Geophysical Journal International*, **133**, 390–406.

AYAZ, E. & GÖKÇE, A. 1998. Geology and genesis of the Sıcak Çermik, Sarıkaya and Uyuz Çermik travertine deposits northwest of Sivas, Turkey. *Bulletin of the Faculty of Engineering, Cumhuriyet University, A*, **15**, 1–12.

BUTLER, R. F. 1992. *Palaeomagnetism: Magnetic Domains to Geologic Terranes.* Blackwell, Oxford.

ERGIN, K., GÜÇLÜ, U. & UZ, Z. 1967. *Türkiye ve Civarının Deprem Kataloğu.* İTÜ, Maden Fakültesi Yayınları, **24**.

GENÇOĞLU, S., İNAN, E. & GÜLER, H. H. 1990. *Tükiye'nin Deprem Tehlikesi.* Chamber of Geophysical Engineering of Turkey, Ankara.

GÜRSOY, H., TEMIZ, H. & POISSON, A. 1992. Recent faulting in the Sivas area (Sivas Basin, Central Anatolia—Turkey). *Bulletin of the Faculty of Engineering, Cumhuriyet University, A*, **9**, 11–17.

GÜRSOY, H., PIPER, J. D. A., TATAR, O. & TEMIZ, H. 1997. A palaeomagnetic study of the Sivas basin, Central Turkey: crustal deformation during lateral extrusion of the Anatolian block. *Tectonophysics*, **271**, 89–105.

HANCOCK, P. L., CHALMERS, R. M. L., ALTUNEL, E. & CAKIR, Z. 1999. Travitonics: using travertines in active fault studies. *Journal of Structural Geology*, **21**, 903–916.

KELLER, E. A. & PINTER, N. 1996. *Active Tectonics.* Prentice–Hall, Englewood Cliffs, NJ.

MARTON, P. 1996. Archaeomagnetic directions: the Hungarian calibration curve. *In*: MORRIS, A & TARLING, D. H. (eds) *Palaeomagnetism and Tectonics of the Mediterranean Region.* Geological Society, London, Special Publications, **105**, 385–399.

MESCI, B. L. 2004. *The development of travertine occurrences around Sıcak Çermik (Sivas) and their relationship with active tectonics.* PhD thesis, Cumhuriyet University, Sivas.

PARLAK, O., DELALOYE, M., DEMIRKOL, C. & ÜNLÜGENÇ, U. C. 2001. Geochemistry of Pliocene–Pleistocene basalts along the Central Anatolian Fault Zone. *Geodinamica Acta*, **14**, 159–167.

PIPER, J. D. A., TATAR, O., GÜRSOY, H., KOÇBULUT, F. & MESCI, B. L. 2007. Paleomagnetic analysis of neotectonic deformation in the Anatolian accretionary collage, Turkey. *In*: DILEK, Y. & PAVLIDES, S. (eds) *Postcollisional tectonics and magmatism in the Mediterranean region and Asia.* Geological Society of America, Special Paper, **409**, 417–439.

SCHWARTZ, D. P. & COPPERSMITH, K. J. 1984. Fault behaviour and characteristic earthquakes: examples from the Wasatch and San Andreas Fault Zones. *Journal of Geophysical Research*, **89**, 5681–5698.

TATAR, O., PIPER, J. D. A., GÜRSOY, H. & TEMIZ, H. 1996. Regional significance of neotectonic anticlockwise rotation in central Turkey. *International Geology Review*, **38**, 692–700.

TATAR, O., TEMIZ, H., GÜRSOY, H. & GUEZOU, J. C. 2004. *Orta ve Kuzey Anadolu Miyosen-Kuvaterner volkanizması ışığında Avrasya-Anadolu levhalarının çarpışması.* TÜBİTAK Projesi, No. **YDABÇAG-198Y092**.

THOMPSON, R. & OLDFIELD, F. 1986. *Environmental Magnetism.* Allen & Unwin, London.

Index

Note: Page numbers for figures are denoted in *italics* and page numbers for tables in **bold**.